Recent Trends of Functional Nanomaterials for Biomedical and Healthcare Applications

Recent Trends of Functional Nanomaterials for Biomedical and Healthcare Applications

Editors

Sudeshna Chandra
Heinrich Lang

Basel • Beijing • Wuhan • Barcelona • Belgrade • Novi Sad • Cluj • Manchester

Editors

Sudeshna Chandra
Technology and Science
Hanse-Wissenschaftskolleg,
Institute of Advanced Study
Delmenhorst
Germany

Heinrich Lang
Research Group
Organometallics
Technische Universität
Chemnitz, MAIN Research
Center
Chemnitz
Germany

Editorial Office
MDPI
St. Alban-Anlage 66
4052 Basel, Switzerland

This is a reprint of articles from the Special Issue published online in the open access journal *Molecules* (ISSN 1420-3049) (available at: www.mdpi.com/journal/molecules/special_issues/3J36W4I3X7).

For citation purposes, cite each article independently as indicated on the article page online and as indicated below:

Lastname, A.A.; Lastname, B.B. Article Title. *Journal Name* **Year**, *Volume Number*, Page Range.

ISBN 978-3-7258-0180-0 (Hbk)
ISBN 978-3-7258-0179-4 (PDF)
doi.org/10.3390/books978-3-7258-0179-4

© 2024 by the authors. Articles in this book are Open Access and distributed under the Creative Commons Attribution (CC BY) license. The book as a whole is distributed by MDPI under the terms and conditions of the Creative Commons Attribution-NonCommercial-NoDerivs (CC BY-NC-ND) license.

Contents

About the Editors vii

Qingpan Bu, Ping Li, Yunfei Xia, Die Hu, Wenjing Li, Dongfang Shi and Kai Song
Design, Synthesis, and Biomedical Application of Multifunctional Fluorescent Polymer Nanomaterials
Reprinted from: *Molecules* **2023**, *28*, 3819, doi:10.3390/molecules28093819 1

Paola Di Matteo, Rita Petrucci and Antonella Curulli
Not Only Graphene Two-Dimensional Nanomaterials: Recent Trends in Electrochemical (Bio)sensing Area for Biomedical and Healthcare Applications
Reprinted from: *Molecules* **2024**, *29*, 172, doi:10.3390/molecules29010172 32

Chien-Hsiu Li, Ming-Hsien Chan, Yu-Chan Chang and Michael Hsiao
Gold Nanoparticles as a Biosensor for Cancer Biomarker Determination
Reprinted from: *Molecules* **2023**, *28*, 364, doi:10.3390/molecules28010364 84

Yujia Cheng and Guang Yu
Application and Research Status of Long-Wavelength Fluorescent Carbon Dots
Reprinted from: *Molecules* **2023**, *28*, 7473, doi:10.3390/molecules28227473 106

Mark Zamansky, Doron Yariv, Valeria Feinshtein, Shimon Ben-Shabat and Amnon C. Sintov
Cannabidiol-Loaded Lipid-Stabilized Nanoparticles Alleviate Psoriasis Severity in Mice: A New Approach for Improved Topical Drug Delivery
Reprinted from: *Molecules* **2023**, *28*, 6907, doi:10.3390/molecules28196907 123

Aishwarya Shetty, Heinrich Lang and Sudeshna Chandra
Metal Sulfide Nanoparticles for Imaging and Phototherapeutic Applications
Reprinted from: *Molecules* **2023**, *28*, 2553, doi:10.3390/molecules28062553 139

Joana Ribeiro, Ivo Lopes and Andreia Castro Gomes
A New Perspective for the Treatment of Alzheimer's Disease: Exosome-like Liposomes to Deliver Natural Compounds and RNA Therapies
Reprinted from: *Molecules* **2023**, *28*, 6015, doi:10.3390/molecules28166015 160

Monica-Cornelia Sardaru, Narcisa-Laura Marangoci, Rosanna Palumbo, Giovanni N. Roviello and Alexandru Rotaru
Nucleic Acid Probes in Bio-Imaging and Diagnostics: Recent Advances in ODN-Based Fluorescent and Surface-Enhanced Raman Scattering Nanoparticle and Nanostructured Systems
Reprinted from: *Molecules* **2023**, *28*, 3561, doi:10.3390/molecules28083561 186

Sergey Tsymbal, Ge Li, Nikol Agadzhanian, Yuhao Sun, Jiazhennan Zhang, Marina Dukhinova, et al.
Recent Advances in Copper-Based Organic Complexes and Nanoparticles for Tumor Theranostics
Reprinted from: *Molecules* **2022**, *27*, 7066, doi:10.3390/molecules27207066 206

Shuangling Chen, Meng Liang, Chengli Wu, Xiaoyi Zhang, Yuji Wang and Ming Zhao
Poly-α, β-D, L-Aspartyl-Arg-Gly-Asp-Ser-Based Urokinase Nanoparticles for Thrombolysis Therapy
Reprinted from: *Molecules* **2023**, *28*, 2578, doi:10.3390/molecules28062578 226

Fatih Yanar, Dario Carugo and Xunli Zhang
Hybrid Nanoplatforms Comprising Organic Nanocompartments Encapsulating Inorganic Nanoparticles for Enhanced Drug Delivery and Bioimaging Applications
Reprinted from: *Molecules* **2023**, *28*, 5694, doi:10.3390/molecules28155694 241

Juan Bian, Nemal Gobalasingham, Anatolii Purchel and Jessica Lin
The Power of Field-Flow Fractionation in Characterization of Nanoparticles in Drug Delivery
Reprinted from: *Molecules* **2023**, *28*, 4169, doi:10.3390/molecules28104169 265

Arina D. Filippova, Madina M. Sozarukova, Alexander E. Baranchikov, Sergey Yu. Kottsov, Kirill A. Cherednichenko and Vladimir K. Ivanov
Peroxidase-like Activity of CeO_2 Nanozymes: Particle Size and Chemical Environment Matter
Reprinted from: *Molecules* **2023**, *28*, 3811, doi:10.3390/molecules28093811 291

Mattia Bartoli, Elena Marras and Alberto Tagliaferro
Computational Investigation of Interactions between Carbon Nitride Dots and Doxorubicin
Reprinted from: *Molecules* **2023**, *28*, 4660, doi:10.3390/molecules28124660 309

About the Editors

Sudeshna Chandra

Dr. Sudeshna Chandra is currently a fellow of the Hanse-Wissenschaftskolleg (HWK) at the University of Oldenburg, Germany. She received her PhD in chemistry from the Indian Institute of Technology, Roorkee, India. Prior to her current position, she was a professor and head of chemistry at the Sunandan Divatia School of Science, SVKM's NMIMS University, Mumbai, India. She has experience spanning over 20 years in research and more than 7 years in academia. She is also a visiting researcher at the University of Regensburg and at the Technical University of Chemnitz, Germany. She is a recipient of the Alexander von Humboldt fellowship and of several awards, including but not limited to the Peter Salamon Young Scientist award, the DST Women Scientist award, and the DST Young Scientist award. She is a member of the Indian Science Congress Association, Alexander von Humboldt Stiftung/Foundation, and the Society of Materials Chemistry.

Her research interests focus mainly on (1) smart multifunctional hybrid nanomaterials and (2) the application of nanomaterials. In terms of the former, her research aims to create novel hybrid nanomaterials based on metals, metal oxides and sulfides, dendrimers, upconversion nanomaterials with controlled morphology, attractive functionalities, and targeted properties. In terms of the latter, her research aims at the application of nanomaterials for drug delivery, biosensing, and energy whilst also seeking to investigate formation mechanisms and process–structure–property relationships.

Heinrich Lang

Heinrich Lang studied chemistry at the University of Constance. He then spent 2 years as postdoctoral fellow (DFG) at M.I.T. and later became a member of the faculty at the University of Heidelberg in 1988, with habilitation in 1992. From 1992 to 1996, he was a Heisenberg fellow of the Deutsche Forschungsgemeinschaft. Between 1996 and 2022, he was full professor (of inorganic chemistry) at TU Chemnitz. A call to TU Kaiserslautern in 2003 was not taken up. Since 2008, prof. Lang has been an adjunct faculty member of the Department of Chemistry, University of Alabama, Tuscaloosa, U.S.A. Since April 2022, he has been group leader at the MAIN (Materials, Architectures and Integration of Nanomembranes) research center at TU Chemnitz. He is a member of various scientific and editorial advisory boards, he has published 860 peer-reviewed papers, and has received multiple honors and awards. Between 2012 and 2017, he was vice-president of TU Chemnitz for research and the promotion of young scientists. His research interests are in organometallic and metal–organic chemistry, including (nano)materials.

Review

Design, Synthesis, and Biomedical Application of Multifunctional Fluorescent Polymer Nanomaterials

Qingpan Bu [1], Ping Li [1], Yunfei Xia [1], Die Hu [1], Wenjing Li [2], Dongfang Shi [3,*] and Kai Song [1,3,*]

1. School of Life Science, Changchun Normal University, Changchun 130032, China; buqingpan@ccsfu.edu.cn (Q.B.); chunlps@163.com (P.L.); xiayunfei1103@163.com (Y.X.); hudie18784775519@163.com (D.H.)
2. School of Education, Changchun Normal University, Changchun 130032, China; liwenj678@163.com
3. Institute of Science, Technology and Innovation, Changchun Normal University, Changchun 130032, China
* Correspondence: shidongfang@ccsfu.edu.cn (D.S.); songkai@ccsfu.edu.cn (K.S.)

Abstract: Luminescent polymer nanomaterials not only have the characteristics of various types of luminescent functional materials and a wide range of applications, but also have the characteristics of good biocompatibility and easy functionalization of polymer nanomaterials. They are widely used in biomedical fields such as bioimaging, biosensing, and drug delivery. Designing and constructing new controllable synthesis methods for multifunctional fluorescent polymer nanomaterials with good water solubility and excellent biocompatibility is of great significance. Exploring efficient functionalization methods for luminescent materials is still one of the core issues in the design and development of new fluorescent materials. With this in mind, this review first introduces the structures, properties, and synthetic methods regarding fluorescent polymeric nanomaterials. Then, the functionalization strategies of fluorescent polymer nanomaterials are summarized. In addition, the research progress of multifunctional fluorescent polymer nanomaterials for bioimaging is also discussed. Finally, the synthesis, development, and application fields of fluorescent polymeric nanomaterials, as well as the challenges and opportunities of structure–property correlations, are comprehensively summarized and the corresponding perspectives are well illustrated.

Keywords: light-emitting polymer nanomaterials; rare earth polymers; semiconducting polymers; organic fluorescent small molecule cell imaging; biomedical imaging

Citation: Bu, Q.; Li, P.; Xia, Y.; Hu, D.; Li, W.; Shi, D.; Song, K. Design, Synthesis, and Biomedical Application of Multifunctional Fluorescent Polymer Nanomaterials. *Molecules* 2023, 28, 3819. https://doi.org/10.3390/molecules28093819

Academic Editors: Sudeshna Chandra and Heinrich Lang

Received: 12 April 2023
Revised: 24 April 2023
Accepted: 27 April 2023
Published: 29 April 2023

Copyright: © 2023 by the authors. Licensee MDPI, Basel, Switzerland. This article is an open access article distributed under the terms and conditions of the Creative Commons Attribution (CC BY) license (https://creativecommons.org/licenses/by/4.0/).

1. Introduction

Bioluminescent imaging is a visual imaging method that detects the intrinsic fluorescence of organisms, or the intensity of fluorescence luminescence after fluorescent materials mark organisms. Compared with traditional medical diagnostic imaging methods, it has the characteristics of fast imaging speed, high resolution, and applying no radiation damage to organisms. Applications can range from micron-sized cells to large-sized living organisms [1–3]. After years of research and development, commercial bioluminescence imaging systems are now widely used in the biomedical field, including laser scanning confocal microscopes, two-photon laser scanning microscopy imaging systems, and in vivo fluorescence imaging techniques. In combination with traditional biological imaging techniques such as magnetic resonance imaging (MRI), ultrasonic imaging (US), and computed tomography (CT), etc., the functions are complementary, and more accurate and effective biological imaging can be achieved, providing a reliable imaging method for the early diagnosis of cancer [4–6].

At present, fluorescent imaging materials mainly include inorganic fluorescent functional materials (quantum dots, rare earth luminescent materials, noble metal nanomaterials, etc.) and organic fluorescent functional materials (organic small molecule fluorescent materials and semiconducting polymers) [7–9]. Among them, quantum dots have good fluorescence quantum efficiency and photostability, but cannot avoid the biological toxicity of heavy metals [10,11]. Organic small molecule fluorescent dyes are the most widely used class of fluorescent materials. However, due to the disadvantages of poor stability, easy photobleaching, small Stokes shift and short fluorescence lifetime, its application range is greatly limited [12]. Rare earth luminescent materials, compared with other fluorescent polymer materials, have a narrow emission band (10–20 nm) and a large Stokes shift. The emission lifetime from microseconds to milliseconds can be achieved, which greatly reduces the interference of the self-luminous background of biological tissues. However, the material itself has poor biocompatibility, and it is prone to fluorescence aggregation quenching or quenching by water in a physiological environment [13,14]. Semiconducting polymers also have many excellent characteristics, for example, good photostability and photothermal performance, easy functional modification of the surface of the material, and good biocompatibility. However, the molecular weight is not easy to control during the preparation process, and the metabolic mechanism in the human body is still unclear [15,16].

Fluorescent polymer nanomaterials are prepared by physically doping or chemically bonding amphiphilic block polymers and fluorescent functional materials [17,18]. The composite material has the characteristics of good optical stability and wide application range of functional fluorescent materials. Additionally, it is also possible to select polymer monomers with different functions; while retaining the good water solubility and excellent biocompatibility of polymer nanomaterials, the controllable synthesis of properties such as size, morphology, stability, and surface properties can be achieved [19,20]. The complementary advantages of the two provide new ideas for expanding the application range of fluorescent functional materials in the field of biomedicine and have attracted the extensive attention of scientific researchers [21–23].

In summary, this paper discusses the functionalization strategy of fluorescent polymers and the preparation methods for fluorescent polymer nanomaterials. It also systematically introduces the research and development status and application prospects of fluorescent polymer nanomaterials based on rare earth luminescent materials, semiconducting polymers, and small organic molecules from recent years. It is emphasized that the design and application of fluorescent polymer nanomaterials should be functionalized from the perspective of synthesis and optimization according to need. Finally, the development direction and challenges of polymer nanomaterials in the fields of optics and medical tumors are prospected.

2. Fluorescent Polymer Functionalization Strategy

Ideal bioluminescent imaging probes, in addition to high fluorescence efficiency and stable luminescent properties, also need to have good monodispersity in aqueous systems. They have low toxicity to biological organisms and active groups on the surface to facilitate the connection of targeting molecules, achieving the effect of targeted imaging [24–26]. However, quantum dots of inorganic functional fluorescent materials, rare earth luminescent materials and noble metal nanoclusters, and semiconducting polymers of organic fluorescent functional materials and organic small molecule fluorescent materials have poor compatibility, poor degradability and long-term biological toxicity in vivo, and other defects. These problems severely limit their application in the biological field [27,28].

Fluorescent polymer nanomaterials are prepared by combining amphiphilic polymers and fluorescent functional materials using physical doping or covalent linkage. They not only have the characteristics of good optical stability and wide application range of fluorescent functional materials but can also be polymerized by selecting different functional monomers [29,30]. While retaining the good water solubility and excellent biocompatibility of polymer nanomaterials, the controllable synthesis of properties such as size, shape, stability, and surface properties can be achieved. As shown in Figure 1, the currently commonly used strategies for functional modification of fluorescent polymers mainly include physical encapsulation and covalent linkage [31–33].

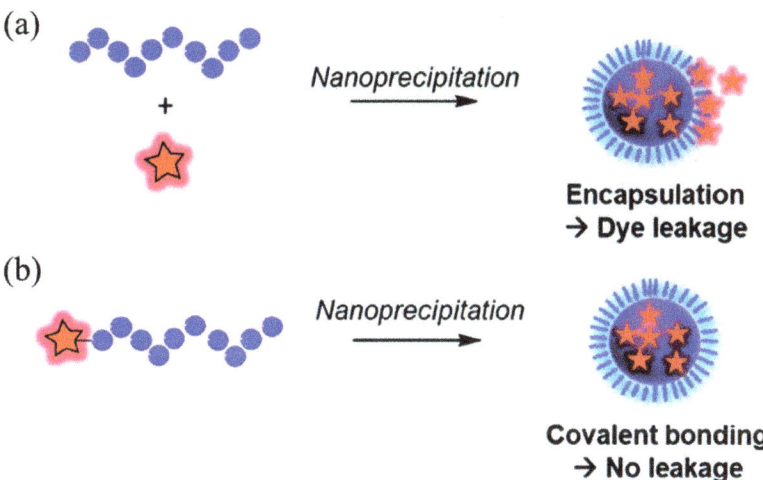

Figure 1. Schematic diagram of the preparation of fluorescent polymer nanoparticles: (**a**) physical encapsulation; (**b**) covalent attachment [31]. Copyright (2021), with permission from Royal Society of Chemistry.

2.1. Physical Package

Physical encapsulation is the most commonly used method at present. Fluorescent functional materials are directly embedded in polymer nanoparticles using physical encapsulation to prepare fluorescent polymer nanoparticles. The nanoparticles belong to a typical core–shell structure, in which the hydrophilic polymer acts as a protective layer in the shell, and the hydrophobic fluorescent material acts as a fluorescent chromophore in the core. It not only retains the luminescent properties of fluorescent materials, but also improves the stability and biocompatibility of hydrophobic fluorescent materials in water systems [34–36]. Yu et al. [37] designed and synthesized a semiconducting polymer PDFT based on diketopyrrolopyrrole (DPP). As shown in Figure 2, the amphiphilic distearoylphosphatidylethanolamine-polyethylene glycol (DSPE-mPEG) was used for physical encapsulation. Self-assembled into an NIR-II fluorescent nanoprobe PDFT1032 with a particle size of 68 nm, the maximum emission wavelength is 1032 nm, and it has excellent photostability, excellent biocompatibility, and extremely low in vivo toxicity. It presents high-resolution, real-time imaging in tumor diagnosis and vascular thrombosis treatment and, more importantly, realizes precise fluorescence imaging "navigation" for in situ tumor surgery and sentinel lymph node biopsy.

If the fluorescent material has hydrophobic properties, the preparation method of physical encapsulation can be used, which has good universality. However, due to the absence of chemical bonds between the polymer and the fluorescent-emitting group, there are situations where the fluorescent material leaks from the fluorescent polymer composite system or the aggregation and quenching of the local fluorescent material occurs, resulting in a decrease in luminescent performance [38,39]. Therefore, how to improve the preparation of fluorescent polymer nanoparticles with stable luminescence using physical encapsulation is still a research hotspot.

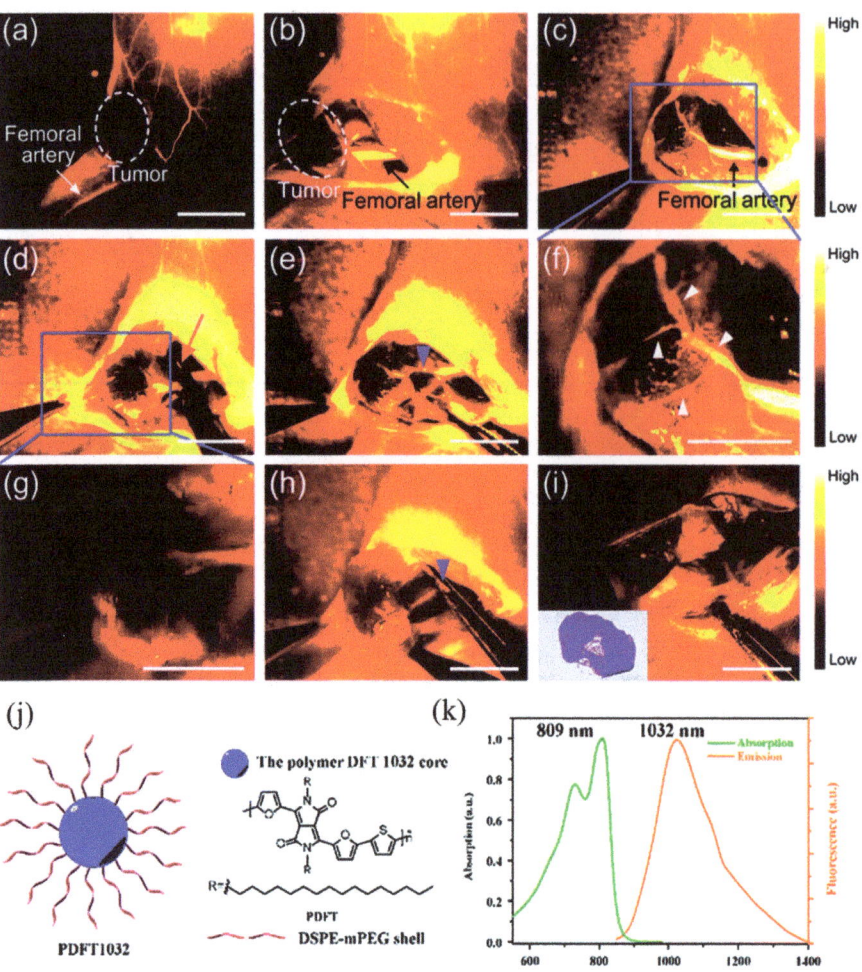

Figure 2. Schematic of DPP-based semiconducting polymer PDFT nanoscale self-assembly and imaging. (**a**) The vascular mapping and the hemodynamic status of the tumor and the femoral artery were determined. The white dashed circle contours the location of the tumor. (**b**,**c**) The branch of the femoral artery that supports the tumor and the vascular network of the tumor (exhibited as a claw shape) were clearly identified. (**d**) A vessel clamp was used to block the blood flow (red arrow) and the signal of the vascular network vanished. (**e**) After 5 min, the clamp was removed and the blood flow of the tumor was still devoid because a temporary thrombus was formed (blue arrowhead). (**f**) Magnification of (**c**). The vascular network of the tumor was clearly identified (white arrowheads). (**g**) Magnification of (**d**). (**h**) The major artery was surgically incised (blue arrowhead). (**i**) NIR-II imaging exhibited the absence of the residual tumor fluorescence and normal circulation (femoral artery) was successfully maintained. Inset is the histological analysis of the osteosarcoma. Scale bar: 8 mm. (**j**) Schematic drawing of a PDFT1032 nanoparticle composed of semiconducting polymer DFT and a hydrophilic DSPE-mPEG shell. (**k**) Absorbance and fluorescence spectrum of PDFT1032 showing an absorption peak at 809 nm and a fluorescence peak at 1032 nm with an 808 nm excitation laser [37]. Copyright (2018), with permission from Royal Society of Chemistry.

2.2. Covalent Linkage

There are two ways to prepare covalently linked fluorescent polymer nanomaterials [31,40]: one is to first copolymerize polymer monomers into polymer chains, and then use covalently linked methods. The fluorescent-emitting group is attached to the polymer chain, and then prepared into nanoparticles. As shown in Figure 3a, it is referred to as "aggregate first and then join". The nanoparticles prepared in this way have good stability, and the desired multifunctional nanoparticles can be customized by selecting different polymer nanoparticles. However, this preparation method requires functional group matching between empty polymer nanoparticles and fluorescent groups, and its universality is slightly worse than physical packaging. In addition, fluorescent group materials are also prone to fluorescence quenching on the surface of nanoparticles [41,42]. The second approach is to prepare fluorescent groups and polymer monomers into fluorescent polymer monomers, then copolymerize and self-assemble them into fluorescent polymer nanomaterials, as shown in Figure 3b, referred to as "connection first and then polymerization". The distribution of light-emitting groups in the fluorescent nanoparticles prepared using this method is relatively more uniform, and the optical stability is good. However, there is also the problem that the size of fluorescent polymer nanoparticles is not easy to control due to the steric hindrance of the luminescent group [43,44].

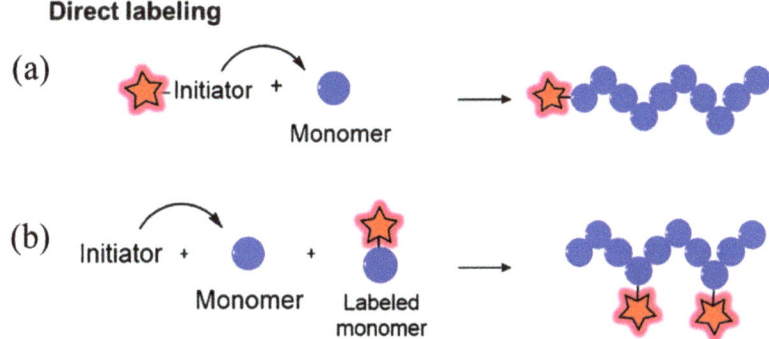

Figure 3. Two forms of covalent linkage: (**a**) polymerization followed by linkage; (**b**) linkage followed by aggregation [31]. Copyright (2021), with permission from Royal Society of Chemistry.

In summary, although the method of physical packaging is used to prepare fluorescent polymer nanomaterials, the distribution of fluorescent materials in nanomaterials is uneven and leaks easily. However, it is still the most commonly used method at present. In order to solve the above problems, designing and synthesizing new fluorescent polymer nanomaterials by means of covalent linkage is still one of the hotspots of scientific research.

3. Preparation Method of Fluorescent Polymer Nanomaterials

3.1. Active/Controllable Synthesis of Amphiphilic Block Polymers

From the preparation strategy of fluorescent polymer nanomaterials in the previous section, it can be seen that the controllable synthesis of amphiphilic block polymers directly determines the monodispersity, morphology, and size of bioluminescent probes in aqueous solution. At present, the most commonly used living/controllable polymerization methods mainly include atom transfer radical polymerization (ATRP) [45,46] and reversible addition–fragmentation chain transfer polymerization (RAFT) [47]. Below, we focus on the introduction of these two technologies, as shown in Table 1.

Table 1. Comparison of different preparation strategies for fluorescent polymer materials.

Synthesis Technology		Advantages	Disadvantages
Living/controllable synthesis of amphiphilic block polymers	ATRP	The reaction temperature is mild, the operation is simple, and it is easy to industrialize.	The intermediate process is completely uncontrolled; the amount of transition metal complex is large; the aging of the polymer.
	RAFT	It has a wide range of applications, good polymerization ability, and the molecular weight of the obtained polymer is uniform.	Few applicable monomers, limited scope of molecular design, expensive.
Physical package	Nanoprecipitation	Simple operation, fast, high reproducibility, good dispersion of colloidal nanoparticles and easy functionalization.	There are fewer types of polymers, and the process of particle growth is not easy to control.
	Microemulsion method	Narrow particle size distribution, controllable, good stability.	Surfactants are difficult to remove and have large particle sizes.
	Self-assembly method	It is very convenient to prepare various exotic three-dimensional structures; it is also possible to prepare porous materials that inherit the original morphology and structure	Unstable under physiological conditions.
Covalent linkage	PISA	The process is simple, the price adjustment is gentle, and nano-medicine can be prepared in one step.	The operation is complicated, the reaction takes a long time, and the concentration of the prepared nanoparticles is low (≤ 1 mg/mL), which makes it impossible to achieve large-scale mass production.
	Precipitation polymerization	The particle size of the polymer is uniform and clean, the viscosity of the polymerization system is low, and no surfactant and stabilizer are needed.	Low microsphere yield and high solvent toxicity.

3.1.1. ATRP

Matyjaszewski et al. [48] and Sawamoto et al. [49] successively proposed the method of atom transfer radical polymerization (ATRP). The low-valence metal complex M_t^n takes an electron from the initiator organic halide R-X to form R free radicals to initiate monomer aggregation. The formation of chain free radicals P can also take the halogen atom X from the high-valence metal halide M_t^{n+1}-X passivation to P-X, and reduce the high-valence metal halide to M_t^n. The reversible transfer equilibrium reaction between the free radical active seeds and the halide dormant seeds of the polymer chains enables effective control of the reactions. Compared with the traditional free radical polymerization, the ATRP method has a wide range of adaptability, can control the molecular weight distribution (PDI) of the polymer between 1.05 and 1.5, and has a mild reaction temperature, simple operation, and is easy to industrialize [50–52]. However, when applied to the synthesis of bioluminescent probes, copper-based catalysts have high biotoxicity, and finding other catalysts to replace copper-based catalysts is still a hotspot in ATRP research [53].

3.1.2. RAFT

Compared with the ATRP reaction, the reversible addition–fragmentation chain transfer polymerization (RAFT) reaction system does not involve the participation of copper-based and other biologically toxic transition metals. Additionally, the source of free radicals is basically from the decomposition of organic initiators. For example, azobisisobutyronitrile (AIBN) or dibenzoyl peroxide (BPO) are more suitable for the controlled synthesis of amphiphilic block polymers for biological use [54,55].

The RAFT method was proposed by Rizzardo [56]. The first is initiation (initiation), where the initiator generates free radicals I, then monomers M are initiated to polymerize with each other to generate extended chain free radicals Pn. The second step is the chain transfer reaction (chain transfer); the extended chain free radical Pn reacts with the dithioester chain transfer agent (1) to form an unstable intermediate (2). The groups on both sides of the intermediate can be broken to form a temporarily inactive thioester dormant (3) and a new free radical R·. The third step is to reinitiate the polymerization between the new free radical R· and the monomer to form Pm· (re-initiation). The fourth step is the process of chain equilibrium (chain equilibration), and the macromolecular chain transfer agent (macro-CTA) plays a controlling role. The free radical concentration is low throughout the reaction. Therefore, the molecular weight distribution of the polymer is relatively uniform. The final termination reaction (termination) generally quenches the reaction directly at low temperature, and the product is a mainly macromolecular chain transfer [57,58].

The RAFT mechanism is applicable to a wide range of monomers. The reaction temperature is 60–70 °C, and it has good polymerization ability for monomers such as acrylic acid (AA), methacrylic acid (MAA), and methyl methacrylate (MMA). The resulting polymers are of uniform molecular weight (PDI typically below 1.3). However, this reaction relies heavily on expensive RAFT reagents, and the development of stable, low-cost, and easy-to-synthesize RAFT reagents that meet different systems is still one of the research hotspots [59–61].

3.2. Preparation Method of Physically Encapsulating Fluorescent Polymer Nanoparticles

From the perspective of preparation strategy, fluorescent polymer nanoparticles wrap luminescent materials into amphiphilic block polymers using physical doping. Nano-precipitation, microemulsion, and self-assembly methods are commonly used, which are similar to the preparation methods of semiconducting polymer nanomaterials, as shown in Figure 4 [62,63].

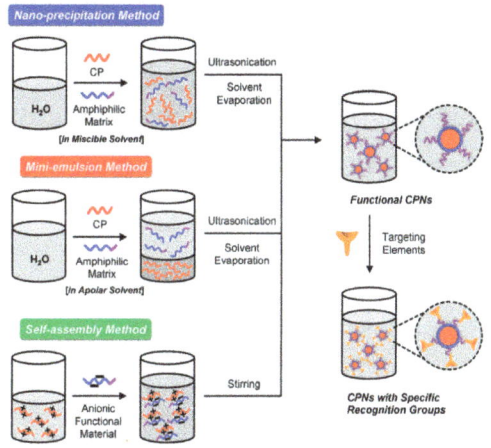

Figure 4. Preparation method for semiconducting polymer nanoparticles [63]. Copyright (2013), with permission from Royal Society of Chemistry.

Nanoprecipitation is a method based on the interfacial deposition of polymers. It was first proposed by Masuhara et al. [64] and then improved by McNeill [65] and Chiu [66]. It is widely used in the preparation of fluorescent nanoparticles in the biological field. Firstly, the fluorescent material and the amphiphilic block polymer material are dissolved in a small amount of good solvent. The material is quickly dropped into a poor solvent (usually deionized water) with vigorous stirring or ultrasound; the huge difference in solubility of the two solvents promotes the aggregation of polymer materials to form nanoparticles, with a particle size of about 15 nm. The process of nanoparticle formation mainly includes several steps: supersaturation, nucleation, coagulation growth, and formation of polymer nanoparticles. The method is simple, fast, and highly reproducible. The nanoparticle colloid has good dispersion and is easy to be functionalized. It is a common method for preparing drug-loaded nanomaterials, but there are also defects such as fewer types of suitable polymers and difficult control of the particle growth process [67].

The microemulsion method is similar to the nanoprecipitation method. The prepared amphiphilic polymer and fluorescent material are first dissolved in a good solvent and then mixed with a poor solvent (usually deionized water) [68]. The huge difference in solubility is used to prepare nanoparticles, but the difference is that a certain concentration of surfactant needs to be added. However, the final surfactant is difficult to remove from the reaction system, which affects the application of fluorescent nanoparticles in the biological field. Additionally, the size of the prepared nanoparticles is relatively large, between 240 and 270 nm [69–71].

The usual preparation of the self-assembly method is to dissolve fluorescent materials and functional materials with opposite charges in an aqueous solution according to a certain ratio. After fully stirring and mixing evenly, the functionalized polymer nanoparticles are prepared using high-speed centrifugation and the particle size is about 100 nm. However, the preparation of the self-assembly method also has the instability of nanoparticles in a physiological environment, which limits its further application [72,73].

3.3. Preparation Method of Covalently Linked Fluorescent Polymer Nanoparticles

By means of covalent connection, fluorescent materials and amphiphilic block polymer materials are connected and self-assembled into polymer nanoparticles with luminescent properties, mainly including two types of nanoparticles: polymer micelles and nanogels.

3.3.1. Preparation of Polymer Micelles Using Aggregation-Induced Self-Assembly (PISA)

The traditional preparation methods of polymer micelles mainly include the solvent induction method, dialysis method, and direct dissolution method [74,75]. Nanoparticles with morphologies such as spherical, worm-like, and vesicular are prepared using the self-assembly of amphiphilic block polymers with different solubility differences in different solvents. However, the operation is complicated, the reaction takes a long time, and the concentration of the prepared nanoparticles is low (≤ 1 mg/mL), which makes it impossible to achieve large-scale mass production [76–78]. In recent years, the polymerization-induced self-assembly (PISA) method can not only prepare micelles and assemblies with different morphological structures (including spherical, worm-like, vesicular, etc.) in one pot, but nanoparticles with solids content up to 50% can also be synthesized in bulk [79,80]. This provides a new idea for the commercial application of preparing polymer nanoparticles, and is also widely used in the fields of drug-controlled release, bioimaging, and catalysis [81,82].

Hawkett et al. [83] first used polyacrylic acid (PAA) as a water-soluble macromolecular RAFT chain transfer agent and induced self-assembly into spherical micelles in aqueous solution. Subsequently, Pan et al. [84,85] utilized poly-4-vinylpyridine (P4VP) as a macromolecular RAFT chain transfer agent for dispersion polymerization in methanol solvent. With the chain growth of PS spheres, the morphology of the polymer gradually changed from spherical to worm-like and vesicle-like. A schematic diagram of nanomaterials prepared using the aggregation-induced self-assembly method [86] is shown in Figure 5. The water-soluble polymer chain transfer agent (macro-CTA) prepared using the RAFT method initiates another hydrophobic polymer monomer, and the newly synthesized diblock polymer can be dissolved in the reaction system at the early stage of the reaction. With the continuous growth of the second hydrophobic chain, the volume of the insoluble polymer continues to increase. When the critical micelle concentration (CMC) is reached, it self-assembles into different morphologies. A series of theoretical studies have shown that the morphology of block polymer self-assembly is determined by the volume ratio P of the polymer at the hydrophobic end. When $P \leq 1/3$, the diblock polymer exhibits spherical nanoparticles. When $1/3 < P \leq 1/2$, the diblock polymer exhibits worm-shaped nanoparticles. When $1/2 < P \leq 1$, the diblock polymer exhibits a vesicle shape.

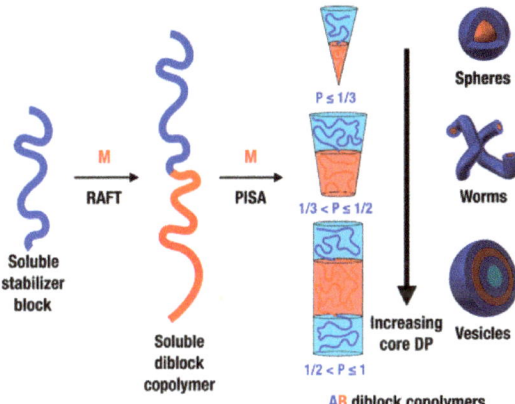

Figure 5. Schematic diagram of the preparation of nanoparticles from diblock polymers using the PISA method [86]. Copyright (2016), with permission from American Chemical Society.

3.3.2. Preparation of Nanogel Polymer Microspheres Using Precipitation Polymerization

Hydrogel is a kind of hydrophilic polymer material with a three-dimensional network structure, which can absorb water several times the weight of the material and has good biocompatibility and degradability [87]. Nanohydrogels are hydrogels with a size between 100 and 1000 nm that have the dual characteristics of hydrogels and nanomaterials. The structure is stable under physiological conditions, and it has a high drug loading rate and a long drug release cycle. It is a drug carrier material that has developed rapidly in recent years [88,89].

The nanogel preparation method [90] is shown in Figure 6. The traditional preparation methods mainly include emulsion polymerization, microemulsion polymerization and dispersion polymerization, all of which need to add stabilizers or surfactants to stabilize the reaction system and avoid aggregation and precipitation. However, it is difficult to remove the stabilizer or surfactant from the reaction system after the reaction, which affects the application of polymer microspheres in the biomedical field [91]. The method of precipitation polymerization was first proposed by Chibante et al. [92]. The stabilizer is replaced by a cross-linking agent, which is added to the reaction system together with the reactive monomer, and polymer microspheres with uniform particle size and clean surface are prepared after heating and polymerization. However, this method is only suitable for the polymerization of hydrophobic monomers and cannot prepare polymer microspheres for biomedicine. To expand the application of precipitation polymerization to hydrophilic monomers, Yang et al. [93] developed a distillation precipitation method, which shortened the precipitation polymerization time to 1–2 h. However, with the decrease in the solvent amount in the reaction system, the late reaction was unstable and the product yield was low. Wang et al. [94] developed the reflux precipitation polymerization method based on the technique of distillation precipitation. A return-shaped condenser was connected to the reaction device to ensure that the volume of the solvent in the reaction system remained unchanged, making the polymerization of the reaction system more stable. Compared with the traditional preparation method, the reflux precipitation polymerization method has the characteristics of short time consumption, no need for stabilizer, clean particle surface, simplicity of device, and less byproducts.

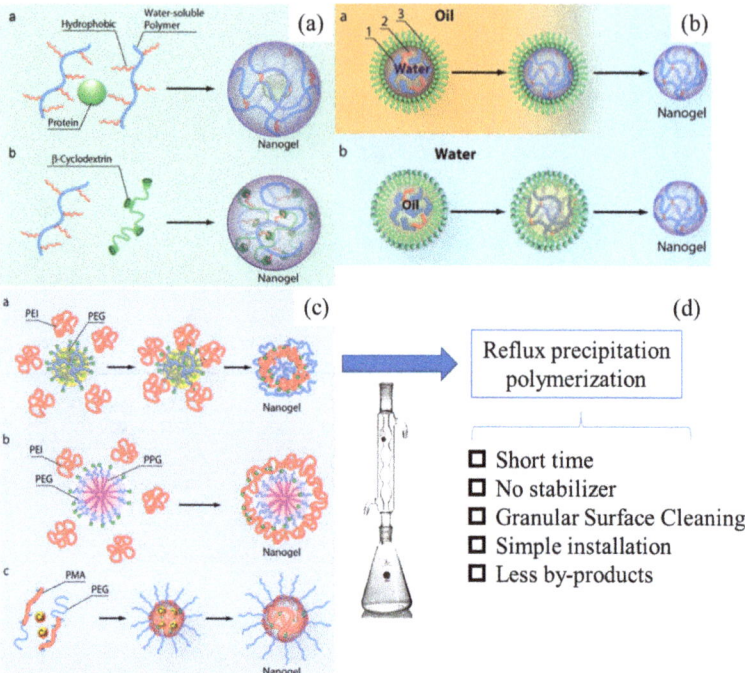

Figure 6. Schematic diagram of the nanohydrogel preparation method: (**a**) polymer self-assembly method; (**b**) inverse emulsion polymerization method; (**c**) precipitation polymerization method; (**d**) reflux precipitation polymerization method [90]. Copyright (2009), with permission from Wiley.

4. Biotoxicity of Fluorescent Polymer Nanomaterials

Due to their unique physical and chemical properties, nanomaterials have broad application prospects in the field of biomedicine. For example, as a drug carrier or a bioimaging probe, the pharmacokinetics and potential toxic effects in the organism need to be tested before practical application. In vivo toxicity studies of nanomaterials involve a variety of exposure methods, such as intravenous, transdermal absorption, subcutaneous, inhalation, intraperitoneal, and oral administration, and various animal models such as mice, rats, dogs, and monkeys. After nanomaterials enter the body and interact with biological components (proteins, cells), they are distributed to different organs of the body. At this point, the particles maintain their original structure or degrade. The slow removal and accumulation of materials, as well as the large number of phagocytes make the liver, spleen, and other organs in the reticuloendothelial system the most important targets of oxidative stress of nanomaterials. In addition, organs with high blood flow, such as lungs and kidneys, are also affected by nanomaterials [95].

Current toxicological mechanisms of nanomaterials mainly focus on the hypothesis of free radical oxidative damage [96]. This hypothesis holds that, under normal conditions, the content of reactive oxygen species in the mitochondria of body cells is very low. Additionally, there are many antioxidant systems in the body, and the active oxygen free radicals produced by normal cell metabolism are easily removed by glutathione reductase and antioxidant enzymes. When nanoparticles enter the body, they can induce the production of a large amount of reactive oxygen species (ROS). ROS mainly activate the inflammatory response by activating the phosphorylation of NF-κB transcription factors and MAPKs. As a result, the antioxidant defense system in the mitochondria is destroyed, causing various damages and further affecting the normal physiological functions of the body.

Li et al. [97] found that nanoparticles can induce an increase in reactive oxygen species in RAW264.7 cells, leading to cell apoptosis. Shvedova et al. [98] summarized the main mechanism of nanoparticle-induced ROS generation in cells, resulting in oxidative damage: (1) oxidation of liposomes in mitochondria; (2) NADPH oxidation leading to cell apoptosis and inflammatory response; (3) depletion of reduced glutathione in the body; (4) activation of peroxidase, leading to degradation of nanoparticles. At the same time, this is precisely the oxidative damage effect of nanoparticles exposed to the body.

In addition to ROS, reactive nitrogen species (RNS) may also be involved in the free radical oxidative damage effect of nanoparticles. Recent studies have proved that RNS play a role in the inflammatory damage caused by nanoparticles. Lanone and Boczkowski suggested that the main molecular mechanism of in vivo toxicity of nanomaterials is the induction of free radicals leading to oxidative damage [99]. Free radicals can not only cause damage to biological components by oxidizing lipids, proteins, and DNA, but also induce and enhance inflammation by upregulating redox-sensitive transcription factors (such as NF-kB) and inflammation-related kinases [100,101]. The composition of some materials, such as iron, cadmium, chromium, and other atoms also affects the toxicity in vivo. In addition, surface modification of nanoparticles can alter their interaction with cell membranes, resulting in their altered cellular uptake, thereby affecting their toxicological effects on targeted cells. The application and toxicity of different fluorescent nanomaterials in biological systems are shown in Table 2.

Table 2. Biotoxicity and biological system applications of different fluorescent nanomaterials.

Fluorescent Nanomaterial Type	Intrinsic Material Toxicity	Materials	Biological System
Carbon dots	Low	C-dots, PEG stabilized	Mice
Carbon nanotubes	Low–medium	Many types of CNTs	Various in vitro/in vivo
Dendrimers	High	Various dendrimer types	Various in vitro/in vivo
Doped graphene QDs	Medium	N-doped graphene quantum dots	Red blood cells
Fluorescent beads	Medium (polymer)	Polystyrene nanoparticles	Endothelial cells
	Low (silica)	Silica nanoparticles	Epithelial cells and fibroblasts
Fluorescent proteins	Medium	Red fluorescent protein	HeLa cells
Graphene oxide	Medium	Graphene oxide	Various in vitro/in vivo
		Graphene oxide	Red blood cells
Organic dyes	Medium	Various organic fluorophores	Various in vitro/in vivo
Metal clusters	Medium	MPA or GSH stabilized Au clusters	Colonic epithelial cells
		GSH and BSA stabilized $Au_2 5$ clusters	Mice
Nanodiamonds	Low	Detonation nanodiamond	Various in vitro/in vivo
		Various diamond types	Human liver cancer and HeLa cells in vitro
		Detonation nanodiamond	Human embryonic kidney cells and Xenopus laevis embryos
P-dots	Medium	Quinoxaline based polymer, STV conjugated	Zebrafish embryo
		Polybutylcyanoacrylate	HeLa and human embryonic kidney cells/rats
Quantum dots	High	CdSe–ZnS; PEG, BSA or polymer stabilized	Rats
		CdTe	Mice
		Several types	Various in vitro/in vivo
		Several types	Various in vitro/in vivo
Rare earth nanoparticles	Medium-high	UCNPs, $NaYF_4$:Yb,Tm, polyacrylic acid coated	Mice
		UCNPs, $NaYF_4$:Yb,Tm	HeLa cells, caenorhabditis elegans
		DCNPs, Gd_2O_2S:Tb^{3+}	Human peripheral blood mononuclear cells, human-derived macrophages, HeLa cells

Because of the small size effect of nanomaterials and the complexity of biological systems, the effects caused by the processing of nanomaterials in biological systems are unpredictable. The interaction between biological components (proteins and cells) and nanostructured materials may cause unique biodistribution and metabolic reactions, making it difficult to predict the metabolism and safety of nanomaterials in biological systems. As evidenced by the above literature review, fluorescent materials have been widely used in biomedical fields such as bioimaging, biosensors, and drug delivery because of their excellent luminescent properties and photoconversion properties. However, due to their particularity, their biological safety cannot be predicted. Therefore, they may cause certain harm to organisms during use, which greatly limits the application of fluorescent nanomaterials. Therefore, it is urgent to find fluorescent nanomaterials with good biocompatibility.

5. Application of Fluorescent Polymer Nanomaterials in Bioimaging

In order to meet the needs of different biological applications, many different types of fluorescent polymer nanomaterials have been developed. This paper focuses on the research direction and focuses on the introduction of polymer nanomaterials based on rare earth luminescent materials, semiconducting polymers, and organic, small molecule luminescent materials [7–9]. The use of nano-fluorescent probes can quickly, accurately, and selectively label and study target molecules on cells. Labels were obtained according to their excitation wavelengths at different emission wavelengths, as shown in Table 3.

Table 3. Common fluorescence excitation and emission wavelengths.

Fluorescent Substance	Excitation Wavelength EX nm	Excitation Wavelength EX (sub)	Emission Wavelength EM nm
Alexa Fluor 532	532		554
Cy3	550		570
DsRed	557		579
EtBr	300	518	605
FITC	490		525
Gel Green	250	500	530
GFP	488		507
mCherry	580		610
SYBR Gold	495		540
SYBR Green I	498		522
SYPRO Red	550	300	630
SYPRO Ruby	280	450	620
TagRFP	555		583
Gel Red	270	510	600

5.1. Polymer Nanomaterials Based on Rare Earth Luminescent Materials

Rare earth elements include lanthanides with atomic numbers 57–71 in the periodic table of chemical elements and scandium (Sc, 21) and yttrium (Y, 39) with similar chemical properties, a total of 17 elements [102]. The electron configuration of lanthanides is $[Xe]4f^{0-14}5d^{0-1}6s^2$, and each element has a 4f electron shell. Compared with other luminescent materials, it has narrow emission band (10–20 nm), high luminous efficiency, large Stokes shift, and long emission lifetime (range μs-ms) features [103]. At present, rare earth luminescent materials used in biological imaging mainly include rare earth organic complexes and rare earth upconversion materials.

5.1.1. Rare Earth Organic Complexes

Rare earth organic complexes are due to the small molar absorptivity of lanthanide trivalent ions and the prohibition of ff transitions in the electron shell [104,105]. As a result, very little energy is directly absorbed by the 4f energy level of lanthanide elements, and organic ligands are required to act as antennas to absorb excitation energy, thereby sensitizing rare earth ions to emit light. This process is called "antenna" (Figure 7a) [106].

As shown in Figure 7b, the main process of energy transfer of rare earth complexes is as follows: (1) After the rare earth complex absorbs energy, electrons transition from the ground state (S0) to the excited state (S1). (2) The energy of the excited state (S1) transfers energy to the lowest excited triplet state (T1) through intersystem crossing (ISC). (3) When the lowest excited triplet state (T1) matches the lowest excited state energy level (5DJ) of rare earth ions, energy transfer occurs between T1 and 5DJ. Eventually, the rare earth ions return to the ground state (7DJ) in the form of radiation, thereby emitting the characteristic fluorescence of rare earth ions [103,107,108].

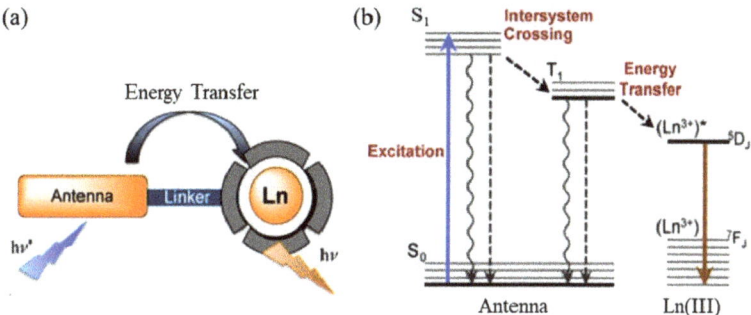

Figure 7. (a) Schematic diagram of the antenna effect of rare earth complexes; (b) schematic diagram of the luminescent principle of rare earth complexes [103]. Copyright (2014), with permission from American Chemical Society.

According to the structure of the ligands, it can be divided into β-diketone ligands, carboxylic acid ligands and macrocyclic ligands such as crown ethers [109]. There are coordination elements such as O, N, and S on the ligand, and a stable six-membered ring structure is formed after coordination with rare earth. It can directly absorb laser energy and effectively transfer energy to rare earth ions through the structure of the six-membered ring, and then emit the characteristic fluorescence of rare earth ions. The coordination ability of the coordinating atoms is O > N > S. When rare earth complexes are used in biological imaging, water molecules easily replace the coordination bonds of organic ligands [110]. The high-frequency O–H bonds of water increase the nonradiative decay of excited states of rare earth ions, which in turn affects the luminescent properties of rare earths [111]. When rare earth organic complexes are used as fluorescent probes, they are usually physically wrapped with biocompatible silica or polyethylene glycol to form a core–shell structure, which improves the chemical stability of the material and avoids the interference of water molecules.

Dos Santos et al. used the polymer PMMA-COOH to physically wrap the rare earth complex Eu(TTA)$_3$phen, and the preparation's schematic diagram is shown in Figure 8. By adjusting the concentration of the precursor, rare earth polymer nanoparticles of 10 nm, 20 nm, and 30 nm were prepared. The fluorescence quantum efficiency exceeded 20%, and the brightness of a single particle was as high as 4.0×10^7 M^{-1} cm^{-1}. A lower laser intensity of 0.24 W/cm^2 can be used to image single particles, and time-resolved imaging microscopy can be used to dynamically observe the progress of nanoparticles into cells [112]. However, the polymer physically wraps the rare earth complex material, and there is also uneven distribution of the fluorescent material. Aggregation quenching is prone to occur, which affects the luminescent properties of fluorescent probes, and the related preparation methods still need further research [113–115].

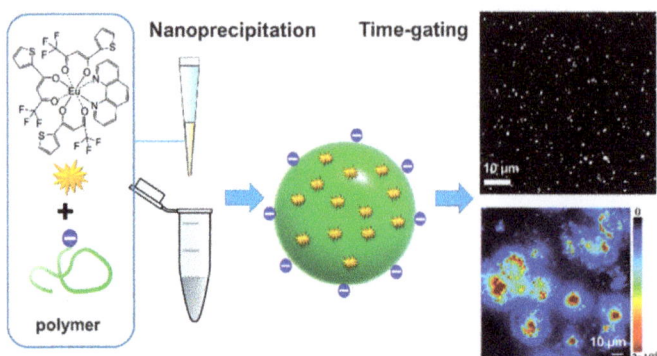

Figure 8. Schematic diagram of rare earth polymer nanoprobes for single particle detection and cell imaging [112]. Copyright (2019), with permission from American Chemical Society.

In order to solve the problem of uneven distribution of fluorescent materials in polymer nanosystems, Xu et al. [116,117] modified hydroxyl and amino groups on Eu(TTA)3phen, respectively. As shown in Figures 9a and 10, two complexes, [Eu(TTA)3phen]-OH and (Eu(TTA)3phen]-NH$_2$, were prepared. They then reacted with hydrophilic polymers PEG2000 and GluEG NCA through covalent connection to generate [Eu(TTA)3phen]-PEG2000 and [Eu(TTA)3phen]-GluEG, respectively, and self-assembled into water-soluble nanoparticles. They emi the 614 nm characteristic peak of Eu(III) ion in aqueous solution, and can be successfully taken up by L929 cells and HeLa cells, and emit strong red light (as shown in Figures 9b and 10).

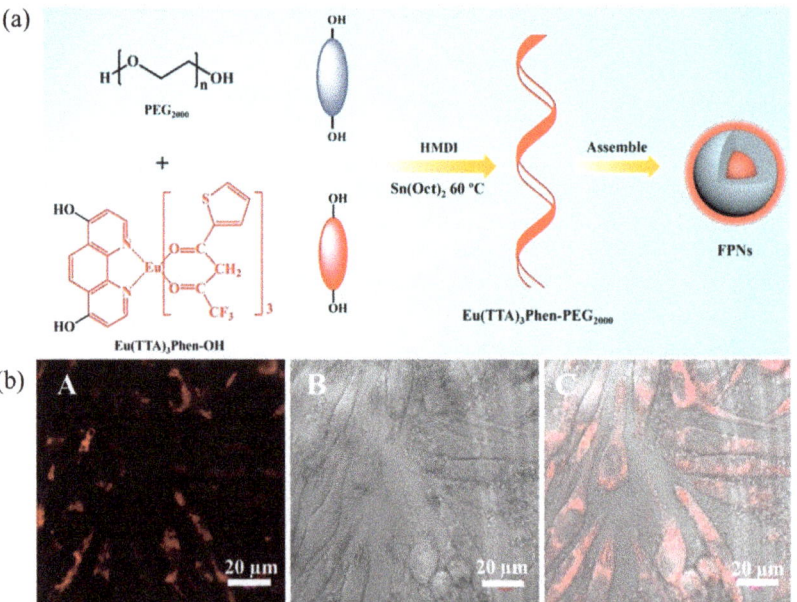

Figure 9. [Eu(TTA)3phen]-PEG2000 nanomaterials: (**a**) schematic diagram of preparation; (**b**) cell imaging diagram. (**A**) fluorescent image 405 nm laser excitation, (**B**) bright field and (**C**) merged image of (**A**,**B**). Scale bar 1/4 20 mm. [116]. Copyright (2018), with permission from Elsevier.

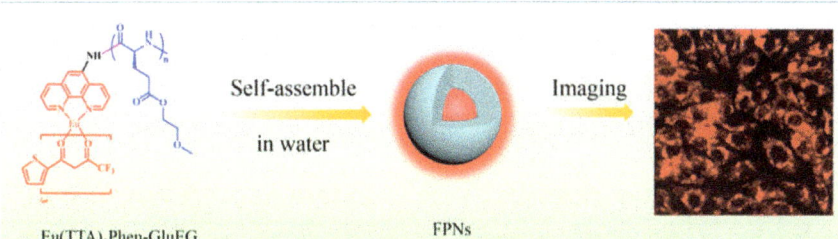

Figure 10. Preparation and cell imaging of [Eu(TTA)₃phen]-GluEG nanomaterials [117]. Copyright (2018), with permission from Elsevier.

5.1.2. Rare Earth-Doped Upconversion Luminescent Materials

Rare earth-doped upconversion materials are doped with trivalent lanthanide ions into a suitable dielectric matrix lattice. The lanthanide ions act as the luminescent center, and the ground state electrons of the sensitized ions are first excited to the excited state under the irradiation of the excitation light of a suitable wavelength. Then, the energy is transferred to the luminescent center, and the luminescent center is excited to an excited state. Finally, the excited state electrons return to the ground state and emit near-infrared fluorescence. Therefore, sensitizing ions are required to have a larger absorption cross-section at NIR-I or NIR-II. For example, the luminescent central ions with NIR-I emission include Nd^{3+}, Yb^{3+}, and Er^{3+}, and the luminescent central ions with NIR-II emission include Nd^{3+}, Ho^{3+}, Pr^{3+}, Tm^{3+}, and Er^{3+} (as shown in Figure 11) [118]. Rare-earth-doped upconversion materials have the advantages of low toxicity, narrow-band emission, long emission lifetime, no photobleaching, and no scintillation. They have broad application prospects in the fields of bioimaging, such as in analytical sensors, PDT, and optical imaging [119].

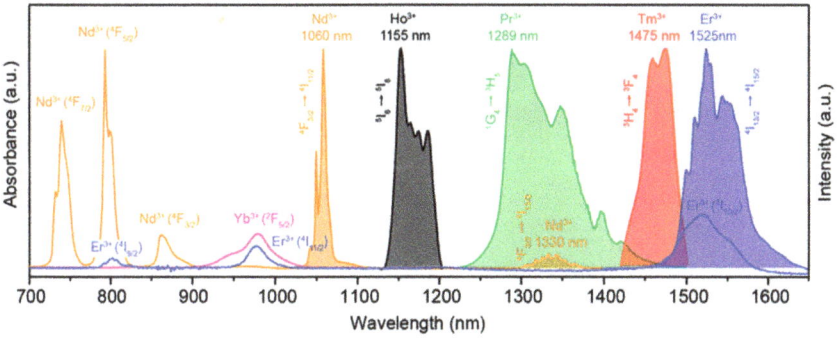

Figure 11. Absorption spectra of Nd^{3+}, Er^{3+}, and Yb^{3+} in NIR-I and emission spectra of Nd^{3+}, Ho^{3+}, Pr^{3+}, Tm^{3+}, and Er^{3+} in NIR-II [118]. Copyright (2019), with permission from Wiley.

Rare earth-doped upconversion luminescent materials are generally hydrophobic. Generally, water-soluble bioluminescent probes are prepared by modifying the surface function of materials with SiO_2 that has better biocompatibility, water-soluble polymers (PEG, PAA, PEI, etc.), and polyols. Tao et al. [120] used the block polymer PEO-b-PCL to wrap NIR-I emitting rare earth nanocrystalline materials $NaYF_4$:Yb/Ho(DiR) and NIR-II-emitting $NaCeF_4$:Er/Yb(LNPs). And luciferase (LUS) and red fluorescent protein (RFP) were doped into the polymer to prepare a multifunctional nanomaterial that can simultaneously generate NIR-I and NIR-II fluorescence spectra (as shown in Figure 12a–c). Then, the nanomaterials were injected intraperitoneally into mice with ovarian cancer, and fluorescence signals of two spectra could be found in the ovarian LUS+/RFP+-responsive cancer cell OVCAR-8 (Figure 12d).

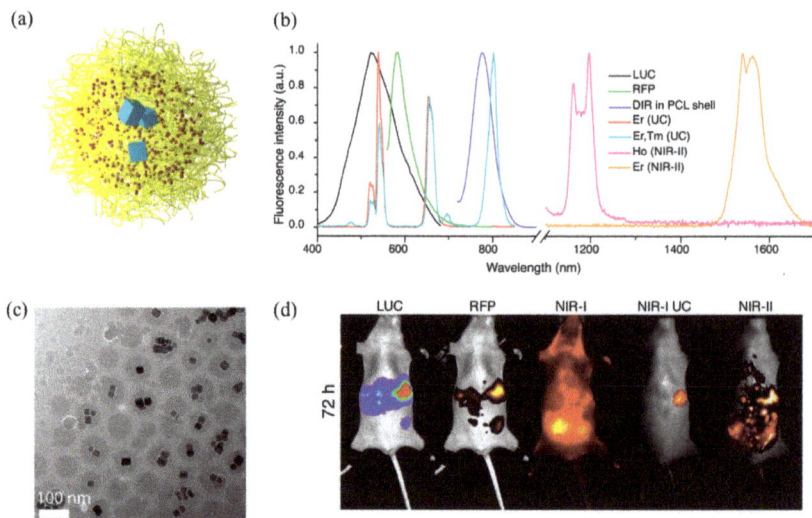

Figure 12. (**a**) Schematic diagram of nanoparticles, in which the yellow block represents the block polymer PEO-b-PCL, the red point represents the NIR-I quantum dot DiR, and the blue block represents the NIR-II quantum dot LNPs; (**b**) the nanoparticle TEM image; (**c**) fluorescence spectrum of multifunctional nanomaterials; (**d**) imaging of multifunctional nanoparticles in mice [120]. Copyright (2017), with permission from Elsevier.

5.2. Polymer Nanomaterials Based on Semiconducting Polymers

Semiconductor polymers (SPs) are a class of polymer materials whose main chain is composed of π–π-conjugated structures [121,122]. Due to the excellent optoelectronic properties and good processing properties of semiconductors, they were first applied in the field of organic optoelectronics. At present, semiconducting polymer materials commonly used in the biomedical field are divided into polyfluorene (PF), polythiophene (PT), poly(phenylene ethylene, PPE) and poly(p-phenylene vinylene, PPV) according to the structure of the main chain. Their derivatives [123,124] and structure are shown in Figure 13.

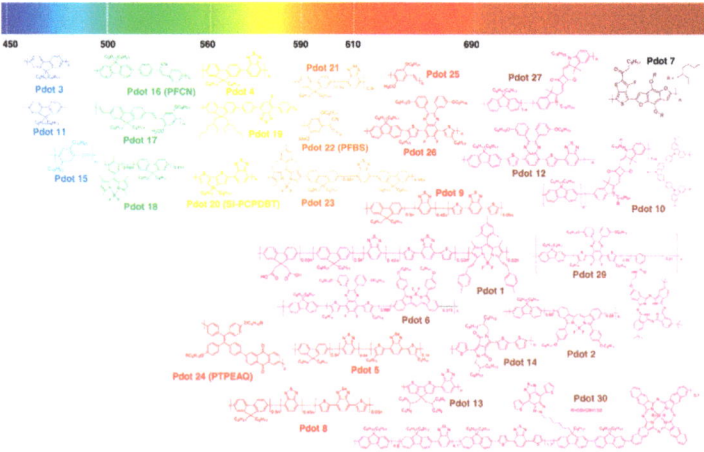

Figure 13. Chemical structures of semiconducting polymers [125]. Copyright (2019), with permission from Wiley.

From the perspective of the main chain structure of the semiconducting polymer, it is a typical rigid structure and hydrophobic, and aggregation occurs in aqueous solution, resulting in fluorescence quenching. At present, the commonly used method is to directly physically encapsulate semiconducting polymer materials in amphiphilic block polymers to prepare water-soluble polymer fluorescent nanoparticles (SPNs). Not only can this effectively solve the problem of monodispersion of semiconducting polymers in water, but the surface can also be easily functionalized. They have been widely used in the fields of tumor diagnosis [126–128] and antibacterials [129–131]. Among them [15], polyethylene glycol (PEG) is widely used to modify semiconducting polymers because of its characteristics of increasing drug solubility, reducing body immunity, and prolonging the residence time of drugs in the body [132].

In order to improve the fluorescence quantum efficiency of semiconducting polymers, Fan et al. used low-energy-band ester-based semiconducting polymers to skillfully control intramolecular charge transfer (ICT) to increase the intensity of NIR-II fluorescence [133]. As shown in Figure 14, as the thiophene group chain lengthened (TT-T to TT-3T), the ICT gradually weakened, and the corresponding NIR-II fluorescence emission gradually increased. TT-3T CPs (51–70 nm) were prepared by physically wrapping TT-3T with amphiphilic block polymer F127. They emit NIR-II light (1050 nm) in aqueous solution with a fluorescence quantum efficiency of 1.75%. Moreover, in vivo cell tracking, vascular system imaging, and lymphatic drainage mapping all had good imaging effects and high NIR-II spatial resolution.

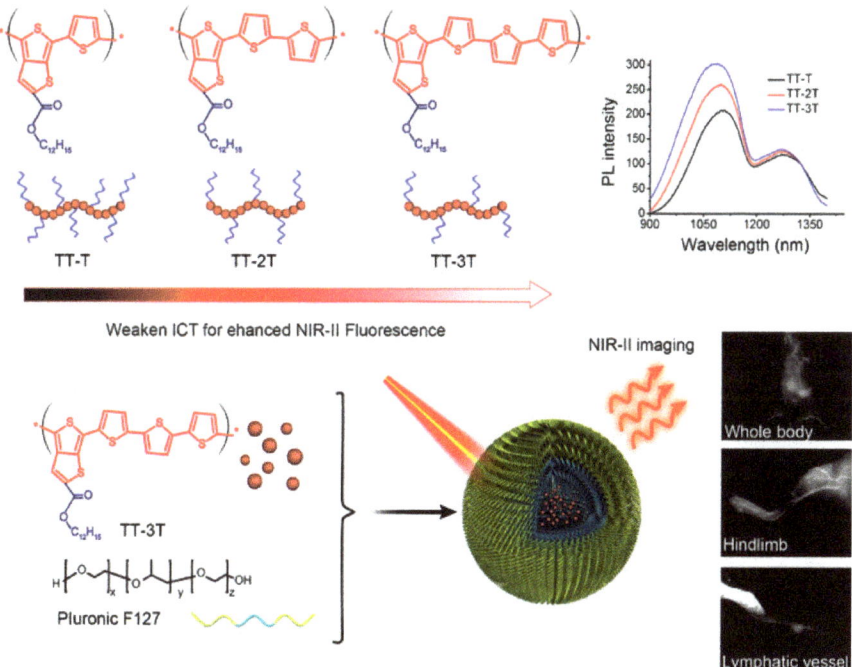

Figure 14. Schematic diagram of the synthesis of ester-based semiconducting polymer TT-3T CPs and their application in bioimaging [133]. Copyright (2019), with permission from American Chemical Society.

In order to further study the NIR-II real-time imaging application of semiconducting polymers in vivo, Hong et al. [134] designed and synthesized a NIR-II semiconducting polymer pDA (Figure 15a). After being physically encapsulated with the amphiphilic block polymer DSPE-MPEG (5kDa), pDA-PEG nanoparticles with a particle size of 2.9 nm were prepared (as shown in Figure 15b,c). The fluorescence emission wavelength is about 1000 nm (Figure 15d), the fluorescence quantum efficiency is about 1.7%, and it has been successfully applied to real-time imaging of vascular diseases.

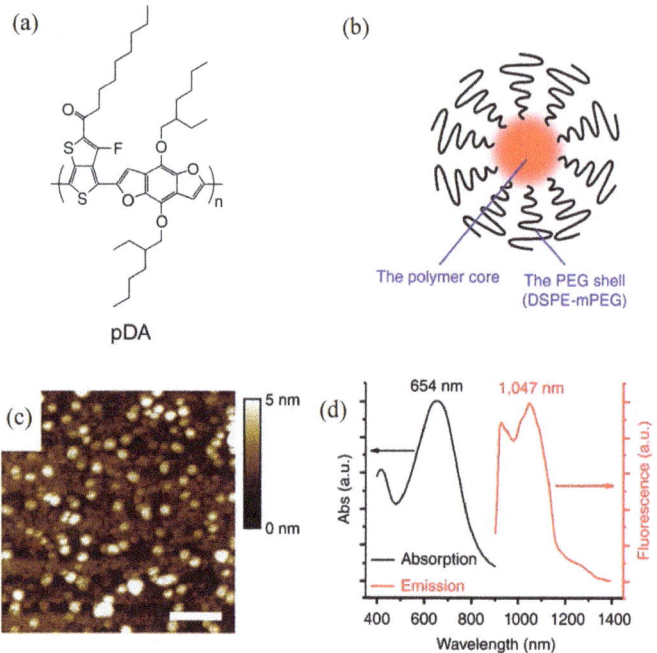

Figure 15. (**a**) Molecular structure diagram of semiconducting polymer pDA; (**b**) schematic diagram of pDA-PEG nanoparticles; (**c**) AFM image of pDA-PEG nanoparticles; (**d**) absorption and emission diagram of pDA-PEG nanoparticles aqueous solution [134]. Copyright (2014), with permission from Springer Nature.

Some semiconducting polymers are prone to the aggregation-caused quenching (ACQ) phenomenon after they are prepared into nanoparticles using physical encapsulation of amphiphilic block polymers [135,136]. Zhang et al. [137] used phenothiazine, which has typical AIE characteristics, as the electron donor. As shown in Figure 16a, different groups were introduced into the side chains to compare the effect of weakening ACQ. The study found that the emission wavelength of P3c modified with 9,10-diphenylanthracene (9,10-diphenylanthracene) was larger than that of P3a modified with hexane, and the emission intensity was high. Then the polymers P3a and P3c were physically encapsulated using the amphiphilic block polymer PS-PEG to prepare P3a NPs and P3c NPs, which were injected into mice. As shown in Figure 16b,c, the mice in the P3c NPs group glowed red, while the mice in the P3a NPs group did not. Additionally, because P3c NPs have strong NIR-II luminescent properties, the skull and cerebral blood vessels of mice can be clearly observed when performing imaging in mice.

Fluorescence brightness is determined using the absorption cross-section and the fluorescence quantum efficiency. Fluorescence quantum efficiency refers to the ratio of the number of emitted photons to the number of absorbed photons, which is one of the important parameters for evaluating the performance of fluorescent probes. The emission wavelength range of polymer quantum dot PFBT is comparable to that of Qdot 565, a commonly used fluorescent probe, inorganic semiconductor quantum dot, and IgG-Alexa 488, which contains approximately six dye molecules per IgG antibody. Therefore, the photophysical properties of the three are summarized and compared (see Table 4).

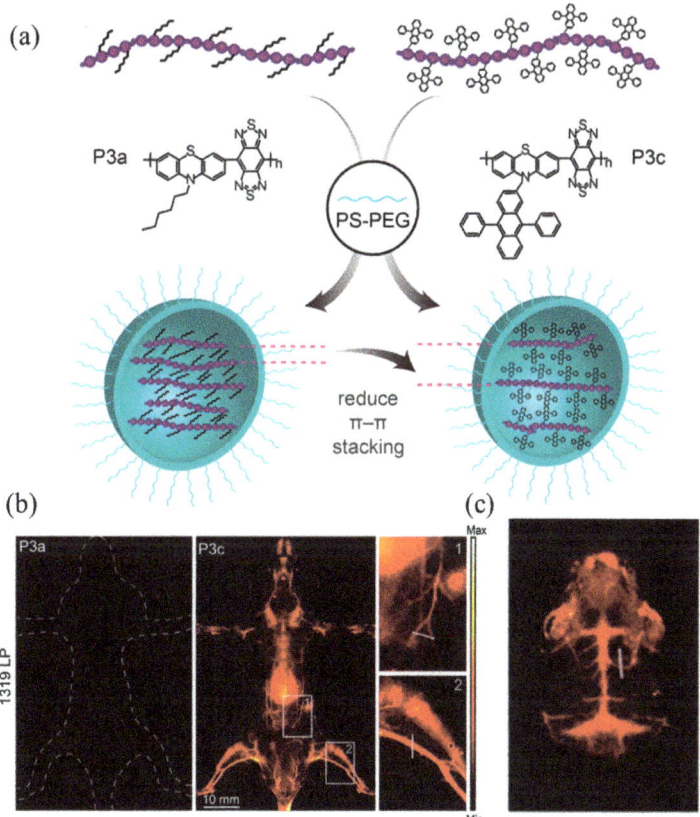

Figure 16. (**a**) Schematic diagram of the preparation of P3a NPs and P3c NPs; (**b**) NIR-II images of mice injected with P3a NPs and P3c NPs, respectively; (**c**) vascular NIR-II images of skull and brain of mice injected with P3c NPs [137]. Copyright (2020), with permission from Wiley.

Table 4. Several photophysical properties of PFBT Pdots, IgG-Alexa 488, and Qdot 565 [138].

Probes Size	PFBT ~10 nm	IgG-Alexa 488 ~1 nm	Qdot 565 ~15 nm
Abs/FL λ_{max} (nm)	460/540	496/519	UV/565
ε (M^{-1}cm^{-1}) λ = 488 nm	1.0×10^7	5.3×10^4	2.9×10^5
Quantum yield (%)	30	90	30~50
Fluorescence lifetime (ns)	0.6	4.2	~20

It can be seen from Table 1 that, when the three nanoparticles are excited by a laser with a wavelength of 488 nm, the single-particle luminescence brightness of PFBT quantum dots with a size of about 10 nm is about 30 times higher than that of IgG-Alexa 488 and Qdot 565. At a wavelength of 488 nm, the absorption cross-section of PFBT quantum dots is approximately half of its own peak absorption cross-section. The luminescence brightness of the three fluorescent probes was compared in parallel using single-particle imaging experiments [139]. Experiments have found that when the probe is excited by a laser with a wavelength of 488 nm, when the excitation power is low, a single PFBT nanoparticle with high luminescence brightness close to the diffraction limit can be observed. Under the same conditions, the luminescence of IgG-Alexa 488 and Qdot 565 probes was found to be very weak. The camera used in the experiment barely captured the fluorescent signal. After counting the fluorescence intensity distribution of thousands of nanoparticles, it was found that the luminescence brightness of PFBT nanoparticles was about 30 times higher than that of IgG-Alexa 488 and Qdot 565 probes. These experimental data are consistent with the results based on the comparison of photophysical parameters.

5.3. Polymer Nanomaterials Based on Organic Fluorescent Small Molecules

The luminescence intensity of fluorescent probes determines the signal-to-noise ratio and imaging depth of the probes. If polymer nanoparticles with higher brightness are to be prepared, more fluorescent materials need to be encapsulated into the polymer nanoparticles. However, in most fluorescent probe materials, as the concentration increases, the fluorescence aggregation-induced quenching (ACQ) phenomenon occurs [140,141]. Tang et al. [142] first proposed the aggregation-induced emission (AIE) phenomenon, and AIE materials are an effective way of solving the above problems. Qi et al. [143] designed and synthesized the AIE compound TQ-BPN, and prepared TQ-BPN nanoparticles with a particle size of 33 nm after physical wrapping with the amphiphilic block polymer Pluronic F-127 (as shown in Figure 17a). The fluorescent quantum effect was as high as 13.9%. Although the maximum emission wavelength was in the NIR-I region (808 nm), there was still a fluorescence quantum efficiency of 2.8% in the NIR-II region (Figure 17b). Additionally, when TQ-BPN nanoparticles were used as fluorescent probes to perform fluorescence imaging on a mouse's brain, it was found that the imaging spatial resolution reached 2.6 μm and the penetration depth reached 150 μm. More importantly, as shown in Figure 17c, clearly identifiable fluorescent signals can be seen at various stages of tumor growth, which can be applied to early diagnosis of cancer. According to the mechanism of aggregation-induced luminescence, more and more AIE molecules have been designed and synthesized by researchers. Typical AIE small molecules include hydrocarbon molecules (1–3 molecules) and heterocyclic small molecules (4–9 molecules) (see Figure 18).

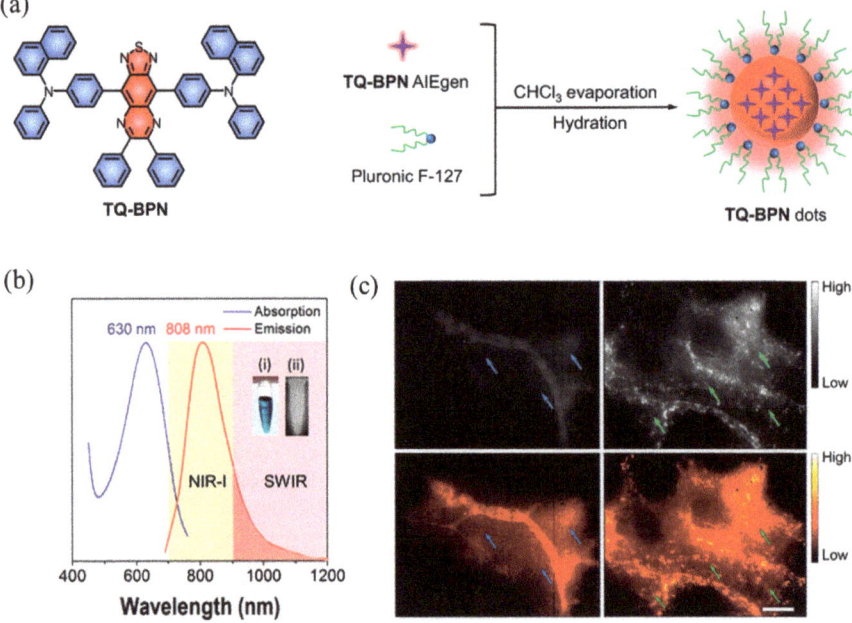

Figure 17. (**a**) Schematic diagram of TQ-BPN dot preparation; (**b**) TQ-BPN dot absorption and generation curves in aqueous dispersion; (**c**) NIR-II of TQ-BPN dots at different stages of tumor growth imaging [143]. Copyright (2018), with permission from Wiley.

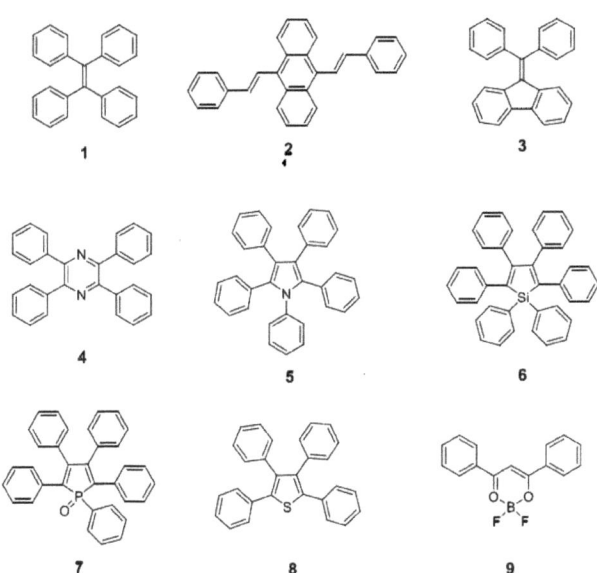

Figure 18. Typical AIE small molecules and their structures. 1. Tetraphenylethene (TPE); 2. Triphenylamine (TPA); 3. Phenothiazine (PTZ); 4. Benzothiazole (BTH); 5. 2-(2′-Hydroxyphenyl)benzoxazole (HBO); 6. 2-(4′-Biphenylyl)-5-(4″-tert-butylphenyl)-1,3,4-oxadiazole (BBD); 7. 1,2,3,4,5-pentaphenylsilole (PPS); 8. 1,2-bis(4′-phenylvinyl)benzene (DPVBi); 9. 1,4-bis(2,2-diphenylvinyl)benzene (DPVBi-Ph).

In order to further improve the fluorescence efficiency of fluorescent probes in NIR-II, Sheng et al. [144] designed and synthesized a new AIE material, TB-1, containing a DA structure. TB-1 dots with a particle size of 32 nm were prepared after physical encapsulation using DSPE-PEG2000, and the schematic diagram of the preparation is shown in Figure 19a. The maximum emission wavelength of the nanomaterial exceeded 1000 nm in the aqueous dispersion system, and the fluorescence efficiency was as high as 6.2%. The blood vessels in the brain of the mouse could be clearly seen without opening the mouse's cranium. In addition, the targeting group c-RGD was further modified to the surface of nanoparticles using a Michael addition reaction, prepared into TB1-RGD dots, and then injected into mice, respectively. From the comparison of Figure 19b,c, it can be seen that the TB1-RGD dot group modified with the targeting group has obvious imaging in the tumor part of the mouse at 24 h. The corresponding TB-1 dots with unmodified targeting groups had no obvious imaging effect.

Although the AIE material is wrapped in the polymer model using physical wrapping, the monodispersity and biocompatibility of the water system of the AIE material improve. However, there are also nanomaterials prepared using physical encapsulation in different batches, showing different particle sizes and encapsulation effects. To prepare AIE nanomaterials with good uniformity, Li et al. [145] alkynylated the AIE material TTB-OH. As shown in Figure 20a, amphiphilic polymers were prepared using amino-alkyne click polymerization with the amino-modified hydrophilic PEG polymer PEG-NH$_2$, which then self-assembled into SA-TTB NPs. At the same time, as a comparison, as shown in Figure 20b, NDP-TTB NPs were prepared by directly encapsulating TTB-OH with the amphiphilic block polymer DSPE-PEG2000. From the TEM comparison images of nanoparticles in Figure 20c III, it can be seen that the two SA-TTB NPs prepared using the self-assembly method are more uniform in particle size and better in monodispersity than the NDP-TTB NPs prepared using the physical encapsulation method. By dispersing the three kinds of nanoparticles into the water system, it was measured that their maximum emission wavelengths were all around 1050 nm, with little difference. However, the fluorescence efficiency of SA-TTB NPs in the water system was as high as 10.3%, which is much higher than that of physically encapsulated NDP-TTB NPs. Additionally, a resolution of 38 μm and a penetration depth of 1 cm can be achieved in mice.

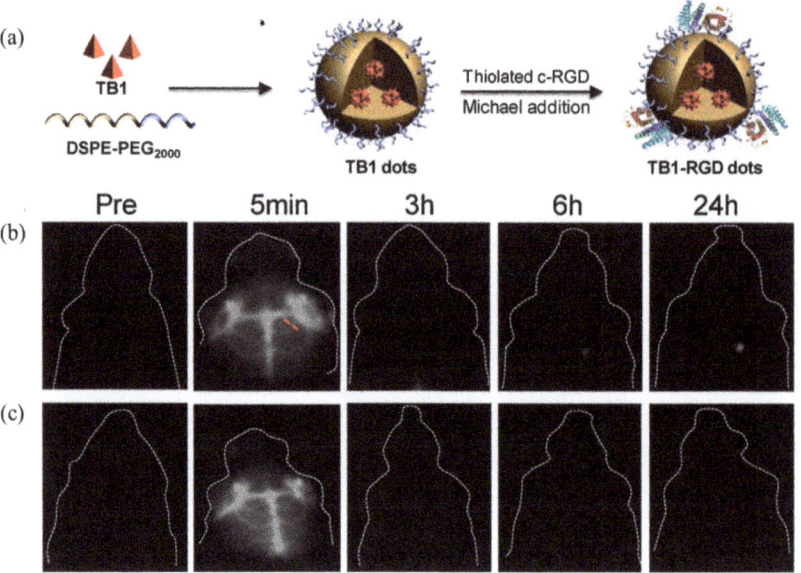

Figure 19. (a) Schematic diagram of the preparation of TB-1 dots and TB-RGD dots.; (b) imaging of TB-RGD dots injected into mice at different times. The red dashed line indicates the location of the intercepted area for calculating intensity of luminescence; (c) images of different time intervals of TB-1 dots injected into mice [144]. Copyright (2018), with permission from Wiley.

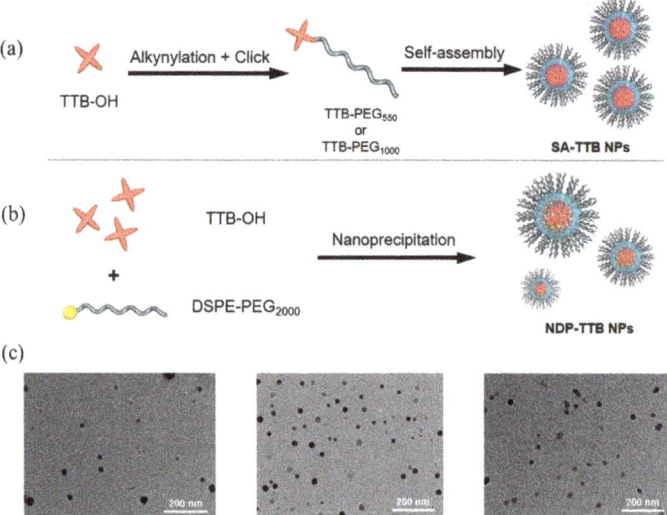

Figure 20. (a) Schematic diagram of TTB-OH self-assembly to prepare nanoparticles; (b) TTB-OH physical encapsulation method to prepare nanoparticles; (c) TEM images of three kinds of nanoparticles [145]. Copyright (2021), with permission from Elsevier.

In addition, based on the typical donor–acceptor–donor (D–A–D) structure, benzobisthiadiazole (BBTD) derivatives are representative organic small molecules [146,147]. With good biocompatibility, low biotoxicity, easy functionalization, and excellent metabolic ability, they show potential application prospects in the field of tumor imaging and treatment. However, BBTD-like derivatives are inherently hydrophobic and have no tumor-targeting ability. Polyethylene glycol (PEG) is often used for functional modification to prepare fluorescent polymer nanoprobes. Targeting groups are then attached to the surface of nanomaterials, so as to achieve partial targeted imaging of tumors and rapid metabolic clearance from organs such as the kidney or liver [148–150]. Dai et al. first used BBTD as the acceptor and triphenylamine (TPA) as the donor to prepare the compound CH1055 with a typical D–A–D structure. CH1055-PEG was prepared after modification with polyethylene glycol (PEG), and its maximum emission wavelength in water was 1055 nm. Experiments have shown that the imaging effect on mouse blood vessels and lymph is better than that of the commercial dye indocyanine green ICG, and about 90% of the material can be metabolically cleared from the kidney within 24 h [151]. In recent years, based on the compound CH1055, it has become possible to prepare a series of CH1055 derivatives by using different group modifications, such as the water-soluble CH-4T [152] prepared using sulfonation modification. With the help of follicle-stimulating hormone (FSH) modification, it is prepared into FSH-CH [153] and so on.

6. Summary and Outlook

Early diagnosis and treatment of cancer can greatly reduce its incidence and mortality. As a noninvasive and visualized diagnosis and treatment method, bioluminescence imaging technology, combined with traditional imaging technology, provides reliable imaging means for the early diagnosis of cancer. This article systematically introduces the research and development status and application prospects of fluorescent polymer nanomaterials based on rare earth luminescent materials, semiconducting polymers, and small organic molecules from recent years. Various types of luminescent probes have been developed for tumor diagnosis. However, due to the hydrophobic nature of the luminescent probe itself, when it is further functionalized, most of the luminescent materials are encapsulated inside the polymer nanoparticles using physical packaging. The method is simple to operate and has good universality. However, there are also defects, for example, the luminescent material leaks easily from the nanoparticles, the particle size of the nanomaterials prepared in different batches is not uniform, and the encapsulation rate of the materials is different. Therefore, searching for efficient functionalization methods for luminescent materials is still one of the core issues in the design and development of new luminescent materials.

In the decades since the development of polymeric nanomaterials in the field of optics, researchers have conducted a lot of work on synthesis control and material selection. While making breakthroughs, there is still demand for both high-performance and multifunctional materials. This also poses a higher challenge to the application of fluorescent polymer nanomaterials and biological fields. This is mainly reflected in the following aspects:

(a) Fluorescent materials should be combined with current scientific theories and technologies. With advancements in science and technology, optical materials, as a traditional research field, can be further developed in the direction of diversification, high technology, and high performance when combined with advanced theory and technology. Thus, new subject areas are created [154,155].

(b) Quantification of the structure–property relationship of multifunctional fluorescent polymer nanomaterials: Just as researchers have studied the "wetting" and "dewetting" of polymer-grafted core–shell particles in the matrix [156,157], essentially, the quantification of the effect of this structure on light transparency has developed from simple conclusions to a formalized theory. This method can be continued and expanded upon for other properties, forming quantitative structure–effect relationships for various properties.

(c) The trade-off between key properties in the design of optical materials: The discussion in this article maintains a relatively consistent train of thought with most of the work. That is, starting from the material with a certain performance, the study on the impact of the structure on the improvement of this single performance, while ignoring the impact on other performances, or even the actual application environment. This is not conducive to the multifunctionalization of optical materials. Starting from key structural factors such as particle size, grafting, and loading in the structure of hybrid materials, the synergistic effects of a certain change in structure on various properties can be discussed to form an application-oriented material performance trade-off strategy.

(d) Attention paid to the development of new optical functional materials: In addition to paying attention to the development of new photofunctional inorganic materials or polymer materials, we also need to pay attention to the development of new hybrid material systems. Carbonized polymer dots (CPDs) are a new type of nanofunctional elementary material that represent a new system of polymer nanohybridization [158,159]. In recent years, the advantages of this material have been reflected in its luminescent properties, and its synthesis process is considered to be a crosslinking carbonization process involving small molecules or polymer precursors. After several years of exploration, a family of CPDs with various luminescent properties such as full-color luminescence and narrow half-peak width emission has been obtained. Recently, Yang et al. further designed and applied this material to a material with both light transmission and mechanical properties, realizing the application of CPDs in the field of transparent optical films [160]. In addition, CPDs have also demonstrated their contributions in multiple fields such as imaging, sensing, and energy [161]. Their advantages such as low toxicity, environmental friendliness, and structural designability [162] lead us to believe that the introduction of new material systems such as CPDs will provide more excellent performance and broader application prospects for multifunctional fluorescent polymer nanomaterials.

(e) In order to promote the further application of fluorescent polymer nanomaterials in biomedicine, future research work can be optimized and expanded in the following aspects. At present, most light-emitting polymers have low light-emitting performance in NIR-II. The development of NIR-II light-emitting polymer materials with higher luminous intensity or photothermal efficiency using DA structure adjustment combined with theoretical calculations is still one of the important research directions for the future. Efficient enrichment of luminescent nanofunctional materials in tumor sites is another key to improving the efficiency of tumor diagnosis and treatment. Using rational molecular design, targeting groups can be effectively bonded to polymer chains to prepare light-emitting polymers with targeting functions, which is of great significance for the precise diagnosis and treatment of tumor sites. A single treatment for tumors is gradually being replaced by multimodal treatment. The therapeutic effect on tumors can be improved by constructing a multifunctional nanodiagnosis and treatment platform with properties such as chemotherapy, photodynamic therapy, or photothermal therapy.

(f) Internal/external stimuli-responsive fluorescent polymer nanoparticles that can be used in theranostics and sensing applications cannot be ignored either [163]. In particular, they respond to internal stimuli, including redox, pH, and enzymes, and external stimuli, including temperature, light, and magnetic fields, for drug delivery and sensing applications [164,165]. In terms of generating stimulus-responsive signals, these signals allow for amplification and easy detection of biologically relevant events. More detailed modeling of the photophysical properties of existing materials and their properties will provide decisive input for designing better performing NPs.

Author Contributions: Conceptualization, D.S. and K.S.; investigation, D.H. and Q.B.; writing—original draft preparation, P.L., W.L. and Y.X. All authors have read and agreed to the published version of the manuscript.

Funding: This work was supported by the Jilin Provincial Department of Education Science and Technology Research Planning Project (JJKH20220833KJ), Natural Science Foundation Project of Jilin Provincial Department of Science and Technology (YDZJ202301ZYTS348, YDZJ202101ZYTS092), Natural Science Foundation Projects of CCNU (CSJJ2022002ZK).

Institutional Review Board Statement: Not applicable.

Informed Consent Statement: Not applicable.

Data Availability Statement: Not applicable.

Conflicts of Interest: The authors declare no conflict of interest.

References

1. Yu, F.; Han, X.; Chen, L. Fluorescent probes for hydrogen sulfide detection and bioimaging. *Chem. Commun.* **2014**, *50*, 12234–12249. [CrossRef]
2. Guo, Z.; Park, S.; Yoon, J.; Shin, I. Recent progress in the development of near-infrared fluorescent probes for bioimaging applications. *Chem. Soc. Rev.* **2014**, *43*, 16–29. [CrossRef]
3. Abelha, T.F.; Dreiss, C.A.; Green, M.A.; Dailey, L.A. Conjugated polymers as nanoparticle probes for fluorescence and photoacoustic imaging. *J. Mater. Chem. B* **2020**, *8*, 592–606. [CrossRef] [PubMed]
4. Kim, S.-B.; Paulmurugan, R. Bioluminescent imaging systems for assay developments. *Anal. Sci.* **2021**, *37*, 233–247. [CrossRef]
5. Kim, S.-B.; Furuta, T.; Ohmuro-Matsuyama, Y.; Kitada, N.; Nishihara, R.; Maki, S.A. Bioluminescent imaging systems boosting near-infrared signals in mammalian cells. *Photochem. Photobiol. Sci.* **2023**, *2023*, 1–12. [CrossRef] [PubMed]
6. Yamada, K.; Nishizono, A. In Vivo Bioluminescent Imaging of Rabies Virus Infection and Evaluation of Antiviral Drug. In *Bioluminescence: Methods and Protocols*; Springer: Berlin/Heidelberg, Germany, 2022; Volume 1, pp. 347–352.
7. Sivaraman, G.; Iniya, M.; Anand, T.; Kotla, N.G.; Sunnapu, O.; Singaravadivel, S.; Gulyani, A.; Chellappa, D. Chemically diverse small molecule fluorescent chemosensors for copper ion. *Coord. Chem. Rev.* **2018**, *357*, 50–104. [CrossRef]
8. Yang, J.; Yang, Y.W. Metal–organic frameworks for biomedical applications. *Small* **2020**, *16*, 1906846. [CrossRef]
9. Liu, X.; Sun, X.; Liang, G. Peptide-based supramolecular hydrogels for bioimaging applications. *Biomater. Sci.* **2021**, *9*, 315–327. [CrossRef]
10. Michalet, X.; Pinaud, F.F.; Bentolila, L.A.; Tsay, J.M.; Doose, S.; Li, J.J.; Sundaresan, G.; Wu, A.; Gambhir, S.; Weiss, S. Quantum dots for live cells, in vivo imaging, and diagnostics. *Science* **2005**, *307*, 538–544. [CrossRef]
11. Shen, M.; Kong, F.; Tong, L.; Luo, Y.; Yin, S.; Liu, C.; Zhang, P.; Wang, L.; Chu, P.K.; Ding, Y. Carbon capture and storage (CCS): Development path based on carbon neutrality and economic policy. *Carbon Neutrality* **2022**, *1*, 37. [CrossRef]
12. Wang, F.; Tan, W.B.; Zhang, Y.; Fan, X.; Wang, M. Luminescent nanomaterials for biological labelling. *Nanotechnology* **2005**, *17*, R1. [CrossRef]
13. Mako, T.L.; Racicot, J.M.; Levine, M. Supramolecular luminescent sensors. *Chem. Rev.* **2018**, *119*, 322–477. [CrossRef]
14. Shen, M.; Ma, H. Metal-organic frameworks (MOFs) and their derivative as electrode materials for lithium-ion batteries. *Coord. Chem. Rev.* **2022**, *470*, 214715. [CrossRef]
15. Wang, Y.; Feng, L.; Wang, S. Conjugated polymer nanoparticles for imaging, cell activity regulation, and therapy. *Adv. Funct. Mater.* **2019**, *29*, 1806818. [CrossRef]
16. Zhu, S.; Tian, R.; Antaris, A.L.; Chen, X.; Dai, H. Near-infrared-II molecular dyes for cancer imaging and surgery. *Adv. Mater.* **2019**, *31*, 1900321. [CrossRef]
17. Li, T.; Liu, J.; Sun, X.-L.; Wan, W.-M.; Xiao, L.; Qian, Q. Boronic acid-containing polymeric nanomaterials via polymerization induced self-assembly as fructose sensor. *Polymer* **2022**, *253*, 125005. [CrossRef]
18. Abraham, J.E.; Balachandran, M. Fluorescent Mechanism in Zero-Dimensional Carbon Nanomaterials: A Review. *J. Fluoresc.* **2022**, *32*, 887–906. [CrossRef] [PubMed]
19. Zhai, S.; Chen, H.; Zhang, Y.; Li, P.; Wu, W. Nanocellulose: A promising nanomaterial for fabricating fluorescent composites. *Cellulose* **2022**, *29*, 7011–7035. [CrossRef]
20. Shen, M.; Tong, L.; Yin, S.; Liu, C.; Wang, L.; Feng, W.; Ding, Y. Cryogenic technology progress for CO_2 capture under carbon neutrality goals: A review. *Sep. Purif. Technol.* **2022**, *2022*, 121734. [CrossRef]
21. Manivasagan, P.; Kim, J.; Jang, E.-S. Recent progress in multifunctional conjugated polymer nanomaterial-based synergistic combination phototherapy for microbial infection theranostics. *Coord. Chem. Rev.* **2022**, *470*, 214701. [CrossRef]
22. Chauhan, N.; Saxena, K.; Jain, U. Smart nanomaterials employed recently for drug delivery in cancer therapy: An intelligent approach. *BioNanoScience* **2022**, *12*, 1356–1365. [CrossRef]
23. Shen, X.; Xu, W.; Ouyang, J.; Na, N. Fluorescence resonance energy transfer-based nanomaterials for the sensing in biological systems. *Chin. Chem. Lett.* **2022**, *33*, 4505–4516. [CrossRef]

24. Yang, X.; Qin, X.; Ji, H.; Du, L.; Li, M. Constructing firefly luciferin bioluminescence probes for in vivo imaging. *Org. Biomol. Chem.* **2022**, *20*, 1360–1372. [CrossRef] [PubMed]
25. Afshari, M.J.; Li, C.; Zeng, J.; Cui, J.; Wu, S.; Gao, M. Self-illuminating NIR-II bioluminescence imaging probe based on silver sulfide quantum dots. *ACS Nano* **2022**, *16*, 16824–16832. [CrossRef] [PubMed]
26. Yoon, S.; Cheon, S.Y.; Park, S.; Lee, D.; Lee, Y.; Han, S.; Kim, M.; Koo, H. Recent advances in optical imaging through deep tissue: Imaging probes and techniques. *Biomater. Res.* **2022**, *26*, 57. [CrossRef]
27. Elsabahy, M.; Heo, G.S.; Lim, S.-M.; Sun, G.; Wooley, K.L. Polymeric nanostructures for imaging and therapy. *Chem. Rev.* **2015**, *115*, 10967–11011. [CrossRef]
28. Smith, B.R.; Gambhir, S.S. Nanomaterials for in vivo imaging. *Chem. Rev.* **2017**, *117*, 901–986. [CrossRef] [PubMed]
29. Wang, B.; Zhou, X.-Q.; Li, L.; Li, Y.-X.; Yu, L.-P.; Chen, Y. Ratiometric fluorescence sensor for point-of-care testing of bilirubin based on tetraphenylethylene functionalized polymer nanoaggregate and rhodamine B. *Sens. Actuators B Chem.* **2022**, *369*, 132392. [CrossRef]
30. Ahumada, G.; Borkowska, M. Fluorescent polymers conspectus. *Polymers* **2022**, *14*, 1118. [CrossRef]
31. Bou, S.; Klymchenko, A.S.; Collot, M. Fluorescent labeling of biocompatible block copolymers: Synthetic strategies and applications in bioimaging. *Mater. Adv.* **2021**, *2*, 3213–3233. [CrossRef]
32. Zhang, P.; Xue, M.; Lin, Z.; Yang, H.; Zhang, C.; Cui, J.; Chen, J. Aptamer functionalization and high-contrast reversible dual-color photoswitching fluorescence of polymeric nanoparticles for latent fingerprints imaging. *Sens. Actuators B Chem.* **2022**, *367*, 132049. [CrossRef]
33. Mohammadi, R.; Naderi-Manesh, H.; Farzin, L.; Vaezi, Z.; Ayarri, N.; Samandari, L.; Shamsipur, M. Fluorescence sensing and imaging with carbon-based quantum dots for early diagnosis of cancer: A review. *J. Pharm. Biomed. Anal.* **2022**, *212*, 114628. [CrossRef]
34. Wan, W.; Li, Z.; Wang, X.; Tian, F.; Yang, J. Surface-fabrication of fluorescent hydroxyapatite for cancer cell imaging and bio-printing applications. *Biosensors* **2022**, *12*, 419. [CrossRef]
35. Jiang, M.-C.; Liu, H.-B.; Wang, J.-Q.; Li, S.; Zheng, Z.; Wang, D.; Wei, H.; Yu, C.-Y. Optimized aptamer functionalization for enhanced anticancer efficiency in vivo. *Int. J. Pharm.* **2022**, *628*, 122330. [CrossRef] [PubMed]
36. Nerantzaki, M.; Michel, A.; Petit, L.; Garnier, M.; Ménager, C.; Griffete, N. Biotinylated magnetic molecularly imprinted polymer nanoparticles for cancer cell targeting and controlled drug delivery. *Chem. Commun.* **2022**, *58*, 5642–5645. [CrossRef] [PubMed]
37. Shou, K.; Tang, Y.; Chen, H.; Chen, S.; Zhang, L.; Zhang, A.; Fan, Q.; Yu, A.; Cheng, Z. Diketopyrrolopyrrole-based semiconducting polymer nanoparticles for in vivo second near-infrared window imaging and image-guided tumor surgery. *Chem. Sci.* **2018**, *9*, 3105–3110. [CrossRef]
38. Zhang, Q.; Wang, X.; Cong, Y.; Kang, Y.; Wu, Z.; Li, L. Conjugated polymer-functionalized stretchable supramolecular hydrogels to monitor and control cellular behavior. *ACS Appl. Mater. Interfaces* **2022**, *14*, 12674–12683. [CrossRef]
39. Hamadani, C.M.; Chandrasiri, I.; Yaddehige, M.L.; Dasanayake, G.S.; Owolabi, I.; Flynt, A.; Hossain, M.; Liberman, L.; Lodge, T.P.; Werfel, T.A. Improved nanoformulation and bio-functionalization of linear-dendritic block copolymers with biocompatible ionic liquids. *Nanoscale* **2022**, *14*, 6021–6036. [CrossRef] [PubMed]
40. Sun, H.; Schanze, K.S. Functionalization of water-soluble conjugated polymers for bioapplications. *ACS Appl. Mater. Interfaces* **2022**, *14*, 20506–20519. [CrossRef]
41. Nothling, M.D.; Bailey, C.G.; Fillbrook, L.L.; Wang, G.; Gao, Y.; McCamey, D.R.; Monfared, M.; Wong, S.; Beves, J.E.; Stenzel, M.H. Polymer grafting to polydopamine free radicals for universal surface functionalization. *J. Am. Chem. Soc.* **2022**, *144*, 6992–7000. [CrossRef]
42. Chatterjee, S.; Lou, X.-Y.; Liang, F.; Yang, Y.-W. Surface-functionalized gold and silver nanoparticles for colorimetric and fluorescent sensing of metal ions and biomolecules. *Coord. Chem. Rev.* **2022**, *459*, 214461. [CrossRef]
43. Shahi, S.; Roghani-Mamaqani, H.; Talebi, S.; Mardani, H. Stimuli-responsive destructible polymeric hydrogels based on irreversible covalent bond dissociation. *Polym. Chem.* **2022**, *13*, 161–192. [CrossRef]
44. Nifant'ev, I.; Besprozvannykh, V.; Shlyakhtin, A.; Tavtorkin, A.; Legkov, S.; Chinova, M.; Arutyunyan, I.; Soboleva, A.; Fatkhudinov, T.; Ivchenko, P. Chain-End Functionalization of Poly (ε-caprolactone) for Chemical Binding with Gelatin: Binary Electrospun Scaffolds with Improved Physico-Mechanical Characteristics and Cell Adhesive Properties. *Polymers* **2022**, *14*, 4203. [CrossRef]
45. Lorandi, F.; Fantin, M.; Matyjaszewski, K. Atom Transfer Radical Polymerization: A Mechanistic Perspective. *J. Am. Chem. Soc.* **2022**, *144*, 15413–15430. [CrossRef] [PubMed]
46. Dworakowska, S.; Lorandi, F.; Gorczyński, A.; Matyjaszewski, K. Toward green atom transfer radical polymerization: Current status and future challenges. *Adv. Sci.* **2022**, *9*, 2106076. [CrossRef]
47. Kim, J.; Cattoz, B.; Leung, A.H.; Parish, J.D.; Becer, C.R. Enabling Reversible Addition-Fragmentation Chain-Transfer Polymerization for Brush Copolymers with a Poly (2-oxazoline) Backbone. *Macromolecules* **2022**, *55*, 4411–4419. [CrossRef]
48. Wang, J.-S.; Matyjaszewski, K. Controlled/"living" radical polymerization. atom transfer radical polymerization in the presence of transition-metal complexes. *J. Am. Chem. Soc.* **1995**, *117*, 5614–5615. [CrossRef]
49. Kato, M.; Kamigaito, M.; Sawamoto, M.; Higashimura, T. Polymerization of methyl methacrylate with the carbon tetrachloride/dichlorotris-(triphenylphosphine) ruthenium (II)/methylaluminum bis (2, 6-di-tert-butylphenoxide) initiating system: Possibility of living radical polymerization. *Macromolecules* **1995**, *28*, 1721–1723. [CrossRef]
50. Matyjaszewski, K.; Xia, J. Atom transfer radical polymerization. *Chem. Rev.* **2001**, *101*, 2921–2990. [CrossRef]

1. Coessens, V.; Pintauer, T.; Matyjaszewski, K. Functional polymers by atom transfer radical polymerization. *Prog. Polym. Sci.* **2001**, *26*, 337–377. [CrossRef]
2. Pan, X.; Fantin, M.; Yuan, F.; Matyjaszewski, K. Externally controlled atom transfer radical polymerization. *Chem. Soc. Rev.* **2018**, *47*, 5457–5490. [CrossRef] [PubMed]
3. Yuan, M.; Cui, X.; Zhu, W.; Tang, H. Development of environmentally friendly atom transfer radical polymerization. *Polymers* **2020**, *12*, 1987. [CrossRef] [PubMed]
4. Chakma, P.; Zeitler, S.M.; Baum, F.; Yu, J.; Shindy, W.; Pozzo, L.D.; Golder, M.R. Mechanoredox Catalysis Enables a Sustainable and Versatile Reversible Addition-Fragmentation Chain Transfer Polymerization Process. *Angew. Chem. Int. Ed.* **2023**, *62*, e202215733. [CrossRef] [PubMed]
5. Bradford, K.G.; Petit, L.M.; Whitfield, R.; Anastasaki, A.; Barner-Kowollik, C.; Konkolewicz, D. Ubiquitous Nature of Rate Retardation in Reversible Addition–Fragmentation Chain Transfer Polymerization. *J. Am. Chem. Soc.* **2021**, *143*, 17769–17777. [CrossRef]
6. Chiefari, J.; Chong, Y.; Ercole, F.; Krstina, J.; Jeffery, J.; Le, T.P.; Mayadunne, R.T.; Meijs, G.F.; Moad, C.L.; Moad, G. Living free-radical polymerization by reversible addition-fragmentation chain transfer: The RAFT process. *Macromolecules* **1998**, *31*, 5559. [CrossRef]
7. Moad, G.; Chiefari, J.; Chong, Y.K.; Krstina, J.; Mayadunne, R.T.A.; Postma, A.; Rizzardo, E.; Thang, S.H. Living free radical polymerization with reversible addition–fragmentation chain transfer (the life of RAFT). *Polym. Int.* **2000**, *49*, 993–1001. [CrossRef]
8. Li, M.; Fromel, M.; Ranaweera, D.; Rocha, S.; Boyer, C.; Pester, C.W. SI-PET-RAFT: Surface-initiated photoinduced electron transfer-reversible addition–fragmentation chain transfer polymerization. *ACS Macro Lett.* **2019**, *8*, 374–380. [CrossRef]
9. Zhang, X.; Yang, Z.; Jiang, Y.; Liao, S. Organocatalytic, stereoselective, cationic reversible addition–fragmentation chain-transfer polymerization of vinyl ethers. *J. Am. Chem. Soc.* **2021**, *144*, 679–684. [CrossRef]
10. Cheng, G.; Xu, D.; Lu, Z.; Liu, K. Chiral self-assembly of nanoparticles induced by polymers synthesized via reversible addition–fragmentation chain transfer polymerization. *ACS Nano* **2019**, *13*, 1479–1489. [CrossRef]
11. Strover, L.T.; Postma, A.; Horne, M.D.; Moad, G. Anthraquinone-mediated reduction of a trithiocarbonate chain-transfer agent to initiate electrochemical reversible addition–fragmentation chain transfer polymerization. *Macromolecules* **2020**, *53*, 10315–10322. [CrossRef]
12. Tuncel, D.; Demir, H.V. Conjugated polymer nanoparticles. *Nanoscale* **2010**, *2*, 484–494. [CrossRef]
13. Feng, L.; Zhu, C.; Yuan, H.; Liu, L.; Lv, F.; Wang, S. Conjugated polymer nanoparticles: Preparation, properties, functionalization and biological applications. *Chem. Soc. Rev.* **2013**, *42*, 6620–6633. [CrossRef]
14. Kurokawa, N.; Yoshikawa, H.; Hirota, N.; Hyodo, K.; Masuhara, H. Size-dependent spectroscopic properties and thermochromic behavior in poly (substituted thiophene) nanoparticles. *ChemPhysChem* **2004**, *5*, 1609–1615. [CrossRef]
15. Wu, C.; Szymanski, C.; McNeill, J. Preparation and encapsulation of highly fluorescent conjugated polymer nanoparticles. *Langmuir* **2006**, *22*, 2956–2960. [CrossRef] [PubMed]
16. Wu, C.; Hansen, S.J.; Hou, Q.; Yu, J.; Zeigler, M.; Jin, Y.; Burnham, D.R.; McNeill, J.D.; Olson, J.M.; Chiu, D.T. Design of highly emissive polymer dot bioconjugates for in vivo tumor targeting. *Angew. Chem.* **2011**, *123*, 3492–3496. [CrossRef]
17. Rivas, C.J.M.; Tarhini, M.; Badri, W.; Miladi, K.; Greige-Gerges, H.; Nazari, Q.A.; Rodríguez, S.A.G.; Román, R.Á.; Fessi, H.; Elaissari, A. Nanoprecipitation process: From encapsulation to drug delivery. *Int. J. Pharm.* **2017**, *532*, 66–81. [CrossRef]
18. Vodyashkin, A.A.; Kezimana, P.; Vetcher, A.A.; Stanishevskiy, Y.M. Biopolymeric nanoparticles–multifunctional materials of the future. *Polymers* **2022**, *14*, 2287. [CrossRef]
19. Kang, S.; Yoon, T.W.; Kim, G.-Y.; Kang, B. Review of Conjugated Polymer Nanoparticles: From Formulation to Applications. *ACS Appl. Nano Mater.* **2022**, *5*, 17436–17460. [CrossRef]
20. Wang, Q.-B.; Zhang, C.-J.; Yu, H.; Zhang, X.; Lu, Q.; Yao, J.-S.; Zhao, H. The sensitive "Turn-on" fluorescence platform of ascorbic acid based on conjugated polymer nanoparticles. *Anal. Chim. Acta* **2020**, *1097*, 153–160. [CrossRef]
21. Fang, F.; Li, M.; Zhang, J.; Lee, C.-S. Different strategies for organic nanoparticle preparation in biomedicine. *ACS Mater. Lett.* **2020**, *2*, 531–549. [CrossRef]
22. MacFarlane, L.R.; Shaikh, H.; Garcia-Hernandez, J.D.; Vespa, M.; Fukui, T.; Manners, I. Functional nanoparticles through π-conjugated polymer self-assembly. *Nat. Rev. Mater.* **2021**, *6*, 7–26. [CrossRef]
23. Ong, S.Y.; Zhang, C.; Dong, X.; Yao, S.Q. Recent advances in polymeric nanoparticles for enhanced fluorescence and photoacoustic imaging. *Angew. Chem. Int. Ed.* **2021**, *60*, 17797–17809. [CrossRef]
24. Sur, S.; Rathore, A.; Dave, V.; Reddy, K.R.; Chouhan, R.S.; Sadhu, V. Recent developments in functionalized polymer nanoparticles for efficient drug delivery system. *Nano-Struct. Nano Objects* **2019**, *20*, 100397. [CrossRef]
25. Indoria, S.; Singh, V.; Hsieh, M.-F. Recent advances in theranostic polymeric nanoparticles for cancer treatment: A review. *Int. J. Pharm.* **2020**, *582*, 119314. [CrossRef] [PubMed]
26. Cölfen, H. Double-hydrophilic block copolymers: Synthesis and application as novel surfactants and crystal growth modifiers. *Macromol. Rapid Commun.* **2001**, *22*, 219–252. [CrossRef]
27. Gaucher, G.; Dufresne, M.-H.; Sant, V.P.; Kang, N.; Maysinger, D.; Leroux, J.-C. Block copolymer micelles: Preparation, characterization and application in drug delivery. *J. Control. Release* **2005**, *109*, 169–188. [CrossRef] [PubMed]
28. Chen, J.Z.; Zhao, Q.L.; Lu, H.C.; Huang, J.; Cao, S.K.; Ma, Z. Polymethylene-b-polystyrene diblock copolymer: Synthesis, property, and application. *J. Polym. Sci. Part A Polym. Chem.* **2010**, *48*, 1894–1900. [CrossRef]

79. Zhou, Y.; Wang, Z.; Wang, Y.; Li, L.; Zhou, N.; Cai, Y.; Zhang, Z.; Zhu, X. Azoreductase-triggered fluorescent nanoprobe synthesized by RAFT-mediated polymerization-induced self-assembly for drug release. *Polym. Chem.* **2020**, *11*, 5619–5629. [CrossRef]
80. Sun, H.; Cao, W.; Zang, N.; Clemons, T.D.; Scheutz, G.M.; Hu, Z.; Thompson, M.P.; Liang, Y.; Vratsanos, M.; Zhou, X. Proapoptotic Peptide Brush Polymer Nanoparticles via Photoinitiated Polymerization-Induced Self-Assembly. *Angew. Chem.* **2020**, *132*, 19298–19304. [CrossRef]
81. Khor, S.Y.; Quinn, J.F.; Whittaker, M.R.; Truong, N.P.; Davis, T.P. Controlling nanomaterial size and shape for biomedical applications via polymerization-induced self-assembly. *Macromol. Rapid Commun.* **2019**, *40*, 1800438. [CrossRef]
82. Ramkumar, R.; Sundaram, M.M. A biopolymer gel-decorated cobalt molybdate nanowafer: Effective graft polymer cross-linked with an organic acid for better energy storage. *New J. Chem.* **2016**, *40*, 2863–2877. [CrossRef]
83. Ferguson, C.J.; Hughes, R.J.; Nguyen, D.; Pham, B.T.; Gilbert, R.G.; Serelis, A.K.; Such, C.H.; Hawkett, B.S. Ab initio emulsion polymerization by RAFT-controlled self-assembly. *Macromolecules* **2005**, *38*, 2191–2204. [CrossRef]
84. Wan, W.-M.; Hong, C.-Y.; Pan, C.-Y. One-pot synthesis of nanomaterials via RAFT polymerization induced self-assembly and morphology transition. *Chem. Commun.* **2009**, *39*, 5883–5885. [CrossRef] [PubMed]
85. Wan, W.-M.; Sun, X.-L.; Pan, C.-Y. Morphology transition in RAFT polymerization for formation of vesicular morphologies in one pot. *Macromolecules* **2009**, *14*, 4950–4952. [CrossRef]
86. Canning, S.L.; Smith, G.N.; Armes, S.P. A critical appraisal of RAFT-mediated polymerization-induced self-assembly. *Macromolecules* **2016**, *49*, 1985–2001. [CrossRef] [PubMed]
87. Buwalda, S.J.; Vermonden, T.; Hennink, W.E. Hydrogels for therapeutic delivery: Current developments and future directions. *Biomacromolecules* **2017**, *18*, 316–330. [CrossRef]
88. Zhang, M.; Huang, Y.; Pan, W.; Tong, X.; Zeng, Q.; Su, T.; Qi, X.; Shen, J. Polydopamine-incorporated dextran hydrogel drug carrier with tailorable structure for wound healing. *Carbohydr. Polym.* **2021**, *253*, 117213. [CrossRef]
89. Ilgin, P.; Ozay, H.; Ozay, O. Synthesis and characterization of pH responsive alginate based-hydrogels as oral drug delivery carrier. *J. Polym. Res.* **2020**, *27*, 251. [CrossRef]
90. Kabanov, A.V.; Vinogradov, S.V. Nanogels as pharmaceutical carriers: Finite networks of infinite capabilities. *Angew. Chem. Int. Ed.* **2009**, *48*, 5418–5429. [CrossRef]
91. Oh, J.K.; Drumright, R.; Siegwart, D.J.; Matyjaszewski, K. The development of microgels/nanogels for drug delivery applications. *Prog. Polym. Sci.* **2008**, *33*, 448–477. [CrossRef]
92. Nishizawa, Y.; Minato, H.; Inui, T.; Uchihashi, T.; Suzuki, D. Nanostructures, thermoresponsiveness, and assembly mechanism of hydrogel microspheres during aqueous free-radical precipitation polymerization. *Langmuir* **2020**, *37*, 151–159. [CrossRef] [PubMed]
93. Bai, F.; Yang, X.; Huang, W. Synthesis of narrow or monodisperse poly (divinylbenzene) microspheres by distillation− precipitation polymerization. *Macromolecules* **2004**, *37*, 9746–9752. [CrossRef]
94. Fan, M.; Wang, F.; Wang, C. Reflux precipitation polymerization: A new platform for the preparation of uniform polymeric nanogels for biomedical applications. *Macromol. Biosci.* **2018**, *18*, 1800077. [CrossRef]
95. Aillon, K.L.; Xie, Y.; El-Gendy, N.; Berkland, C.J.; Forrest, M.L. Effects of nanomaterial physicochemical properties on in vivo toxicity. *Adv. Drug Deliv. Rev.* **2009**, *61*, 457–466. [CrossRef]
96. Rama Narsimha Reddy, A.; Narsimha Reddy, Y.; Himabindu, V.; Rama Krishna, D. Induction of oxidative stress and cytotoxicity by carbon nanomaterials is dependent on physical properties. *Toxicol. Ind. Health* **2011**, *27*, 3–10. [CrossRef] [PubMed]
97. Mahmoudi, M.; Shokrgozar, M.A.; Sardari, S.; Moghadam, M.K.; Vali, H.; Laurent, S.; Stroeve, P. Irreversible changes in protein conformation due to interaction with superparamagnetic iron oxide nanoparticles. *Nanoscale* **2011**, *3*, 1127–1138. [CrossRef]
98. Liu, J.-Y.; Zhao, L.-Y.; Wang, Y.-Y.; Li, D.-Y.; Tao, D.; Li, L.-Y.; Tang, J.-T. Magnetic stent hyperthermia for esophageal cancer: An in vitro investigation in the ECA-109 cell line. *Oncol. Rep.* **2012**, *27*, 791–797.
99. Lanone, S.; Boczkowski, J. Biomedical applications and potential health risks of nanomaterials: Molecular mechanisms. *Curr. Mol. Med.* **2006**, *6*, 651–663. [CrossRef]
100. Rahman, I. Regulation of nuclear factor-κB, activator protein-1, and glutathione levels by tumor necrosis factor-α and dexamethasone in alveolar epithelial cells. *Biochem. Pharmacol.* **2000**, *60*, 1041–1049. [CrossRef]
101. Rahman, I.; Biswas, S.K.; Jimenez, L.A.; Torres, M.; Forman, H.J. Glutathione, stress responses, and redox signaling in lung inflammation. *Antioxid. Redox Signal.* **2005**, *7*, 42–59. [CrossRef]
102. Guanming, Q.; Xikum, L.; Tai, Q.; Haitao, Z.; Honghao, Y.; Ruiting, M. Application of rare earths in advanced ceramic materials. *J. Rare Earths* **2007**, *25*, 281–286. [CrossRef]
103. Heffern, M.C.; Matosziuk, L.M.; Meade, T.J. Lanthanide probes for bioresponsive imaging. *Chem. Rev.* **2014**, *114*, 4496–4539. [CrossRef]
104. Zhang, C.; Zhang, C.; Zhang, Z.; He, T.; Mi, X.; Kong, T.; Fu, Z.; Zheng, H.; Xu, H. Self-suspended rare-earth doped up-conversion luminescent waveguide: Propagating and directional radiation. *Opto-Electron. Adv.* **2020**, *3*, 190045. [CrossRef]
105. Zhu, X.; Zhang, J.; Liu, J.; Zhang, Y. Recent progress of rare-earth doped upconversion nanoparticles: Synthesis, optimization, and applications. *Adv. Sci.* **2019**, *6*, 1901358. [CrossRef]
106. Werts, M.H. Making sense of lanthanide luminescence. *Sci. Prog.* **2005**, *88*, 101–131. [CrossRef]

107. Montgomery, C.P.; Murray, B.S.; New, E.J.; Pal, R.; Parker, D. Cell-penetrating metal complex optical probes: Targeted and responsive systems based on lanthanide luminescence. *Acc. Chem. Res.* **2009**, *42*, 925–937. [CrossRef] [PubMed]
108. Moore, E.G.; Samuel, A.P.; Raymond, K.N. From antenna to assay: Lessons learned in lanthanide luminescence. *Acc. Chem. Res.* **2009**, *42*, 542–552. [CrossRef]
109. Ding, Z.; He, Y.; Rao, H.; Zhang, L.; Nguyen, W.; Wang, J.; Wu, Y.; Han, C.; Xing, C.; Yan, C. Novel Fluorescent Probe Based on Rare-Earth Doped Upconversion Nanomaterials and Its Applications in Early Cancer Detection. *Nanomaterials* **2022**, *12*, 1787. [CrossRef] [PubMed]
110. Li, H.; Wei, R.; Yan, G.-H.; Sun, J.; Li, C.; Wang, H.; Shi, L.; Capobianco, J.A.; Sun, L. Smart self-assembled nanosystem based on water-soluble pillararene and rare-earth-doped upconversion nanoparticles for ph-responsive drug delivery. *ACS Appl. Mater. Interfaces* **2018**, *10*, 4910–4920. [CrossRef]
111. Sun, L.-D.; Wang, Y.-F.; Yan, C.-H. Paradigms and challenges for bioapplication of rare earth upconversion luminescent nanoparticles: Small size and tunable emission/excitation spectra. *Acc. Chem. Res.* **2014**, *47*, 1001–1009. [CrossRef]
112. Cardoso Dos Santos, M.; Runser, A.; Bartenlian, H.; Nonat, A.M.; Charbonnière, L.J.; Klymchenko, A.S.; Hildebrandt, N.; Reisch, A. Lanthanide-complex-loaded polymer nanoparticles for background-free single-particle and live-cell imaging. *Chem. Mater.* **2019**, *31*, 4034–4041. [CrossRef]
113. Ai, K.; Zhang, B.; Lu, L. Europium-based fluorescence nanoparticle sensor for rapid and ultrasensitive detection of an anthrax biomarker. *Angew. Chem.* **2009**, *121*, 310–314. [CrossRef]
114. Tan, H.; Liu, B.; Chen, Y. Lanthanide coordination polymer nanoparticles for sensing of mercury (II) by photoinduced electron transfer. *ACS Nano* **2012**, *6*, 10505–10511. [CrossRef]
115. Binnemans, K. Lanthanide-based luminescent hybrid materials. *Chem. Rev.* **2009**, *109*, 4283–4374. [CrossRef] [PubMed]
116. Xu, D.; Zhou, X.; Huang, Q.; Tian, J.; Huang, H.; Wan, Q.; Dai, Y.; Wen, Y.; Zhang, X.; Wei, Y. Facile fabrication of biodegradable lanthanide ions containing fluorescent polymeric nanoparticles: Characterization, optical properties and biological imaging. *Mater. Chem. Phys.* **2018**, *207*, 226–232. [CrossRef]
117. Xu, D.; Liu, M.; Huang, Q.; Chen, J.; Huang, H.; Deng, F.; Wen, Y.; Tian, J.; Zhang, X.; Wei, Y. One-step synthesis of europium complexes containing polyamino acids through ring-opening polymerization and their potential for biological imaging applications. *Talanta* **2018**, *188*, 1–6. [CrossRef]
118. Fan, Y.; Zhang, F. A new generation of NIR-II probes: Lanthanide-based nanocrystals for bioimaging and biosensing. *Adv. Opt. Mater.* **2019**, *7*, 1801417. [CrossRef]
119. Cai, Y.; Wei, Z.; Song, C.; Tang, C.; Han, W.; Dong, X. Optical nano-agents in the second near-infrared window for biomedical applications. *Chem. Soc. Rev.* **2019**, *48*, 22–37. [CrossRef]
120. Tao, Z.; Dang, X.; Huang, X.; Muzumdar, M.D.; Xu, E.S.; Bardhan, N.M.; Song, H.; Qi, R.; Yu, Y.; Li, T. Early tumor detection afforded by in vivo imaging of near-infrared II fluorescence. *Biomaterials* **2017**, *134*, 202–215. [CrossRef]
121. Dimov, I.B.; Moser, M.; Malliaras, G.G.; McCulloch, I. Semiconducting polymers for neural applications. *Chem. Rev.* **2022**, *122*, 4356–4396. [CrossRef]
122. Liu, C.; Wang, K.; Gong, X.; Heeger, A.J. Low bandgap semiconducting polymers for polymeric photovoltaics. *Chem. Soc. Rev.* **2016**, *45*, 4825–4846. [CrossRef]
123. Scharber, M.C.; Sariciftci, N.S. Low band gap conjugated semiconducting polymers. *Adv. Mater. Technol.* **2021**, *6*, 2000857. [CrossRef]
124. Yan, X.; Xiong, M.; Deng, X.-Y.; Liu, K.-K.; Li, J.-T.; Wang, X.-Q.; Zhang, S.; Prine, N.; Zhang, Z.; Huang, W. Approaching disorder-tolerant semiconducting polymers. *Nat. Commun.* **2021**, *12*, 5723. [CrossRef]
125. Tsai, W.K.; Chan, Y.H. Semiconducting polymer dots as near-infrared fluorescent probes for bioimaging and sensing. *J. Chin. Chem. Soc.* **2019**, *66*, 9–20. [CrossRef]
126. Xu, L.; Cheng, L.; Wang, C.; Peng, R.; Liu, Z. Conjugated polymers for photothermal therapy of cancer. *Polym. Chem.* **2014**, *5*, 1573–1580. [CrossRef]
127. Chen, P.; Ma, Y.; Zheng, Z.; Wu, C.; Wang, Y.; Liang, G. Facile syntheses of conjugated polymers for photothermal tumour therapy. *Nat. Commun.* **2019**, *10*, 1192. [CrossRef] [PubMed]
128. Li, S.; Wang, X.; Hu, R.; Chen, H.; Li, M.; Wang, J.; Wang, Y.; Liu, L.; Lv, F.; Liang, X.-J. Near-infrared (NIR)-absorbing conjugated polymer dots as highly effective photothermal materials for in vivo cancer therapy. *Chem. Mater.* **2016**, *28*, 8669–8675. [CrossRef]
129. Chen, J.; Wang, F.; Liu, Q.; Du, J. Antibacterial polymeric nanostructures for biomedical applications. *Chem. Commun.* **2014**, *50*, 14482–14493. [CrossRef] [PubMed]
130. Chong, H.; Nie, C.; Zhu, C.; Yang, Q.; Liu, L.; Lv, F.; Wang, S. Conjugated polymer nanoparticles for light-activated anticancer and antibacterial activity with imaging capability. *Langmuir* **2012**, *28*, 2091–2098. [CrossRef]
131. Li, J.; Zhao, Q.; Shi, F.; Liu, C.; Tang, Y. NIR-mediated nanohybrids of upconversion nanophosphors and fluorescent conjugated polymers for high-efficiency antibacterial performance based on fluorescence resonance energy transfer. *Adv. Healthc. Mater.* **2016**, *5*, 2967–2971. [CrossRef] [PubMed]
132. Zhang, K.; Tang, X.; Zhang, J.; Lu, W.; Lin, X.; Zhang, Y.; Tian, B.; Yang, H.; He, H. PEG–PLGA copolymers: Their structure and structure-influenced drug delivery applications. *J. Control. Release* **2014**, *183*, 77–86. [CrossRef] [PubMed]

133. Zhang, W.; Huang, T.; Li, J.; Sun, P.; Wang, Y.; Shi, W.; Han, W.; Wang, W.; Fan, Q.; Huang, W. Facial control intramolecular charge transfer of quinoid conjugated polymers for efficient in vivo NIR-II imaging. *ACS Appl. Mater. Interfaces* **2019**, *11*, 16311–16319. [CrossRef] [PubMed]
134. Hong, G.; Zou, Y.; Antaris, A.L.; Diao, S.; Wu, D.; Cheng, K.; Zhang, X.; Chen, C.; Liu, B.; He, Y. Ultrafast fluorescence imaging in vivo with conjugated polymer fluorophores in the second near-infrared window. *Nat. Commun.* **2014**, *5*, 4206. [CrossRef] [PubMed]
135. Qi, J.; Hu, X.; Dong, X.; Lu, Y.; Lu, H.; Zhao, W.; Wu, W. Towards more accurate bioimaging of drug nanocarriers: Turning aggregation-caused quenching into a useful tool. *Adv. Drug Deliv. Rev.* **2019**, *143*, 206–225. [CrossRef]
136. Wang, B.W.; Jiang, K.; Li, J.X.; Luo, S.H.; Wang, Z.Y.; Jiang, H.F. 1, 1-Diphenylvinylsulfide as a Functional AIEgen Derived from the Aggregation-Caused-Quenching Molecule 1, 1-Diphenylethene through Simple Thioetherification. *Angew. Chem. Int. Ed.* **2020**, *59*, 2338–2343. [CrossRef]
137. Zhang, Z.; Fang, X.; Liu, Z.; Liu, H.; Chen, D.; He, S.; Zheng, J.; Yang, B.; Qin, W.; Zhang, X. Semiconducting Polymer Dots with Dual-Enhanced NIR-IIa Fluorescence for Through-Skull Mouse-Brain Imaging. *Angew. Chem. Int. Ed.* **2020**, *59*, 3691–3698. [CrossRef]
138. Wu, C.; Chiu, D.T. Highly fluorescent semiconducting polymer dots for biology and medicine. *Angew. Chem. Int. Ed.* **2013**, *52*, 3086–3109. [CrossRef]
139. Wu, C.; Schneider, T.; Zeigler, M.; Yu, J.; Schiro, P.G.; Burnham, D.R.; McNeill, J.D.; Chiu, D.T. Bioconjugation of ultrabright semiconducting polymer dots for specific cellular targeting. *J. Am. Chem. Soc.* **2010**, *132*, 15410–15417. [CrossRef]
140. Gao, M.; Tang, B.Z. Fluorescent sensors based on aggregation-induced emission: Recent advances and perspectives. *ACS Sens.* **2017**, *2*, 1382–1399. [CrossRef]
141. Tsai, W.-K.; Wang, C.-I.; Liao, C.-H.; Yao, C.-N.; Kuo, T.-J.; Liu, M.-H.; Hsu, C.-P.; Lin, S.-Y.; Wu, C.-Y.; Pyle, J.R. Molecular design of near-infrared fluorescent Pdots for tumor targeting: Aggregation-induced emission versus anti-aggregation-caused quenching. *Chem. Sci.* **2019**, *10*, 198–207. [CrossRef]
142. Luo, J.; Xie, Z.; Lam, J.W.; Cheng, L.; Chen, H.; Qiu, C.; Kwok, H.S.; Zhan, X.; Liu, Y.; Zhu, D. Aggregation-induced emission of 1-methyl-1, 2, 3, 4, 5-pentaphenylsilole. *Chem. Commun.* **2001**, *18*, 1740–1741. [CrossRef] [PubMed]
143. Qi, J.; Sun, C.; Zebibula, A.; Zhang, H.; Kwok, R.T.; Zhao, X.; Xi, W.; Lam, J.W.; Qian, J.; Tang, B.Z. Real-time and high-resolution bioimaging with bright aggregation-induced emission dots in short-wave infrared region. *Adv. Mater.* **2018**, *30*, 1706856. [CrossRef] [PubMed]
144. Sheng, Z.; Guo, B.; Hu, D.; Xu, S.; Wu, W.; Liew, W.H.; Yao, K.; Jiang, J.; Liu, C.; Zheng, H. Bright aggregation-induced-emission dots for targeted synergetic NIR-II fluorescence and NIR-I photoacoustic imaging of orthotopic brain tumors. *Adv. Mater.* **2018**, *30*, 1800766. [CrossRef] [PubMed]
145. Li, Y.; Hu, D.; Sheng, Z.; Min, T.; Zha, M.; Ni, J.-S.; Zheng, H.; Li, K. Self-assembled AIEgen nanoparticles for multiscale NIR-II vascular imaging. *Biomaterials* **2021**, *264*, 120365. [CrossRef]
146. Fang, Y.; Shang, J.; Liu, D.; Shi, W.; Li, X.; Ma, H. Design, synthesis, and application of a small molecular NIR-II fluorophore with maximal emission beyond 1200 nm. *J. Am. Chem. Soc.* **2020**, *142*, 15271–15275. [CrossRef] [PubMed]
147. Ye, F.; Huang, W.; Li, C.; Li, G.; Yang, W.C.; Liu, S.H.; Yin, J.; Sun, Y.; Yang, G.F. Near-Infrared Fluorescence/Photoacoustic Agent with an Intensifying Optical Performance for Imaging-Guided Effective Photothermal Therapy. *Adv. Ther.* **2020**, *3*, 2000170. [CrossRef]
148. Rao, R.S.; Singh, S.P. Near-Infrared (>1000 nm) Light-Harvesters: Design, Synthesis and Applications. *Chem. Eur. J.* **2020**, *26*, 16582–16593. [CrossRef]
149. Wang, Y.; Wang, M.; Xia, G.; Yang, Y.; Si, L.; Wang, H.; Wang, H. Maximal Emission beyond 1200 nm Dicyanovinyl-Functionalized Squaraine for in vivo Vascular Imaging. *Chem. Commun.* **2023**, *59*, 3598–3601. [CrossRef]
150. Tu, L.; Xu, Y.; Ouyang, Q.; Li, X.; Sun, Y. Recent advances on small-molecule fluorophores with emission beyond 1000 nm for better molecular imaging in vivo. *Chin. Chem. Lett.* **2019**, *30*, 1731–1737. [CrossRef]
151. Zhang, X.D.; Wang, H.; Antaris, A.L.; Li, L.; Diao, S.; Ma, R.; Nguyen, A.; Hong, G.; Ma, Z.; Wang, J. Traumatic brain injury imaging in the second near-infrared window with a molecular fluorophore. *Adv. Mater.* **2016**, *28*, 6872–6879. [CrossRef]
152. Antaris, A.L.; Chen, H.; Diao, S.; Ma, Z.; Zhang, Z.; Zhu, S.; Wang, J.; Lozano, A.X.; Fan, Q.; Chew, L. A high quantum yield molecule-protein complex fluorophore for near-infrared II imaging. *Nat. Commun.* **2017**, *8*, 15269. [CrossRef] [PubMed]
153. Feng, Y.; Zhu, S.; Antaris, A.L.; Chen, H.; Xiao, Y.; Lu, X.; Jiang, L.; Diao, S.; Yu, K.; Wang, Y. Live imaging of follicle stimulating hormone receptors in gonads and bones using near infrared II fluorophore. *Chem. Sci.* **2017**, *8*, 3703–3711. [CrossRef] [PubMed]
154. Kalay, S.; Stetsyshyn, Y.; Donchak, V.; Harhay, K.; Lishchynskyi, O.; Ohar, H.; Panchenko, Y.; Voronov, S.; Çulha, M. pH-Controlled fluorescence switching in water-dispersed polymer brushes grafted to modified boron nitride nanotubes for cellular imaging. *Beilstein J. Nanotechnol.* **2019**, *10*, 2428–2439. [CrossRef] [PubMed]
155. Nese, A.; Lebedeva, N.V.; Sherwood, G.; Averick, S.; Li, Y.; Gao, H.; Peteanu, L.; Sheiko, S.S.; Matyjaszewski, K. pH-responsive fluorescent molecular bottlebrushes prepared by Atom Transfer Radical polymerization. *Macromolecules* **2011**, *44*, 5905–5910. [CrossRef]
156. Liu, S.; de Beer, S.; Batenburg, K.M.; Gojzewski, H.; Duvigneau, J.; Vancso, G.J. Designer Core–Shell Nanoparticles as Polymer Foam Cell Nucleating Agents: The Impact of Molecularly Engineered Interfaces. *ACS Appl. Mater. Interfaces* **2021**, *13*, 17034–17045. [CrossRef] [PubMed]

57. Chen, Y.; Xu, H.; Ma, Y.; Liu, J.; Zhang, L. Diffusion of polymer-grafted nanoparticles with dynamical fluctuations in unentangled polymer melts. *Phys. Chem. Chem. Phys.* **2022**, *24*, 11322–11335. [CrossRef]
58. Xia, C.; Zhu, S.; Feng, T.; Yang, M.; Yang, B. Evolution and synthesis of carbon dots: From carbon dots to carbonized polymer dots. *Adv. Sci.* **2019**, *6*, 1901316. [CrossRef]
59. Ru, Y.; Ai, L.; Jia, T.; Liu, X.; Lu, S.; Tang, Z.; Yang, B. Recent advances in chiral carbonized polymer dots: From synthesis and properties to applications. *Nano Today* **2020**, *34*, 100953. [CrossRef]
60. Pan, K.; Liu, C.; Zhu, Z.; Feng, T.; Tao, S.; Yang, B. Soft–Hard Segment Combined Carbonized Polymer Dots for Flexible Optical Film with Superhigh Surface Hardness. *ACS Appl. Mater. Interfaces* **2022**, *14*, 14504–14512. [CrossRef]
61. Liu, J.; Li, R.; Yang, B. Carbon dots: A new type of carbon-based nanomaterial with wide applications. *ACS Cent. Sci.* **2020**, *6*, 2179–2195. [CrossRef]
62. Zeng, Q.; Feng, T.; Tao, S.; Zhu, S.; Yang, B. Precursor-dependent structural diversity in luminescent carbonized polymer dots (CPDs): The nomenclature. *Light Sci. Appl.* **2021**, *10*, 142. [CrossRef] [PubMed]
63. Lou, K.; Hu, Z.; Zhang, H.; Li, Q.; Ji, X. Information Storage Based on Stimuli-Responsive Fluorescent 3D Code Materials. *Adv. Funct. Mater.* **2022**, *32*, 2113274. [CrossRef]
64. Battistelli, G.; Cantelli, A.; Guidetti, G.; Manzi, J.; Montalti, M. Ultra-bright and stimuli-responsive fluorescent nanoparticles for bioimaging. *Wiley Interdiscip. Rev. Nanomed. Nanobiotech.* **2016**, *8*, 139–150. [CrossRef] [PubMed]
65. Mazrad, Z.A.I.; Lee, K.; Chae, A.; In, I.; Lee, H.; Park, S.Y. Progress in internal/external stimuli responsive fluorescent carbon nanoparticles for theranostic and sensing applications. *J. Mater. Chem. B* **2018**, *6*, 1149–1178. [CrossRef] [PubMed]

Disclaimer/Publisher's Note: The statements, opinions and data contained in all publications are solely those of the individual author(s) and contributor(s) and not of MDPI and/or the editor(s). MDPI and/or the editor(s) disclaim responsibility for any injury to people or property resulting from any ideas, methods, instructions or products referred to in the content.

Review

Not Only Graphene Two-Dimensional Nanomaterials: Recent Trends in Electrochemical (Bio)sensing Area for Biomedical and Healthcare Applications

Paola Di Matteo [1], Rita Petrucci [1] and Antonella Curulli [2,*]

[1] Dipartimento Scienze di Base e Applicate per l'Ingegneria, Sapienza University of Rome, 00161 Rome, Italy; p.dimatteo@uniroma1.it (P.D.M.); rita.petrucci@uniroma1.it (R.P.)
[2] Consiglio Nazionale delle Ricerche (CNR), Istituto per lo Studio dei Materiali Nanostrutturati (ISMN), 00161 Rome, Italy
* Correspondence: antonella.curulli@cnr.it

Abstract: Two-dimensional (2D) nanomaterials (e.g., graphene) have attracted growing attention in the (bio)sensing area and, in particular, for biomedical applications because of their unique mechanical and physicochemical properties, such as their high thermal and electrical conductivity, biocompatibility, and large surface area. Graphene (G) and its derivatives represent the most common 2D nanomaterials applied to electrochemical (bio)sensors for healthcare applications. This review will pay particular attention to other 2D nanomaterials, such as transition metal dichalcogenides (TMDs), metal–organic frameworks (MOFs), covalent organic frameworks (COFs), and MXenes, applied to the electrochemical biomedical (bio)sensing area, considering the literature of the last five years (2018–2022). An overview of 2D nanostructures focusing on the synthetic approach, the integration with electrodic materials, including other nanomaterials, and with different biorecognition elements such as antibodies, nucleic acids, enzymes, and aptamers, will be provided. Next, significant examples of applications in the clinical field will be reported and discussed together with the role of nanomaterials, the type of (bio)sensor, and the adopted electrochemical technique. Finally, challenges related to future developments of these nanomaterials to design portable sensing systems will be shortly discussed.

Keywords: electrochemical (bio)sensors; 2D nanomaterials; TMDs; MOFs; COFs; MXenes; biomedical analysis; healthcare

Citation: Di Matteo, P.; Petrucci, R.; Curulli, A. Not Only Graphene Two-Dimensional Nanomaterials: Recent Trends in Electrochemical (Bio)sensing Area for Biomedical and Healthcare Applications. *Molecules* **2024**, *29*, 172. https://doi.org/10.3390/molecules29010172

Academic Editors: Sudeshna Chandra and Heinrich Lang

Received: 20 November 2023
Revised: 20 December 2023
Accepted: 21 December 2023
Published: 27 December 2023

Copyright: © 2023 by the authors. Licensee MDPI, Basel, Switzerland. This article is an open access article distributed under the terms and conditions of the Creative Commons Attribution (CC BY) license (https://creativecommons.org/licenses/by/4.0/).

1. Introduction

The development of (bio)sensors for the quick and cost-effective determination of low-level analytes within a wide linearity range can be considered a key goal to be achieved in the field of biomedical analysis and healthcare [1–3], among others.

The main requirement to compete with conventional analytical methods is the realization of an accurate and robust analysis system through the development of sensing platforms capable of stabilizing the biorecognition element, if present, and effectively interweaving with the signal transduction system. Consequently, the selection of appropriate sensing materials is fundamental for achieving the required performance.

Among various (bio)sensors, the electrochemical ones can represent smart detection tools for biomedical analysis and healthcare as part of accurate, sensitive, specific, and rapid analysis systems [4,5].

Recently, the interest in two-dimensional (2D) nanomaterials has grown considerably, and one-atom-thick graphene (G) is definitely the best known and most studied [6].

Due to its peculiar physicochemical properties, the so-called 'graphene rush' has triggered the study of atomically thin sheets of other layered materials, such as transition metal dichalcogenides (TMDs), metal–organic frameworks (MOFs), covalent organic frameworks (COFs), and MXenes, among others.

The chemical and structural heterogeneity of these new 2D nanomaterials, whose properties are actually controlled and tuned by their dimensions, provides huge possibilities for basic and applied research. In particular, the high surface area-to-volume ratio together with their optical/electrical properties and biocompatibility make such materials extremely attractive for their application in sensing and biosensing areas, including electrochemical sensors and biosensors such as immunosensors, genosensors, enzyme-based biosensors, and aptasensors.

This review focuses on the application of various 2D nanomaterials beyond graphene in electrochemical biosensing and sensing systems for the analysis of significant analytes of clinical and biomedical interest, such as glucose, neurotransmitters, hormones, viruses, cancer biomarkers, and drugs, among others.

Several reviews reported in the literature are focused on the application of 2D nanomaterials in the (bio)sensing area [1,2,6–13], but no special attention has been paid to electrochemical (bio)sensors for biomedical and healthcare applications.

This review aims to provide an up-to-date overview of which 2D nanomaterials beyond G are applied to electrochemical (bio)sensors for biomedical and healthcare applications, considering the literature of the last five years (2018–2022), evidencing strengths, limits, and future perspectives. For this purpose, it is organized in two parts.

In the first part, the synthetic approach and properties of 2D nanomaterials used in the herein reported (bio)sensor examples are briefly described; the second part introduces the most significant examples regarding the determination of analytes of clinical and biomedical interest.

Special attention is paid to the role of the involved nanomaterial, type of receptor, recognition mechanism, and selectivity, beyond the possibility of using them on real samples, and to the comparison with official validation methods, if provided.

Brief comments and/or observations are reported at the end of each subsection, but a more detailed and in-depth discussion can be found in Section 4.

2. Two-Dimensional Nanomaterials

It is well-known that dimensions of 2D nanomaterials are not within the nanoscale and that they are considered single- or few-layer nanocrystalline materials with a planar morphology. Electronic conduction is restricted by the nanostructure's thickness and it is assumed delocalized on the plan of the sheet [2,4]. They make up a big family involving different elements, from transition metals to carbon, nitrogen, and/or sulfur [12]. Among them, G is the most popular for designing and developing electrochemical (bio)sensors, drawing attention to other 2D nanomaterials.

More recently, graphene-like 2D-layered nanomaterials such as MXenes, TMDs, COFs, and MOFs have evidenced strong mechanical strength, high surface areas, fast electron transfer kinetics, significant biocompatibility, and easy functionalization; for these reasons they have begun to be used to assemble and develop electrochemical sensors.

2.1. MXenes

MXenes display peculiar properties, including metallic electrical conductivity [14,15] and hydrophilicity, which are not always available in other 2D family members.

MXenes are 2D transitional metal carbides/nitrides/carbonitrides, obtained by the selective etching of the so-called MAX phases (Figure 1). MAX phases are conductive 2D layers of transition metal carbides/nitrides interconnected by the 'element A' (Group 13–16) through strong ionic, metallic, and covalent bonds [14,15]. In chemical exfoliation, method hydrofluoric acid (HF) or a mixture of lithium fluoride and hydrochloric acid (HCl) are used, while the electrochemical selective exfoliation procedure is assumed as a fluoride-free etching procedure [14].

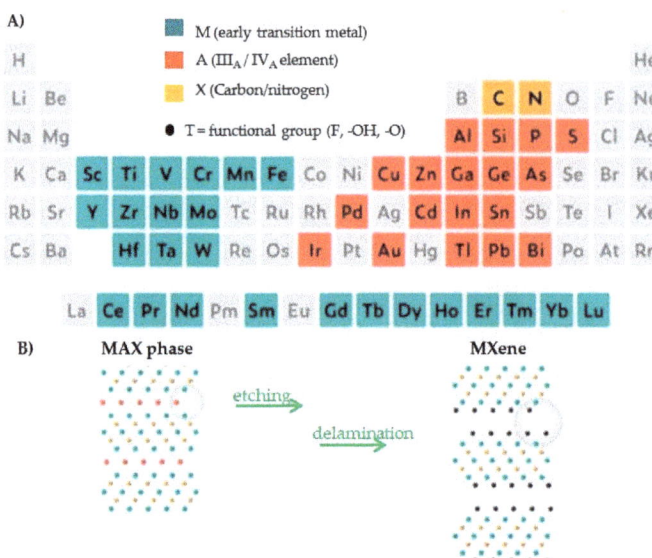

Figure 1. MXenes: (**A**) chemical composition starting from MAX and (**B**) schematics of the synthetic path.

$M_{n+1}X_nT_x$ is the general formula, M being an early transition metal (Mo, W, Ti, V, Sc, Y, Zr, Hf, Nb, Ta, or Cr) and X carbon or nitrogen with n between 1 and 3. T indicates surface functional groups, such as -O, -F, -OH, and rarely -Cl and x in T_x their number. MXenes are classified as layered sheet-like nanomaterials and their thicknesses, starting from 1 nm, can be modulated to vary in value in MXenes' general formula to provide different chemical compositions, using both computational and experimental methods.

MXenes have been intensively investigated for designing and assembling electrochemical (bio)sensors because of their high electrochemical activity and conductivity, and for their large surface area [16–20]. Notably, MXenes can be easily functionalized with other materials and their performance in the electrochemical sensing area relies on the type and number of functional groups [16,20]. MXenes nanohybrids can be synthesized including, for instance, metal nanoparticles and other nanomaterials, with a synergistic effect enhancing the sensing properties.

Notably, novel 2D monoelemental materials called Xenes, (e.g., borophene, silicene, germanene, stanene, phosphorene, arsenene, antimonene, bismuthene, and tellurene) have recently attracted considerable attention as promising 2D nanomaterials for biosensors, bioimaging, therapeutic delivery, theranostics, and other new bio-applications, due to their interesting physical, chemical, electronic, and optical properties [21].

When the lateral dimension of 2D nanomaterial is lower than 10 nm, 2D quantum dots (2D-QDs) can be produced owing to the strong quantum confinement effect [22,23]. Thus, 2D-QDs are generally classified as quantum dots derived from 2D materials. Although 2D-QDs might be considered zero-dimensional (0D) nanomaterials, they can actually be assumed as miniaturized/scaled-down structures of 2D-layered materials retaining their two-dimensional lattices. Compared with conventional forms, they present better solubility and dispersibility, beyond a larger surface area-to-volume ratio, so they can be easily doped and functionalized. In addition, they generally present electrochemical activity and high electrical conductivity. All these properties make 2D-QDs suitable for applications in various fields, including electrochemical sensing and biosensing areas, and appealing for the development of electrochemical sensors.

Two-dimensional quantum dots are generally synthesised by top-down approaches, involving the cleavage of bulk precursors such as the 2D nanomaterial, and by bottom-up approaches, including the aggregation and growth of small organic or inorganic molecules [22,23].

Regarding 2D-QDs derived from MXenes, the ultra-thin size involves a large specific surface area and a high number of functional groups, thereby allowing the binding of a high number of biomolecules and/or interacting more effectively with different analytes.

2.2. TMDs

TMDs represent a very interesting and unique 2D nanomaterial's family. Their general formula is MX_2, where M is a transition metal such as Ti, Zr, V, Nb, Mo, W, Hf, and Ta and X is a chalcogen such as Se, S, or Te [24,25]. TMDs are characterized by an M-X-M sandwich structure, in which the chalcogen X and the two metal atoms M in the same layer are covalently linked. Different layers are held together only by weak van der Waals interactions (Figure 2).

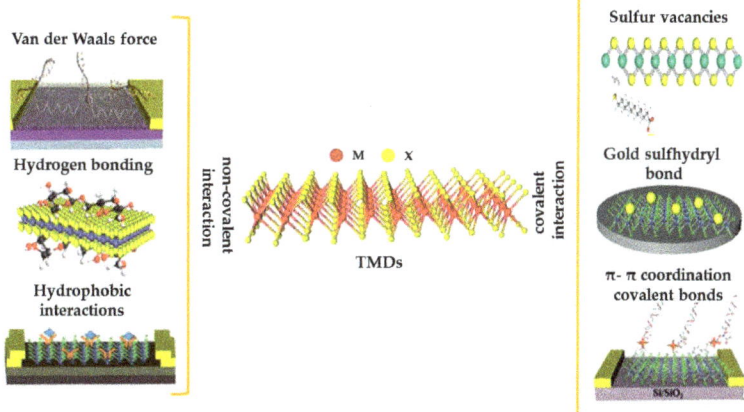

Figure 2. TMDs: structure (center), M = transition metal and X = chalcogen; structures and surface modification: non-covalent (**left**) and covalent (**right**) interactions.

According to their electronic band structure [26], TMDs can be defined as insulators (such as HfS_2), semiconductors (such as MoS_2 and WS_2), semi metals (such as WTe_2 and $TiSe_2$), and metals (such as NbS_2 and VSe_2).

Due to their physical, chemical, optical, and electronic features, such as high bandgap, large surface area-to-volume ratio, large number of active sites available for further functionalization, good electrical conductivity, and fast electron transfer, TMDs can be considered promising nanomaterials to be applied in the electrochemical sensing area [26,27].

Different synthetic approaches have been reported, such as mechanical cleavage, liquid and chemical exfoliation, chemical vapour deposition (CVD), and hydrothermal/solvothermal processes. They can be classified as top-down or bottom-up methods [26].

Starting from the top-down approaches, the mechanical cleavage is the most common method, involving mechanical exfoliation by using scotch tape or adhering polymeric thin films to obtain TMDs single-layer or multi-layer structures from bulky materials [28,29]. It is an easy and inexpensive technique, but it is time demanding, and quality and thickness are not properly managed.

In the chemical exfoliation, intercalators (organometallic compounds) are introduced inside the bulk material via ultrasound. This method develops different ultrathin TMDs without involving toxic solvents. However, it is time demanding and requires a high reaction temperature and high environmental awareness.

The liquid intercalation and exfoliation involve a stabilizer/dispersing agent and solvent. The liquid exfoliation requires three typical steps: bulk starting the material dispersion in solvent, exfoliation, and purification for isolating TMD-exfoliated layers from the unexfoliated ones. It is an easy and cost-effective method, and single-layer or multi-layer nanostructures can be produced on a large scale. Conversely, using toxic solvents can be considered a real disadvantage.

Moving to the bottom-up techniques, the hydrothermal and solvothermal approaches are conventional methods for the synthesis of colloidal TMDs.

The solvothermal method requires high boiling organic solvents and efficient nucleation and growth procedures. The presence of organic ligands is mandatory for controlling TMDs' morphology and size, and for guaranteeing dispersibility. The difference with the hydrothermal method is the precursor is not in an aqueous solution. Hydrothermal and solvothermal approaches are simple and easy to apply but not suitable for large-scale production.

Finally, the bottom-up CVD method under high temperature and pressure conditions produces TMDs' high-quality layers on different substrates. Chalcogenide atoms and transition metals are provided by the corresponding precursors. TMD films with peculiar electronic properties and optimized thicknesses were synthesized by means of the bottom-up CVD method. However, high-temperature and high-vacuum conditions are required for the synthetic protocol.

Two-dimensional quantum dots derived from TMDs [22,23] show large specific surface areas and a high number of functional groups, so they can be used for electrochemical sensors and biosensors, as already reported in Section 2.1 for 2D-QDs derived from MXenes.

2.3. MOFs

Metal–organic frameworks (MOFs) are assumed as organic–inorganic hybrid crystalline porous coordination polymers. They are assembled with inorganic and organic units, acting as ligands through coordination binding, to realize a network together with a flexible and tunable structure [30,31]. MOFs can represent very promising electrode materials for different application fields, since the selection of MOFs' "ingredients" will determine the geometry, form, and size of the pores, as well as the surface area and functionality of the structure. However, the most common MOFs have low electrical conductivity and relatively poor stability under electrochemical experimental conditions. To solve this critical issue, MOF-hybrid electrochemical sensing materials have been developed introducing metal nanoparticles (MNPs) or metal oxide nanoparticles (MONPs) in the corresponding reticulated structure.

Different strategies have been applied for the synthesis of MOFs (Figure 3). In addition, post-synthetic methods for surface functionalization and the modification of MOFs have been implemented during the last years [32,33]. The most common methods are reported and discussed below.

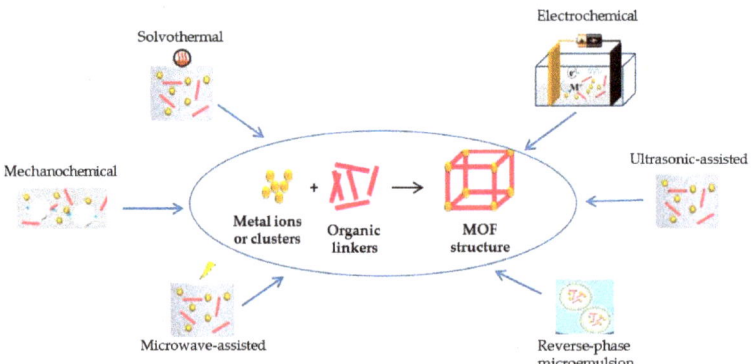

Figure 3. MOFs: synthetic strategies.

The solvothermal (or hydrothermal) method is a conventional procedure to synthesize MOFs crystalline structures, performed at the solvent boiling temperature, as previously reported for TMDs in Section 2.2. A direct coordination between organic ligands and metal clusters is obtained and the crystals' nucleation/growth rate is controlled by the applied temperature, assuring coordination between the inorganic and organic units, irrespective of the generated pressure. However, this method is time and energy demanding.

In the microwave-assisted method, the microwave irradiation supports MOF crystals' nucleation: growth rate, different morphologies, and size, together with faster crystallization, can be achieved by controlling the key parameters such as solvent, energy power, reactants concentration, and reaction time.

With regard to the electrochemical strategy, the anode and cathode are immersed in the electrolytic solution in the presence of the organic linker and the inorganic metal salt. Cathodic electrodeposition (CED), anodic electrodeposition (AED), and electrophoretic deposition (EPD) are the electrochemical synthetic strategies. The cathodic reduction of metal cations corresponds to CED, so MOFs are deposited onto the cathode. In AED, the anode is oxidized producing metal cations for MOF. EPD is a two-step protocol, where MOFs are deposited on the electrode surface in the presence of an applied electrical field. Different parameters can affect the synthesis of MOFs via EPD, such as conductivity, particle size, and so on. The cathodic reduction of metal cations is considered a possible drawback and the use of solvents, such as acrylonitrile, acrylates, and maleates, is strongly suggested for minimizing it.

The ultrasonic (ULS)-assisted approach involves acoustic cavitation phenomena, inducing the increase in temperature and pressure in the reaction system. MOFs at the nanoscale level are obtained with crystallization times faster than those obtained by more conventional approaches.

The mechanochemical method is used to prepare MOFs at room temperature without applying solvents, using a mechanical force for breaking the intramolecular bond in metal salts and creating new bonds between organic ligands and metal cations. On the other hand, this method can be used only for the synthesis of particular MOFs [32].

The reverse microemulsion technique is currently used to synthesize MOFs at the nanoscale level and includes an emulsified liquid phase (water drops in organic solvent or droplets of organic solvent in water). Surfactant molecules act as stabilizers of the emulsified solution and dispersing agents of the resulting nanostructure, thereby increasing MOFs' stability.

The topology-guided design includes ligands with increasing steric hindrance and changes in synthetic conditions to avoid interpenetration in MOFs. Using this approach, it is possible to tune structure and functionality of MOFs.

Different methods of post synthetic modifications are also reported in the literature [32]. They are not actual synthetic routes, but modification strategies for further functionalizing MOFs' structures, introducing proper functional groups.

2.4. COFs

Two-dimensional covalent organic frameworks (COFs) are porous polymers connecting organic molecules in two dimensions via covalent bonds [34,35]. Their structure and pores size are tunable and characterized by high surface area, thermal stability, and long-lasting porosity.

Several COFs have been designed and synthesized with different sizes, symmetries, and geometries of the structural units, such as COFs based on boric anhydride and borate (BCOFs), on Schiff base (SCOFs), on triazine groups (TCOFs), and on polyimide (PCOFs) [36].

The most common approaches are solvothermal, mechanochemical, solvent-free, microwave-assisted, and sonochemical synthesis [37] (Figure 4).

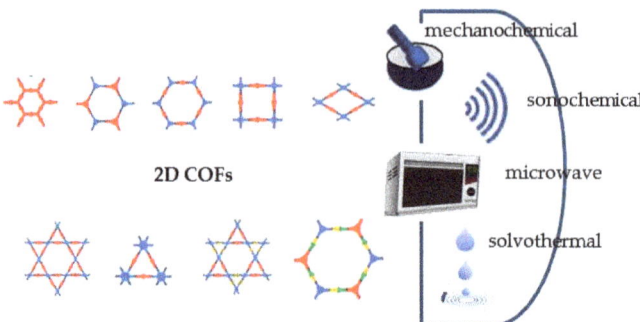

Figure 4. Two-dimensional COFs: structures and synthetic strategies.

The solvothermal method is the most common approach, obtaining highly crystalline COFs. As usual, this method is performed at the solvent boiling temperature, also involving high-pressure conditions, as previously reported for TMDs and MOFs (see Sections 2.2 and 2.3). The main drawbacks are the long-time synthetic procedure and the use of organic solvents.

The mechanochemical synthesis represents an alternative strategy for synthesizing COFs, being easy to perform, efficient, solvent-free, and operating at room temperature. A mechanical force is used for breaking intramolecular bonds in organic molecules, placed in a mortar, and for connecting them in the porous polymeric structure [38]. The addition of small amounts of catalyst in the mortar can accelerate the reaction rate, homogenizing the reagents and thus improving polymer crystallinity [38]. In fact, COFs obtained through this method usually present a lower surface area and poorer crystallinity compared with those prepared by the solvothermal method.

The solvent-free synthesis method is considered an alternative route for COFs, because it is assumed to be environmentally friendly, easy to perform, low-cost, and suitable for large-scale synthesis. The experimental conditions, in terms of temperature and pressure, are comparable with those of the solvothermal approach, involving, in this case, a solid-state catalyst to enhance COFs' crystallinity and yield. The main disadvantages are due to the use of a solid-state catalyst: high pressure and high temperature.

The microwave-assisted approach has gained increased attention because it is fast and environmentally friendly, involves higher yields, and consumes less energy. Microwave irradiation supports the polymerization process, and different morphologies and sizes, along with faster crystallization, can be achieved by controlling key parameters such as solvency, energy power, reactants' concentration, and reaction time, as already reported for MOFs in Section 2.3.

The sonochemical method can improve the homogeneity of COFs and accelerate the rate of polymerization and crystallization, due to the significant increase in pressure and temperature induced by the ultrasonic wave.

However, COFs usually have poor electrical conductivity, which is a real disadvantage for their application in electrochemical sensors and biosensors. Combining COFs with different conducting materials, such as carbon-based nanomaterials, metal and metallic oxides nanoparticles, and/or conducting polymers, can be considered an effective strategy to improve COFs' application in the electrochemical (bio)sensing area.

3. Applications of 2D Nanomaterials to Electrochemical (Bio)Sensors for Healthcare and Biomedical Field

Electrochemical (bio)sensors represent a well-known class of diagnostic systems for the healthcare and biomedical field [1,39–41]. In fact, they enable the simultaneous analysis of analytes, also at low concentrations and under different conditions, by using simple and low-cost instrumentation. Moreover, portability and miniaturization are possible in many cases [39].

The critical analytical parameters of electrochemical (bio)sensors, such as sensitivity, selectivity, and stability, are mainly controlled by the design of the sensing interface and platform. Different electrochemical techniques are generally involved, such as amperometry (A), chronoamperometry (CA), cyclic voltammetry (CV), linear sweep voltammetry (LSV), differential pulse voltammetry (DPV), square-wave voltammetry (SWV), chronocoulometry (CC), and electrochemical impedance spectroscopy (EIS) [38,39], and their applicability has been improved with the introduction of new functional materials such as 2D nanomaterials [13,42–46]. Several examples of electrochemical (bio)sensors for clinical analysis and healthcare using 2D nanomaterials such as MXenes, TMDs, MOFs, and COFs are reported and discussed below.

3.1. Glucose

Diabetes mellitus represents a worldwide public health problem since it is regarded as the seventh cause of death in the world, according to the World Health Organization (WHO) [39,47]. It is a metabolic disorder caused by an altered insulin action involving blood glucose concentration higher than the normal range (4.4–6.6 mol L^{-1}). Also, a glucose level decrease represents a pathological condition, called hypoglycaemia, causing a loss of consciousness in the most serious cases. For these reasons, glucose monitoring is fundamental for diabetics, and fast and reliable glucose sensors for healthcare monitoring are required. Considering the great improvements in nanomaterials over the last decade, an increasing interest in nanomaterial-based electrochemical glucose sensors has been highlighted. Combining glucose biosensors with the particular properties of 2D nanomaterials has significantly improved sensitivity and selectivity. Reviews on electrochemical glucose sensors based on 2D nanomaterials have been recently published. However, they are mainly focused on the application of graphene and its derivatives [48,49].

The first example herein reported is a non-enzymatic glucose electrochemical sensor based on a nanocomposite including a Co-based porous metal–organic framework (MOF) such as ZIF-67 and Ag nanoparticles (Ag@ZIF-67) [50]. Poor conductivity is a real disadvantage for most MOFs, because it can imply a decrease in sensitivity. On the other hand, MOFs' high porosity and flexibility allow the introduction of functional metal nanoparticles (MNPs) within the pores' cages/channels to form metal@MOF nanocomposites [50], in which MOF acts as a supporter to wrap MNPs, and the incorporated MNPs are uniformly dispersed in the framework. Ag nanoparticles (AgNPs) were selected because of their high conductivity and biocompatibility, and they were encapsulated within ZIF-67 by a sequential deposition–reduction method [50]. In this case, we have to note that, considering the convention that a nanomaterial is only one such if one or more external dimensions range from 1 to 100 nm, ZIF-67 is not a proper nanostructure because it is a dodecahedron with a size of about 200 nm. A glassy carbon electrode (GCE) was then modified with the resulting Ag@ZIF-67 nanocomposite by drop-casting the composite suspension onto the electrode surface. Glucose was analysed using amperometry, and a linearity range from 2 to 1000 μM with a limit of detection (LOD) of 0.66 μM was obtained. Selectivity and stability were evaluated. Selectivity was evaluated in the presence of usual interferences for glucose such as uric acid (UA), ascorbic acid (AA), and acetaminophen (AP) that did not affect the amperometric response. Regarding the operational stability, a slight activity loss of 6.5% was observed after 25 consecutive analysis cycles; considering the long-term stability, a decrease of 5.8% in the electrochemical response was found after one month at room temperature (RT). Reproducibility, repeatability, and application to real samples were not considered.

Another non-enzymatic glucose sensor based on ZIF-67 including Ag-doped TiO_2 nanoparticles (Ag@TiO_2NPs) was developed [51]. The hybrid nanocomposite was synthesized using the solvothermal method and drop-casted on GCE. In this case, the poor conductivity of MOF was improved by introducing metal-doped nanoparticles to obtain a nanocomposite, in which MOF acts as a supporter to encapsulate Ag@TiO$_2$NPs uniformly dispersed in the ZIF-67 nanostructure. Glucose was amperometrically detected, and a linearity range from 48μM to 1mM with an LOD of 0.99 μM was obtained. Selectivity was evaluated in the presence of UA, AA, AP, and dopamine (DA) as interferences that did not affect the amperometric response. Reproducibility, repeatability, stability, and application to real samples were not examined.

A flexible Ni–Co MOF/Ag/reduced graphene oxide/polyurethane (Ni–Co MOF/Ag/rGO/PU) fiber-based wearable electrochemical sensor was assembled for monitoring glucose in sweat [52]. rGO/PU fiber was produced by wet spinning technology, and a Ni–Co MOF nanosheet was deposited on its surface to set up the Ni–Co MOF/Ag/rGO/PU (NCGP) fiber electrode. Due to Ni–Co MOF's large specific surface area and high catalytic activity, the fiber sensor showed good electrochemical performances, with a wide linear range of 10 μM–0.66 mM and an LOD of 3.28 μM. In addition, the NCGP fiber electrode displayed significant stretching and bending stability under mechanical deformation. Selectivity and stability were investigated. Selectivity was evaluated in the presence of interferences such as lactic acid (LA), sodium chloride (NaCl), UA, cysteine (Cys), and DA, which did not affect the amperometric response. Long-term stability was evaluated after 7 days at RT, with a 2.90% decrease in the electrochemical response. As the human body temperature can change after exercise, the sensor was also tested at different temperatures, i.e., 24, 30, and 38 °C, and experimental data evidenced that temperature had little influence on glucose detection. An NCGP fiber-based three-electrode system, including an NCGP fiber as a working electrode, Ag/AgCl/rGO/PU as a reference electrode, and a Pt wire as a counter electrode, was sewn with a sweat absorbent cloth and set on a stretchable polydimethylsiloxane film substrate to provide a non-enzymatic sweat glucose wearable sensor, for the real-time monitoring of glucose in human sweat. Data obtained from two different patients were comparable with those coming from a commercial blood glucometer. In addition, a good correlation between glucose level changes in sweat and in blood was found.

A dual-confinement strategy in MOFs' nanocage-based structure was involved in realizing a sensor for glucose in sweat [53]. The enzyme retained its activity and stability, even if incorporated in an MOF nanocage, thereby avoiding the enzyme leakage and maintaining its conformational structure. Moreover, the mass transport resulted enhanced modifying the MOF-nanocage mesoporosity. ZIF-67 was firstly synthesized through a bottom-up method. Subsequently, GOX and Hemin were uniformly adsorbed on the surface of ZIF-67 (ZIF-67@GOX/Hemin) via coordination between metal cations and the Hemin carbonyl group. ZIF-67@GOX/Hemin acted as a template for the synthesis of a well-defined core-shell composite (ZIF-67@GOX/Hemin@ZIF-8).

The sensor stability was investigated under different conditions, denaturing for the free-enzyme, such as high temperature or treatment with urea, DMF, or DMSO. The sensor retained its bioactivity at 72.1% after exposure at 80 °C, at 84.6% after the urea treatment, at 70.7% after DMF exposure, and at 90.1% after DMSO exposure. Reproducibility was also investigated with satisfactory results in terms of RSD% (2.4%). Fructose, AA, DA, UA, maltose, and lactose were selected as possible interfering molecules, and their presence did not affect the glucose electrochemical signal. Furthermore, a GOX/Hemin@NC-ZIF-based sensor and printed circuit board (PCB) were integrated into a sweatband for the continuous real-time monitoring of glucose in human sweat, and a good correlation between glucose level changes in sweat and in blood was evidenced. The miniaturized, portable and all-integrated glucose sensor included GOX/Hemin@NC-ZIF as a working electrode, carbon paste, and Ag/AgCl paste as a counter electrode and reference electrode, respectively. A linearity range was obtained amperometrically from 50 to 600 μM with LOD of 2 μM.

A flexible carbon cloth (CC) was used as a matrix to grow NiCo$_2$O$_4$ nanorods using the hydrothermal method, and then to synthesize in situ the composite ZIF-67@GO, including ZIF-67 and graphene oxide (GO). A ZIF-67@GO/NiCo$_2$O$_4$/CC hybrid system provides a peculiar structure and it was used as a sensing platform for glucose detection [54].

NiCo$_2$O$_4$ nanorod arrays offer growth sites for ZIF-67 nanocubes and fast electron transport pathways for CC and ZIF-67@GO. The presence of GO in MOF provides high conductivity and stability to the composite ZIF-67@GO. Considering the electrochemical non-enzymatic sensing platform, ZIF-67@GO/NiCo$_2$O$_4$/CC acted as a working electrode, a platinum plate as a counter electrode, and Ag/AgCl as a reference electrode. After the morphological, compositional, and electrochemical characterization of the working electrode, glucose was detected using chronoamperometry, achieving a linear range from 0.3 µM to 5.407 mM with LOD of 0.16 µM and a response time within 2 s. Selectivity, repeatability, and stability were investigated. AA, UA, DA, and fructose, sucrose, lactose, maltose, as sugars structurally similar to glucose, were selected as possible interfering compounds, and they did not affect the current value even in large amounts. Repeatability was acceptable in terms of RSD% (1.65%). Long-term stability was tested, and the electrochemical response decreased by 4.20% after 30 days, under not better specified experimental storage conditions. Any application to real samples was not addressed.

Bimetallic CuCo-MOFs were synthesized in situ on nickel foam (NF) electrodes through a simple one-pot hydrothermal procedure and the NF-modified electrode was used as a non-enzymatic sensing platform for glucose detection [55]. Ni foam as a supporting material provides a large surface area, stability, and good electrical conductivity, so MOF criticalities of poor stability and electrical conductivity might be solved. A CuCo-MOF/NF electrode, a platinum sheet, and a standard saturated calomel electrode (SCE) acted as working, counter, and reference electrodes, respectively. Chronoamperometry was used to detect glucose, and a linearity range of 0.05–0.5 mM with LOD of 0.023 mM was obtained. AA, UA, DA, and NaCl were identified as possible interferences, and the response was negligible compared with glucose. Reproducibility was addressed with satisfactory results in terms of RSD% (2.17%). Stability and applications to real samples were not considered.

Ni-based metal–organic framework (Ni-MOF) nanosheets were used as precursors to create Ni-MOF@Ni-2,3,6,7,10,11-hexahydroxytriphenylene (HHTP) core@shell structures by introducing HHTP as a π–conjugated molecule [56]. HHTP interacted with free Ni^{2+} ions on a Ni-MOF surface and etched Ni-MOF NSs at the same time. Ni-MOF@Ni-HHTP nanocomposite was drop-casted onto a GCE surface and the modified GC electrode was used to detect glucose using amperometry in an alkaline aqueous solution. Linear concentrations ranging from 0.5 to 2665.5 µM with LOD of 0.0485 µM were obtained. Selectivity, reproducibility, repeatability, stability, and application to real samples were not examined.

In the next example, MOFs are regarded as sacrificial templates for synthesizing hollow metal oxide/carbon architectures with particular properties such as separate inner voids, low density, structural stability, and large specific surface areas [57,58]. Consequently, the mobility of analyte/ions/electrons at the solid/liquid interphase results in accelerated enhancing of the electron transfer from and to the electrode surface. In particular, NiO/Co$_3$O$_4$/C was developed as hierarchical hollow architecture using Ni-Co MOF as a sacrificial template. The proximity of Ni(II)/Co(II) ions with the organic ligand in the bimetallic MOF structure accelerates the formation of NiO/Co$_3$O$_4$ nanostructures closely linked to carbon. DFT studies have defined the role of the ingredients of nanocatalysts in the charge-transfer process and their structural and electronic relationship. In fact, the hollow architecture of NiO/Co$_3$O$_4$/C used both inner and outer surfaces for glucose interaction, while the presence of carbon linked to NiO/Co$_3$O$_4$ nanostructures enhanced the electron transfer rate. NiO/Co$_3$O$_4$/C was incorporated into a biodegradable corn-starch bag (BCSB) representing a free-standing, disposable, bendable, low-cost, and fast electrochemical probe for glucose detection.

A scheme of Ni-Co MOF synthesis, a non-enzymatic glucose-sensor-assembling and glucose-detection mechanism is reported in Figure 5.

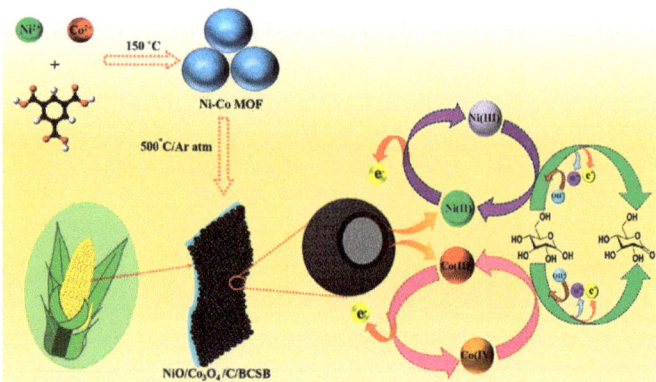

Figure 5. Schematic representation of Ni-Co MOF synthesis, a non-enzymatic glucose-sensor-assembling and glucose-detection mechanism, reprinted with permission by [57] Copyright 2022, Elsevier. The arrows indicate the well-known glucose detection mechanism.

A linear concentration, ranging from 0.0002 to 10 mM with an LOD of 45 nM, was achieved using amperometry. AP, AA, urea (U), DA, NaCl, citric acid (CiA), UA, and KCl were considered as possible interferences, with the sensor response to interferences resulting as negligible compared with glucose. Long-term stability was tested after 60 days under not better specified experimental storage conditions, and the electrochemical response decreased by 7.90%. The reproducibility and repeatability were investigated with satisfactory results in terms of RSD%, resulting in 2.1% and 2.4%, respectively. BCSB enzyme-free glucose sensor performance was evaluated in human serum real samples, obtaining interesting results in terms of recovery (98.3–102.4%) and RSD (2.19–2.72%).

A microelectrode based on a flexible nanocomposite including carbon fiber (CF) wrapped in rGO-supporting Ni-MOF nanoflake arrays was designed and developed for non-enzymatic glucose detection [59]. Combing the flexibility and conductivity of carbon substrate and highly dense Ni-MOF nanoflake arrays, interesting analytical performances for glucose detection were evidenced for the corresponding Ni-MOF/rGO/CF electrode. In particular, Ni-MOF nanoflake nanoarrays resulted uniformly supported on rGO/CF. The nanocomposite synthesis involved GO wrapping on the CF surface, the electrochemical reduction of GO to rGO, and the growth of Ni-MOF nanoflake nanoarrays on the rGO/CF surface using the solvothermal method. In addition, using conductive rGO/CF-supporting Ni-MOF nanostructures can accelerate the electron transfer from and to the CF electrode. The flexible hybrid CF electrode represented an alternative to more conventional electrode typologies such as GCEs or carbon paste electrodes (CPS) because it can be easily incorporated in miniaturized sensing systems. Glucose detection was performed using amperometry and a linearity range from 6 µM to 2.09 mM with LOD of 0.6 µM was found. KCl, NaCl, U, UA, AA, and DA were investigated as possible interferences and no significant changes to the electrochemical response of glucose were evidenced. Reproducibility was analysed, obtaining an RSD value of less than 5%. Concerning stability, the amperometric signal decreased by 7.40% after five weeks at RT. The sensing system was applied to detect glucose in orange juice samples, obtaining an RSD value of 3.6%.

A N-doped-Co-MOF@polydopamine nanocomposite (N-Co-MOF@PDA), including Ag nanoparticles (N-Co-MOF@PDA-Ag), was synthesized and then applied to assemble electrochemical sensors for glucose non-enzymatic determination [60]. DA polymerizes on the surface of N-Co-MOF, with N-Co-MOF acting as a catalyst. In addition, the N-Co-MOF@PDA nanocomposite was the reducing agent of Ag^+ ions to Ag nanoparticles with homogeneous size and good dispersion and conductivity. The combination of N-Co-MOFs and AgNPs assured good analytical performance with the non-enzymatic sensor. The N-Co-MOF@PDA-Ag nanocomposite was drop-casted onto a GCE and the resulting modified electrode was used to electrochemically detect glucose via amperometry, obtaining a linear concentration range from 1 µM to 2 mM and LOD of 0.5 µM. Fructose, AA, and UA were used as potential interferences, evidencing a negligible current response relative to glucose. Reproducibility was analysed, obtaining satisfactory results in terms of RSD (4.6%). Stability was investigated evaluating the sensor response after 3600 s from the first injection of glucose solution: the amperometric signal decreased by 3.8%. The sensor was applied to human serum samples, obtaining recoveries ranging from 96% to 110% with RSDs from 3.54% to 4.95%.

A wearable electrochemical sweat sensor including Ni–Co MOF nanosheets deposited on Au/polydimethylsiloxane (PDMS) film was realized for the continuous checking of glucose in sweat [61]. A flexible three-electrode system based on Au/PDMS film was set up by the chemical deposition of a gold layer on PDMS. Then, Ni–Co MOF nanosheets were synthesized by the solvothermal method and deposited on the Au/PDMS electrode surface. A Ni–Co MOF/Au/PDMS (NCAP) electrode was applied to determine glucose amperometrically and showed a wide linear range from 20 µM to 790 µM, with LOD of 4.25 µM. Selectivity and stability were evaluated. Lactose, U, DA, UA, AA, and NaCl were tested as possible interfering molecules that did not affect the electrochemical response of glucose. Long-term stability was tested and the amperometric response decreased by 2.00% after one month, but storage conditions were not indicated. Repeatability and reproducibility were not analysed. Performances under a stretching state were comparable to those under a not-stretching state. Since body temperature changes during sporting activity, as discussed above, the effect of temperature was studied and similar good performances were evidenced at different temperatures. The flexible sensor was then applied to monitor glucose level in volunteers. A sweat-absorbent cloth was used to cover the working area of the sensing system. Changes in glucose level before and after meals were evidenced, and the results were comparable with those obtained in sweat by commercial glucometers. Moreover, changes in sweat glucose concentration were strictly correlated to the values measured in blood.

A skin-attachable and flexible electrochemical biosensor based on ZnO tetrapods (TPs) and MXene (Ti_3C_2Tx) was developed for the continuous monitoring of glucose in sweat [62]. Skin-attachable electrodes included a carbon working electrode and were produced by a conventional screen-printing method, using thermoplastic polyurethane (TPU). Glucose oxidase (GOX) was immobilized using glutaraldehyde (GA) as a cross-linking agent on the working electrode modified with a ZnO TPs/MXene nanocomposite, with the last one showing a high specific area and electrical conductivity. A linear concentration range from 0.05 to 0.7 mM with LOD of 17 µM was obtained by chronoamperometry, including a satisfactory mechanical stability (up to 30% stretching) of the template. Reproducibility was acceptable in terms of RSD% (10%). Stability was investigated, and the amperometric response decreased by 10% of its initial value, after 10 days storage at 4 °C. The developed skin-attachable flexible biosensor was used to check glucose levels in sweat before and after a meal, and during physical activity, by continuous in vivo monitoring, and data were correlated with those collected by a conventional amperometric glucometer in blood.

A glucose biosensor was developed based on an electrode surface functionalized with MXene (Ti_3C_2Tx), providing binding sites for enzyme immobilization and proper conductivity [63]. Consequently, a transfer channel for electrons produced by the enzymatic redox reaction between GOX and glucose was guaranteed. A linear concentration range from 0.1 to 10 mM with LOD of 12.1 µM was obtained via amperometry. UA and AA, tested as interfering compounds, did not affect the electrochemical response of glucose. Repeatability, reproducibility, and stability were not investigated; the biosensor was not applied to real samples.

An electrochemical glucose biosensor was assembled immobilizing GOX onto GCE modified with poly(3,4-ethylenedioxythiophene):4-sulfocalix [4]arene (PEDOT:SCX)/MXene nanocomposite [64]. PEDOT was synthesized by chemical oxidation using SCX as a counter ion and MXene at high temperature under an inert gas atmosphere. Next, a PE-

DOT:SCX/MXene hybrid film was obtained by the ultrasonication of PEDOT:SCX/MXene (1:1) dispersion. GOX was then immobilized onto chitosan-modified PEDOT:SCX/MXene/ GCE. The biosensor was electrochemically characterized and a stable redox peak of FAD-GOX was observed at −0.435 V, showing a direct electron transfer between the enzyme and the electrode surface. A linear concentration range from 0.5 to 8 mM was achieved, with LOD of 0.0225 mM. Selectivity and repeatability were analysed. UA, AA, oxalic acid (OA), L-alanine (L-ala), and L–tyrosine (L-tyr) were tested as possible interferences and no changes were evidenced in the electrochemical signal after the addition of all of them. Repeatability was satisfactory with an RSD of 2.1%. The electrochemical signal decreased by 13% after 20 days at 4 °C. The biosensor was applied to commercial fruit juice samples with satisfactory recoveries, ranging from 96% to 99%.

A scheme of the biosensor assembling is reported in Figure 6.

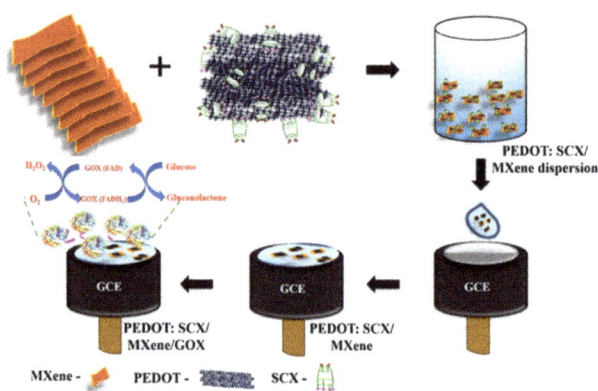

Figure 6. Scheme of GCE modification for assembling glucose biosensor reprinted from [64]. The arrows indicate the several steps of the glucose sensor assembling.

An advanced butterfly-inspired hybrid epidermal biosensing (bi-HEB) patch is now reported [65]. The bi-HEB patch included a glucose biosensor provided with pH and temperature sensors for precise quantitation and two biopotential electrodes for the real-time monitoring of electrophysiological signals. Nanoporous carbon and MXene (NPC@MXene) nanocomposite including platinum nanoparticles (PtNPs) were used to modify a Au electrode and support GOX immobilization via an EDC/NHS approach for assembling a glucose biosensor. Glucose levels were determined using chronoamperometry and a linearity range of 3 μM–21 mM with LOD of 7 μM was obtained. Reproducibility resulted <5% in terms of RSD. The bi-HEB patch integrated in a wearable system was used to control the sweat glucose and electrophysiological (EP) parameters of human subjects participating in indoor physical activities.

A non-enzymatic sensor based on a nanostructured electrode including MXene, chitosan (CHI), and Cu_2O nanoparticles was assembled for the simultaneous detection of glucose and cholesterol [66]. MXene and CHI act as a nanostructured sensing platform, while Cu_2O nanoparticles provide catalytic active edges to improve sensor performances. Glucose and cholesterol were determined simultaneously by CV: a linear concentration range from 52.4 to 2000 μM with LOD of 52.4 μM was achieved for glucose, and a linearity range from 49.8 to 200 μM with LOD of 49.8 μM was found for cholesterol. Sucrose (SC), UA, AP, lactose, NaCl, AA, and L-Cys were tested as possible interferences that did not affect the electrochemical signal of glucose. The sensor was applied to human serum real samples for quantifying glucose and cholesterol simultaneously and acceptable recoveries ranging from 98.04% to 102.94% were evidenced.

A glucose sensor was realized using a GCE modified with MXene/MOFs nanohybrid [67]. The nanocomposite with high conductivity was synthesized by depositing ZIF-67 as an MOF onto two-dimensional Ti$_3$C$_2$Tx nanosheets as MXene. Ti$_3$C$_2$Tx nanosheets improved electrical conductivity, while ZIF-67 improved electrocatalytic activity. Glucose was determined amperometrically, and a linearity range from 5 µM to 7.5 mM with LOD of 3.81 µM was obtained. UA, AP, and AA were tested as possible interferences that did not affect the electrochemical response. Repeatability was investigated with satisfactory results in terms of RSD (1.18%). The sensor was not applied to real samples.

An MXene nanocomposite, presenting a combination of MXene with TiO$_2$ nanocrystals (Ti$_2$C-TiO$_2$), was applied to modify GCE for assembling a non-enzymatic glucose sensor [68]. It was synthesized through the oxidation of Ti$_2$C-MXene nanosheets. The oxidative opening of the nanosheets produced TiO$_2$ nanocrystals on their surface. The combination of MXene nanosheets and TiO$_2$ nanocrystals accelerated the electron transfer from and to the sensing surface. The nanocomposite was then casted on the electrode surface and the corresponding modified GCE was used for determining glucose via DPV and CA. A linearity range from 0.1 to 200 µM and LOD of 0.12 µM were evidenced. AA, UA, and DA were selected as possible interferences, evidencing an insignificant electrochemical response compared with glucose. Reproducibility was acceptable in terms of RSD (4%). After 15 days, the signal response decreased by 6.5%, but storage conditions were not indicated. The sensor was applied to human serum samples with recoveries ranging from 99.80% to 100.23%, comparable with those obtained from a commercial glucometer.

A glucose sensor based on a carbon fiber electrode (CFE) modified with cobalt oxide Co$_3$O$_4$ nanocubes directly grown on a conducting MXene layer was realized [69], where 2D nanosheets deposited on the electrode surface supported the uniform growth of nanocubes. Conductive MXene and Co$_3$O$_4$ nanocubes as catalytic active sites acted synergistically to support sensor performances. Glucose was determined via amperometry, achieving a linear range from 0.05 µM to 7.44 mM with LOD of 10 nM. AA, DA, AP, catechol, L-Cys, resorcinol, UA, and hydrogen peroxide (H$_2$O$_2$) were considered as possible interferences that did not affect the glucose response. After 30 days, the signal response decreased by 2.0%, but storage conditions were not indicated. Reproducibility was satisfactory in terms of RSD (2.52%). The sensor was applied to spiked real samples of human serum, urine, and blood, with recoveries from 97.8% to 101.6% with results comparable to those obtained by the conventional colorimetric method.

An MXenes (Ti$_3$C$_2$T$_x$)-based nanocomposite with Cu$_2$O nanoparticles (Ti$_3$C$_2$T$_x$-Cu$_2$O) was developed to assemble a glucose sensor by modifying GCE [70], as shown in Figure 7. The morphological characterization of Ti$_3$C$_2$T$_x$-Cu$_2$O evidenced that the micro-octahedral Cu$_2$O nanoparticles were distributed uniformly on the MXene surface.

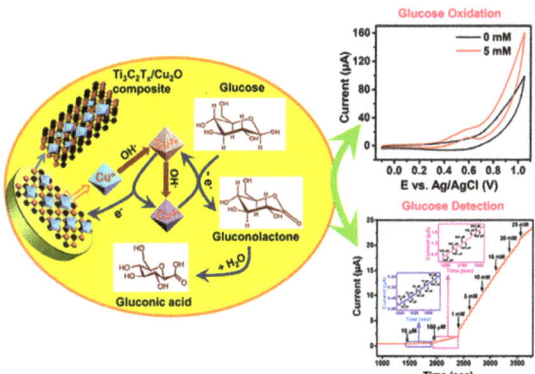

Figure 7. Schematic representation of electrode modification and glucose sensing mechanism, reprinted with permission from [70] Copyright 2022 Elsevier. The arrows indicate the cyclic volyammetry of the glucose oxidation and the glucose detection by chrono amperometry.

Glucose was determined using CA and a linear range from 0.01 to 30 mM with an LOD of 2.83 mM was evidenced. AA, NaCl, U, LT, fructose, and SC were considered as interfering molecules, without affecting the sensor response. After 30 days at RT, the analyte response decreased by 5.00%. The sensor was applied to human serum real samples and the data resulted as comparable with those coming from a commercial glucometer.

As a last example, a glucose sensor based on a two-dimensional (2D) conjugated metal–organic framework (c-MOF) film, such as 2D Cu_3 $(HHTP)_2$ (HHTP = 2,3,6,7,10, 11- hexahydroxytriphenylene) c-MOF, is reported [71]. The MOF film was deposited onto a Au electrode, improving conductivity and the electrocatalytic response. Glucose was detected amperometrically and a linear range from 0.2 µM to 7 mM with LOD of 10 µM was achieved. NaCl, UA, lactose, U, AA, and DA were tested as possible interferences without affecting the electrochemical response of glucose. Long-term stability was investigated and the amperometric signal remained almost unchanged after 60 days at room temperature. Reproducibility was not considered, and the sensor was not applied to real samples.

As a conclusive comment regarding the reported examples of sensors for the determination of glucose, it can be observed that LOD values are generally micromolar, and nanomolar values were reported for two examples only [57,69].

Notably, the complexity of materials does not always correspond to better performances in terms of linearity range or LOD.

Considering enzyme-based biosensors, only a few examples are reported in the literature [53,62–64,66] compared with the total. It is well-known that electrochemical enzyme-based biosensors are easy to assemble and generally reusable, but their major drawback is the stability of the enzyme over time.

Selectivity, applicability to real samples, and sensors' validation with analytical conventional methods have not always been adequately analyzed and addressed.

The glucose sensors' analytical performances, together with the corresponding sensor formats, are summarized in Table 1.

Table 1. Analytical performances and format of electrochemical (bio)sensors for glucose determination.

Electrode	2D Nano-material	Format	Technique	Sample	Linearity (µM)	LOD (µM)	Recovery %	Reference Method	Ref.
GCE	ZIF-67	sensor based on Ag@ZIF-67 nanocomposite	A	-	2–1000	0.66	-	-	[50]
GCE	ZIF-67	sensor based on Ag@TiO$_2$@ZIF-67 nanocomposite	A	-	48–1000	0.99	-	-	[51]
rGO/PU fiber	Ni–Co MOF	sensor based on Ni–Co MOF/Ag nanocomposite	A	Human sweat	10–660	3.28	-	Glucometer blood test	[52]
SPCPE	NC-ZIF	biosensor based on GOX/Hemin@NC-ZIF	A	Human sweat	50–600	2	-	Glucometer blood test	[53]
CC	ZIF-67	sensor based on ZIF-67@GO/NiCo$_2$O$_4$	A	-	0.3–157.4	0.16	-	-	[54]
NF	Cu$_1$Co$_2$-MOF	sensor based on Cu$_1$Co$_2$-MOF	CA	-	50–500	23	-	-	[55]
GCE	Ni-MOF	sensor based on Ni-MOF@Ni-HHTP-5 NSs core@shell structures	A	-	500–2,665,500	48.5	-	-	[56]
BCSB	Ni-Co MOF as sacrificial template	sensor based on NiO/Co$_3$O$_4$/C nanocomposite	A	Human blood serum	0.2–10,000	0.045	98.3–102.4	-	[57]
CFE	Ni-MOF	sensor based on Ni-MOF/rGO nanocomposite	A	Orange juice	6–2090	0.6	-	-	[59]

Table 1. Cont.

Electrode	2D Nano-material	Format	Technique	Sample	Linearity (μM)	LOD (μM)	Recovery %	Reference Method	Ref.
GCE	N-doped-Co-MOF	sensor based on N-Co-MOF@PDA-Ag nanocomposite	A	Human serum	1–2000	0.5	96–110	-	[60]
AuE	Ni-Co-MOF	sensor based on Ni–Co MOF/Au/PDMS nanocomposite	A	Human sweat	20–790	4.25	-	Glucometer blood test	[61]
SPCE	MXene (Ti_3C_2Tx)	biosensor based on ZnO TPs/MXene/GOX	CA	Human sweat	50–700	17	-	Glucometer blood test	[62]
Cr/AuE	MXene (Ti_3C_2Tx)	biosensor based on MXene/GOX	A	-	100–10,000	12.1	-	-	[63]
GCE	MXene (Ti_3C_2Tx)	biosensor based on PEDOT:SCX/MXene/GOX	A	Fruit juice	500–8000	22.5	96–99	-	[64]
AuE	MXene (Ti_3C_2Tx)	biosensor based on GOX/PtNPs/NPC/MXene	CA	Human sweat	3–21,000	7	-	Glucometer blood test	[65]
GCE	MXene	sensor based on MXene/CHI/Cu_2O	CV	Human serum	52.4–2000	52.4	98.04–102.94	-	[66]
GCE	MXene (Ti_3C_2Tx) ZIF-67	sensor based on Ti_3C_2Tx/ZIF-67 nanocomposite	A	-	5–7500	3.81	-	-	[67]
GCE	Ti_2C-MXene	sensor based on Ti_2C-TiO_2-MXene nanocomposite	DPV	Human serum	0.1–200	0.12	99.80–100.23	Glucometer blood test	[68]
CFE	MXene	sensor based on Co_3O_4/MXene nanocomposite	A	Human blood serum, urine	0.05–7440	0.010	97.80–101.60	Glucometer blood test	[69]
GCE	MXene (Ti_3C_2Tx)	sensor based on MXene-Cu_2O nanocomposite	CA	Human serum	10–30,000	2.83	-	Glucometer blood test	[70]
AuE	c-MOF ($Cu_3(HHTP)_2$)	sensor based on c-MOF	A	-	0.2–7 mM	10	-	-	[71]

Abbreviations: A: amperometry; BCSB: biodegradable corn starch bag; CA: chronoamperometry; CC: carbon cloth; CFE: carbon fiber electrode; CHI: chitosan; c-MOF: conjugated MOF; COF: covalent organic framework; DPV: differential pulse voltammetry; GCE: glassy carbon electrode; GO: graphene oxide; GOX: glucose oxidase; HHTP: 2,3,6,7,10,11-hexahydroxytriphenylene; MOF: metal–organic framework; NF: nickel foam; NPC: nanoporous carbon; NSs: nanosheets; PDMS: polydimethylsiloxane; PEDOT: poly(3,4-ethylenedioxythiophene; PtNps: platinum nanoparticles; rGO: reduced graphene oxide; SCX: 4-sulfocalix [4]arene; ZnO TPs: ZnO tetrapods.

3.2. Neurotransmitters

Neurotransmitters are endogenous chemical molecules able to send, improve, and exchange specific signals between neurons and other cells.

They can be classified into two groups, considering their electrochemical activity. Electroactive neurotransmitters include dopamine, serotonin, epinephrine, norepinephrine, among others, while glutamate, acetylcholine, and choline are assumed electroinactive [72].

In this review, attention was focused on electrochemical sensors based on 2D nanomaterials for the determination of two important electroactive neurotransmitters such as dopamine and serotonin, where the nanostructured interface acts to amplify the signal and subsequently to improve the performance of the sensor in terms of stability, sensitivity, and sustainability.

3.2.1. Dopamine

Dopamine is a catecholamine neurotransmitter widely present in the central nervous system (CNS). It influences attention skills and brain plasticity, i.e., the ability of neural networks in the brain to change through growth and reorganization. In addition, DA plays a crucial role in memory and learning. Several neurological disorders such as Parkinson's disease and schizophrenia are associated to abnormal levels of DA [72].

Several studies evidenced that physiological levels of DA in human biological fluids are significantly different. For example, 5 nM DA is reported in urine and cerebrospinal fluid [73], while normal values of DA in blood range from 10 to 480 pM.

The monitoring of DA in the presence of other molecules such as AA, UA, and tyrosine, among others, is essential for the diagnosis of neurological diseases and for understanding mechanisms underlying human neuropathologies.

A $Ti_3C_2T_x$ (MXene)/Pt nanoparticle-modified GCE ($Ti_3C_2T_x$/PtNPs/GCE) was developed for determining AA, DA, UA, and APAP [74]. GCE was modified by drop-casting using a dispersion of MXene and PTNPs. MXene acted as a nanostructured sensing platform, while PtNPs provided catalytic active edges to improve the sensor performance. Target analytes were determined using amperometry: the linear response range was observed up to 750 µM for all the analytes, and an LOD of 10 nM was found for DA. In order to make DA detection selective in samples containing AA, an outer layer from an ethanolic solution of Nafion or from chitosan was deposited onto the modified electrode surface. In fact, they did not affect the electrochemical behaviour of DA and UA, their redox behaviour being similar to the one evidenced in the modified GCE without any additional outer layers. Selectivity, reproducibility, repeatability, stability, and application to real samples were not examined.

Another electrochemical sensor for DA was realized using MoS_2 (TMD) electrode-posited onto pyrolytic graphite sheets (PGSs) and doped with Mn [75]. The presence of Mn as a dopant improved the electrochemical behaviour of DA compared with bare PGS and undoped MoS_2. In fact, Mn-MoS_2 enhanced selectivity toward DA in the presence of other common electroactive interferences such as AA and UA. DA was determined through DPV, and two linearity ranges were observed: one was from 50 pM to 5 nM and the other one was from 5 nM to 5 mM, with LOD of 50 pM. Concerning selectivity, the electrochemical response was not affected by the presence of AA and UA. The sensor was tested in artificial samples of human serum and sweat, with LOD values of 50 nM and 5 nM, respectively. Reproducibility, repeatability and stability data were not provided as well as applications to real samples.

A titanium carbide (MXene) (Ti-C-T_x)-modified GCE was assembled for the simultaneous determination of AA, DA, and UA [76]. It was demonstrated that Ti-C-Tx/GCE evidenced satisfactory electrocatalytic activity and separated oxidation peaks for AA (0.01 V), DA (0.21 V), and UA (0.33 V). Their simultaneous determination was performed at physiological pH, obtaining a linearity range of 100–1000 µM for AA, 0.5–50 µM and 0.5–4 µM for DA, and 100–1500 µM for UA. The LOD values were 4.6 µM, 0.06 µM, and 0.075 µM for AA, DA, and UA, respectively. Reproducibility was investigated involving three different independent sensors, which evidenced no significant changes in the electrochemical response for AA, DA, and UA in PBS. Considering stability, the electrochemical response decreased by 10.4% after 25 days at RT even after repeated use. Selectivity was also tested in the presence of U, nicotine, L-Cys, and APAP, and the electrochemical response was not affected by the presence of interferences. Any application to real samples was not provided.

An electrochemical flexible sensor for DA determination was realized by the direct growth of molybdenum disulphide nanosheets (MoS_2N_S) on carbon cloth (CC) [77], as reported in Figure 8.

In MoS_2N_S/CC flexible electrodes, CC acts as a conductive and flexible support, and MoS_2Ns promotes uniform DA adsorption and oxidation. DA was determined by CV, and a linear concentration range from 250 to 4000 µM with LOD of 0.3 µM was evidenced. Repeatability and reproducibility were considered satisfactory in terms of RSD, 1.87% and 1.35%, respectively. The sensor, stored under ambient conditions, was tested against DA every 4 days for 16 days. RSD was 2.9%, indicating good long-term stability. Data concerning real samples application were not provided.

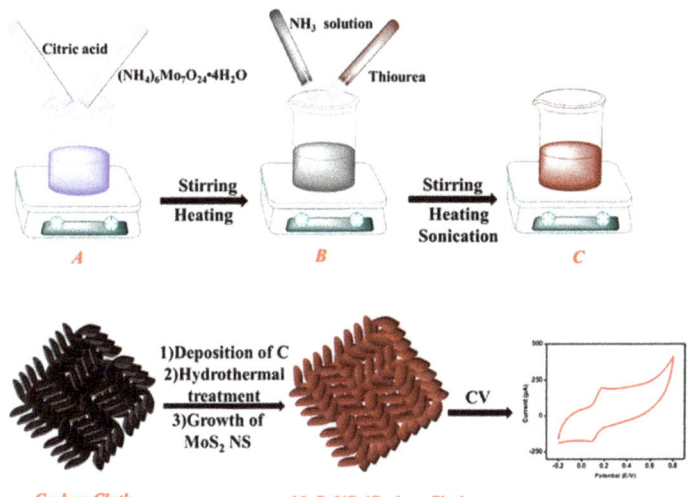

Figure 8. Scheme of TMD synthesis, CC electrode modification and electrochemical characterization, reprinted with permission from [77] Copyright 2022 Elsevier.

A novel triazine-based covalent organic framework (TS-COF) was synthesized by condensation of 1, 3, 5-tris-(4-aminophenyl) triazine (TAPT) and squaric acid (SA) via a solvothermal approach. Next, AuNPs were grown on a TS-COF surface and rGO was incorporated in AuNPs@TS-COF composite. AuNPs@TS-COF/rGO nanocomposite was then used for modifying GCE and the resulting sensor was used to assay DA, UA, and AA [78]. TS-COF bounded to rGO by π–π stacking promoted electron transfer and electron mobility, while TS-COF acted as a dispersing agent for AuNPs that improved both electrical conductivity and electrocatalytic activity. AA, UA, and DA were determined via DPV, and linearity ranges of 8–900 µM, 25–80 µM, and 20–100 µM were obtained for AA, UA, and DA, respectively, with LODs of 4.30 µM for AA, 0.07 µM for UA, and 0.03 µM for DA. Glucose, L-cys, glycine, CA, tyrosine, lysine, tryptophan, leucine, and L-glutamic acid were used as possible interfering molecules, and they did not affect the determination of AA, DA, and UA, the relative errors being under ± 5%. Reproducibility was considered acceptable in terms of RSD: 4.76% for UA, 3.53% for DA, and 3.15% for AA. Concerning repeatability, satisfactory RSD values were obtained: 2.19%, 4.48%, and 3.26% for AA, DA, and UA, respectively. The sensor was applied to human urine real samples, with recoveries ranging within 97.0% and 104.4%.

A composite was prepared including $PMo_{12}O_{40}{}^{3-}$ (PMo_{12}) as polyoxmetalate (POM), $C_9H_5FeO_7$ (MIL-100(Fe) as an Fe-based MOF, and polyvinylpyrrolidone (PVP), and then it was deposited on a GCE for the simultaneous determination of DA and UA [79]. MIL-100(Fe) can encapsulate POMs within its mesoporous architecture. These POM-based MOF composites not only evidenced large surface areas, but their multi-component and multi-interface architectures also improved electron transfer, thereby enhancing analytical performances and the stability of the sensor. PVP can prevent hybrid particles agglomeration, thereby supporting conductivity and increasing electrochemical sensor performance. DA and UA were determined using DPV, and linear concentration ranges of 1–247 µM for DA and 5–406 µM for UA, with corresponding LODs of 1 µM and 5 µM, were obtained. AA, tyrosine, guanosine, isoleucine, alanine, tryptophan, KCl, xanthine, glucose, hypoxanthine, and NaCl were chosen to check selectivity, and no significant response to these interfering molecules was found. Considering long-term stability, DA response decreased by 10.31%, while the one of UA decreased by 11.31%, after 42 days at RT. Repeatability was acceptable in terms of RSD: 2.54% (DA) and 3.18% (UA). Similarly, reproducibility was found to be satisfactory, with RSDs of 2.08% and 5.19% for DA and UA, respectively. The sensor was applied to human serum real samples and recoveries of 97.67–102.16% and 97.81–102.89% for DA and UA, respectively, were found.

An electrochemical sensor for DA determination was realized, modifying a low-cost lead pencil graphitic electrode (LP) by dip-coating using a ZIF-67/PEDOT composite [80], as shown in Figure 9.

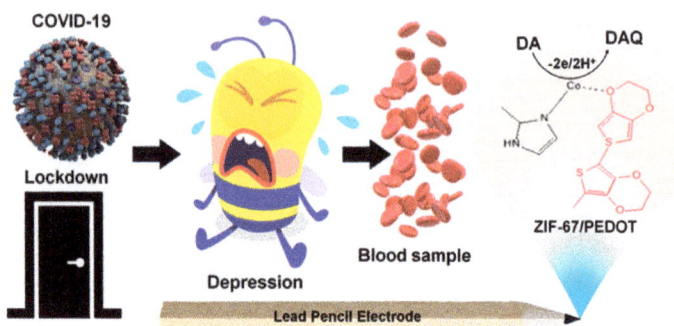

Figure 9. Scheme of the sensing mechanism involved in DA assay at ZIF-67/PEDOT-modified LPE, reprinted with permission from [80] Copyright 2022 Elsevier.

ZIF-67 as an MOF has interesting properties such as a crystalline structure, chemical durability, high surface area, high thermal stability, and tunable pore size, while PEDOT improves conductivity and accelerates electrode transfer at the ZIF-67 surface, preserving the peculiar MOF properties such as large surface area and stability. DA was determined using amperometry, and a linearity range from 15 to 240 µM with LOD of 0.04 µM was found. AA, UA, L-Cyst, glucose, and salbutamol were assumed as possible interfering molecules and negligible interfering impact towards DA was evidenced, even in the presence of several-fold higher concentrations of interfering molecules. Reproducibility was acceptable in terms of RSD (0.92%); concerning stability, the amperometric response decreased by 2% after 10 days, but storage conditions were not identified. The sensor was then applied to real blood samples of a COVID-19 quarantine patient, since neurological disorders such as anxiety, depression, and agitation were connected to the lockdown condition, but the results were not accurately explained.

A conductive graphitic pencil electrode (GPE) was modified using a nanocomposite including $Ti_3C_2Cl_2$ as MXene and 1-methyl imidazolium acetate as ionic liquid (IL) for preparing a DA electrochemical sensor [81]. As discussed in Section 2.1., MXenes are characterized by good conductivity, a high surface area, biocompatibility, hydrophilicity, and resistivity against electrode surface fouling and passivation. However, low flexibility and poor stability in aqueous media and air, due to the presence of hydrophilic functional groups, make those materials easy to oxidize; consequently, their working potential range is limited. For this reason, the stabilization of an MXene surface by IL is an effective strategy to minimize its oxidative degradation without compromising MXene conductivity. An IL/MXene composite was drop-casted onto PGE, and DA was determined via amperometry. A linearity range from 100 µM to 2 mM with LOD of 702 nM was obtained. Repeatability, reproducibility and stability were investigated, and acceptable results in terms of RSD were found: 2.3% (repeatability) and 1.9% (reproducibility), besides an RSD of 1.3% obtained after 14 days storage at RT (stability). UA, AA, glucose, fructose, L-cys, and U were tested as interfering molecules, without affecting the DA amperometric signal. The sensor was applied to spiked real samples of human serum with recoveries ranging from 98.3% to 100.0%.

MOF are considered promising nanomaterials in the sensing area because of their high porosity, adsorption capability, film-forming ability, and tunable synthesis protocol, and nanocomposites including MOF and MXene produce a hybrid with a synergistic combination of the properties of both 2D nanomaterials.

A Ti_3C_2 membrane was synthesized by doping UIO-66-NH2 (MOF) with Ti_3C_2 (MXene) through a hydrogen bond and used to modify GCE by drop-casting, for assembling a DA electrochemical sensor [82]. DA was determined via DPV, and a linear concentration range from 1 to 250 fM with LOD of 0.81 fM was obtained. Glucose, bovine serum albumin (BSA), AA, and UA were investigated as possible interferences, but they did not affect the DA signal response. After 15 days at 4 °C, the electrochemical response decreased by 7.9%. Reproducibility was found to be good in terms of RSD (3.16%). The sensor was applied to spiked human serum real samples, with recoveries in the range 101.2–103.5%.

A monolayer titanium carbide ($Ti_3C_2T_x$) material was prepared, and an MXene-based nanocomposite was developed by self-assembling ZnO nanoparticles and a $Ti_3C_2T_x$ monolayer. It was applied to modify a gold electrode for determining DA [83]. Ti_3C_2Tx presents good conductivity, hydrophilicity, a high surface area, biocompatibility, and its integration with another nanomaterial such as ZnO nanoparticles could improve the analytical performances of the sensor. DA was analysed via CA, and a linear concentration range from 0.1 to 1200 µM with LOD of 0.076 µM was evidenced. Glucose, U, serotonin, AA, and UA were investigated as possible interferences that did not interfere with the analysis of DA. Reproducibility and repeatability were considered satisfactory in terms of RSD, at 4.24% and 1.04%, respectively. The sensor was applied to human serum samples, with recoveries ranging from 97.8% to 102.2%.

A hybrid nanomaterial based on Ti_3C_2 as MXene, graphitized multi-walled carbon nanotubes (g-MWCNTs), and ZnO nanospheres (ZnO NSPs) was realized for developing an electrochemical sensor for DA assay [84]. Ti_3C_2 was combined with g-MWCNTs to enhance stability and electrochemical properties, and the addition of ZnO NSPs could further improve the catalytic and electrochemical properties of the Ti_3C_2/g-MWCNTs nanocomposite. Finally, Ti_3C_2/g-MWCNTs/ZnO NSPs was deposited on GCE for assembling the DA sensor. DA was determined using DPV, with a linearity range from 0.01 to 30 µM and LOD of 3.2 nM. Glucose, alanine, glycine, leucine, OA, U, AA, and UA were assumed as possible interferences that did not affect the DPV response of DA. Long-term stability was investigated, and a decreased response of 7.4% was evidenced after 25 days at 4 °C. Reproducibility and repeatability were considered acceptable in terms of RSD, 0.54% and 1.16%, respectively. The sensor was applied to human serum samples with recoveries ranging from 98.6% to 105.9%.

A screen-printed electrode (SPCE) modified with Ti_3C_2 nanolayers was assembled for the simultaneous determination of DA and tyrosine [85]. SPCE was modified by the drop-casting of MXene suspension. DA and tyrosine were determined using DPV, but only in the linearity range (0.5–600.0 µM,) and LOD (0.15 µM) of DA were reported and commented. The sensor was then applied to determine DA and tyrosine in pharmaceutical drugs and in human urine samples, with recoveries ranging from 97.1% to 104.0% for DA and from 96.7% to 102.5% for tyrosine.

A DA electrochemical sensor based on flower-like MoS_2 (TMD) nanomaterial, decorated with single Ni site catalyst (Ni-MoS_2), was realized by modifying GCE [86]. Single atom catalysts (SACs) are a new class of electrocatalysts where isolated metal atoms are distributed and docked on solid supports [87]. SACs have been proposed as single-atom nanozymes (SAzymes) mimicking natural enzymes, due to high catalytic stability, tunable activity, low cost, and high storage stability [88]. MoS_2 has been investigated as a support material, because of its surface area-to-volume ratio, conductivity, and capability in biomarker detection [89]. Moreover, considering Ni electronegativity and the redox properties of the couple Ni(II)/Ni(III), a flower-like MoS_2 modified by a Ni atom (Ni-MoS_2) was developed as a proper and suitable sensing nanohybrid for the selective and sensitive determination of DA. The neurotransmitter was detected using DPV in a linear range from 1 pM to 1 mM, with LOD of 1 pM. AA, UA, glucose, and U were assumed as interfering molecules, and a negligible interfering effect was observed. Repeatability was found to be acceptable in terms of RSD (3.2%), while after 7 days at RT in air, a decreased electrochemical response of 11.4% was evidenced. The sensor was applied to bovine serum samples with recoveries ranging from 97.00% to 105.00%.

MoS_2 screen-printed electrodes (SPEs) were developed for DA electrochemical detection [90]. MoS_2 SPEs were developed utilising high viscosity screen printable inks containing MoS_2 particles in different concentrations and with various sizes, and ethylcellulose as a binder. MoS_2 inks were printed onto conductive FTO (Fluorine-doped Tin Oxide) substrate. DA was determined via DPV in a linear range up to 300 μM, with LOD of 260 nM. Selectivity, reproducibility, repeatability, stability, and application to the real samples were not examined.

3.2.2. Serotonin

Serotonin is also called 5-hydroxytryptamine (5-HT) and is the most important monoamine neurotransmitter and neuromodulator. Normal values of 5-HT in blood vary in the range 0.6–1.6 μM [91]. It plays a fundamental role in several biological and psychopathological processes, such as sleep regulation, depression, eating disorders, irritable bowel syndrome, anxiety disorders, and psychosis, among others. Consequently, the development of rapid and sensitive methods for detecting 5-HT in body fluids is crucial for supporting the correct diagnosis of neuropathologies and psychopathological disorders.

As a first example, an electrochemical sensor based on GCE modified with a nanocomposite including L-cysteine-terminated triangular silver nanoplates (Tri-AgNPs/L-Cys) and $Ti_3C_2T_x$ (MXene) nanosheets [92] is reported, and shown in Figure 10. AgNPs are biocompatible and have good electrocatalytic activity but tend to aggregate if any system supporting their dispersion is not used. As the multilayer structure of MXene can prevent nanoparticles' aggregation, AgNPs can be incorporated between MXene layers, promoting electron transfer and increasing surface area, with improved performances of the sensor as consequence. 5-HT was determined using DPV in a linear range of 0.5–150 μM, with LOD of 0.08 μM. Selectivity was tested using different interfering molecules such as UA, L-Cys, L-Arginine (L-Arg), L-Phenylalanine (L-Phe), and L-Glycine (L-Gly), and little peak current changes were observed. The sensor was used to assay 5-HT in human serum samples, with recoveries ranging from 95.38% to 102.3%.

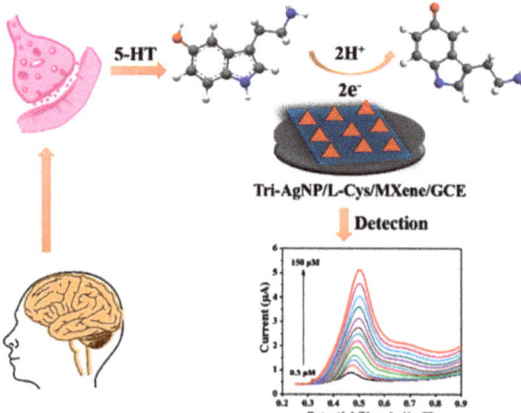

Figure 10. Sensing mechanism of 5-HT determination at MXene-modified GCE, reprinted with permission from [92] Copyright 2021 American Chemical Society. The colored lines correspond to the different electrochemical signals using different concentration of 5-HT.

A flexible sensing platform was developed with a WS_2/graphene nanostructured hybrid supported on polyimide (WGP) for the simultaneous and selective determination of

DA and 5-HT [93]. WS$_2$ (TMD) and graphene acted synergistically, producing a nanocomposite with unique electrical conductivity and several electrochemically active sites. DA and 5-HT were detected simultaneously using DPV with linear ranges of 1.21–13.37 µM (DA) and 249 nM–9.9 µM (5-HT) and LODs of 1240 nM (DA) and 240 nM (5-HT). Stability and selectivity were investigated. DA and 5-HT electrochemical responses were recorded every five days over 30 days, with decreases of 16.3% (DA) and 17.5% (5-HT). Storage conditions were not specified. UA, AA, glucose, and epinephrine were identified as possible interferences, and negligible DPV responses occurred for all of them. The sensor was applied to artificial cerebrospinal fluid (CSF) samples because abnormal levels of DA and 5-HT in the CSF are correlated to the malfunctioning of dopaminergic and serotonergic neurons in CNS. Satisfactory recoveries ranging from 95.2% to 104.1% were obtained.

An electrochemical cell-sensing platform for 5-HT detection based on MXene/SWCNTs nanocomposites was assembled onto GCE [94]. The nanohybrid included SWCNTs uniformly distributed on the surface and among the MXene layers. The conductivity and electrocatalytic properties of both the starting nanomaterials improved in the synthesized MXene-SWCNTs composite. 5-HT was determined using amperometry in a linearity range from 4 nM to 103.2 µM with LOD of 1.5 nM. Reproducibility was considered acceptable in terms of RSD (3.8%). In terms of stability, the signal response decreased by 10% after 3 weeks, but storage conditions were not specified. Concerning selectivity, AA, DA, UA, and tyrosine were assumed as possible interfering molecules, yielding negligible current signals. Since the real-time monitoring of 5-HT produced from living cells is crucial for the early diagnosis and rapid treatment of several neuropathogical disorders, the sensor was applied to different cell lines to assess its applicability under physiological conditions.

The same research group had proposed a Ti$_3$C$_2$T$_x$-reduced oxide graphene (Ti$_3$C$_2$Tx-rGO) nanocomposite for modifying GCE to determine 5-HT in biological fluids [95]. The nanostructured hybrid material provided a large surface area and effective active sites for enhancing sensing layer electrochemical performances and, compared with [94], SWCNTs were substituted by rGO. 5-HT was quantified using DPV, in a linear range of 0.025–147 µM with LOD of 10 nM. Concerning stability, the sensor response decreased by 8.3% after 10 days at 4 °C. DA, AA, and UA were tested as possible interferences, and they did not affect the analyte signal response. Reproducibility was acceptable in terms of RSD (5.2%). The sensor was applied to real samples of human blood plasma and recoveries ranged from 96.4% to 107%.

As a conclusive comment regarding the reported examples of sensors for the detection of neurotransmitters, it can be observed that LOD values are generally micromolar, even if several examples reported lower values ranging from nanomolar to femtomolar [74,75,81,82,84,86,90,93–95].

Questionable points concern the validation of the proposed methods. In fact, in all cases, results were not validated with conventional and standard analytical methods.

The data of sensor formats and analytical performances concerning the herein reported examples for the determination of neurotransmitters are summarized in Table 2.

Table 2. Analytical performances and format of electrochemical sensors for neurotransmitters determination.

Electrode	2D Nano-material	Format	Technique	Sample	Linearity	LOD	Recovery (%)	Ref.
GCE	MXene (Ti$_3$C$_2$T$_x$)	sensor based on Ti$_3$C$_2$T$_x$/PtNPs	A	-	Up to 750 µM	10 nM	-	[74]
PGS	TMD (MoS$_2$)	sensor based on MoS$_2$	DPV	DA	0.05–5 nM 5 nM–5 mM	50 pM	-	[75]
GCE	MXene (Ti-C-T$_x$)	sensor based on Ti-C-T$_x$	DPV	DA,AA,UA	DA: 0.5–4 µM AA: 0.5–50 µM UA: 0.1–1.5 µM	DA: 0.06 µM AA: 4.6 µM UA: 0.075 µM	-	[76]

Table 2. Cont.

Electrode	2D Nano-material	Format	Technique	Sample	Linearity	LOD	Recovery (%)	Ref.
CC	TMD (MoS$_2$)	sensor based on MoS$_2$ nanosheets	CV	DA	250–4000 µM	0.3 µM	-	[77]
GCE	TS-COF	sensor based on AuNPs@TS-COF/rGO	DPV	DA,AA,UA in human urine	DA: 20–100 µM AA: 8–900 µM UA: 25–80 µM	DA: 0.03 µM AA: 4.30 µM UA: 0.07 µM	97–104.4	[78]
GCE	Fe-based MOF MIL-100(Fe)	sensor based on POM- MOF/PVP	DPV	DA,UA in human serum	DA: 1–247 µM UA: 5–406 µM	DA: 1 µM UA: 5 µM	DA: 97.67–102.16 UA: 97.81–102.89	[79]
LPE	MOF (ZIF-67)	sensor based on ZIF-67/PEDOT	A	DA in human blood	15–240 µM	0.04 µM	-	[80]
GPE	MXene (Ti$_3$C$_2$Cl$_2$)	sensor based on IL-MXene	A	DA in human serum	100–2000 µM	702 nM	98.3–100	[81]
GCE	MXene (Ti$_3$C$_2$) MOF (UIO-66-NH$_2$)	sensor based on Ti$_3$C$_2$/UIO-66-NH$_2$	DPV	DA in human serum	1–250 fM	0.81 fM	101.2–103.5	[82]
AuE	MXene (Ti$_3$C$_2$T$_x$)	sensor based on ZnO NPs/Ti$_3$C$_2$T$_x$	CA	DA in human serum	0.1–1200 µM	0.076 µM	97.8–102.2	[83]
GCE	MXene (Ti$_3$C$_2$)	sensor based on Ti$_3$C$_2$/g-MWCNTs/ZnO NSPs	DPV	DA in human serum	0.01–30 µM	3.2 nM	98.6–105.9	[84]
SPCE	MXene (Ti$_3$C$_2$)	sensor based on Ti$_3$C$_2$	DPV	DA in human urine Tyr in drug	0.5–600 µM	0.15 µM	DA: 97.1–104.0 Tyr: 96.7–102.5	[85]
GCE	TMD (MoS$_2$)	sensor based on Ni-MoS$_2$	DPV	DA in bovine serum	1 pM–1 mM	1 pM	97.0–105.0	[87]
FTO SPE	TMD (MoS$_2$)	sensor based on MoS$_2$	DPV	DA	Up to 300 µM	260 nM	-	[90]
GCE	MXene (Ti$_3$C$_2$T$_x$)	sensor based on Tri-AgNPs/L-Cys/MXene	DPV	5-HT in human serum	0.5–150 µM	0.08 µM	95.38–102.3	[92]
WGPE	TMD (WS$_2$)	sensor based on WGP	DPV	DA, 5-HT in artificial CSF	DA: 1.21–13.37 µM 5-HT: 9.9–0.249 µM	DA: 1.24 µM 5-HT: 0.24 µM	95.2–104.1	[93]
GCE	MXene (Ti$_3$C$_2$)	sensor based on MXene/SWCNTs	CA	5-HT produced by living cells	0.004–103.2 µM	1.5 nM	-	[94]
GCE	MXene (Ti$_3$C$_2$)	sensor based on MXene/rGO	DPV	5-HT in human plasma	0.025–147 µM	10 nM	96.4–107	[95]

Abbreviations: A: amperometry; AuE: gold electrode; AuNPs: gold nanoparticles; CA: chronoamperometry; CC: carbon cloth; COF: covalent organic framework; CSF: cerebrospinal fluid; CV: cyclic voltammetry; DA: dopamine; DPV: differential pulse voltammetry; FTO: fluorine-doped tin oxide; GCE: glassy carbon electrode; GPE: graphite pencil electrode; 5-HT: 5-hydroxytryptamine (serotonine); IL: ionic liquid; LPE: lead pencil graphitic electrode; MOF: metal–organic framework; g-MWCNTs: graphitized multi-walled carbon nanotubes; NSPs: nanospheres; PGS: pyrolytic graphitic sheets; POM: polioxametalate; rGO: reduced graphene oxide; SPE: screen-printed electrode; SWCNTs: single-walled carbon nanotubes; TMD: transition metal dichalcogenide; Tri-AgNP: T triangular silver nanoplate; TS-COF: triazine-based covalent organic framework; Tyr: tyrosine; WGP: WS$_2$/graphene heterostructure on polyimide.

3.3. Hormones

According to Starling's original definition (1905), "a hormone is a substance produced by glands with internal secretion, which serve to carry signals through the blood to target organs" [96]. In other words, hormones are secreted generally by glands, and circulate in blood by simple diffusion before reaching the target cell. Hormones can induce responses of target cells either by interaction with a specific receptor on the cell membrane without entering the cell or by interaction with a receptor within the target cell. Notably, hormones circulate in the blood at very low concentrations (nanomolar amounts or even less); this is the reason why accurate and sensitive approaches are needed for their determination, such as the use of electrochemical sensors integrated with nanomaterials or nanocomposites [97].

Herein, some recent examples of electrochemical sensors based on 2D nanomaterials for hormone determination are reported and discussed.

Human cortisol (11b,17a, 21-Trihydroxypregn-4-ene-3,20-dione) is a steroid hormone, one of the major glucocorticoids synthesized in the zona fasciculata of adrenal glands, and it plays a vital role in emotional responses like stress or depression [98].

A thread-based electrochemical immunosensor was developed for cortisol determination in sweat immobilizing anti-cortisol on L-cys/AuNPs/MXene-modified conductive thread electrode [99], as shown in Figure 11. MXene and AuNPs increase the surface area and promote antibody immobilization, thereby improving sensitivity. The antibody was immobilized on the sensing layer by using EDC and NHS as coupling agents.

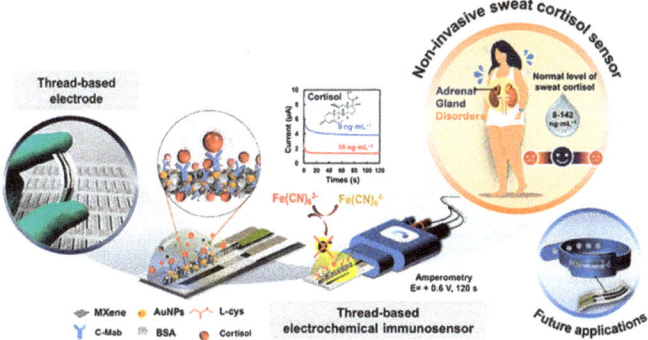

Figure 11. Schematic representation of L-cys/AuNPs/MXene-modified conductive thread electrode assembling and sensing mechanism of cortisol determination, reprinted with permission from [99] Copyright 2022 Elsevier.

Cortisol was analysed using amperometry and a linearity range from 5 to 180 ng mL^{-1} and LOD of 0.54 ng mL^{-1} were observed under optimized conditions. Corticosterone, cortisone, AA, UA, and creatinine were investigated as possible interfering molecules, but their amperometric response did not affect the cortisol response. Considering reproducibility at different cortisol concentrations, the results were acceptable in terms of RSD, ranging from 2.60% to 3.46%. Long-term stability was addressed, and the initial value decreased by 20% after 6 weeks at 4 °C. The sensor was applied to artificial sweat samples, with recoveries in the range 94.47–102%.

A wearable electrochemical impedimetric immunosensor based on a $Ti_3C_2T_x$-decorated laser-burned graphene (LBG) flake 3D electrode network including a microfluidic channel and chamber was assembled for cortisol determination in human sweat [100]. Polydimethylsiloxane (PDMS) was selected as a substrate for the flexible and stretchable patch sensor. Then, LBG was deposited on it and a laser line gap induced a disconnection between laser-burned graphene flakes and conductivity, so that the electrochemical activity of the LBG electrode was reduced. For this reason, highly conductive $Ti_3C_2T_x$ was deposited onto the electrode. In addition, this flexible microfluidic system was prepared using 3D-printed mold and PDMS. Finally, anti-cortisol was immobilized on a $Ti_3C_2T_x$ MXene/LBG/PDMS 3D sensing network and cortisol was determined using EIS. A linear concentration range between 10 pM and 100 nM with LOD of 3.88 pM was achieved. Aldosterone, corticosterone, prednisolone, and progesterone were tested as possible interferences that did not significantly affect the response of cortisol. Reproducibility with an RSD of 4.6% was considered satisfactory. The wearable sensor was applied to quantify the cortisol in freshly collected spiked sweat samples to evaluate its applicability in clinical diagnostics, and recoveries ranging from 93.68% to 99.1% were obtained.

Epinephrine (EP) or adrenaline is a hormone belonging to the catecholamine family secreted by the suprarenal gland. Catecholamines stimulate CNS and cardiac contraction [101]. Adrenaline is present in human brain fluid, blood, and body fluids at nanomolar levels. Alterations in EP levels can be correlated to several disorders, such as Alzheimer's or Parkinson's disease, and multiple sclerosis.

A nanocomposite including reduced graphene oxide (rGO) and $Ti_3C_2T_x$ (GMA) was used to modify an indium tin oxide (ITO) electrode for determining EP [102]. $Ti_3C_2T_x$ was uniformly deposited on the 3D G layer and the 3D structure of the GMA supported the interaction with biomolecules. EP was determined using DPV, and a linear concentration range of 1–60 µM with LOD of 3.5 nM was observed. UA was considered a possible interfering compound and any appreciable changes in analytical parameters were evidenced. Repeatability was acceptable in terms of RSD (1%); concerning stability, the voltammetric response retained 96.2% of its initial value after 16 days at RT. The GMA sensor was applied to human urine real samples for verifying its application in clinical analysis and recoveries ranging from 95.7% to 105.7% were evidenced.

An electrochemical sensor based on GCE modified by CoMn-based porous MOF deposited on carbon nanofiber (CNF) was prepared for EP determination [103]. CoMn-ZIF/CNF combined CNF's good conductivity and CoMn-ZIF's electrocatalytic properties. EP was determined using DPV, and a linear range from 5 to 1000 µM with LOD of 1.667 µM was achieved. Reproducibility was satisfactory in terms of RSD (1.95%); the voltammetric response retained 87.81% of its initial value after 20 days, under not specified storage conditions. Different amino acids such as tyrosine and leucine, UA, DA, norepinephrine, and glucose and drugs such as ibuprofen and amoxicillin were selected as possible interferences, and they exhibited negligible influence on the EP electrochemical signal. The sensor was applied to human urine samples for evaluating its applicability to a real matrix, with recoveries from 97.51% to 102.53%.

Insulin is an important hormone secreted by β-cells in the pancreas. This hormone regulates the glucose level in the blood through the control of the metabolism of proteins, carbohydrates, and lipids. The normal value of insulin concentration in blood is 50 pmol L^{-1} [104]. Nevertheless, inflammatory autoimmune and/or metabolic disorders can undermine β-cells inducing type I (insulin-dependent) diabetes.

An aptasensor for insulin assay in human serum was assembled modifying disposable Au electrodes (DGEs) with a nanocomposite including copper (II) benzene-1,3,5-tricarboxylate (Cu-BTC) nanowires and a leaf-like zeolitic imidazolate framework (ZIF-L) as MOF [105], as illustrated in Figure 12.

Aptamers were immobilized via physical adsorption on a Cu-BTC/ZIF-L composite that enhanced the aptasensor electrochemical performance due to the integration of electrocatalytic properties and the larger surface area of MOF and Cu-BTC nanowires. Insulin was analysed using DPV, obtaining a linear concentration range from 0.1 pM to 5 µM with LOD of 0.027 pM. Glucose, DA, AA, and melatonin (MELA) were used as interferences for evaluating aptasensor selectivity: the error percentage of 10.5% evidenced the good selectivity of the sensor. Reproducibility was investigated and acceptable results in terms of RSD (5.5%) were obtained. Regarding operative stability, the voltammetric signal decreased a little after continuous measurements up to 20 cycles, evidencing a stability rate of 91.7%. The aptasensor was applied to human serum samples with recoveries ranging from 97.2% to 98.5%, and the data were comparable to those coming from the ELISA standard method The aptasensor was used to monitor in vivo insulin in non-diabetic and diabetic mice, and data were comparable to those coming from the ELISA standard method, as shown in Figure 13. Biofouling was observed due to the presence of proteins, so the use of antibiological attachment polymers, such as polyethylene glycol and polyhydroxy ethylmethacrylate, was suggested to prevent biofouling and to improve aptasensor durability.

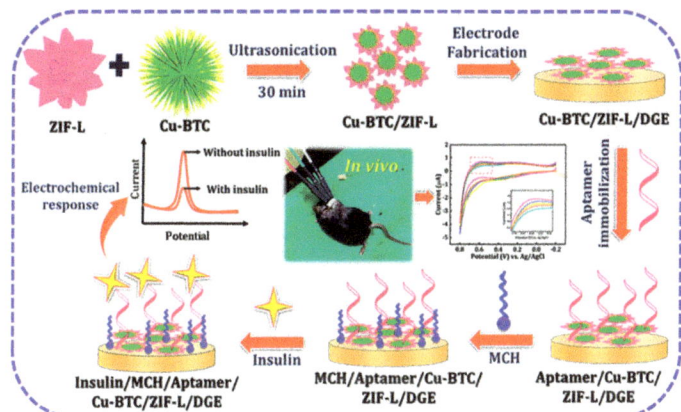

Figure 12. Assembling of Cu-BTC/ZIF-L composite-modified DGE and sensing mechanism of insulin assay, reprinted with permission from [105] Copyright 2022 American Chemical Society.

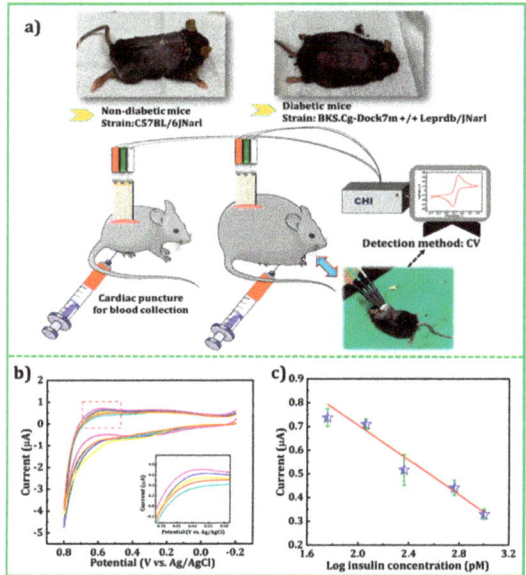

Figure 13. (a) Schematic of real-time in vivo insulin monitoring using aptamer/Cu-BTC/ZIF-L/DGE. (b) CV curves of insulin concentration in serum samples determined by insulin aptasensor. (c) Calibration plot of anodic peak current vs. insulin concentration logarithm, reprinted with permission from [105] Copyright 2022 American Chemical Society.

As a conclusive comment on the reported sensors for the detection of hormones, linearity ranges were sufficiently wide, with LOD values being generally nanomolar as well as picomolar, independently from analyte, and according to hormonal levels in biological fluids such as blood, urine, or sweat.

Notably, an example involving 2D-based biosensor nanomaterials for biomedical applications was used for the in vivo determination of mice for the first time, with important results for the possible commercialization of the device [105].

Analytical performances of the electrochemical sensors reported for the determination of hormones as well as the corresponding sensor format are summarized in Table 3.

Table 3. Analytical performances and format of electrochemical (bio)sensors for hormones determination.

Electrode	2D Nano-material	Format	Technique	Sample	Linearity	LOD	Recovery %	Reference Method	Ref.
Thread Conductive E	MXene	immunosensor immobilizing anti-cortisol on L-cys/AuNPs/MXene	A	Cortisol in sweat	5–180 ng mL^{-1}	0.54 ng mL^{-1}	94.47–102	-	[99]
LBG/PDMS	MXene (Ti$_3$C$_2$T$_x$)	immunosensor immobilizing anti-cortisol on Ti$_3$C$_2$T$_x$/LBG/PDMS	EIS	Cortisol in sweat	0.01–100 nM	3.88 pM	93.68–99.1	-	[100]
ITOE	MXene (Ti$_3$C$_2$T$_x$)	sensor based on GMA	DPV	EP in human urine	1–60 µM	3.5 nM	95.7–105.7	-	[102]
GCE	MOF (CoMn-ZIF)	sensor based on CoMnZIF-CNF	DPV	EP in human urine	5–1000 µM	1.667 µM	97.51–102.53	-	[103]
DGE	MOF (ZIF-L)	aptasensor immobilizing insulin aptamer on Cu-BTC/ZIF-L	DPV	Insulin in human serum	0.1 pM–5 µM	0.027 pM	97.2–98.5	ELISA	[105]

Abbreviations: A: amperometry; AuNPs: gold nanoparticles; CNF: carbon nanofiber; Cu-BTC: copper(II) benzene-1,3,5-tricarboxylate; DGE: disposable gold electrode; DPV: differential pulse voltammetry; EIS: electrochemical impedance spectroscopy; ELISA: enzyme-linked immunosorbent assay; EP: epinephrine; GCE: glassy carbon electrode; GMA: reduced graphene oxide/MXene Ti$_3$C$_2$T$_x$; ITOE: indium tin oxide electrode; L-cys: L-cysteine; LBG: laser graphene burned; MOF: metal–organic framework; PDMS: polydimethylsiloxane.

3.4. Pathogens

A pathogen is defined as an organism causing disease to its host, and virulence is the severity of the disease symptoms [106]. Pathogens are biologically different organisms and comprise viruses and bacteria as well as unicellular and multicellular eukaryotes. In this review, examples of (bio)sensors based on 2D nanomaterials for the assay of bacteria and viruses are reported.

3.4.1. Bacteria

Bacteria are the most common cause of foodborne diseases and present different shapes, types, and properties. Pathogenic bacteria directly or indirectly infect food and water sources, and the ingestion of contaminated foods can induce intestinal infectious diseases and/or food poisoning [107].

In this review, recent examples of electrochemical (bio)sensors based on 2D nanomaterials for the determination of *Escherichia coli* (*E. coli*), *Salmonella Mycobacterium tuberculosis*, and two vibrio bacteria such as *Vibrio vulnificus* (VV) and *Vibrio parahaemolyticus* (VP) have been considered.

The first two examples concern the determination of Gram-negative bacterium *E. coli*, a facultative anaerobic rod present in the intestinal tract of animals and humans from birth. A wide and differentiated class of bacteria includes *E. coli*: most strains are not pathogenic while a few induce diseases in humans and animals.

An electrochemical immunosensor was developed based on magnetic COF [108]. A specific egg yolk antibody (IgY) with affinity for *E. coli* was labelled with a porous magnetic covalent organic framework (m-COF) microbeads to assemble a capture probe (m-COF@IgY) for efficiently recognizing *E. coli*, as shown in Figure 14.

m-COF@IgY was then used with ferrocene boronic acid (FBA) as signal tag to combine with *E. coli* in a sandwich complex, dropped on a screen-printed electrode (SPE). The voltammetric signal generated by FBA in the sandwich complex was used for the quantitation of bacterium. After experimental conditions had been optimized, *E. coli* was determined by SVW in the linear range of 10–10^8 CFU mL^{-1} with LOD of 3 CFU mL^{-1}.

Immunosensor specificity was investigated using *Vibrio parahaemolyticus* (VP), *Salmonella typhimurium* (ST), and *Listeria monocytogenes* and even at concentration 10 times higher than *E. coli*, the sensor had no significant response to them compared with that from *E. coli*. SPE maintained 97.0% of the initial signal response after being reused 60 times for various samples by controlling the sandwich complex through the magnet, so that the sensing platform could be considered reusable. Long-term stability was also addressed, and the voltammetric signal decreased by less than 10% after 3 months at 4 °C. The sensor was applied to determine *E. coli* in several spiked food samples such as milk, beef, and shrimps, with recoveries ranging from 90% to 103%. The results were validated and compared with those obtained by the ELISA standard method.

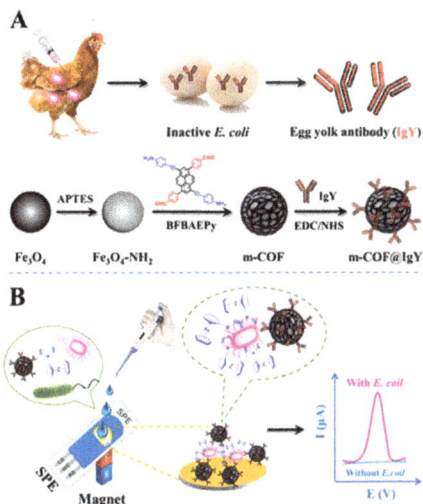

Figure 14. (**A**) Scheme for the synthesis of m-COF@IgY and IgY for *E. coli* and (**B**) sensing mechanism of the immunosensor for *E. coli* based on m-COF@IgY and ferrocene boric acid (FBA) using magnetic control screen-printed electrode (SPE) as detection platform, reprinted with permission from [108], Copyright 2022 Elsevier.

A cationic covalent organic polymer (COP), named CATN, was employed to develop an impedimetric sensor for *E. coli* cells [109]. COPs are a class of porous organic materials, generally including polycyclic aromatic hydrocarbons linked by C-C σ bonds and several π electronic systems, thereby guaranteeing good thermal and chemical stability and representing a proper sensing platform [110]. CATN was deposited via electrophoresis onto the interdigitated electrode array (IDEA) and EIS response showed a linear logarithmic relationship with increasing concentrations of *E. coli* up to 10 CFU mL^{-1}, with LODs of 2 CFU mL^{-1}. Repeatability, reproducibility and stability were not analysed. The sensor was not applied to real samples.

A two-dimensional porphyrin-based covalent organic framework (Tph-TDC-COF) was used for developing an electrochemical aptasensor to determine *E. coli* [111]. Tph-TDC-COF was synthesized starting from 5,10,15,20-tetrakis(4-aminophenyl)-21H, 23H-porphine (Tph), and [2,2′-bithiophene]-2,5′-dicarbaldehyde (TDC) and a highly conjugated structure was obtained that had high conductivity, a large specific surface area, and was ideal for immobilizing aptamers through π-π stacking, hydrogen bonding, and electrostatic interactions. In particular, the specific recognition between the aptamer and *E. coli* results in the formation of the G-quadruplex. Tph-TDC-COF was deposited on AuE and *E. coli* was determined using EIS and DPV. The same linearity range was obtained using EIS and DPV (10–10^8 CFU mL^{-1}), while two different LOD values were found: 0.17 CFU mL^{-1} (EIS) and 0.38 CFU mL^{-1} (DPV). *Staphylococcus aureus*, Basophils (Bas), *Staaue* (Sta), and

Salmonella typhimurium (ST) were considered as possible interferences: the EIS response obtained by a mixture of interferents and *E. coli* was about 8% higher than the EIS signal obtained by detecting *E. coli* alone. Reproducibility was acceptable in terms of RSD (4.92%). The electrochemical response remained almost stable after 15 days at 4 °C. The aptasensor was applied to spiked real samples of bread and milk, with recoveries ranging from 100.09 to 103.97% (bread) and from 99.61 to 100.71% (milk).

Salmonella is a flagellated Gram-negative, non-spore-forming bacillus, growing at temperatures between 35 °C and 37 °C. It is a foodborne pathogen because most infections are due to the ingestion of contaminated food. *Salmonella* induces salmonellosis, the most significant symptoms of which are nausea, vomiting, abdominal pain, and diarrhoea, among others.

An MXene/poly (pyrrole) (PPy)-based bacteria-imprinted polymer (MPBIP) sensor was assembled to modify GCE for determining *Salmonella* [112]. MPBIP was prepared via the one-step electropolymerization of pyrrole, and a *Salmonella* template was then eluted. The interaction of *Salmonella* surface groups with MXene functional groups enhanced the connection between MPBIP and the target, assisting the biorecognition process. A linear relationship with the logarithmic concentration of *Salmonella* was found from 10^3 to 10^7 CFU mL^{-1} and the corresponding LOD was 23 CFU mL^{-1}. Repeatability was acceptable in terms of RSD (0.91%), and the sensor maintained 94.5% of its initial response after 7 days at 4 °C. *Staphilococcus aureus*, *E. coli*, and *Listeria monocytogenes* were tested at the same concentration of *Salmonella* for evaluating specificity and selectivity, without any significant response compared to that from *Salmonella*. The impedimetric sensor was applied to quantify *Salmonella* in drinking water samples, with recoveries in the range 96–109.4%.

Mycobacterium tuberculosis (*M. tb*) is a human pathogen that causes tuberculosis (TB), a serious infectious disease among the top 10 causes of death worldwide.

A genosensor was assembled for the determination of a DNA target inside an IS6110 sequence of the *M. tb* genome, using MXene (Ti$_3$C$_2$) nanosheets and PPY as modifiers of GCE, ssDNA as Probe DNA (p IS6110), and methylene blue (MB) as a redox indicator [113]. ssDNA was immobilized onto MXene/PPY/GCE via covalent bonding. The combination of MXene and PPY improved the electron transfer rate, conductivity, and sensitivity of the genosensing platform. The analytical performances of the genosensor were evaluated by recording the MB electrochemical response as a function of DNA target concentration using DPV. The MB signal intensity decreased as the DNA target concentration increased, since the redox probe electron transfer decreased because of the hybridization among the ssDNA immobilized onto the electrode and the target DNA. Under optimized conditions, a linearity range of 100 fM–25 nM and LOD of 11.24 fM were observed. Two-base mismatch DNA (2 m-DNA), five-base mismatch DNA (5 m-DNA), complementary target DNA, non-complementary DNA (nc-DNA), and, also, the genomic single-strand of other bacteria similar in sequence to *M. tb*., including the *M. bovis* BCG strain GL2 and *M. simiae* strain TMC 1226, were tested for evaluating specificity. It was evidenced that the MB current intensity increased with the number of mismatches. Reproducibility was acceptable in terms of RSD (6.05%). The biosensor applicability was investigated by analysing *M. tb*-extracted DNA from clinical patient sputum samples. The same samples were also analysed using a conventional PCR method. Recoveries were in the range of 90.52–100.8% and results were comparable with those coming from the PCR method.

The genus *Vibrio*, belonging to the family *Vibrionaceae*, includes more than 35 species, and more than one-third are pathogenic to humans [114]. *Vibrios* are Gram-negative straight or curved rods. Among the pathogenic *Vibrios*, *Vibrio cholerae*, *Vibrio parahaemolyticus*, and *Vibrio vulnificus* have to be mentioned.

As one of the most dangerous *Vibrios*, *Vibrio vulnificus* (VV) could cause acute gastroenteritis, primary sepsis, necrotizing wound infection, and even death.

A dual-mode immunoassay based on electrochemiluminescence (ECL) coupled with Surface Enhanced Raman Spectroscopy (SERS) was designed for the determination of VV [115]. A multifunctional sensing platform (R6G-Ti$_3$C$_2$T$_x$ @AuNRs-Ab$_2$/ABEI), including $Ti_3C_2T_x$ MXene, Rhodamine 6G (R6G), gold nanorods (AuNRs), detection antibodies (Ab$_2$), and N-(4-aminobutyl)-N-ethylisoluminol (ABEI), was developed as a signal unit. The dual-mode immunosensor included a capture unit, i.e., Fe_3O_4@Ab$_1$ where Fe_3O_4 supported the immunosensor assembling and Ab$_1$ acted as capture antibody. The signal unit R6G-$Ti_3C_2T_x$@AuNRs-Ab$_2$/ABEI evidenced good conductivity and large specific area. AuNRs and SERS signal R6G molecules were deposited on the surface of $Ti_3C_2T_x$ by electrostatic adsorption, while R6G, detection antibody Ab$_2$ and ECL signal tags ABEI were immobilized onto AuNRs. After capture unit Fe_3O_4@Ab$_1$ was assembled onto a magnetic glassy carbon electrode (m-GCE), VV was recognized and captured by Fe_3O_4@Ab$_1$, followed by the formation of the capture unit-target-signal unit immunocomplex Fe_3O_4@Ab$_1$-VV-R6G-$Ti_3C_2T_x$@AuNRs-Ab$_2$/ABEI after immobilizing the signal unit R6G$Ti_3C_2T_x$@AuNRs-Ab2/ABEI. Consequently, ABEI produced an ECL signal and R6G produced a SERS signal separately. Under optimized conditions, ECL intensity increased with VV concentration. A linear relationship between ECL and the logarithm of VV concentration was found in the range 1–10^8 CFU mL^{-1} with limit of quantitation of 1 CFU mL^{-1}. Considering SERS determination, there was a linear relationship between SERS intensity and the logarithm of VV concentration in the range 10^2–10^8 CFU mL^{-1} with a limit of quantitation of 10^2 CFU mL^{-1}. Considering operational stability, ECL response was stable after 16 consecutive scan cycles, with an RSD of 2.0%. In addition, SERS intensity was stable after 10 consecutive measurements, with an RSD of 2.9%. Reproducibility was also investigated, with acceptable results in terms of RSD: 3% for ECL determination and 2.9% for SERS determination. *Vibrio parahaemolyticus* (VP), *Shewanella marisflavi* (SM), *Vibrio harveyi* (VH), and *Enterobacter cloacae* (EC) were analysed as common interfering bacteria in a mixture with VV, providing signal intensities roughly similar to those of blank. The immunosensor was applied to VV-spiked real samples of seawater with recoveries ranging from 94.8% to 110.3% for ECL, and from 92.3% to 112.1% for SERS.

The consumption of raw or undercooked seafood such as crabs, shrimps, scallops, seaweed, oysters, and clams can induce gastroenteritis due to the presence of *Vibrio parahaemolyticus* (VP). In severe cases, the bacterium can cause dysentery, primary septicaemia, or cholera-like illness with the possibility of death [114].

A dual-mode electrochemical and colorimetric aptasensor was developed for the on-site detection of (VP) in shrimps [116]. Mercapto-phenylboronic acid (PBA), ferrocene (Fc), and Pt nanoparticles were assembled on an MXenes layer to develop the nanoprobe PBAFc@Pt@MXenes that acted as dual-signal probe, evidencing peroxidase mimic features and electrochemical activity. A screen-printed Au electrode functionalized with an aptamer acted as a capture probe for VP. A sandwich complex was produced after the conjugation of PBAFc@Pt@MXenes with the capture probe. The colorimetric determination of VP was related to the chromogenic reaction of tetramethylbenzidine (TMB)-H_2O_2, catalysed by PBA on the probe, while the electrochemical response of Fc on the probe was further used for the determination of VP. Under optimal conditions, VP was determined electrochemically using DPV with a linearity range of 10^1–10^8 CFU mL^{-1} and LOD of 5 CFU mL^{-1}, while a linearity range of 10^2–10^8 CFU mL^{-1} with LOD of 30 CFU mL^{-1} was obtained by the colorimetric method. *Staphylococcus aureus*, *Salmonella typhimurium*, *Listeria monocytogenes*, and *E. coli* were analysed as common interfering bacteria in a mixture with VP, the signal intensities resulting as roughly similar to those of blank. The electrochemical signal retained 89.6% of the initial value after 2 weeks under not clearly specified storage conditions. The aptasensor was applied to real samples of shrimps with recoveries ranging from 95.0% to 104.3%, in good agreement with those obtained by the colorimetric mode (94.2% to 102.5%).

3.4.2. Viruses

Viruses are everywhere and can induce life-threatening diseases in humans and animals. Indeed, it is well-known that viral infections cause a third of deaths worldwide [117,118].

Virus biosensors have been described in recent reviews, also providing a comparison with the conventional analytical methods [118–121]. In this review, attention was focused on electrochemical biosensors based on 2D nanomaterials, considering the literature of the last five years (2018–2022) and integrating the data of a previous review [117].

Almost four years ago, the worldwide coronavirus 2019 (CoV-2) pandemic was announced. It is well known that the CoV-2 virus induces Severe Acute Respiratory Coronavirus Syndrome SARS-CoV-2 [118]. Coronaviruses are enveloped viruses, and they can infect humans and animals. COVID-19 contains four structural proteins and protein S on the virus surface is responsible for infection transmission.

The first three examples include MOF-based biosensors but, notably, there is an interesting review focused on the use of MOF for the determination of viruses [122] comprising the literature up to 2019.

An electrochemical dual-aptamer biosensor based on NH_2-MIL-53(Al) as MOF, Au@Pt nanoparticles, horseradish peroxidase (HRP), and hemin/Gquadruplex DNAzyme (GQH DNAzyme) as a signal nanoprobe was developed for the detection of SARS-CoV-2 nucleocapsid protein (NP) [123]. Au@Pt/NH_2-MIL-53 was modified with HRP and with the thiolated aptamers (SH-2G-N48 and SH-2G-N61) including a double G-quadruplex sequence, for amplifying the aptasensor response and for catalyzing the oxidation of hydroquinone (HQ). In the presence of NP, HQ was determined using DPV, obtaining a linear correlation with the logarithmic concentration in the range of 0.025–50 ng mL^{-1} with LOD of 8.33 pg mL^{-1}. Several proteins, such as cTnI, Her2, and MPT64, were selected as interferences, and they did not affect the SARS-CoV-2 NP response. Repeatability was analyzed with acceptable results in terms of RSD%, ranging from 2.6% to 5.0. Results were comparable to those coming from the ELISA method, and recoveries were in the range 92.0–110%. A scheme of the nanoprobe's assembling and sensing mechanism are shown in Figure 15.

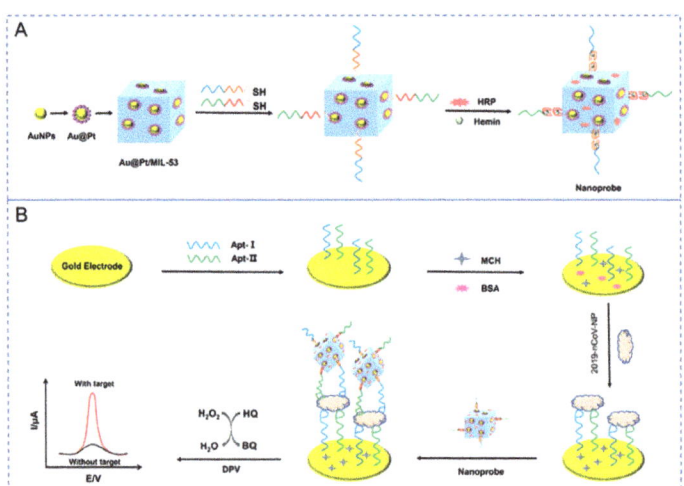

Figure 15. (A) Scheme of nanoprobe assembling and (B) sensing mechanism of the aptasensor for the detection of SARS-CoV-2 NP, reprinted with permission from [123], Copyright 2021 Elsevier.

A label-free electrochemical immunosensor for the determination of the SARS-CoV-2 S-protein [124] was assembled modifying SPCE with SiO_2@UiO-66 nanocomposite, including UiO-66 and a Zr-MOF nanostructure. The nanocomposite showed high surface area and porosity, good thermal conductivity, and chemical stability. SiO_2 nanoparticles improved the electron transfer and the conductivity of Zr-MOF. Angiotensin-converting enzyme 2 (ACE2) has been used as receptor for the S-protein [124]. S-protein determination was performed by EIS, and a linear concentration range from 100.0 fg mL^{-1} to 10.0 ng mL^{-1} with LOD of 100.0 fg mL^{-1} was obtained. Human coronavirus HCOV, L-glucose, L-Cys, L-Arg,

UA, DA, AA, vitamin D, ribavirin, zanamivir, favipiravir, remdesiver, and tenofovir were selected as possible interferences, and they did not affect the determination of S-proteins, except favipiravir, remdesiver, and tenofovir, since they are antiviral drugs. Reproducibility and repeatability were acceptable in terms of RSD (4.85%). The immunosensor was considered reusable, [124] and was applied to nasal fluid samples, with satisfactory recoveries ranging from 91.6% to 93.2%. Results were validated with the PCR test.

An aptasensor was developed using an aptamer and an imprinting polymer (MIP) for the determination of an intact SARS-CoV-2 virus [125]. The aptasensor was based on SPCE modified with nickel-benzene tricarboxylic acid-MOF ($Ni_3(BTC)_2$ MOF), SARSCoV-2 S-protein aptamer, and polydopamine synthesized via electropolymerization (ePDA). MIP synthesis was performed via PDA electropolymerization on the aptamer [SARS-CoV-2 virus] complex, immobilized on the modified SPCE. The template virus was removed just after the electropolymerization ended. Analytical performances of the MIP-aptamer nanohybrid sensor were evaluated by measuring EIS response to different concentrations of SARS-CoV-2 virus, and under optimized experimental conditions, a linear relationship with logarithmic concentration in the range of $10-10^8$ PFU mL^{-1} with LOD of 3.3 PFU mL 1 was achieved. SARS-CoV, MERS-CoV, influenza A H1N1, and influenza A H3N2 were selected as possible interfering viruses, and no significant response was observed for all of them. Reproducibility and repeatability were considered satisfactory in terms of RSD, at 4.2% and 1.4%, respectively. Considering long-term stability, EIS response did not show significant changes after 14 days at 4 °C. The aptasensor was applied to real saliva and nasopharyngeal swab samples in a viral transport medium (VTM) of sick and healthy patients for the qualitative and quantitative analysis of the virus. Results in both cases were comparable to those obtained via PCR, with recoveries ranging from 98% to 104%.

The last example includes an MXenes-based genosensor, reminding that a recent review focusing on the use of MXenes for the determination of the SARSCoV-2 virus has been reported in the literature [126].

A $Ti_3C_2T_x$ MXene was functionalized with a single-stranded DNA (ssDNA) through noncovalent adsorption, which facilitates the sequence-specific detection via hybridization with a target SARS-CoV-2 gene [127]. Consequently, ssDNA/$Ti_3C_2T_x$ was used for developing a chemoresistive biosensing platform for the determination of the SARS-CoV-2 N gene. The hybridization of the SARS-CoV-2 N gene with complementary DNA probes induced the detachment of dsDNA from the $Ti_3C_2T_x$ layer, so an increase in $Ti_3C_2T_x$ conductivity occurred. LOD below 10^5 copies mL^{-1} in saliva was obtained, with a linear concentration range from 10^5 to 10^9 copies mL^{-1}. Reproducibility, repeatability, selectivity, and stability data were not provided.

The hepatitis B (HBV) virus is widely spread worldwide and is transmitted through blood and body fluids. HBV belongs to the *Hepadnavirus* family; it is an enveloped icosahedral DNA virus with a spherical shape [118]. The virus's outer layer is a denominated surface antigen (HBS Ag) and it is a lipid envelope containing viral proteins responsible for the host cells attack.

An electrochemical genosensor, based on a modified GCE, was prepared for detecting HBV DNA by combining the electroactive Cu-MOF as signal nanoprobe and electroreduced GO (ErGO) as signal amplification material [128], without any enzymes, labels, or other redox indicators. Cu-MOF with a strong π-conjugate system can interact with ErGO through π–π interaction, enhancing the analytical performances of the sensor. Moreover, the strong interaction between Cu-MOF and DNA through covalent bonding improves stability. The genosensor was used to detect HBV using DPV, and a linear concentration range between 50.0 fM and 10.0 nM with LOD of 5.2 fM was evidenced. Reproducibility was evaluated satisfactory with RSD 3.02%. The voltammetric signal decreased by 4.5% after 2 weeks at 4 °C. The genosensor was applied to spiked human serum and urine samples, obtaining recoveries from 95.2% to 99.8%.

An electrochemical immunosensor based on GCE modified with Cu-MOF was developed for HBS Ag detection [129]. In particular, amine-functionalized Cu-MOF nanospheres were synthesized, for immobilizing Ab via covalent interaction between the Ab carboxyl group and the Cu-MOF amino groups. In addition, the nanospheres acted as electrocatalysts and electrochemical signal amplifiers. After the optimization of the experimental conditions, HBS Ag was determined by means of DPV and a linearity range from 1 ng·mL^{-1} to 500 ng·mL^{-1} with LOD of 730 pg·mL^{-1} was found. Considering reproducibility, RSD was 3.24%. Selectivity was tested by comparing the electrochemical response of the target antigen with those coming from other hepatitis virus biomarkers such as HAV, HDV, and HCV. HBS Ag, and the electrochemical response was much higher compared with other hepatitis virus markers. The biosensor was applied to spiked clinical samples, obtaining recoveries from 79.63% to 92.18%.

As a final remark regarding the biosensors for pathogen analysis herein reported, LOD values achieved fM or fg mL^{-1} in different cases. Concerning the biosensor format, the number of immunosensors, genosensors, and aptasensors is very similar.

Analytical data are generally sufficiently accurate, evidencing, in many cases, applicability to real samples. In some cases, results obtained from real samples were validated with external or standard methods.

Analytical performance and sensor format of the reported electrochemical (bio)sensors for pathogens determination are summarized in Table 4.

3.5. Cancer Biomarkers

According to the literature [130], biomarkers can be defined as "a characteristic that is objectively measured and evaluated as an indicator of normal biological processes, pathogenic processes, or pharmacologic responses to a therapeutic intervention". In particular, cancer biomarkers (CBs) are specific proteins and/or oligonucleotides spread into body fluids during the early stages of cancer at abnormal levels with respect to those found in healthy people. Moreover, they are important in providing data indicating the type and phase of the cancer and to monitor and evaluate treatment efficacy [131].

Herein, recent examples of electrochemical biosensors based on 2D nanomaterials for the determination of some relevant CBs are reported.

Carcinoembryonic antigen (CEA) is a cell membrane structural protein produced by embryonic gastrointestinal tissue and epithelial tumours and is one of the most important clinical CBs for the diagnosis of colon and breast cancers, ovarian carcinoma, colorectal cancer, and cystadenocarcinoma. Its normal level should be below 5 ng mL^{-1} in human serum. Concentration of 5 ng mL^{-1} is assumed as a threshold for differentiating abnormal from normal expression. In fact, CEA concentration is higher than 20 ng mL^{-1} in cancer patients [132].

An integrated microfluidic electrochemical platform was organized for the determination of CEA [133]. It included two functional parts: a herringbone-embedded microfluidic chip and an electrochemical aptasensor. The electrochemical aptasensor was based on SPCE modified with nanocomposite including MXene (Ti$_3$C$_2$) nanosheets and functionalized carbon nanotubes (CCNTs) with hemin for electrochemical signal amplification. Hemin supported the aptamer immobilization via EDC-NHS covalent coupling. The herringbone-embedded chip produced local mixed flow and enhanced the interaction between CEA and the sensing interface. All the analytical processes involving sample injection, efficient enrichment, target capture, and detection was performed at one integrated platform. CEA was determined using DPV and a dynamic concentrations range of 10–1 × 10^6 pg mL^{-1} with LOD of 2.88 pg mL^{-1} was obtained. Human serum albumin (HSA), immunoglobulin G (IgG), and glucose were tested in mixture with CEA as possible interfering molecules, and no significant response was observed for all of them. Reproducibility was considered satisfactory in terms of RSD (3.6%). The integrated platform was applied to spiked human serum samples, obtaining acceptable recoveries in the range of 95.29–105.19%.

Table 4. Analytical performances and format of electrochemical (bio)sensors for pathogen determination.

Electrode	2D Nanomaterial	Format	Technique	Sample	Linearity	LOD	Recovery (%)	Reference Method	Ref.
SPCE	m-COF	immunosensor immobilizing IgY antibody on m-COF	SWV	E. coli in milk, beef, shrimps	$10–10^8$ CFU mL^{-1}	3 CFU mL^{-1}	90–103	ELISA	[108]
IDEA	COP (CATN)	sensor based on CATN	EIS	E. coli	Up to 10 CFU mL^{-1}	2 CFU mL^{-1}	-	-	[109]
AuE	COF (Tph-TDC-COF)	aptasensor based on E. coli aptamer immobilized on Tph-TDC-COF	EIS/DPV	E. coli in bread, milk	$10–10^8$ CFU mL^{-1}	0.17 CFU mL^{-1} (EIS) 0.38 CFU mL^{-1} (DPV)	100.09–103.97 (bread) 99.61–100.71 (milk).	-	[111]
GCE	MXene	sensor based on MPBIP	EIS	Salmonella in drinking water	$10^3–10^7$ CFU mL^{-1}	23 CFU mL^{-1}	96–109.4	-	[112]
GCE	MXene (Ti$_3$C$_2$)	genosensor immobilizing ssDNA on PPY/MXene	DPV	M.tb in human sputum	100 fM-25 nM	11.24 fM	90.52–100.8	PCR	[113]
MGCE	MXene (Ti$_3$C$_2$T$_x$)	ECL immunosensor based on Fe$_3$O$_4$@Ab$_1$-VV-R6G-Ti$_3$C$_2$T$_x$@AuNRs-Ab$_2$/ABEI	ECL	VV in seawater	$1–10^8$ CFU mL^{-1}	1 CFU mL^{-1}	94.8–110.3	-	[115]
AuSPE	MXene	sandwich-type aptasensor involving PBA-Fc@Pt@MXenes	DPV	VP in shrimps	$10–10^8$ CFU mL^{-1}	5 CFU mL^{-1}	95.0–104.3	-	[116]
AuE	MOF (MIL-53 (Al))	sandwich-type aptasensor using 2 aptamers immobilized on AuE and Au@Pt/MIL-53 (Al) nanocomposite modified with HRP and hemin/Gquadruplex DNAzyme as signal nanoprobe	DPV	SARS-CoV-2 NP	0.025–50 ng mL^{-1}	8.33 pg mL^{-1}	92–110	ELISA	[123]
SPCE	MOF (SiO$_2$@UiO-66)	label-free immunosensor including SiO$_2$@UiO-66 nanocomposite	EIS	SARS-CoV-2 SP in nasal fluid samples	100.0 fg·mL^{-1}–10.0 ng·mL^{-1}	100.0 fg mL^{-1}	91.6–93.2	PCR	[124]
SPCE	MOF Ni$_3$(BTC)$_2$	label-free aptasensor using Ni$_3$(BTC)$_2$, SARS-CoV-2 aptamer, ePDA	EIS	SARS-CoV-2 virus in saliva, oropharyngeal swab	$10–10^8$ PFU mL^{-1}	3.3 PFU mL^{-1}	98–104	PCR	[125]

Table 4. Cont.

Electrode	2D Nanomaterial	Format	Technique	Sample	Linearity	LOD	Recovery (%)	Reference Method	Ref.
IGE	MXene ($Ti_3C_2T_x$)	genosensor using ssDNA/Ti_3C_2Tx	EIS	SARS-CoV-2 N gene	10^5–10^9 copies mL^{-1}	10^3 copies mL^{-1}	-	-	[127]
GCE	Cu-MOF	genosensor including Cu-MOF/ErGO	DPV	HBV in human serum, urine	50.0 fM–10.0 nM	5.2 fM	95.2–99.8	-	[128]
GCE	Cu-MOF	label-free immunosensor using Cu-MOF nanospheres	DPV	HBV in human serum	1–500 ng mL^{-1}	730 pg mL^{-1}	76.93–92.18	-	[129]

Abbreviations: Ab$_1$: capture antibody; Ab$_2$: detection antibody; ABEI: N-(4-aminobutyl)-N-ethylisoluminol; AuE: gold electrode; AuNRs: gold nanorods; AuSPE: gold screen-printed electrode; CATN: cationic network; COF: covalent organic framework; m-COF: magnetic COF; CFU: colony forming unit; COP: covalent organic polymer; DPV: differential pulse voltammetry; ELISA: enzyme-linked immunosorbent assay; ECL: electrochemiluminescence; EIS: electrochemical impedance spectroscopy; ePDA: electropolymerized poly(dopamine); ErGO: electroreduced graphene oxide; Fc: ferrocene; GCE: glassy carbon electrode; HBV: *Hepatitis B virus*; HRP: horseradish peroxidase; IDEA: interdigitated electrode arrays; IGE: interdigitated gold electrode; L-cys: L-cysteine; LGB: laser graphene burned; MGCE: magnetic glassy carbon electrode; MOF: metal-organic framework; MPBIP: MXene/PPY-based bacterial imprinted polymer; M.Tb: *Mycobacterium tuberculosis*; PBA: phenylboronic acid; PCR: protein C reactive; PFU: plaque forming unit; PPY: poly(pyrrole); SPCE: screen-printed carbon electrode; ssDNA: single-stranded DNA; SPCE: screen printed carbon electrode; SWV: square wave voltammetry; Tph: 10,15,20-tetrakis(4-aminophenyl)-21H,23H-porphine); TDC: [2,2′-bithiophene]-2,5′-dicarbaldehyde; UiO: Universitetet Oslo; VP: *Vibrio parahaemolyticus*; VV: *Vibrio vulnificus*.

A label-free immunosensor for CEA assay was realized including β-cyclodextrin (β-CDs) and gold nanoparticles (AuNPs) (Au-β-CD) deposited on the surface of FTO modified with a composite consisting of PANI and MXene (MXene@PANI) [134]. MXene@PANI improved the electrocatalytic activity and conductivity of the immunosensing platform, while Au-β-CD supported anti-CEA immobilization, as shown in Figure 16.

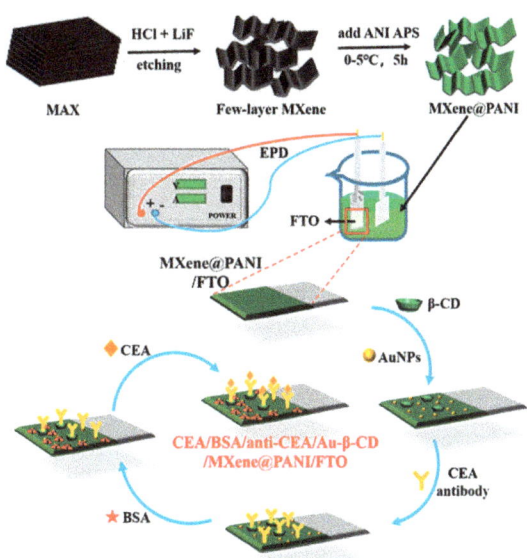

Figure 16. Scheme of the electrochemical immunosensor assembling steps for label-free detection of CEA, reprinted from [134].

The resulting BSA/anti-CEA/Au-β-CD/MXene@PANI/FTO immunosensor was electrochemically and morphologically characterized and, under optimized conditions, CEA was determined using DPV. The electrochemical signal decreased as the concentration of CEA increased, because the electron transfer was hindered after immunoreaction between anti-CEA, immobilized on the sensing interface, and CEA. There was a linear relationship between current intensity and the logarithm of CEA concentrations in the range 0.5~350 ng mL^{-1} with LOD of 0.0429 ng mL^{-1}. Reproducibility was considered satisfactory in terms of RSD (3.61%). AA, glucose, BSA, human immunoglobulin G (IgG), and cancer antigen 15-3 (CA15-3) were selected as possible interferences, without affecting the CEA DPV response. Concerning long-term stability, the electrochemical response decreased by only 9.4% after 10 days at 4 °C. The applicability of the CEA immunosensor was investigated by analysing several spiked real samples of human serum and recoveries ranging from 97.52% to 103.98% were obtained.

A nanocomposite including trimetallic nanoparticles (Au-Pd-Pt NPs) and MXene ($Ti_3C_2T_x$) nanosheets was drop-casted onto a GCE for assembling a CEA aptasensor [135]. In particular, CEA detection was performed via exonuclease III (Exo III)-supported recycling amplifications using triple-helix complex probes (THC). A CEA target interacted with the aptamer containing hairpin probes to cause cyclic cleavage of the secondary hairpins supported by Exo III to release ssDNAs. The obtained ssDNA hybridized with G-quadruplex-integrated triple-helix complex (THC) signal probes onto the electrode. Then, Exo III cyclically cleaved dsDNAs to release G-quadruplex sequences able to constrain hemin on the electrodic surface. Then, hemin catalysed H_2O_2 reduction at Au-Pd-Pt/$Ti_3C_2T_x$/GCE thus enhancing CEA electrochemical response. A linearity range from 1 fg mL^{-1} to 1 ng mL^{-1} with LOD of 0.32 fg mL^{-1} was obtained using DPV. Considering

BSA, α-fetoprotein (AFP), and platelet-derived growth factor (PDGF-BB) as possible interfering proteins, their DPV current intensities showed no significant differences in a blank test. Reproducibility was acceptable in terms of RSD (4.74%). Then, aptasensor was applied to spiked real samples of human serum with recoveries ranging from 98.8 to 104.1%.

The next two examples reported the application of COF nanomaterials for CB determination. Notably, a review regarding COF application for determining disease biomarkers, including CBs, can be found in the literature. There are generally reported applications of COFs to all types of sensors for CB analysis, from optics to fluorescence sensors, but not necessarily to the electrochemical ones [136].

Cancer antigen 125 (CA-125) or Mucin 16 (MUC16) is a protein and biomarker produced by ovarian cancer cells. CA 125 normal values found in healthy women are in the range 0–35 U mL^{-1}. Values higher than 35 U mL^{-1} are linked to ovarian cancer occurrence and development. Therefore, a CA-125 test could be useful to follow the ovarian cancer evolution during and after its treatment [137].

A sandwich immunosensor based on COF-LZU1 and multilayer reduced graphene oxide frame (MrGOF) was developed for assaying CA 125 [137]. COF-LZU1 was casted on GCE and acted as a platform for immobilizing CA 125 first antibody, while MrGOF, functionalized with an amino group and decorated with silver nanoparticles (AgNPs), acted as a probe to label CA 125 s antibody. A scheme of the immunosensor assembling is reported in Figure 17.

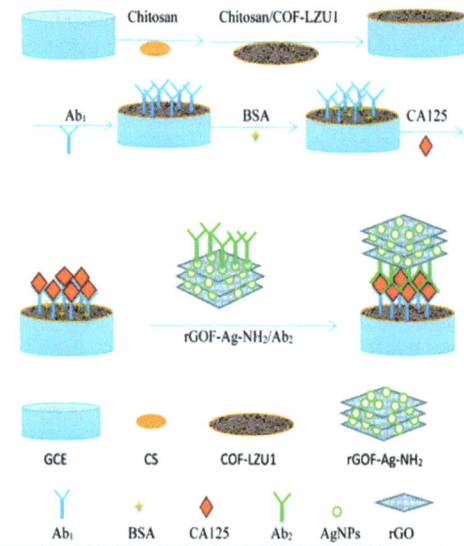

Figure 17. Scheme of the electrochemical sandwich immunosensor assembling for the determination of CA 125, reprinted with permission from [137] Copyright 2022 Elsevier.

Under optimized conditions, CA 125 was determined using DPV, and a linear concentration range from 0.001 to 40 U mL^{-1} and LOD of 0.00023 U mL^{-1} were found. CA19-9 antigen, CA72-4 antigen, horseradish peroxidase (HRP), AA, and BSA were analysed as possible interferences, evidencing that they did not affect CA 125 determination. Long-term stability was investigated: the DPV signal decreased by 19.11% after 7 days at 4 °C. The immunosensor was applied to spiked samples of human serum, obtaining recoveries ranging from 91.54% to 105.21%.

Prostate-specific antigen (PSA) is a well-known biomarker for prostate cancer (PCa) diagnosis. PSA normal values found in healthy men are in the range 0–4 ng mL^{-1}. Values higher than 4 ng mL^{-1} are related to PCa occurrence and development. Therefore, the PSA method to determine PCa accurately is very important [138].

A peptide-PSA-antibody sandwich immunosensor was developed using polydopamine-coated boron-doped carbon nitride, including AuNPs, (Au@PDA@BCN) and covalent organic frameworks (COF) functionalized with AuPt bimetallic nanoparticles and manganese dioxide (MnO$_2$) (AuPt@MnO$_2$@COF) [131]. Boron-doped carbon nitride (BCN) is a heteroatom-doped 2D carbon material with low toxicity and high stability, while PDA is a conducting polymer. A Au@PDA@BCN nanocomposite was deposited on a GCE and acted as a sensing platform to immobilize PSA primary antibodies. On the other hand, AuPt@MnO$_2$@COF served as an electrocatalyst and signal-amplifier. PSA affinity peptide was immobilized on it to form a Pep/MB/AuPt@MnO$_2$@COF nanocomposite, including MB as a redox indicator. PSA was determined using DPV, with a linear response in the range of 0.00005–10 ng mL^{-1} with LOD of 16.7 fg mL^{-1}. AA, BSA, lysine, ovalbumin, glucose, lysozyme, sucrose, and lipase were considered as interferences because they are present together with PSA in the same complex matrices or structural analogues. The DPV response of a mixture of PSA and all the interfering molecules was similar to that produced by PSA. Repeatability was satisfactory in terms of RSD (1.6%); DPV response decreased by 5% with respect to its initial value after 14 days at 4 °C. Spiked serum samples were analysed with recoveries ranging from 98.9% to 100.2%.

An aptasensor based on a two-dimensional porphyrin-based covalent organic framework (p-COF) was developed, immobilizing epidermal growth factor receptor (EGFR)-targeting aptamer strands for determining trace EGFR and living Michigan Cancer Foundation-7 (MCF-7) cells [139].

Epidermal growth factor receptor (EGFR) is a transmembrane protein and its abnormal values (>75.3 mg·L^{-1}) may be linked to different cancer typologies, such as lung, breast, prostate, bladder, colorectal, pancreatic, and ovarian.

MCF-7 is a human breast cancer cell line with estrogen, progesterone, and glucocorticoid receptors. It was isolated for the first time from the pleural effusion of a 69-year-old Caucasian metastatic breast cancer (adenocarcinoma) at the Michigan Cancer Foundation.

The two-dimensional p-COF structure can provide several binding sites for aptamers or biomolecules and its conjugated structure can improve electrochemical activity. In addition, the presence of large pore channels supports the aptamer immobilization inside the p-COF structure, thereby increasing the number of biomolecules adsorbed and further enhancing the sensor analytical performances. P-COF was deposited on AuE and EGFR was determined using EIS and DPV. The same linearity range was obtained via EIS and DPV (0.05–100 pg mL^{-1}), while two different LOD values were found: 7.54×10^{-3} pg mL^{-1} (EIS) and 5.64×10^{-3} pg mL^{-1} (DPV). CEA, PSA, Mucin-1 (MUC1), human epidermal growth factor receptor 2 (HER2), vascular endothelial growth factor (VEGF), immunoglobulin G (IgG), platelet-derived growth factor-BB (PDGF-BB), and BSA, present together with EGFR in human serum, were tested as interfering biomolecules at a concentration 1000 times higher than that of EGFR. Their electrochemical response was negligible compared with that of EGFR. After 10 days at 4 °C in dry state, the electrochemical response remained almost stable. Reproducibility was acceptable in terms of RSD (2.29%). The aptasensor was applied to spiked human serum samples to determine EGFR with recoveries ranging from 96.2% to 103.2%. After testing the biocompatibility of the aptasensor with MCF-7 cells, the sensor was applied for determination in artificial samples MCF-7 cells, with LOD of 61 cells·mL^{-1} and a linear detection range of 5×10^2–1×10^5 cell·mL^{-1}.

As final considerations, LOD values achieved generally pg mL^{-1} and the corresponding linearity ranges seem to be wide, considering the application field.

There is not a preferred sensor format and all the (bio)sensors have been applied to real samples. Comparison and validation with standard methods are, however, missing.

Analytical performance and format of the reported electrochemical biosensors for CBs determination are summarized in Table 5.

Table 5. Analytical performances and format of electrochemical (bio)sensors for cancer biomarkers determination.

Electrode	2D Nano-material	(Bio)Sensor Format	Technique	Sample	Linearity	LOD	Recovery%	Ref.
SPCE	MXene (Ti_3C_2)	Aptasensor including He@CCNT/Ti_3C_2	DPV	CEA in human serum	$10–10^6$ pg mL^{-1}	2.88 pg mL^{-1}	95.29–105.19	[133]
FTOE	MXene (Ti_3C_2)	Immunosensor including Au-β-CD/MXene@PANI	DPV	CEA in human serum	0.5–350 ng mL^{-1}	42.9 pg mL^{-1}	97.52–103.98	[134]
GCE	Mxene ($Ti_3C_2T_x$)	Aptasensor involving Au-Pd-Pt/$Ti_3C_2T_x$	DPV	CEA in human serum	1 fg mL^{-1} –1 ng mL^{-1}	0.32 fg mL^{-1}	98.8–104.1	[135]
GCE	COF-LZU1	Sandwich-type Immunosensor involving COF-LZU1, MrGOF, AgNPs	DPV	CA 125 in human serum	0.001–40 U mL^{-1}	0.00023 U mL^{-1}	91.54–105.21	[137]
GCE	COF BCN	Sandwich-type immunosensor involving MnO_2@COF, AuPtNPs, BCN, PDA	DPV	PSA in human serum	0.00005–10 ng mL^{-1}	16.7 fg mL^{-1}	98.9–100.2	[138]
AuE	COF (p-COF)	Aptasensor based on EGFR aptamer immobilized on p-COF	DPV EIS	EGFR in human serum	0.05–100 pg mL^{-1}	7.54×10^{-3} pg mL^{-1} (EIS) 5.64×10^{-3} pg mL^{-1} (DPV)	96.2–103.2	[139]
AuE	COF (p-COF)	Aptasensor based on EGFR aptamer immobilized on p-COF	EIS	MCF-7/-	$5 \times 10^2–1 \times 10^5$ cell·mL^{-1}	61 cells·mL^{-1}	-	[139]

Abbreviations: AgNPs: silver nanoparticles; AuPtNPs: gold platinum nanoparticles; AuE: gold electrode; β-CD: β-cyclodestrin; BCN: boron-doped carbon nitride; CEA: carcinoembryonic antigen; CNT: carbon nanotube; COF: covalent organic framework; DPV: differential pulse voltammetry; EGFR: epidermal growth factor receptor; FTOE: fluorine-doped tin oxide electrode; GCE: glassy carbon electrode; He: herringbone-embedded; MCF-7: Michigan Cancer Foundation-7; MrGOF: multi-layer reduced graphene oxide frame; p-COF: porphyrin-COF; PDA: polydopamine; PSA: prostate-specific antigen; SPCE: screen-printed carbon electrode; U: enzymatic unit.

3.6. Antibiotics

Antibiotics are drugs used for treating bacterial diseases in animals and plants because they can destroy bacterial cells by either preventing cell reproduction or modifying necessary cellular function or processes within the cell [140]. Antibiotics are non-biodegradable compounds, and they can induce endocrine disorders, anaemia, mutagenicity, etc., if they remain in the human body. Consequently, it is mandatory to have effective, responsive, and fast antibiotic trace detection methods.

A recent review reported some examples of electrochemical sensors based on MOFs for the determination of antibiotics [140], describing MOFs' synthetic methods and the different typologies of the included sensors. The most recent examples of electrochemical sensors for antibiotics have been herein reviewed, also considering other 2D nanomaterials.

Enrofloxacin (ENR) is a fluoroquinolone antibiotic used to fight bacterial diseases in livestock and aquaculture. Ampicillin (AMP) is a broad-spectrum antibiotic classified as β-lactam, widely used because of its ability to kill Gram-positive and -negative bacteria by destroying the cell wall.

Electrochemical aptasensors were developed for ENR and AMP analysis using AuE modified with COF synthesized from the condensation-polymerization of 1,3,6,8-tetrakis(4-formylphenyl)pyrene and melamine through imine bonds (Py-M-COF) [141]. Py-M-COF presented an extensive π-conjugation framework, a large specific surface area, a nanosheet-like structure, different functional groups, and good conductivity, so a stable and firm aptamers immobilization was performed via π-π stacking and electrostatic interactions. ENR was determined using EIS and a linear concentration range of 0.01 pg mL^{-1}–2 ng mL^{-1} with LOD of 6.07 fg mL^{-1} was found. Similarly, AMP was quantified with a linearity range of 0.001–1000 pg mL^{-1} and LOD of 0.04 fg mL^{-1}. Tetracycline, kanamycin (Kana), tobramycin (TOB), streptomycin, and oxytetracycline (OTC) were tested as possible interfering antibiotics and the electrochemical response of all those vs. both ENR and AMP was negligible. The reproducibility of the two aptasensors was acceptable considering the RSD values, 1.25% and 1.44% for ENR and AMP, respectively. Long-term stability was investigated: the EIS response remained almost stable after 14 days at 4 °C. ENR and AMP were assayed in spiked samples of human serum and recoveries were in the range of 101.0–112.4% for ENR and 99.5–103.0% for AMP.

Tetracycline (TC) is a broad-spectrum antibiotic for the treatment of bacterial infections in humans and animals and is used as animal feed-additive to prevent animal infections because of its low cost and high antibacterial activity. Notably, antibiotic residues in food can cause health problems and induce bacterial resistance to TC in humans and animals.

A molecularly imprinted tetracycline electrochemiluminescence (ECL) sensor was assembled based on Zr-coordinated amide porphyrin-based 2D COF (Zr-amide-Por-based 2D COF) [142]. Zr-amide-Por-based 2D COF was dropped onto GCE and then the TC-molecularly imprinted electrochemiluminescence sensor (TC-MIECS) was assembled to perform the electropolymerization of o-phenylendiamine (o-PD), acting as a functional monomer, and using TC as a molecular template. After removing the template molecule, imprinted cavities serving as TC recognition elements were realized. Under optimized experimental conditions, TC-MIECS showed a linear relationship with tetracycline in the concentration range 5–60 pM, with LOD of 2.3 pM. The sensor evidenced an acceptable repeatability in terms of RSD (6.7%). Operative stability was considered satisfactory because, after 12 measurements, the RSD value of the ECL intensity variation was 0.94%. Chloramphenicol (CAP), OTC, AMP, and penicillin were investigated as interfering antibiotics and the ECL intensity response of all the interferences vs. that of TC was negligible because of the high specificity of the molecularly imprinted cavities. TC-MIECS were applied to spiked real samples of milk and recoveries ranging from 94.0% to 103.5% were obtained.

Chloramphenicol (CAP) is a broad-spectrum antibiotic used against the main species of Gram-positive and -negative bacteria, as well as other groups of microorganisms such as *Salmonella*. Since CAP can trigger several collateral effects, such as anaemia or mutagenicity in humans, its use is limited and controlled or even forbidden in animal husbandry for food production in many countries [143].

An electrochemical aptasensor based on a nanocomposite including Zirconium-porphyrin MOF (PCN-222) and graphene oxide (PCN-222/GO) was prepared for the determination of CAP [144]. The high conductivity of GO, together with MOF mesoporous channels and metal sites, facilitated a stable aptamer immobilization owing to a π−π stacking interaction and the connection between the aptamer phosphate group and Zr(IV) sites of PCN-222. PCN-222/GO was dropped on AuE and then the aptamer was adsorbed on the nanocomposite. CAP was determined using EIS and a linear concentration range from 0.01 to 50 ng mL^{-1} and LOD of 7.04 pg mL^{-1} were found. OTC, TC, kanamycin sulphate (KS), metronidazole (MDZ), and nitrofurantoin (NFT) were used as interfering compounds, with the EIS response of all of them being insignificant compared with that of CAP. Reproducibility was acceptable in terms of RSD (6.29%), and long-term stability was defined as satisfactory without indicating storage conditions. The aptasensor was applied to spiked real samples of milk, human serum, and urine with recoveries ranging from 94.6 to 107.2%.

Norfloxacin (NF) presents a broad-spectrum antibacterial activity and, for this reason, it has been used for the treatment of human and animal diseases. However, its excessive use can trigger antibiotic resistance, collateral effects, and toxicity. Residual amounts or traces of NF can remain in foods and tissues because of its uncompleted metabolism. Therefore, the production and use of NF have been limited and controlled.

β-CD porous polymers (P-CDPs) functionalized with COF (PCDPs/COFs) and combined with Pd^{2+} via electrostatic interaction (Pd^{2+}@P-CDPs/COFs) were synthesized and casted onto GCE to assemble a non-enzymatic electrochemical sensor for the assay of NF [145], as shown in Figure 18.

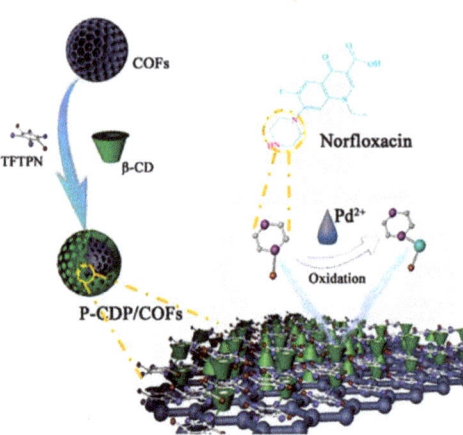

Figure 18. Scheme of non-enzymatic electrochemical sensor assembling for the determination of NF, reprinted with permission from [145] Copyright 2022 Elsevier.

COF was prepared starting from 1,3,5-tris(4-aminophenyl)benzene (TAPB) and 2,5-dimethoxyterephaldehyde (DMTP) via the simple solution infiltration method. β-CD entrapped NF through a host–guest interaction, while COF interacted electrostatically via -NH_2 functionalities with the negatively charged functional group of NF; consequently, the adsorption of NF onto the modified electrode improved. Moreover, Pd^{2+} increased the nanocomposite catalytic performances and acted as an electrochemical signal amplifier. After experimental conditions optimization, the NF was determined using DPV, and two linear concentrations in the ranges of 0.08–7.0 µM and 7.0–100.0 µM with LOD of 0.031 µM were evidenced. Precision was verified and the results in terms of RSD were in the range of 2.91–3.46% for intra-day precision and 3.13–3.47% for inter-day precision. Concerning accuracy, the results obtained for intra-day and inter-day accuracy were < 3.00%. Glucose, AA, UA, DA, and Levofloxacin (LF) were evaluated as interference molecules and none of them affected the electrochemical response of the NF. Repeatability and reproducibility were considered acceptable in terms of RSD, 2.9% and 3.1%, respectively. The sensor was applied to determine NF in the spiked samples of Norfloxacin Eye-drops. The results were comparable to those coming from a standard method using HPLC, with recoveries ranging from 97.8% to 101.7%.

The next three examples introduce electrochemical sensors for the determination of three antibiotics belonging to the sulphonamide family with a wide spectrum of antimicrobial activities against protozoa and bacteria. The first one is sulfathiazole (STZ), well-known as being low cost and a broad-spectrum antibiotic, widely employed for the treatment of animal and human infections. On the other hand, its accumulation in humans and livestock causes serious effects on hematopoietic systems by blocking the dihydrofolic acid synthesis [146], so its use is limited and controlled especially to guarantee food safety.

A GCE-modified electrochemical MIP sensor for the determination of STZ was designed to exploit CuS microflowers as redox probe polypyrrole as MIP, imprinted with STZ, and AuNPs incorporated in a COF structure (Au@COF) for conductivity [147]. The scheme for MIP-sensor assembling is shown in Figure 19.

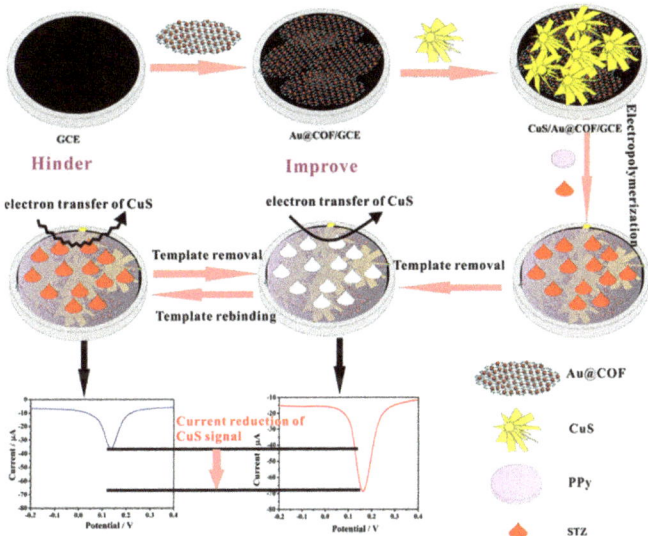

Figure 19. Scheme for MIP/CuS/Au@COF/GCE-sensor assembling and the STZ-detection strategy, reprinted with permission from [147] Copyright 2020 Elsevier.

STZ was quantified using DPV, and a decreasing electrochemical response was observed for increasing STZ concentration, owing to selective STZ adsorption on the MIP film. Consequently, the corresponding CuS electron transfer at the sensing interface was hampered. The DPV current decrease was directly proportional to the logarithm of STZ concentration over a concentration range from 0.001 to 100,000 nM with LOD of 0.0043 nM. Structural analogues of STZ, such as sulfadimidine, sulfacetamide, and sulfadiazine (SDZ), and glucose, glutamate, and AA were tested as possible interfering molecules, evidencing that they did not interfere with the STZ assay. Reproducibility was considered, with acceptable results in terms of RSD (4.5%); DPV response decreased by 15.4% after 30 days, without specifying storage conditions. Mutton and fodder spiked samples were used to investigate accuracy and the applicability to real samples, with recoveries in the range of 83.0–107.3%. The MIP sensor was applied to samples of chicken liver and pig liver and the results were comparable to those coming from HPLC as a reference method, indicating no significant difference in amounts determined by the two methods ($p > 0.05$).

Sulfamethoxazole as STZ can cause haematological problems, as well as hypersensitivity and gastrointestinal diseases. Moreover, sulfamethoxazole traces in food with animal origin can induce thyroid cancer and several other diseases [148].

An electrochemical sensor based on CPE modified with a Fe_3O_4/ZIF-67 nanocomposite and ionic liquid (IL) 1-Butyl-3-methylimidazolium hexafluorophosphate (Fe_3O_4/ZIF-67/ILCPE) was prepared for the determination of sulfamethoxazole [149]. ZIF-67 is a well-known MOF with peculiar catalytic properties and chemical and thermal stability, as described above, while Fe_3O_4 can improve conductivity and electron-transfer to and from the electrode surface. Under the optimized experimental conditions, the antibiotic was determined using DPV, with the current intensity directly proportional to the concentration over the range 0 0.01–520.0 µM with LOD of 5 nM. River and tap water and urine-spiked samples were used to investigate the sensor applicability to real samples, and recoveries in the range of 97.1–103.3% were found. The reproducibility, repeatability, and stability data of the sensor were not provided.

Sulfadiazine (SDZ) is an antibiotic used for the treatment of several infections in animals and humans. SDZ inhibits bacteria growth by interfering with folate metabolism, so its use is regulated and controlled especially for food safety [150].

An electrochemical sensor for the simultaneous determination of SDZ and AP was developed based on an MIP for the recognition of SDZ and AP, and a GO@COF nanocomposite for signal amplification [150]. The nanocomposite was first deposited on GCE; a polypyrrole MIP was then electropolymerized on the modified electrode. MIP synthesis was performed via pyrrole electropolymerization in the presence of the template molecules. Then, they were removed just after the electropolymerization ended. Under optimized experimental conditions, SDZ and AP were determined simultaneously through DPV. Linear concentration ranges of 0.5–200 µM for SDZ and 0.05–20 µM for AP, with LODs of 0.16 µM and 0.032 µM, respectively, were found. Molecules structurally similar to SDZ, such as sulfamerazine and sulfacetamide, and to AP, such as AA and p-nitrophenol, were assumed as possible interferences. The electrochemical response of the modified GCE towards SDZ and AP were higher than those of the respective structural analogues. Reproducibility was satisfactory in terms of RSD: 5.5% (SDZ) and 6.7% (AP). Concerning repeatability, RSD values of 2.7% for SDZ and 5.8% for AP were found. The DPV response decreased by 17.6% for SDZ and 14.6% for AP, after 30 days, without specifying the storage conditions. The sensor was applied to spiked samples of fodder and beef, extracted with an organic solvent, for investigating its accuracy. Recoveries ranging from 82.0% to 108.0% were obtained. The sensor was applied to real samples of pork and chicken, with data comparable to those coming from the HPLC analysis.

Tobramycin (TOB) is an effective antibiotic commonly used in the treatment of various systemic and ocular infections. As an aminoglycoside antibiotic, tobramycin is active against aerobic gram-negative bacteria. Like other aminoglycosides, several side effects for tobramycin can be evidenced such as ototoxicity, nephrotoxicity, and neuromuscular toxicity.

An impedimetric aptasensor based on a Zirconium-porphyrin MOF PCN-222 (Fe) nanosheet (PCN-222(Fe)-NS) was assembled to analyse TOB [151]. PCN-222(Fe)-NS was prepared, starting from [5,10,15,20-tetrakis(4-carboxyphenyl) porphyrinato]-Fe(III) chloride and Zr clusters, and supported the aptamer immobilization via π–π interaction and coordination bond. Under optimized conditions, a linear range of 1.1×10^{-4}–10.7 nM and LOD of 1.3×10^{-4} nM were determined. Considering the aptasensor reproducibility, RSD values smaller than 4.92% were obtained. Penicillin G (PCL), oxytetracycline (OTC), doxycycline (DOX), streptomycin (STR), and kanamycin (KAN) were investigated as possible interfering antibiotics, without affecting the impedimetric response of TOB. The electrochemical signal was almost stable after 7 days of storage at $-20\,°C$. The aptasensor was applied to spiked real samples of milk and recoveries ranging from 97.5% to 104.8% were found.

A covalent organic framework nanosheet (COF NS) synthesized from 2,4,6-triformylphloroglucinol (Tp) and 5,5′-diamino-2,2′-bipyridine (Bpy) (Tp-Bpy COF NS) was used as functional material to immobilize the aptamer for assembling an impedimetric aptasensor to determine TOB [152]. Tp-Bpy COF NS showed a high surface area and stability and different functional sites to support the aptamer immobilization. TOB was assayed by EIS and a linear concentration range of 2.1×10^{-4}–10.7 nM and LOD of 6.57 fM were obtained. Reproducibility was investigated by using five Tp-Bpy COF NS aptasensors, with RSD values lower than 4.17%. Tetracycline (TET), ampicillin (AMP), oxytetracycline (OTC), kanamycin (KAN), penicillin G (PCL), doxycycline (DOX), and ornidazole (ODZ) were tested as possible interfering antibiotics, which did not affect the EIS response of the TOB. The electrochemical signal was almost stable after 10 days storage at $-20\,°C$. The aptasensor was applied to spiked real samples of milk and river water, with recoveries in the range of 95.4–105.2% in milk and of 96.1–104.4% in river water.

Considering all the sensors described, the LOD values achieved concentration levels pM or nM, and the corresponding linearity ranges seem to be wide enough considering the application field. The preferred (bio)sensor format seems to be aptasensor. Validation with standard methods was performed for three examples only [145,147,150].

The analytical performance and format of the reported electrochemical biosensors for antibiotics determination are summarized in Table 6.

Table 6. Analytical performances and format of electrochemical (bio)sensors for antibiotics determination.

Electrode	2D Nanomaterial	Format	Technique	Sample	Linearity	LOD	Recovery %	Reference Method	Ref.
AuE	Py-M-COF	Aptasensor including Py-M-COF	EIS	ENR in human serum	0.01–2000 pg mL^{-1}	6.07 fg mL^{-1}	101.0–112.4	-	[141]
AuE	Py-M-COF	Aptasensor including Py-M-COF	EIS	AMP in human serum	0.0001–1000 pg mL^{-1}	0.04 fg mL^{-1}	99.5–103.0	-	[141]
GCE	Zr-amide-Por-based 2D COF	TC-MIECS	ECL	TC in milk	5–60 pM	2.3 pM	94.0–103.5	-	[142]
AuE	MOF (PCN-222)	Aptasensor including PCN-222/GO	EIS	CAP in milk, human serum, urine	0.01–50 ng mL^{-1}	7.04 pg mL^{-1}	94.6–107.2	-	[144]
GCE	COF	Electrochemical sensor based on Pd^{2+}@P-CDPs/COFs	DPV	NF in eye drops	0.08–7.0 μM 7.0–100.0 μM	0.031 μM	97.8–101.7	HPLC	[145]
GCE	COF	Electrochemical sensor based on MIP/CuS/Au@COF	DPV	STZ in mutton, fodder, chicken liver, pig liver	0.001–100,000 nM	0.0043 nM	83.0–107.3	HPLC	[147]
CPE	MOF (ZIF-67)	Electrochemical sensor based on Fe$_3$O$_4$/ZIF-67/IL	DPV	Sulfamethoxazole in river, tap water, urine	0.01–520.0 μM	5.0 nM	97.1–103.3	-	[149]
GCE	COF	Electrochemical sensor based on GO@COF nanocomposite and MIP	DPV	SDZ/beef, pork, chicken, fodder	0.5–200 μM	0.16 μM	82.0–108.0	HPLC	[150]
AuE	MOF (PCN-222(Fe) NS)	Aptasensor including PCN-222 (Fe) NS	EIS	TOB/milk	1.1×10^{-4}–10.7 nM	1.3×10^{-4} nM	97.5–104.8	-	[151]
AuE	COF (Tp-Bpy COF NS)	Aptasensor including Tp-Bpy COF NS	EIS	TOB/milk, river water	2.1×10^{-4}–10.7 nM	6.57 fM	95.4–105.2% milk 96.1–104.4% river	-	[152]

Abbreviations: AMP: ampicillin; AuE: gold electrode; Bpy: 5,5′-diamino-2,2′-bipyridine; CAP: chloramphenicol; COF: covalent organic framework; CPE: carbon paste electrode; DPV: differential pulse voltammetry; ECL: electrochemiluminescence; EIS: electrochemical impedance spectroscopy; ENR: enrofloxacin; GCE: glassy carbon electrode; GO: graphene oxide; HPLC: high performance liquid chromatography; IL: ionic liquid; M: melamine; MOF: metal–organic framework; NF: norfloxacin; P-CDPs: β-cyclodextrin porous polymers; PCN-222 (Fe) NS: zirconium metal–organic framework nanosheet; Py: 1,3,6,8-tetrakis(4-formylphenyl)pyrene); SDZ: sulfadiazine; STZ: sulfathiazole; TC: tetracycline; TC-MIECS: TC-molecularly imprinted electrochemiluminescence sensor; TOB: tobramycin; Tp: 2,4,6-triformylphloroglucinol; Zr-amide-Por-based 2D COF: Zr-coordinated amide porphyrin-based 2D COF.

4. Conclusions

In this section, some considerations and comments regarding the role of 2D nanomaterials and the different types of electrodes and biosensors are summarized.

In addition, some comments concerning biosensors' analytical performances, such as linearity range, detection limits, selectivity, and validation with standard methods of analysis, along with the possibility of determining several analytes at the same time or the applicability in real matrices, are introduced. Finally, critical issues, challenges, and future perspectives on the electrochemical (bio)sensing approach for biomedical applications involving 2D nanomaterials are outlined.

In most cases, the most used nanomaterial was found to be MXenes because of their high electrochemical activity and conductivity, as well as large surface area and well-established synthetic procedures [16–20]. In addition, MXenes can be easily functionalized and, consequently, their performance in the electrochemical sensing area relies on the type and number of these functional groups. Finally, MXenes nanohybrids including different nanomaterials or polymers can be easily synthesized, and the sensing properties of the nanocomposite can be enhanced with respect to those of the starting material.

MOFs are also largely employed in (bio)sensors for biomedical applications, probably because it is possible to tune their structure and functionality [30,31], and include in the MOF's structure different conducting nanomaterials such as metal nanoparticles can improve the poor conductivity and sensing properties of the starting MOFs.

COFs and TMDs are much less likely to be used because COFs' synthesis methods still need to be optimized, while synthetic approaches for TMDs require drastic conditions in terms of temperature and pressure, which do not facilitate their use to design (bio)sensors for biomedical applications.

The integration of nanomaterials in the development of electrochemical (bio)sensors is crucial for improving their analytical performances.

Nanocomposites and/or nanohybrids represent the preferred sensing interface, including different nanomaterials or synthetic polymers, and sometimes the resulting nanostructure with tailored architecture can be very complex.

Considering the electrode typologies reported in this review (see Tables 1–6), different types of electrodes are employed for the determination of analytes of clinical and biomedical interest, from more conventional bulk electrodes such as GCE, AuE, and ITOE, to IDEA (interdigitated electrode array), IGE (interdigitated gold electrode), CCE (carbon cloth electrode), and SPEs, among others. GCE was the preferred option probably because of its well-known chemical–physical properties [3].

Gold electrode (AuE) and SPEs were employed in several examples because Au is a biocompatible, stable, easy to functionalise, and a conducting material, and SPEs represent low-cost sensing platforms, able to move the transition from conventional laboratory equipment to portable devices.

Regarding different types of (bio)sensors, it is difficult to make general considerations, because examples with different types of analytes have been examined. In the case of glucose, neurotransmitters and hormones as target molecules, chemosensors, are prevailing, the key recognition element being generally a nanocomposite including 2D nanomaterials [see Tables 1–3]. In the case of glucose, only a few examples are represented by classical enzymatic biosensors based on glucose oxidase [see Table 1].

On the other hand, when pathogens, CBs, or antibiotics are the target analyte, immunosensor or aptasensor are the prevalent type [see Tables 4–6].

The number of examples using molecularly imprinted polymer as a key recognition element was limited [see Tables 3, 4 and 6], even if it could represent a promising synthetic receptor to be used instead of more conventional ones.

Analytical performances in terms of linearity range and/or LOD seem to be very promising: in fact, several examples involved nanomolar concentrations with the LOD at a femtomolar level, depending on the target. Regarding selectivity, this issue is generally addressed, but the criterion of choice of interfering compounds is often not clear, even for the same target. It would be useful to indicate this criterion: interferences might have been chosen because they have similar structure or function or because they are present in the same complex matrices to be analysed where the target is also present, just to suggest some examples.

The reproducibility, repeatability, and stability of the sensors have been generally investigated. However, it is difficult to compare data from different sensors, even when regarding the same target, because comparable experimental procedures are not used. In particular, referring to long-term stability, storage conditions in terms of temperature and wet or dry storage are not always indicated.

Validation with standard methods is rarely performed, even if this step is mandatory in order to have a clear indication of the sensors' analytical performances in comparison with those of more conventional approaches.

Moreover, it should be stressed once again that electrochemical (bio)sensors are generally not commercially available except for the glucometer [see Table 1]. Different problems and challenges are involved, such as complex and expensive material synthesis, samples preparation and stability, unforeseen interfering molecules, collateral chemical reactions in real matrices, or the fouling and biofouling of the electrode surface.

Finally, coming back to 2D nanomaterials, greater attention should be paid to the relationship among structure and material properties, and which 2D nanomaterial properties really affect the (bio)sensor performances.

All these challenges should be addressed and solved in view of the electrochemical (bio)sensor introduction into the market, because the industrial interest in their commercialization is connected to market demand considering all the costs associated with new technologies. Consequently, an interdisciplinary approach and the close collaborations of analytical chemists, materials scientists, biochemists, engineers, and physicians can effectively support solving these problems.

Author Contributions: Writing—original draft preparation, A.C.; writing—review and editing, A.C., R.P. and P.D.M. All authors have read and agreed to the published version of the manuscript.

Funding: This research received no external funding.

Data Availability Statement: Not applicable.

Acknowledgments: The author would thank Alessandro Trani for the technical support.

Conflicts of Interest: The author declares no conflict of interest.

References

1. Khan, R.; Radoi, A.; Rashid, S.; Hayat, A.; Vasilescu, A.; Silvana Andreescu, S. Two-Dimensional Nanostructures for Electrochemical Biosensor. *Sensors* **2021**, *21*, 3369. [CrossRef] [PubMed]
2. Wongkaew, N.; Simsek, M.; Griesche, C.; Baeumner, A.J. Functional Nanomaterials and Nanostructures Enhancing Electrochemical Biosensors and Lab-on-a-Chip Performances: Recent Progress, Applications, and Future Perspective. *Chem. Rev.* **2019**, *119*, 120–194. [CrossRef] [PubMed]
3. Curulli, A. Electrochemical Biosensors in Food Safety: Challenges and Perspectives. *Molecules* **2021**, *26*, 2940. [CrossRef]
4. Curulli, A. Functional Nanomaterials Enhancing Electrochemical Biosensors as Smart Tools for Detecting Infectious Viral Diseases. *Molecules* **2023**, *28*, 3777. [CrossRef]
5. Brazaca, L.C.; dos Santos, L.P.; de Oliveira, P.R.; Rocha, D.P.; Stefano, J.S.; Kalinke, C.; Abarza Munoz, R.A.; Alves Bonacin, J.; Janegitz, B.C.; Carrilho, E. Biosensing strategies for the electrochemical detection of viruses and viral diseases—A review. *Anal. Chim. Acta* **2021**, *1159*, 338384. [CrossRef] [PubMed]
6. Anichini, C.; Czepa, W.; Pakulski, D.; Aliprandi, A.; Ciesielski, A.; Samorì, P. Chemical sensing with 2D materials. *Chem. Soc. Rev.* **2018**, *47*, 4860. [CrossRef]
7. Varghese, S.S.; Varghese, S.H.; Swaminathan, S.; Singh, K.K.; Mittal, V. Two-Dimensional Materials for Sensing: Graphene and Beyond. *Electronics* **2015**, *4*, 651–687. [CrossRef]
8. Wang, L.; Xiong, Q.; Xiao, F.; Duan, H. 2D nanomaterials based electrochemical biosensors for cancer diagnosis. *Biosens. Bioelectron.* **2017**, *89*, 136–151. [CrossRef]
9. Campuzano, S.; Pedrero, M.; Nikoleli, G.-P.; Pingarrón, J.M.; Nikolelis, D.P. Hybrid 2D-nanomaterials-based electrochemical immunosensing strategies for clinical biomarkers determination. *Biosens. Bioelectron.* **2017**, *89*, 269–279. [CrossRef]

10. Bolotsky, A.; Butler, D.; Dong, C.; Gerace, K.; Glavin, N.R.; Muratore, C.; Robinson, J.A.; Ebrahimi, A. Two-Dimensional Materials in Biosensing and Healthcare: From In Vitro Diagnostics to Optogenetics and Beyond. *ACS Nano* 2019, *13*, 9781–9810. [CrossRef]
11. Derakhshi, M.; Daemi, S.; Shahini, P.; Habibzadeh, A.; Mostafavi, E.; Ashkarran, A.A. Two-Dimensional Nanomaterials beyond Graphene for Biomedical Applications. *J. Funct. Biomater.* 2022, *13*, 27. [CrossRef] [PubMed]
12. Zhu, S.; Liu, Y.; Gu, Z.; Zhao, Y. Research trends in biomedical applications of two-dimensional nanomaterials over the last decade—A bibliometric analysis. *Adv. Drug Deliv. Rev.* 2022, *188*, 114420. [CrossRef] [PubMed]
13. Sakthivel, R.; Keerthi, M.; Chung, R.-J.; He, J.-H. Heterostructures of 2D materials and their applications in biosensing. *Prog. Mater. Sci.* 2023, *132*, 101024. [CrossRef]
14. Arulraj, A.; Mangalaraja, R.V.; Khalid, M. MXene: Pioneering 2D Materials. In *Fundamental Aspects and Perspectives of Mxenes*, 1st ed.; Khalid, M., Nirmala Grace, A., Arulraj, A., Numan, A., Eds.; Springer Nature: Berlin, Germany, 2022; pp. 17–37.
15. Koyappayil, A.; Ganpat Chavan, S.; Roh, Y.-G.; Lee, M.-H. Advances of MXenes; Perspectives on Biomedical Research. *Biosensors* 2022, *12*, 454. [CrossRef] [PubMed]
16. Alwarappan, S.; Nesakumar, N.; Sun, D.; Hu, T.Y.; Li, C.-Z. 2D metal carbides and nitrides (MXenes) for sensors and biosensors. *Biosens. Bioelectron.* 2022, *205*, 113943. [CrossRef] [PubMed]
17. Rajeev, R.; Thadathil, D.A.; Varghese, A. New horizons in surface topography modulation of MXenes for electrochemical sensing toward potential biomarkers of chronic disorders. *Crit. Rev. Solid State Mater. Sci.* 2023, *48*, 580–622. [CrossRef]
18. Zhu, S.; Wang, D.; Li, M.; Zhou, C.; Yu, D.; Lin, Y. Recent advances in flexible and wearable chemo and bio-sensors based on two-dimensional transition metal carbides and nitrides MXenes. *J. Mater. Chem. B* 2022, *10*, 2113. [CrossRef]
19. Ab Latif, F.E.; Numan, A.; Mubarak, N.M.; Khalid, M.; Abdullah, E.C.; Manaf, N.A.; Walvekar, R. Evolution of MXene and its 2D heterostructure in electrochemical sensor applications. *Coord. Chem. Rev.* 2022, *471*, 214755. [CrossRef]
20. Sajid, M. MXenes: Are they emerging materials for analytical chemistry applications?—A review. *Anal. Chim. Acta* 2021, *1143*, 267–280. [CrossRef]
21. Tao, W.; Kong, N.; Ji, X.; Zhang, Y.; Sharma, A.; Ouyang, J.; Qi, B.; Wang, J.; Xie, N.; Kang, C.; et al. Emerging two-dimensional monoelemental materials (Xenes) for biomedical applications. *Chem. Soc. Rev.* 2019, *48*, 2891. [CrossRef]
22. Zhang, J.; Zhang, X.; Bi, S. Two-Dimensional Quantum Dot-Based Electrochemical Biosensors. *Biosensors* 2022, *12*, 254. [CrossRef] [PubMed]
23. Xu, Y.; Wang, X.; Zhang, W.L.; Lv, F.; Guo, S. Recent progress in two-dimensional inorganic quantum dots. *Chem. Soc. Rev.* 2018, *47*, 586–625. [CrossRef] [PubMed]
24. Tajik, S.; Dourandish, Z.; Garkani Nejad, F.; Beitollahi, H.; Jahani, P.M.; Di Bartolomeo, A. Transition metal dichalcogenides: Synthesis and use in the development of electrochemical sensors and biosensors. *Biosens. Bioelectron.* 2022, *216*, 114674. [CrossRef] [PubMed]
25. Sun, H.; Li, D.; Yue, X.; Hong, R.; Yang, W.; Liu, C.; Xu, H.; Lu, J.; Dong, L.; Wang, G.; et al. A Review of Transition MetalDichalcogenides-Based Biosensors. *Front. Bioeng. Biotechnol.* 2022, *10*, 941135.
26. Zhao, Y.; Wang, S.-B.; Chen, A.-Z.; Kankala, R.K. Nanoarchitectured assembly and surface of two-dimensional (2D) transition metal dichalcogenides (TMDCs) for cancer therapy. *Coord. Chem. Rev.* 2022, *472*, 214765. [CrossRef]
27. Lam, C.Y.C.; Zhang, Q.; Yin, B.; Huang, Y.; Wang, H.; Yang, M.; Wong, S.H.D. Recent Advances in Two-Dimensional Transition Metal Dichalcogenide Nanocomposites Biosensors for Virus Detection before and during COVID-19 Outbreak. *J. Compos. Sci.* 2021, *5*, 190. [CrossRef]
28. Mohammadpour, Z.; Hossein Abdollahi, S.; Safavi, A. Sugar-Based Natural Deep Eutectic Mixtures as Green Intercalating Solvents for High-Yield Preparation of Stable MoS$_2$ Nanosheets: Application to Electrocatalysis of Hydrogen Evolution Reaction. *ACS Appl. Energy Mater.* 2018, *1*, 5896–5906. [CrossRef]
29. Mohammadpour, Z.; Hossein Abdollahi, S.; Omidvar, A.; Mohajeri, A.; Safavi, A. Aqueous solutions of carbohydrates are new choices of green solvents for highly efficient exfoliation of two-dimensional nanomaterials. *J. Mol. Liq.* 2020, *309*, 113087. [CrossRef]
30. Gonçalves, J.M.; Martins, P.R.; Rocha, D.P.; Matias, T.A.; Juliao, M.S.S.; Munoz, R.A.A.; Angnes, L. Recent trends and perspectives in electrochemical sensors based on MOF-derived materials. *J. Mater. Chem. C* 2021, *9*, 8718. [CrossRef]
31. Chang, Y.; Lou, J.; Yang, L.; Liu, M.; Xia, N.; Liu, L. Design and Application of Electrochemical Sensors with Metal–Organic Frameworks as the Electrode Materials or Signal Tags. *Nanomaterials* 2022, *12*, 3248. [CrossRef]
32. Haider, J.; Shahzadi, A.; Akbar, M.U.; Hafeez, I.; Shahzadi, I.; Khalid, A.; Ashfaq, A.; Ahmad, S.O.A.; Dilpazir, S.; Imran, M.; et al. A review of synthesis, fabrication, and emerging biomedical applications of metal-organic frameworks. *Biomater. Adv.* 2022, *140*, 213049. [CrossRef] [PubMed]
33. Dourandish, Z.; Tajik, S.; Beitollahi, H.; Jahani, P.M.; Garkani Nejad, F.; Sheikhshoaie, I.; Di Bartolomeo, A. A Comprehensive Review of Metal–Organic Framework: Synthesis, Characterization, and Investigation of Their Application in Electrochemical Biosensors for Biomedical Analysis. *Sensors* 2022, *22*, 2238. [CrossRef] [PubMed]
34. Pirzada, M.; Altintas, Z. Nanomaterials for Healthcare Biosensing Applications. *Sensors* 2019, *19*, 5311. [CrossRef] [PubMed]
35. Martínez-Periñán, E.; Martínez-Fernández, M.; Segura, J.L.; Lorenzo, E. Electrochemical (Bio)Sensors Based on Covalent Organic Frameworks (COFs). *Sensors* 2022, *22*, 4758. [CrossRef] [PubMed]
36. Zhu, J.; Wen, W.; Tian, Z.; Zhang, X.; Wang, S. Covalent organic framework: A state-of-the-art review of electrochemical sensing applications. *Talanta* 2023, *260*, 124613.

17. Lu, Z.; Wang, Y.; Li, G. Covalent Organic Frameworks-Based Electrochemical Sensors for Food Safety Analysis. *Biosensors* **2023**, *13*, 291. [CrossRef] [PubMed]
18. Geng, K.; He, T.; Liu, R.; Dalapati, S.; Tan, K.T.; Li, Z.; Tao, S.; Gong, Y.; Jiang, Q.; Jiang, D. Covalent Organic Frameworks: Design, Synthesis, and Functions. *Chem. Rev.* **2020**, *120*, 8814–8933. [CrossRef]
19. Labib, M.; Sargent, E.H.; O'Kelley, S. Electrochemical Methods for the Analysis of Clinically Relevant Biomolecules. *Chem. Rev.* **2016**, *116*, 9001–9090. [CrossRef]
20. Hosseinzadeh, B.; Rodriguez-Mendez, M.L. Electrochemical Sensor for Food Monitoring Using Metal-Organic Framework Materials. *Chemosensors* **2023**, *11*, 357. [CrossRef]
21. Bertok, T.; Lorencova, L.; Chocholova, E.; Jane, E.; Vikartovska, A.; Kasak, P.; Tkac, J. Electrochemical Impedance Spectroscopy Based Biosensors: Mechanistic Principles, Analytical Examples and Challenges towards Commercialization for Assays of Protein Cancer Biomarkers. *ChemElectroChem* **2019**, *6*, 989–1003. [CrossRef]
22. Rohaizad, N.; Mayorga-Martinez, C.C.; Fojtů, M.; Latiff, N.M.; Pumera, M. Two-dimensional materials in biomedical, biosensing and sensing applications. *Chem. Soc. Rev.* **2021**, *50*, 619. [CrossRef] [PubMed]
23. Meng, Z.; Stolz, R.M.; Mendecki, L.; Mirica, K.A. Electrically-Transduced Chemical Sensors Based on Two-Dimensional Nanomaterials. *Chem. Rev.* **2019**, *119*, 478–598. [CrossRef] [PubMed]
24. Yuan, R.; Lib, H.-K.; He, H. Recent advances in metal/covalent organic framework-based electrochemical aptasensors for biosensing applications. *Dalton Trans.* **2021**, *50*, 14091. [CrossRef] [PubMed]
25. Uçar, A.; Aydogdu Tığ, G.; Er, E. Recent advances in two dimensional nanomaterial-based electrochemical (bio)sensing platforms for trace-level detection of amino acids and pharmaceuticals. *Trends Anal. Chem.* **2023**, *162*, 117027. [CrossRef]
26. Ganesan, S.; Ramajayam, K.; Kokulnathan, T.; Palaniappan, A. Recent Advances in Two-Dimensional MXene-Based Electrochemical Biosensors for Sweat Analysis. *Molecules* **2023**, *28*, 4617. [CrossRef]
27. Lee, C.W.; Suh, J.M.; Jang, H.W. Chemical Sensors Based on Two-Dimensional (2D) Materials for Selective Detection of Ions and Molecules in Liquid. *Front. Chem.* **2019**, *7*, 708. [CrossRef]
28. Balkourani, G.; Damartzis, T.; Brouzgou, A.; Tsiakaras, P. Cost Effective Synthesis of Graphene Nanomaterials for Non-Enzymatic Electrochemical Sensors for Glucose: A Comprehensive Review. *Sensors* **2022**, *22*, 355. [CrossRef]
29. Radhakrishnan, S.; Lakshmy, S.; Santhosh, S.; Kalarikkal, N.; Chakraborty, B.; Rout, C.S. Recent Developments and Future Perspective on Electrochemical Glucose Sensors Based on 2D Materials. *Biosensors* **2022**, *12*, 467. [CrossRef]
30. Meng, W.; Wen, Y.; Dai, L.; He, Z.; Wang, L. A novel electrochemical sensor for glucose detection based on Ag@ZIF-67 nanocomposite. *Sens. Actuators B* **2018**, *260*, 852–860. [CrossRef]
31. Arif, D.; Hussain, Z.; Sohail, M.; Liaqat, M.A.; Khan, M.A.; Noor, T. A Non-enzymatic Electrochemical Sensor for Glucose Detection Based on Ag@TiO$_2$@ Metal-Organic Framework (ZIF-67) Nanocomposite. *Front. Chem.* **2020**, *8*, 573510. [CrossRef]
32. Shu, Y.; Su, T.; Lu, Q.; Shang, Z.; Qin Xu, Q.; Hu, X. Highly Stretchable Wearable Electrochemical Sensor Based on Ni-Co MOF Nanosheet-Decorated Ag/rGO/PU Fiber for Continuous Sweat Glucose Detection. *Anal. Chem.* **2021**, *93*, 16222–16230. [CrossRef] [PubMed]
33. Wang, Q.; Chen, M.; Xiong, C.; Zhu, X.; Chen, C.; Zhou, F.; Dong, Y.; Wang, Y.; Xu, J.; Li, Y.; et al. Dual confinement of high–loading enzymes within metal–organic frameworks for glucose sensor with enhanced cascade biocatalysis. *Biosens. Bioelectron.* **2022**, *196*, 113695. [CrossRef] [PubMed]
34. Tao, B.; Li, J.; Miao, F.; Zang, Y. Carbon Cloth Loaded NiCo$_2$O$_4$ Nano-Arrays to Construct Co-MOF@GO Nanocubes: A High-Performance Electrochemical Sensor for Non-Enzymatic Glucose. *IEEE Sens. J.* **2022**, *22*, 13898–13907. [CrossRef]
35. Du, Q.; Liao, Y.; Shi, N.; Sun, S.; Liao, X.; Yin, G.; Huang, Z.; Pu, X.; Wang, J. Facile synthesis of bimetallic metal–organic frameworks on nickel foam for a high performance non-enzymatic glucose sensor. *J. Electroanal. Chem.* **2022**, *904*, 115887. [CrossRef]
36. Wang, F.; Hu, J.; Liu, Y.; Yuan, G.; Zhang, S.; Xu, L.; Xue, H.; Pang, H. Turning coordination environment of 2D nickel-based metal-organic frameworks by π-conjugated molecule for enhancing glucose electrochemical sensor performance. *Mater. Today Chem.* **2022**, *24*, 100885. [CrossRef]
37. Vignesh, A.; Vajeeston, P.; Pannipara, M.; Al-Sehemi, A.G.; Xia, Y.; Gnana Kumar, G. Bimetallic metal-organic framework derived 3D hierarchical NiO/Co$_3$O$_4$/C hollow microspheres on biodegradable garbage bag for sensitive, selective, and flexible enzyme-free electrochemical glucose detection. *Chem. Eng. J.* **2022**, *430*, 133157. [CrossRef]
38. Wei, X.; Liu, N.; Chen, W.; Qiao, S.; Chen, Y. Three-phase composites of NiFe$_2$O$_4$/Ni@C nanoparticles derived from metal-organic frameworks as electrocatalysts for the oxygen evolution reaction. *Nanotechnology* **2021**, *32*, 175701. [CrossRef]
39. Dong, S.; Niu, H.; Sun, L.; Zhang, S.; Wub, D.; Yang, Z.; Xiang, M. Highly dense Ni-MOF nanoflake arrays supported on conductive graphene/carbon fiber substrate as flexible microelectrode for electrochemical sensing of glucose. *J. Electroanal. Chem.* **2022**, *911*, 116219. [CrossRef]
40. Zhai, X.; Cao, Y.; Sun, W.; Cao, S.; Wang, Y.; Hea, L.; Yao, N.; Zhao, D. Core-shell composite N-doped-Co-MOF@polydopamine decorated with Ag nanoparticles for non-enzymatic glucose sensors. *J. Electroanal. Chem.* **2022**, *918*, 116491. [CrossRef]
41. Shu, Y.; Shang, Z.; Su, T.; Zhang, S.; Lu, Q.; Xu, Q.; Hu, X. A highly flexible Ni–Co MOF nanosheet coated Au/PDMS film based wearable electrochemical sensor for continuous human sweat glucose monitoring. *Analyst* **2022**, *147*, 1440–1448. [CrossRef]

62. Myndrul, V.; Coy, E.; Babayevska, N.; Zahorodna, V.; Balitskyi, V.; Baginskiy, I.; Gogotsi, O.; Bechelany, M.; Giardi, M.T.; Iatsunskyi, I. MXene nanoflakes decorating ZnO tetrapods for enhanced performance of skin-attachable stretchable enzymatic electrochemical glucose sensor. *Biosens. Bioelectron.* **2022**, *207*, 114141. [CrossRef] [PubMed]
63. Huang, Y.; Long, Z.; Zou, J.; Luo, L.; Zhou, X.; Liu, H.; He, W.; Shen, K.; Wu, J. A Glucose Sensor Based on Surface Functionalized MXene. *IEEE Trans. Nanotechnol.* **2022**, *21*, 399–405. [CrossRef]
64. Murugan, P.; Annamalai, J.; Atchudan, R.; Govindasamy, M.; Nallaswamy, D.; Ganapathy, D.; Reshetilov, A.; Sundramoorthy, A.K. Electrochemical Sensing of Glucose Using Glucose Oxidase/PEDOT:4-Sulfocalix [4]arene/MXene Composite Modified Electrode. *Micromachines* **2022**, *13*, 304. [CrossRef] [PubMed]
65. Zahed, M.A.; Sharifuzzaman, M.; Yoon, H.; Asaduzzaman, M.; Kim, D.K.; Jeong, S.; Pradhan, G.B.; Shin, Y.D.; Yoon, S.H.; Sharma, S.; et al. A Nanoporous Carbon-MXene Heterostructured Nanocomposite-Based Epidermal Patch for Real-Time Biopotentials and Sweat Glucose Monitoring. *Adv. Funct. Mater.* **2022**, *32*, 2208344. [CrossRef]
66. Hu, T.; Zhang, M.; Dong, H.; Li, T.; Zang, X.; Li, X.; Ni, Z. Free-standing MXene/chitosan/Cu_2O electrode: An enzyme-free and efficient biosensor for simultaneous determination of glucose and cholesterol. *J. Zhejiang Univ.-Sci. A Appl. Phys. Eng.* **2022**, *23*, 579–586.
67. Han, X.; Cao, K.; Yao, Y.; Zhao, J.; Chai, C.; Dai, P. A novel electrochemical sensor for glucose detection based on a Ti_3C_2Tx/ZIF-67 nanocomposite. *RSC Adv.* **2022**, *12*, 20138. [CrossRef]
68. Kumar, V.; Shukla, S.K.; Choudhary, M.; Gupta, J.; Chaudhary, P.; Srivastava, S.; Kumar, M.; Kumar, M.; Kumar Sarma, D.; Yadav, B.C.; et al. Ti_2C-TiO_2 MXene Nanocomposite-Based High-Efficiency Non-Enzymatic Glucose Sensing Platform for Diabetes Monitoring. *Sensors* **2022**, *22*, 5589. [CrossRef]
69. Manoj, D.; Aziz, A.; Muhammad, N.; Wang, Z.; Xiao, F.; Asif, M.; Sun, Y. Integrating Co_3O_4 nanocubes on MXene anchored CFE for improved electrocatalytic activity: Freestanding flexible electrode for glucose sensing. *J. Environ. Chem. Eng.* **2022**, *10*, 108433. [CrossRef]
70. Gopal, T.S.; Jeong, S.K.; Alrebdi, T.A.; Pandiaraj, S.; Alodhayb, A.; Muthuramamoorthy, M.; Grace, A.N. MXene-based composite electrodes for efficient electrochemical sensing of glucose by non-enzymatic method. *Mater. Today Chem.* **2022**, *24*, 100891. [CrossRef]
71. Liu, Y.; Liu, M.; Shang, S.; Gao, W.; Wang, X.; Hong, J.; Hua, C.; You, Z.; Liu, Y.; Chen, J. Recrystallization of 2D C-MOF Films for High-Performance Electrochemical Sensors. *ACS Appl. Mater. Interfaces* **2023**, *15*, 16991–16998. [CrossRef]
72. Venkata Ratnam, K.; Manjunatha, H.; Janardan, S.; Chandra Babu Naidu, K.; Ramesh, S. Nonenzymatic electrochemical sensor based on metal oxide, MO (M¼ Cu, Ni, Zn, and Fe) nanomaterials for neurotransmitters: An abridged review. *Sens. Int.* **2020**, *1*, 100047. [CrossRef]
73. Lakshmanakumar, M.; Nesakumar, N.; Jayalatha Kulandaisamy, A.; Balaguru Rayappan, J.B. Principles and recent developments in optical and electrochemical sensing of dopamine: A comprehensive review. *Measurement* **2021**, *183*, 109873. [CrossRef]
74. Lorencova, L.; Bertok, T.; Filip, J.; Jerigova, M.; Velic, D.; Kasak, P.; Mahmoud, K.A.; Tkac, J. Highly stable Ti_3C_2Tx (MXene)/Pt nanoparticles-modified glassy carbon electrode for H_2O_2 and small molecules sensing applications. *Sens. Actuators B* **2018**, *263*, 360–368. [CrossRef]
75. Lei, Y.; Butler, D.; Lucking, M.C.; Zhang, F.; Xia, T.; Fujisawa, K.; Granzier-Nakajima, T.; Cruz-Silva, R.; Endo, M.; Terrones, H.; et al. Single-atom doping of MoS_2 with manganese enables ultrasensitive detection of dopamine: Experimental and computational approach. *Sci. Adv.* **2020**, *6*, abc4250. [CrossRef] [PubMed]
76. Murugan, N.; Jerome, R.; Preethika, M.; Sundaramurthy, A.; Sundramoorthy, A.K. 2D-titanium carbide (MXene) based selective electrochemical sensor for simultaneous detection of ascorbic acid, dopamine and uric acid. *J. Mater. Sci. Technol.* **2021**, *72*, 122–131. [CrossRef]
77. Sabar, M.; Amara, U.; Riaz, S.; Hayat, A.; Nasir, M.; Nawaz, M.H. Fabrication of MoS_2 enwrapped carbon cloth as electrochemical probe for non-enzymatic detection of dopamine. *Mater. Lett.* **2022**, *308*, 131233. [CrossRef]
78. Wang, M.; Guo, H.; Wu, N.; Zhang, J.; Zhang, T.; Liu, B.; Pan, Z.; Liping Peng, L.; Yang, W. A novel triazine-based covalent organic framework combined with AuNPs and reduced graphene oxide as an electrochemical sensing platform for the simultaneous detection of uric acid, dopamine and ascorbic acid. *Colloids Surf. A Physicochem. Eng. Asp.* **2022**, *634*, 127928. [CrossRef]
79. Liu, X.; Cui, G.; Dong, L.; Wang, X.; Zhen, Q.; Sun, Y.; Ma, S.; Zhang, C.; Pang, H. Synchronous electrochemical detection of dopamine and uric acid by a PMo_{12}@MIL-100(Fe)@PVP nanocomposite. *Anal. Biochem.* **2022**, *648*, 114670. [CrossRef]
80. Masood, T.; Asad, M.; Riaz, S.; Akhtar, N.; Hayat, A.; Shenashen, M.A.; Rahman, M.M. Non-enzymatic electrochemical sensing of dopamine from COVID-19 quarantine person. *Mater. Chem. Phys.* **2022**, *289*, 126451. [CrossRef]
81. Amara, U.; Sarfraz, B.; Mahmood, K.; Taqi Mehran, M.; Muhammad, N.; Hayat, A.; Hasnain Nawaz, M. Fabrication of ionic liquid stabilized MXene interface for electrochemical dopamine detection. *Microchim. Acta* **2022**, *189*, 64. [CrossRef]
82. Wen, M.; Xing, Y.; Liu, G.; Hou, S.; Hou, S. Electrochemical sensor based on Ti_3C_2 membrane doped with UIO-66-NH_2 for dopamine. *Microchim. Acta* **2022**, *189*, 141. [CrossRef] [PubMed]
83. Cao, M.; Liu, S.; Liu, S.; Tong, Z.; Wang, X.; Xu, X. Preparation of ZnO/Ti_3C_2Tx/Nafion/Au electrode. *Microchem. J.* **2022**, *175*, 107068. [CrossRef]
84. Ni, M.; Chen, J.; Wang, C.; Wang, Y.; Huang, L.; Xiong, W.; Zhao, P.; Xie, Y.; Fei, J. A high-sensitive dopamine electrochemical sensor based on multilayer Ti_3C_2 MXene, graphitized multi-walled carbon nanotubes and ZnO nanospheres. *Microchem. J.* **2022**, *178*, 107410. [CrossRef]

85. Arbabi, N.; Beitollahi, H. Ti$_3$C$_2$ Nano Layer Modified Screen Printed Electrode as a Highly Sensitive Electrochemical Sensor for the Simultaneous Determination of Dopamine and Tyrosine. *Surf. Eng. Appl. Electrochem.* **2022**, *58*, 13–19. [CrossRef]
86. Sun, X.; Chen, C.; Xiong, C.; Zhang, C.; Zheng, X.; Wang, J.; Gao, X.; Yu, Z.-Q.; Wu, Y. Surface modification of MoS$_2$ nanosheets by single Ni atom for ultrasensitive dopamine detection. *Nano Res.* **2023**, *16*, 917–924. [CrossRef]
87. Ji, S.F.; Chen, Y.J.; Wang, X.L.; Zhang, Z.D.; Wang, D.S.; Li, Y.D. Chemical synthesis of single atomic site catalysts. *Chem. Rev.* **2020**, *120*, 11900–11955. [CrossRef] [PubMed]
88. Zhang, X.L.; Li, G.L.; Chen, G.; Wu, D.; Zhou, X.X.; Wu, Y.N. Single-atom nanozymes: A rising star for biosensing and biomedicine. *Coord. Chem. Rev.* **2020**, *418*, 213376. [CrossRef]
89. Barua, S.; Dutta, H.S.; Gogoi, S.; Devi, R.; Khan, R. Nanostructured MoS$_2$-based advanced biosensors: A review. *ACS Appl. Nano Mater.* **2018**, *1*, 2–25. [CrossRef]
90. Pavlíčková, M.; Lorencová, L.; Hatala, M.; Kováč, M.; Tkáč, J.; Gemeiner, P. Facile fabrication of screen-printed MoS$_2$ electrodes for electrochemical sensing of dopamine. *Sci. Rep.* **2022**, *12*, 11900. [CrossRef]
91. Peaston, R.T.; Weinkove, C. Measurement of catecholamines and their metabolites. *Ann. Clin. Biochem.* **2004**, *41*, 17–38. [CrossRef]
92. Chen, J.; Li, S.; Chen, Y.; Yang, J.; Dong, J.; Lu, X. L-Cysteine-Terminated Triangular Silver Nanoplates/MXene Nanosheets are Used as Electrochemical Biosensors for Efficiently Detecting 5-Hydroxytryptamine. *Anal. Chem.* **2021**, *93*, 16655–16663. [CrossRef]
93. Kim, H.-U.; Koyappayil, A.; Seok, H.; Aydin, K.; Kim, C.; Park, K.-Y.; Jeon, N.; Seok Kang, W.; Lee, M.-H.; Kim, T. Concurrent and Selective Determination of Dopamine and Serotonin with Flexible WS$_2$/Graphene/Polyimide Electrode Using Cold Plasma. *Small* **2021**, *17*, 2102757. [CrossRef]
94. Jiang, M.; Tian, L.; Su, M.; Cao, X.; Jiang, Q.; Huo, X.; Yu, C. Real-time monitoring of 5-HT release from cells based on MXene hybrid single-walled carbon nanotubes modified electrode. *Anal. Bioanal. Chem.* **2022**, *414*, 7967–7976. [CrossRef]
95. Su, M.; Lan, H.; Tian, L.; Jiang, M.; Cao, X.; Zhu, C.; Yu, C. Ti$_3$C$_2$Tx-reduced graphene oxide nanocomposite-based electrochemical sensor for serotonin in human biofluids. *Sens. Actuators B Chem.* **2022**, *367*, 132019. [CrossRef]
96. Stárka, L.; Dušková, M. What Is a Hormone? *Physiol. Res.* **2020**, *69*, S183–S185. [CrossRef]
97. Burcu Bahadır, E.; Sezgintürk, M.K. Electrochemical biosensors for hormone analyses. *Biosens. Bioelectron.* **2015**, *68*, 62–71. [CrossRef]
98. Zea, M.; Bellagambi, F.C.; Halima, H.B.; Zine, N.; Jaffrezic-Renault, N.; Villa, R.; Gabriel, G.; Errachid, A. Electrochemical sensors for cortisol detections: Almost there. *Trends Anal. Chem.* **2020**, *132*, 116058. [CrossRef]
99. Laochai, T.; Yukird, J.; Promphet, N.; Qin, J.; Chailapakul, O.; Rodthongkum, N. Non-invasive electrochemical immunosensor for sweat cortisol based on L-cys/AuNPs/MXene modified thread electrode. *Biosens. Bioelectron.* **2022**, *203*, 114039. [CrossRef]
100. San Nah, J.; Chandra Barman, S.; Abu Zahed, M.; Sharifuzzaman, M.; Yoon, H.; Park, C.; Yoon, S.; Zhang, S.; Yeong Park, J. A wearable microfluidics-integrated impedimetric immunosensor based on Ti$_3$C$_2$Tx MXene incorporated laser-burned graphene for noninvasive sweat cortisol detection. *Sens. Actuators B Chem.* **2021**, *329*, 129206. [CrossRef]
101. Hernandez, P.; Sanchez, I.; Paton, F.; Hernandez, L. Cyclic voltammetry determination of epinephrine with a carbon fiber ultramicroelectrode. *Talanta* **1998**, *46*, 985–991. [CrossRef]
102. Li, Z.; Guo, Y.; Yue, H.; Gao, X.; Huang, S.; Zhang, X.; Yu, Y.; Zhang, H.; Zhang, H. Electrochemical determination of epinephrine based on Ti$_3$C$_2$Tx MXene-reduced graphene oxide/ITO electrode. *J. Electroanal. Chem.* **2021**, *895*, 115425. [CrossRef]
103. Lin, J.; Lu, H.; Duan, Y.; Yu, C.; Li, L.; Ding, Y. Fabrication of bimetallic ZIF/carbon nanofibers composite for electrochemical sensing of adrenaline. *J. Mater. Sci.* **2022**, *57*, 6629–6639. [CrossRef]
104. Singh, V.; Krishnan, S. An Electrochemical Mass Sensor for Diagnosing Diabetes in Human Serum. *Analyst* **2014**, *139*, 724–728. [CrossRef]
105. Sakthivel, R.; Lin, L.-Y.; Duann, Y.-F.; Chen, H.-H.; Su, C.; Liu, X.; He, J.-H.; Chung, R.-J. MOF-Derived Cu-BTC Nanowire-Embedded 2D Leaf-like Structured ZIF Composite-Based Aptamer Sensors for Real-Time In Vivo Insulin Monitoring. *ACS Appl. Mater. Interfaces* **2022**, *14*, 28639–28650. [CrossRef]
106. Balloux, F.; van Dorp, L. Q&A: What are pathogens, and what have they done to and for us? *BMC Biol.* **2017**, *15*, 91.
107. Wang, B.; Wang, H.; Lu, X.; Zheng, X.; Yang, Z. Recent Advances in Electrochemical Biosensors for the Detection of Foodborne Pathogens: Current Perspective and Challenges. *Foods* **2023**, *12*, 2795. [CrossRef]
108. Xiao, S.; Yang, X.; Wu, J.; Liu, Q.; Li, D.; Huang, S.; Xie, H.; Yu, Z.; Gan, N. Reusable electrochemical biosensing platform based on egg yolk antibody-labeled magnetic covalent organic framework for on-site detection of Escherichia coli in foods. *Sens. Actuators B Chem.* **2022**, *369*, 132320. [CrossRef]
109. Skorjanc, T.; Mavrič, A.; Nybo Sørensen, M.; Mali, G.; Wu, C.; Valant, M. Cationic Covalent Organic Polymer Thin Film for Label-free Electrochemical Bacterial Cell Detection. *ACS Sens.* **2022**, *7*, 2743–2749. [CrossRef]
110. Wang, S.; Li, H.; Huang, H.; Cao, X.; Chen, X.; Cao, D. Porous organic polymers as a platform for sensing applications. *Chem. Soc. Rev.* **2022**, *51*, 2031. [CrossRef] [PubMed]
111. Cui, J.; Zhang, Y.; Lun, K.; Wu, B.; He, L.; Wang, M.; Fang, S.; Zhang, Z.; Zhou, L. Sensitive detection of *Escherichia coli* in diverse foodstuffs by electrochemical aptasensor based on 2D porphyrin-based COF. *Microchim. Acta* **2023**, *190*, 421–432. [CrossRef] [PubMed]
112. Wang, O.; Jia, X.; Liu, J.; Sun, M.; Wu, J. Rapid and simple preparation of an MXene/polypyrrole-based bacteria imprinted sensor for ultrasensitive Salmonella detection. *J. Electroanal. Chem.* **2022**, *918*, 116513. [CrossRef]

113. Salimiyan Rizi, K.; Hatamluyi, B.; Darroudi, M.; Meshkat, Z.; Aryan, E.; Soleimanpour, S.; Rezayi, M. PCR-free electrochemical genosensor for Mycobacterium tuberculosis complex detection based on two-dimensional Ti_3C_2 Mxene-polypyrrole signal amplification. *Microchem. J.* **2022**, *179*, 107467. [CrossRef]
114. Bintsis, T. Foodborne pathogens. *AIMS Microbiol.* **2017**, *3*, 529–563. [CrossRef]
115. Wei, W.; Lin, H.; Hao, T.; Su, X.; Jiang, X.; Wang, S.; Hu, Y.; Guo, Z. Dual-mode ECL/SERS immunoassay for ultrasensitive determination of Vibrio vulnificus based on multifunctional MXene. *Sens. Actuators B Chem.* **2021**, *332*, 129525. [CrossRef]
116. Wang, W.; Xiao, S.; Jia, Z.; Xie, H.; Li, T.; Wang, Q.; Gan, N. A dual-mode aptasensor for foodborne pathogens detection using Pt, phenylboric acid and ferrocene modified Ti_3C_2 MXenes nanoprobe. *Sens. Actuators B Chem.* **2022**, *351*, 130839. [CrossRef]
117. Ménard-Moyon, C.; Bianco, A.; Kalantar-Zadeh, K. Two-Dimensional Material-Based Biosensors for Virus Detection. *ACS Sens.* **2020**, *5*, 3739–3769. [CrossRef]
118. Sheikhzadeh, E.; Beni, V.; Zourob, M. Nanomaterial application in bio/sensors for the detection of infectious diseases. *Talanta* **2021**, *230*, 122026. [CrossRef]
119. Viana Ribeiro, B.; Reis Cordeiro, T.A.; Ramos Oliveira Freitas, G.; Ferreira, L.F.; Leoni Franco, D. Biosensors for the detection of respiratory viruses: A review. *Talanta Open* **2020**, *2*, 100007. [CrossRef] [PubMed]
120. Yugender Goud, K.; Koteshwara Reddy, K.; Khorshed, A.; Sunil Kumar, V.; Mishra, R.K.; Oraby, M.; Hatem Ibrahim, A.; Kim, H.; Vengatajalabathy Gobid, K. Electrochemical diagnostics of infectious viral diseases: Trends and challenges. *Biosens. Bioelectron.* **2021**, *180*, 113112.
121. Manring, N.; Ahmed, M.M.N.; Tenhoff, N.; Smeltz, J.L.; Pathirathna, P. Recent Advances in Electrochemical Tools for Virus Detection. *Anal. Chem.* **2022**, *94*, 7149–7157. [CrossRef] [PubMed]
122. Wang, Y.; Hu, Y.; He, Q.; Yan, J.; Xiong, H.; Wen, N.; Cai, S.; Peng, D.; Liu, Y.; Liu, Z. Metal-organic frameworks for virus detection. *Biosens. Bioelectron.* **2020**, *169*, 112604. [CrossRef]
123. Tian, J.; Liang, Z.; Hu, O.; He, O.; Sun, D.; Chen, Z. An electrochemical dual-aptamer biosensor based on metal-organic frameworks MIL-53 decorated with Au@Pt nanoparticles and enzymes for detection of COVID-19 nucleocapsid protein. *Electrochim. Acta* **2021**, *387*, 138553. [CrossRef]
124. Mehmandoust, M.; Gumus, Z.P.; Soylak, M.; Erk, N. Electrochemical immunosensor for rapid and highly sensitive detection of SARS-CoV-2 antigen in the nasal sample. *Talanta* **2022**, *240*, 123211. [CrossRef]
125. Rahmati, Z.; Roushani, M. SARS-CoV-2 virus label-free electrochemical nanohybrid MIP-aptasensor based on $Ni_3(BTC)_2$ MOF as a high-performance surface substrate. *Microchim. Acta* **2022**, *189*, 287. [CrossRef]
126. Panda, S.; Deshmukh, K.; Mustansar Hussain, C.; Khadheer Pasha, S.K. 2D MXenes for combatting COVID-19 Pandemic: A perspective on latest developments and innovations. *FlatChem* **2022**, *33*, 100377. [CrossRef]
127. Yenyu Chen, W.; Hang Lin, H.; Kumar Barui, A.; Ulloa Gomez, A.M.; Wendt, M.K.; Stanciu, L.A. DNA-Functionalized Ti_3C_2Tx MXenes for Selective and Rapid Detection of SARS-CoV-2 Nucleocapsid Gene. *ACS Appl. Nano Mater.* **2022**, *5*, 1902–1910. [CrossRef]
128. Lin, X.; Lian, X.; Luo, B.; Huang, X.-C. A highly sensitive and stable electrochemical HBV DNA biosensor based on ErGO-supported Cu-MOF. *Inorg. Chem. Commun.* **2020**, *119*, 108095. [CrossRef]
129. Rezki, M.; Septiani, N.L.W.; Iqbal, M.; Harimurti, S.; Sambegoro, P.; Damar Rastri Adhika, D.R.; Yuliarto, B. Amine-functionalized Cu-MOF nanospheres towards label-free hepatitis B surface antigen electrochemical immunosensors. *J. Mater. Chem. B* **2021**, *9*, 5711–5721. [CrossRef]
130. Biomarkers Definition Working Group. Biomarkers and surrogate endpoints: Preferred definitions and conceptual framework. *Clin. Pharmacol. Ther.* **2001**, *69*, 89–95. [CrossRef] [PubMed]
131. Sohrabi, H.; Bolandi, N.; Hemmati, H.; Eyvazi, S.; Ghasemzadeh, S.; Baradaran, B.; Oroojalian, F.; Reza Majidi, M.; de la Guardia, M.; Mokhtarzadeh, A. State-of-the-art cancer biomarker detection by portable (Bio) sensing technology: A critical review. *Microchem. J.* **2022**, *177*, 107248. [CrossRef]
132. Hammarström, S. The carcinoembryonic antigen (CEA) family: Structures, suggested functions and expression in normal and malignant tissues. *Semin. Cancer Biol.* **1999**, *9*, 67–81. [CrossRef] [PubMed]
133. Zhao, P.; Zheng, J.; Liang, Y.; Tian, F.; Peng, L.; Huo, D.; Hou, C. Functionalized Carbon Nanotube-Decorated MXene Nanosheet-Enabled Microfluidic Electrochemical Aptasensor for Carcinoembryonic Antigen Determination. *ACS Sustain. Chem. Eng.* **2021**, *9*, 15386–15393. [CrossRef]
134. Wang, Q.; Xin, H.; Wang, Z. Label-Free Immunosensor Based on Polyaniline-Loaded MXene and Gold-Decorated β-Cyclodextrin for Efficient Detection of Carcinoembryonic Antigen. *Biosensors* **2022**, *12*, 657. [CrossRef] [PubMed]
135. Song, X.; Gao, H.; Yuan, R.; Xiang, Y. Trimetallic nanoparticle-decorated MXene nanosheets for catalytic electrochemical detection of carcinoembryonic antigen via Exo III-aided dual recycling amplifications. *Sens. Actuators B Chem.* **2022**, *359*, 131617. [CrossRef]
136. Wang, L.; Xie, H.; Lin, Y.; Wang, M.; Sha, L.; Yu, X.; Yang, J.; Zhao, J.; Li, G. Covalent organic frameworks (COFs)-based biosensors for the assay of disease biomarkers with clinical applications. *Biosens. Bioelectron.* **2022**, *217*, 114668. [CrossRef]
137. Qiu, R.; Mu, W.; Wu, C.; Wu, M.; Feng, J.; Rong, S.; Ma, H.; Chang, D.; Pan, H. Sandwich-type immunosensor based on COF-LZU1 as the substrate platform and graphene framework supported nanosilver as probe for CA125 detection. *J. Immunol. Methods* **2022**, *504*, 113261. [CrossRef]

38. Zheng, J.; Zhao, H.; Ning, G.; Sun, W.; Wang, L.; Liang, H.; Xu, H.; He, C.; Zhao, H.; Li, C.-P. A novel affinity peptide–antibody sandwich electrochemical biosensor for PSA based on the signal amplification of MnO_2-functionalized covalent organic framework. *Talanta* **2021**, *233*, 122520. [CrossRef]
39. Yan, X.; Song, Y.; Liu, J.; Zhou, N.; Zhang, C.; He, L.; Zhang, Z.; Liu, Z. Two-dimensional porphyrin-based covalent organic framework: A novel platform for sensitive epidermal growth factor receptor and living cancer cell detection. *Biosens. Bioelectron.* **2019**, *126*, 734–742. [CrossRef]
40. Yin, M.; Zhang, L.; Wei, X.; Sun, J.; Xu, D. Detection of antibiotics by electrochemical sensors based on metal-organic frameworks and their derived materials. *Microchem. J.* **2022**, *183*, 107946. [CrossRef]
41. Wang, M.; Hu, M.; Liu, J.; Guo, C.; Peng, D.; Jia, Q.; He, L.; Zhang, Z.; Du, M. Covalent organic framework-based electrochemical aptasensors for the ultrasensitive detection of antibiotics. *Biosens. Bioelectron.* **2019**, *132*, 8–16. [CrossRef]
42. Ma, X.; Pang, C.; Li, S.; Xiong, Y.; Li, J.; Luo, J.; Yang, Y. Synthesis of Zr-coordinated amide porphyrin-based two-dimensional covalent organic framework at liquid-liquid interface for electrochemical sensing of tetracycline. *Biosens. Bioelectron.* **2019**, *146*, 111734. [CrossRef]
43. Yan, C.; Zhang, J.; Yao, L.; Xue, F.; Lu, J.; Li, B.; Chen, W. Aptamer-mediated colorimetric method for rapid and sensitive detection of chloramphenicol in food. *Food Chem.* **2018**, *260*, 208–212. [CrossRef] [PubMed]
44. Li, H.-K.; Ye, H.-L.; Zhao, X.-X.; Sun, X.-L.; Zhu, Q.-Q.; Han, Z.-Y.; Yuan, R.; He, H. Artful union of a zirconium-porphyrin MOF/GO composite for fabricating an aptamer-based electrochemical sensor with superb detecting performance. *Chin. Chem. Lett.* **2021**, *32*, 2851–2855. [CrossRef]
45. Zhang, C.; Fan, L.; Ren, J.; Cui, M.; Li, N.; Zhao, H.; Qu, Y.; Ji, X. Facile synthesis of surface functionalized Pd^{2+}@P-CDP/COFs for highly sensitive detection of norfloxacin drug based on the host-guest interaction. *J. Pharm. Biomed. Anal.* **2022**, *219*, 114956. [CrossRef] [PubMed]
46. Cháfer-Pericás, C.; Maquieira, A.; Puchades, R. Fast screening methods to detect antibiotic residues in food samples. *TrAC Trends Anal. Chem.* **2010**, *29*, 1038–1049. [CrossRef]
47. Sun, Y.; Gao, H.; Xu, L.; Waterhouse, G.I.N.; Zhang, H.; Qiao, X.; Xu, Z. Ultrasensitive determination of sulfathiazole using a molecularly imprinted electrochemical sensor with CuS microflowers as an electron transfer probe and Au@COF for signal amplification. *Food Chem.* **2020**, *332*, 127376. [CrossRef]
48. Balasubramanian, P.; Settu, R.; Chen, S.M.; Chen, T.W. Voltammetric sensing of sulfamethoxazole using a glassy carbon electrode modified with a graphitic carbon nitride and zinc oxide nanocomposite. *Microchim Acta* **2018**, *185*, 396. [CrossRef]
49. Shahsavari, M.; Tajik, S.; Sheikhshoaie, I.; Beitollahi, H. Fabrication of Nanostructure Electrochemical Sensor Based on the Carbon Paste Electrode (CPE) Modified With Ionic Liquid and Fe_3O_4/ZIF-67 for Electrocatalytic Sulfamethoxazole Detection. *Top. Catal.* **2022**, *65*, 577–586. [CrossRef]
50. Sun, Y.; He, J.; Waterhouse, G.I.; Xu, L.; Zhang, H.; Qiao, X.; Xu, Z. A selective molecularly imprinted electrochemical sensor with GO@COF signal amplification for the simultaneous determination of sulfadiazine and acetaminophen. *Sens. Actuators B Chem.* **2019**, *300*, 126993. [CrossRef]
51. Guo, F.; Tian, G.; Fan, C.; Zong, Z.; Wang, J.; Xu, J. A zirconium–organic framework nanosheet-based aptasensor with outstanding electrochemical sensing performance. *Inorg. Chem. Commun.* **2022**, *145*, 109970. [CrossRef]
52. Li, H.-K.; An, Y.-X.; Zhang, E.-H.; Zhou, S.-N.; Li, M.-X.; Li, Z.-J.; Li, X.; Yuan, R.; Zhang, W.; He, H. A covalent organic framework nanosheet-based electrochemical aptasensor with sensitive detection performance. *Anal. Chim. Acta* **2022**, *1223*, 340204. [CrossRef]

Disclaimer/Publisher's Note: The statements, opinions and data contained in all publications are solely those of the individual author(s) and contributor(s) and not of MDPI and/or the editor(s). MDPI and/or the editor(s) disclaim responsibility for any injury to people or property resulting from any ideas, methods, instructions or products referred to in the content.

Review

Gold Nanoparticles as a Biosensor for Cancer Biomarker Determination

Chien-Hsiu Li [1,†], Ming-Hsien Chan [1,†], Yu-Chan Chang [2] and Michael Hsiao [1,3,*]

1. Genomics Research Center, Academia Sinica, Taipei 115, Taiwan
2. Department of Biomedical Imaging and Radiological Sciences, National Yang Ming Chiao Tung University, Taipei 112, Taiwan
3. Department and Graduate Institute of Veterinary Medicine, School of Veterinary Medicine, National Taiwan University, Taipei 106, Taiwan
* Correspondence: mhsiao@gate.sinica.edu.tw
† Both authors contributed equally to this manuscript.

Abstract: Molecular biology applications based on gold nanotechnology have revolutionary impacts, especially in diagnosing and treating molecular and cellular levels. The combination of plasmonic resonance, biochemistry, and optoelectronic engineering has increased the detection of molecules and the possibility of atoms. These advantages have brought medical research to the cellular level for application potential. Many research groups are working towards this. The superior analytical properties of gold nanoparticles can not only be used as an effective drug screening instrument for gene sequencing in new drug development but also as an essential tool for detecting physiological functions, such as blood glucose, antigen-antibody analysis, etc. The review introduces the principles of biomedical sensing systems, the principles of nanomaterial analysis applied to biomedicine at home and abroad, and the chemical surface modification of various gold nanoparticles.

Keywords: gold nanoparticles; biosensing; surface plasmon resonance; cancer marker; surface modification

1. Introduction

The unique optical properties of gold nanomaterials were revealed as early as the Middle Ages, when the scientist Michael Faraday prepared a gold colloid solution via the wet chemical synthesis method and demonstrated its extraordinary optical properties in 1857 [1], suggesting that gold particles (AuNPs) appear red in the nanoscale and gradually change to dark blue as the particle size increases.

It has been demonstrated that the incident light's electromagnetic field can excite the surface-free electrons of metallic materials on the nanoscale, as the particle size is much smaller than the incident wavelength. The frequency of the incident light produces a collective oscillation motion called surface plasmon resonance (SPR), as shown in Figure 1a [2]. The uneven distribution of electrons illustrates the electric field at different cross-sections when free electrons are affected by electromagnetic radiation. For example, deviation from an electric field causes negatively charged electrons to move in the direction of the electric field. Transiently induced dipoles are generated, resulting in the separation of free electrons from the metal nucleus. This separation leads to a Coulomb restoring force in the opposite direction and, finally, causes the metal's free electrons to generate a collective back-and-forth oscillatory motion at both ends of the particle. Surface plasmon absorption bands appear in the ultraviolet and visible wavelengths of light as a result of this resonance phenomenon. This explains the different colors of metallic nanomaterials under white light. As shown in Equation (1), there is a strong correlation between the peak of surface plasmon resonance (ω_{sp}) and the surface charge density of the material. In contrast, the charge density of the particle is affected by particle size, shape, structure, dielectric constant, etc. [3].

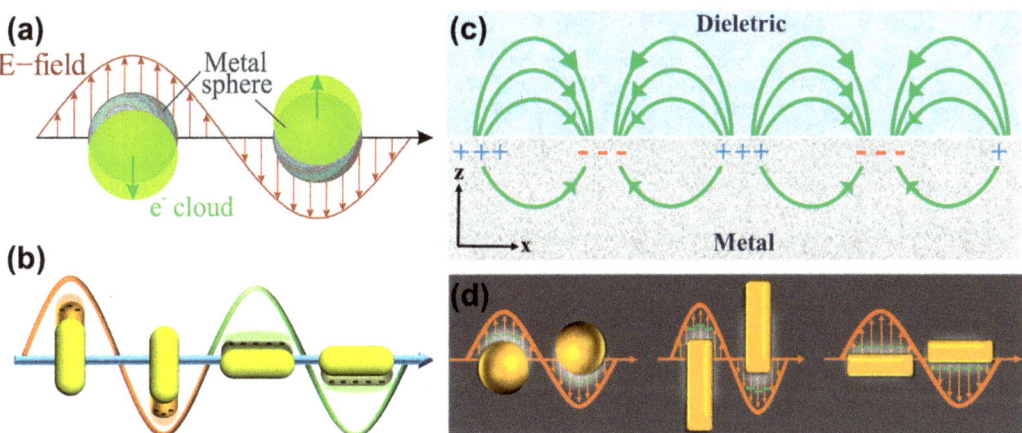

Figure 1. Schematic diagrams: (**a**) surface plasmon polariton. (**b**) Localized surface plasmon resonance. (**c**) A dipole centered on a metal sphere with a radius r smaller than the incident wavelength was used to oscillate metallic nanoparticles that were subjected to an electromagnetic field and oscillated by the dipole. (**d**) Excitation of AuNPs by incident light results in surface plasmon resonance. The surface plasmon resonance oscillating along different axes comprises the short-axis absorption peak and the long-axis absorption peak.

In addition to the above advantages, AuNPs combined with composite materials of upconversion nanoparticles can solve some of the shortcomings and increase the diagnostic and therapeutic value of single materials. For instance, AuNPs need light energy to promote SPR, which upconversion nanoparticles can provide. As shown in Figure 1b, Sun et al. [4] used a AuNP dimer as a core material. They combined it with a grafted photosensitizer to form a core-satellite type composite material for diagnostic and mouse tumor treatment. A 980 nm laser was used to excite the upconverted nanoparticles to emit fluorescence to drive the photosensitizer for photodynamic therapy. An 808 nm laser was then used to excite the AuNP dimer for photothermal treatment. Using near-infrared light as the excitation source can increase the penetration of biological tissues and reduce unnecessary heat damage. A combination of photothermal and photodynamic therapy is more effective than a single light treatment. In addition to its therapeutic function, the above-converted emitted light can be used as a fluorescent probe for cellular calibration. In nuclear magnetic resonance imaging, gadolinium (Gd^{3+}) can serve as a contrast agent (NMRI) to enhance the bio-imaging function. Due to the photothermal conversion effect, the excited AuNPs can be used for photoacoustic imaging (PAI), which provides an alternative way to image tumors. The composite material can also be used as a contrast agent for computed tomography (CT) imaging to enhance image contrast. Therefore, the composite material has excellent therapeutic effects and high application value for four types of bioimaging calibrations. Based on the advantages mentioned above, AuNPs are often used in biomedical research to carry certain substances into organisms. The surface of AuNPs can be coated to modify the specific molecules and sensors expected to be held on their surface when the AuNPs are used carry biomarker receptors into a body. Optical biomarkers are produced by SPR using AuNPs adhered to high-refractive-index surfaces. The AuNPs absorb laser light and generate electron waves on their surface to generate measurable signals for any analyte bound to the AuNPs. Therefore, this review is focused on AuNPs, specifically considering the changes in signals caused by the characteristics of AuNPs that can be used to detect molecules, such as DNA, RNA, and protein, flowing into an organism. This review further discusses the synthesis and properties of AuNPs, as well as their potential applications as sensors.

2. Synthesis of Gold Nanoparticles

Depending on the method of preparation, AuNPs can be classified into two major categories: (i) top–down and (ii) bottom–up. The top–down approach includes the template method [5], lithographic methods [6], and catalytic methods [7]; the bottom–up process consists of the electrochemical method [8], the seedless growth method [9], and the seed-mediated growth method [10]. It is most commonly used to synthesize AuNPs by seed-mediated growth.

2.1. Surfactant-Preferential-Binding-Directed Growth

The seed-mediated growth method was first proposed by Wiesner and Wokaun [11] in 1989 to obtain the crystals of AuNPs by reducing the tetrachloroauric acid ($HAuCl_4$) with phosphorus elements using the crystals as the nuclei to grow AuNPs. In the beginning, Turjevich's team used sodium citrate as a reducing agent and an interfacial reactive agent to reduce $HAuCl_4$ from Au^{3+} to Au by hydrothermal method to synthesize many gold nanoparticles of different sizes under different reaction parameters. In the aqueous phase, the $AuCl_4^-$ atoms are reduced by sodium citrate to form gold atoms, which then aggregate to form gold nanoparticles. The negatively charged citrate ions play the reducing agent and capping agent. The size of the particles is controlled by the ratio of gold ions to sodium citrate with the heating time of the reaction. At this time, the gold nanoparticles are protected by the negatively charged citrate on the outside of the particles and are stably stored in the aqueous solution. In 2003, Nikoobakht and El-Sayed [12] proposed two modifications of this synthesis method: sodium citrate, the protective agent, was replaced by hexadecyltrimethylammonium bromide (CTAB) in the synthesis of the seeds and Ag^+ was used to regulate the aspect ratio of the AuNPs. After modification, the yield of AuNPs produced by the crystal growth method could reach 99%, and the aspect ratio of AuNPs could be adjusted from 1.5 to 4.5 by changing the silver ion concentration. There are two significant steps in the crystal growth process. The first step is the preparation of a crystal solution; a tetrachloroauric acid solution containing the surfactant CTAB is reduced to crystals with the reductant sodium borohydride ($NaBH_4$), in which the surfactant CTAB is used as a protective agent to stabilize the crystal. In the second step, reductant ascorbic acid is added to a growth solution containing the surfactant CTAB, silver nitrate ($AgNO_3$), and tetrachloroauric acid, and the trivalent gold ions of tetrachloroauric acid in the solution are reduced to monovalent gold ions. Then, the crystalline solution synthesized in the previous step is added to the growth solution to produce AuNPs. Currently, Murphy et al. propose a mechanism in which surfactant-preferential-binding-directed growth is responsible for the growth of AuNPs [13,14]. This team observed the structure of AuNPs and the effects of various reaction conditions on AuNP generation with high-resolution penetrating electron microscopy. The growth of AuNPs begins with the change of the intrinsic structure of the crystalline species, and the disruption of the crystalline species symmetry leads to non-isotropic growth, in which the {100} lattice of the AuNPs has a higher surface energy. Hence, CTAB preferentially adsorbs on the {100} surface. In addition, the adsorption of CTAB stabilizes the {100} surface and restricts the growth of the {100} surface so that the AuNPs grow towards the ends of the {111} surface to form a rod, as shown in Figure 2a.

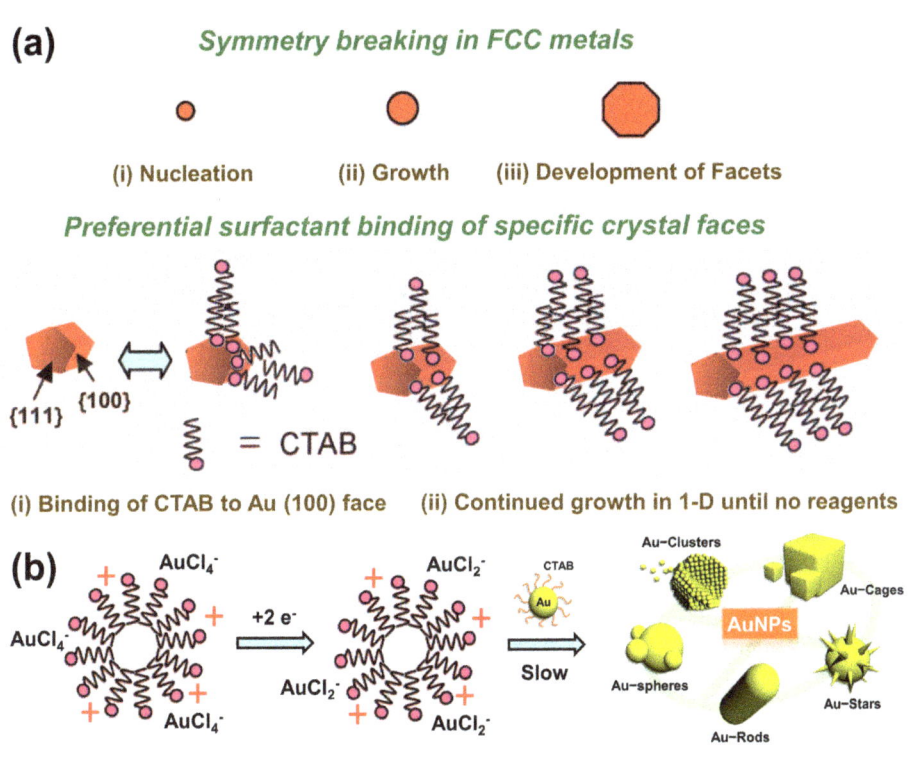

Figure 2. AuNP synthesis method. (**a**) In the surfactant-preferential-binding-directed growth mechanism, because CTAB protects the {100} side of the crystalline species, the subsequently added gold ions can only be stacked and grown on the unprotected {111} side, resulting in AuNPs. (**b**) Electric-field-directed growth mechanism, in which AuCl₄-substituted Br is connected to the CTAB microcell and reacts with the CTAB-coated crystalline species.

2.2. Electric-Field-Directed Growth Mechanism

According to Mulvaney et al., the electric-field-directed is the second growth mechanism [15]. This mechanism involves the addition of $AuCl_4^{2-}$ to the CTAB microbattery, followed by its reduction to an $AuCl_2^-$-CTAB microbattery by a reducing agent, and the reaction equation is as follows. The negatively charged $AuCl_2^-$ at the surface of the CTAB microcell attracts the CTAB on the surface of the crystal, causing the $AuCl_2^-$-CTAB microcell to collide with the CTAB-protected crystal. In contrast, the electrons on the surface of the crystal are transferred to the monovalent gold ions to produce the reduction reaction. Finally, the collision probability of the two ends of the crystal becomes greater than that of the side, resulting in the generation of the rod-like structure. In the synthesis of AuNPs, the silver ions of silver nitrate are used to regulate the aspect ratio of AuNPs by forming silver bromide (AgBr) with the bromine ions (Br-) of CTAB to reduce the charge density and the negative electric repulsion between the functional groups at the head end of the CTAB; therefore, regulating the concentration of silver ions can regulate the arrangement of the soft templates of CTAB and change the aspect ratio of AuNPs. As shown in Figure 2b, As part of the growth mechanisms of AuNPs, CTAB plays a crucial role in preventing AuNPs from aggregating due to electrostatic repulsion between positive surface charges [16].

3. Biological Characteristics of AuNPs

Metal materials have localized surface plasmon resonance (LSPR) properties, and gold nanomaterials have become a popular research material because of their unique optical properties. To understand the LSPR phenomenon, the SPR effect first needed to be understood. In 1902, Wood [17] designed a grating experiment on polarizing light. He found an abnormal phenomenon of diffraction on the flat metal surface when the grating was placed near a metal surface. Later, Fano provided a theory to explain this, i.e., light is a kind of electromagnetic wave that could interact with the free conduction electrons on the surface of the metal. In this case, plasma polaritons propagate along the metal-dielectric interface in both directions. This is called surface plasma polarization (SPP) or surface plasma resonance (SPR), as shown in Figure 1c [18]. Furthermore, LSPR occurs at the nanoscale because the incident light is far greater than the nanomaterials, resulting in all the free electrons on the surface reacting together. Briefly, when light is irradiated on a metallic nanoparticle, the free conductive electrons on the surface interact with the incident light via the electromagnetic interaction, followed by all polarized electrons oscillating together at a specific frequency, as shown in Figure 1d. In 1908, this phenomenon in spherical nanomaterials was investigated and resolved by Mie using Maxwell's equations to explain the relationship of the extinction (SPR) spectra (extinction = scattering + absorption) in any particle size [19].

3.1. Localized Surface Plasmon Resonance (LSPR)

As a result of some elemental species inside organisms, UV and visible light have limited penetration abilities regarding skin, such as water and oxy/deoxyhemoglobin, which show tremendous individual absorption ranges from ~900 to 1000 nm and from ~400 to 690 nm, respectively [20]. Accordingly, the suitable SPR peaks at the greatest wavelength are between 700 and 900 nm, which is intended to mean that the energy is more efficiently absorbed when moving away from the red-shifted region to the near-infrared (NIR) region. Gold nanorods with an anisotropic structure generate two bands of SPR at 520 nm or 650~1200 nm, denoted as the transverse and longitudinal axis, respectively. It is easy to red-shift the long axis of an SPR band by increasing the aspect ratio for gold nanorods. When AuNPs are excited by incident light to generate surface plasmon resonance, the oscillation directions of surface free electrons are divided into two types due to different polarization modes, one of which is the oscillation motion along the long axis of the material, and the other is the oscillation motion along the short axis of the material [21,22]. The short-axis surface plasmon oscillates as a result of the short-axis surface plasmon, and AuNPs exhibit a characteristic absorption peak at about 520 nm in the visible light range. This is called the short-axis absorption peak (transverse band). The long-axis plasmon oscillations produce stronger absorptions and fall in the longitudinal band, which corresponds to the long-wavelength region. The short-axis absorption peaks do not significantly change with the size change of AuNPs. On the contrary, the long-axis absorption peaks drastically change with the aspect ratio (length/width, aspect ratio) of AuNPs. When the aspect ratio increases, the long-axis absorption peak shifts from the visible light band to the near-infrared light band, resulting in a red-shift phenomenon.

$$\omega_{sp} = \sqrt{\frac{ne^2}{m_0\varepsilon_0(\varepsilon_\infty + \kappa\varepsilon_m)} - \Gamma^2} \qquad (1)$$

ω_{sp}: surface plasmon resonance wavelength; n: free carrier density; m_0: effective mass of charge carrier; ε_0: dielectic constant offree space; ε_m: medium dielectric constant; ε_∞ high dielectric constant of medium; κ: geometry of nanomaterials; and Γ: damping constant

When considering using AuNPs for biotherapeutic applications, it is essential to use appropriate-length AuNPs because red blood cells and water occupy a large proportion of the biological body. In the visible wavelength (<650 nm) and infrared region (>900 nm), the absorption intensity of red blood cells and water molecules is the highest. Still, the absorption coefficients of red blood cells and water are the lowest in the near-infrared wavelength between 650 nm and 900 nm. This distance is also called the near-infrared (NIR) window [20]. Therefore, when near-infrared light is used as the excitation source the biological tissues absorb the least, which also means that the penetration depth of near-infrared light in biological tissues is higher than that of other wavelengths and is most suitable for biomedical applications. When the aspect ratio is about 4, the position of the long-axis absorption peak of AuNPs is red-shifted to about 800 nm, which is the near-infrared region (650–900 nm) where the incident light penetrates deepest into human tissues. AuNPs can convert near-infrared light energy into heat when stimulated by near-infrared light [23–25]. As a result of the scattering and absorption of light, the scattered light is converted into thermal energy when the incident light absorbs it, AuNPs can be used for the detection of diseased tissues and the diagnosis and treatment of tumor cells at the same time; therefore, AuNPs have great potential in the biomedical field due to their unique optical properties.

3.2. Surface Modification of Gold Nanoparticles

AuNPs are often used as emerging nanomaterials with unique and excellent optical properties. They are widely used in the biotechnology and biomedicine fields, including biomedical imaging, drug transportation, disease diagnosis, and treatment, as shown in Figure 3a. Whether AuNPs are synthesized by the crystal-free growth or crystal growth methods, their surface must be adsorbed with a CTAB double-layer protective agent so that the AuNPs are monodisperse in aqueous solutions due to electrostatic repulsive forces. In addition, the CTAB can protect AuNPs in environments with high concentrations of salts. However, CTAB causes a high degree of biotoxicity when it is detached from the surface of AuNPs. Therefore, the surface modification will become an essential issue if we want to apply AuNPs in the biomedical field further. Currently, there are three types of surface modification methods for AuNPs.

3.2.1. Ligand Exchange

The differences between carboxylic acids and amines can be observed when compared to other functional groups and alcohols, and there is a strong covalent bonding between thiol and gold in AuNPs. The molecules to be modified on the surface of AuNPs can be functionalized with thiol functional groups, and the thiol groups can generate Au-S bonds on the surface of the AuNPs to replace the interfacial reactive agent CTAB, deoxyribonucleic acid (DNA) [26], or other small molecules [27]. For example, Maltzahn et al. [28] replaced CTAB with PEG of high biocompatibility on the surface of AuNPs to enhance their stability, improve the shortcomings of CTAB-coated AuNPs that tend to aggregate in serum environments, and increase the circulation time (t1/2) of the material in an organism for up to 17 h, as shown in Figure 3b. Due to the close arrangement of CTAB on AuNP surfaces, some biomolecules, including antibodies and proteins, cannot bind to them after they have been modified with thiols. Therefore, small molecules, such as 3-mercaptopropionic acid (MPA) and 11-mercaptoundecanoic acid (11-MPA), must be used first. Then, the functionalized ends of these molecules can be used to covalently bond with the large molecules, such as antibodies or proteins, to be modified. For example, Yu et al. [29] first replaced CTAB on the surface of AuNPs with MUDA and AMTAZ, two small molecules of organic thiol, to significantly reduce the biotoxicity of the composite, which then reacted with the amine or carboxylic acid groups of MUDA and AMTAZ to form an amide bond with the antibodies. Antibody-modified AuNPs can be applied to nanosensors and target therapy.

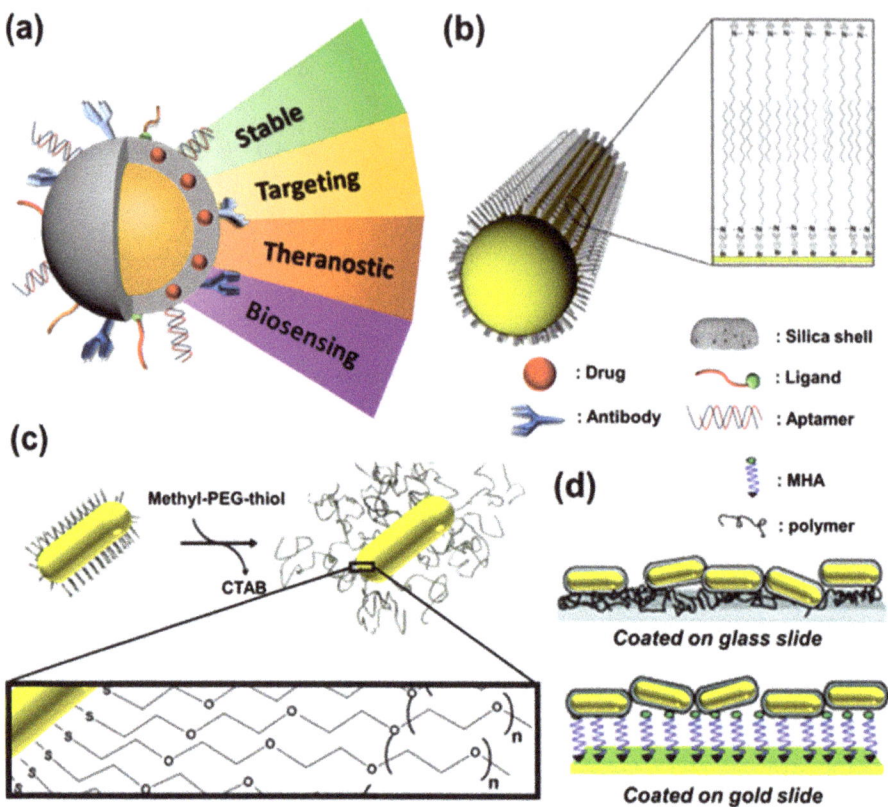

Figure 3. Surface modification of AuNPs. (**a**) Application of AuNPs in the biomedical field after surface modification. (**b**) Schematic diagram of the interface active agent CTAB adsorbed on the surface of AuNPs with a bilayer structure. (**c**) Schematic diagram of a modified polymer on AuNPs' surface (n means the numbers of polymer molecules). (**d**) Layer-by-layer modification of charged polymer on AuNPs' surface for selective adhesion to the substrate.

3.2.2. Electrostatic Adsorption

Furthermore, gold surfaces can be modified by forming gold-sulfur bonds between thiol groups and gold surfaces, as well as electrostatic adsorption between negative molecules and positive CTAB, while antibodies or proteins may also be directly adsorbable onto the surface of AuNPs [30] or via layer-by-layer modification [31]. The layer-by-layer modification of polymers on the surface of AuNPs can easily manipulate their surface properties. For example, Gole et al. [32] performed the layer-by-layer modification of negatively and positively charged sodium polystyrene sulfonate (PSS) and poly (dimethyl diallyldimethylammonium chloride) (PDADMAC), respectively. The polymers on the surface of the AuNPs were selectively attached to the substrates with different charges, as shown in Figure 3c. In addition, the modification of polymers with further charges also affects the phagocytosis behavior of cells [33]. Compared with the gold–sulfur bond modification strategy, electrostatic adsorption does not require ligand substitution and is relatively simple, fast, and efficient.

3.2.3. Electrostatic Adsorption

In addition to using the two abovementioned methods to improve the stability and applicability of AuNPs, Sendroiu et al. [34] proposed covering the surface of AuNPs with a layer of silica to strengthen their application value in living organisms through silica's excellent biocompatibility and ease of modification, which also enables the use of application strategies. In addition, it is possible to increase the surface area of AuNPs by using mesoporous silica as a carrier, which can facilitate the loading of more drug molecules and provide the material with more functions. For example, Zhang et al. [35] used mesoporous silica-coated AuNPs to load the anticancer drug doxorubicin (Dox) and modulate the long-axis absorption peak of the AuNPs to the near-infrared region so that the AuNPs could undergo photothermal conversion after near-infrared laser excitation to produce hyperthermia while also promoting the release of Dox loaded into the mesopores to enhance chemotherapy (chemotherapy). The two-photon fluorescence generated by the Dox release and the AuNPs could be used for cell imaging. This nanoplatform could simultaneously realize the dual effect of cell imaging and therapy, as shown in Figure 3d.

3.3. Biosensing Assays

3.3.1. Förster Resonance Energy Transfer (FRET)

An energy transfer mechanism based on Förster resonance energy transfer (FRET) involves energy transfer between two chromophores. FRET is similar to near-field transport, i.e., compared with the wavelength of the excitation light, the reaction distance of action is much smaller. Near-field regions are characterized by the emission of virtual photons by excitations of donor chromophores, which are then absorbed by acceptor chromophores. Due to the fact that these photons violate energy conservation and momentum, these particles are not detectable. Therefore, FRET is considered a radiation-free process. From quantum electrodynamics calculations, we can determine the short-range and long-range approximations of radiation-free and radiative energy transfers under a unified field [36,37], respectively. When both chromophores are fluorophores, the mechanism is called fluorescence resonance energy transfer; however, in reality, the energy transfer is not conducted through fluorescence [38,39]. Since this phenomenon is based on non-radiative energy transfer, we prefer to use FRET terminology to avoid misleading names. It should also be noted that FRET is not limited to fluorescence as it can also be related to phosphorescence. For instance, a combination of AuNPs and upconversion nanoparticles was first applied to enhance the fluorescence intensity of upconversion nanoparticles with the unique SPR optical properties of AuNPs. After continuous research, in 2015, Zhan [40] and his research team adjusted the aspect ratio of AuNPs so that the absorption peaks of the long and short axes overlapped with the excitation source and emission position of the upconversion nanoparticles, and controlled the particle size of the upconversion nanoparticles to 4 nm to ensure that all the doped ions were affected by the electromagnetic field of the SPR to enhance the intensity of the upconversion emission. Dadmehr [41] et al. conjugated the aptamers to deposit on the surface of graphene oxide decorated with AuNPs, and the following formation of a hetero-duplex stem-loop structure led to fluorescence quenching. Moreover, Hu et al. [42] also reported a sensitive fluorescent probe with off–on FRET response for harmful thiourea that can be detected by fluorescent carbon nanodots combined with AuNPs via electrostatic interaction. As shown in Figure 4a, the electromagnetic field generated by the long-axis absorption peak of AuNPs very strongly absorbs photons, which consequently enhances the absorption rate of photons via the surrounding upconversion nanoparticles, thus increasing the intensity of the emitted light. The short-axis absorption peak of AuNPs enhances the density of the optical states of upconversion nanoparticles through the Purcell effect, which regulates the radiative decay rate to strengthen the intensity of the upconversion.

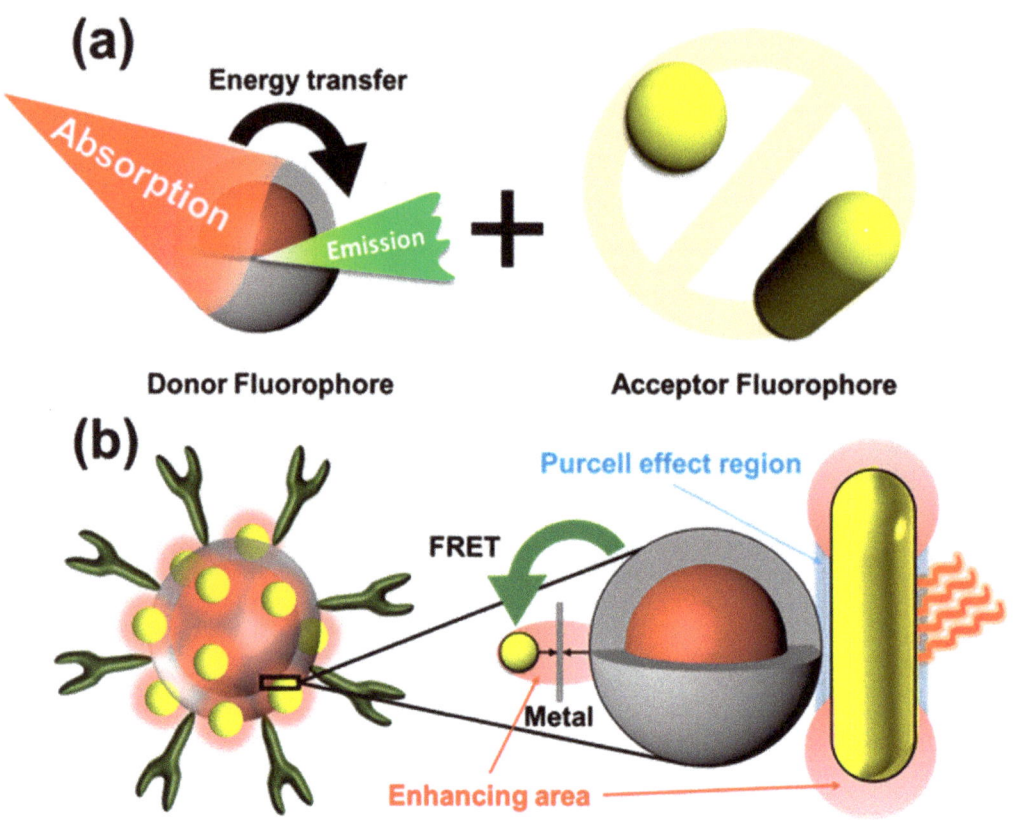

Figure 4. (**a**) The upconverted nanoparticles (donor fluorophore) are affected by the surface plasmon resonance (SPR) of AuNPs. (**b**) The FRET effect enhances the absorption of the excitation source, and the Purcell effect enhances the emission intensity.

3.3.2. LSPR Electric Field Enhancement

AuNPs enhance the light emission of organic fluorophores by means of the SPR effect. In this article, we discuss this effect using an example of plasmon-enhanced fluorescence emission [43]. As a result of the abundance of conductive electrons in AuNPs, both the intensity and the rate at which electromagnetic fields are radiated increase as an electromagnetic field is excited. However, AuNPs could simultaneously quench or enhance the emitting intensity because it also has a high absorption coefficient. Therefore, the distance between AuNPs and another emitter, which we call upconversion nanoparticles (UCNPs), is the key point to control which behavior in the significant part. In Figure 4b, there is another viewpoint that an emitter with a low quantum yield could be significantly enhanced compared with an emitter with a higher quantum yield, even following radiative rate enhancement, meaning that UCNPs have a high potential for enhancing upconverted luminescence efficiency based on the low quantum yield of UCNPs. Among the plasmonic nanomaterials, silver and gold have been studied extensively for their ability to enhance upconversion. Silver has a more substantial SPR effect than gold because overlapping interband transitions do not dampen its plasmon resonance. In spite of this, silver does not consistently result in a greater enhancement of upconversion emissions, suggesting that other factors such as overlap between spectral bands or overlap between fields may have a greater influence on the observed enhancements. Additionally, the maximum enhancement for a core-shell or flat gold film structure is 10 times lower than that of a gold hole array-patterned system with 450 times enhancement because SPR displays a specific region with a higher resonance strength for two reasons: the coupling phenomena between each AuNP and SPR's focus on sharp tip structures.

Additionally, AuNPs be used as energy absorbers because of the significantly different order of extinction coefficient between AuNPs and organic fluorophores. The value of the extinction coefficient for the commonly used organic dye indocyanine green (ICG) is only 10^5 (cm^{-1}/M) at 790 nm, but AuNPs have an extinction coefficient of 10^{11}~10^{13} (cm^{-1}/M), which is higher than the ICG 6~8 order. Therefore, AuNPs can be used to improve ROS production from ICG by accepting the energy transferred from the dye [44].

3.3.3. Fluorescent and Colorimetric Assay

AuNPs have unique physical and chemical properties, which make them easier to be novel chemical and biological sensors. Over the past ten years of research, AuNPs have been widely used as sensors and in detecting heavy metals. Among them, the colorimetric method is a common and convenient detection method. The content of the component to be tested is determined through the solution's color depth, and the sample's concentration is analyzed by optical and photoelectric colorimetry. Both methods are calculated through the Beer–Lambert law (Equation (2)). For example, DNA-functionalized gold nanoparticles can be used to detect lead ions with a colorimetric method: DNase-based functionalized gold nanoparticles; the sensor's detection range can be from 3 nM to 1 µM [45]. However, DNA molecule synthesis and chemical modification are very complex and expensive.

$$A = \varepsilon \cdot l \cdot c \tag{2}$$

A is the absorbance; ε is the molar attenuation coefficient or absorptivity of the attenuating species; l is the optical path length in cm; c is the concentration of the attenuating species.

However, colorimetric techniques tend to be less sensitive and less selective. To effectively improve its sensitivity and resolution, recent studies have used composite materials to improve the accuracy of colorimetry. In current literature, AuNPs are used as a heavy metal sensor at different pH values. Mercury ions and lead ions will form mercury-gold and lead-gold alloys on the surface of AuNPs [46]. In the presence \ of hydrogen peroxide (H_2O_2), the fluorescent substrate is catalyzed with peroxidase-like properties. In order to make the sensor more stable to heat, pH, and salt, Li [47] et al. designed a fluorescent and colorimetric dual-modal sensor based on AuNPs prepared using carbon dots (CDs) to detect the concentration of Cu^{2+} and Hg^{2+} in water. In addition to environmental heavy metal detection, AuNPs can also be used to analyze cancer biomarkers through dual-mode fluorescence colorimetry. Dadmehr [48] et al. used AuNPs grafted with gelatin and combined with gold nanoclusters to detect matrix metalloproteinase. Food samples can also be detected by AuNPs colorimetric method. Shahi [49] et al. used gelatin-functionalized AuNPs (AuNPs@gelatin) and applied a competitive colorimetric assay to detect aflatoxin B1 (AFB1) and measure the concentration of this carcinogen in food samples.

4. Clinical Biomarkers for Early Cancer Detection

Crick proposed the central dogma in 1957 and demonstrated that the transmission of genetic information relies on DNA, RNA, and proteins [50]. With the development of molecular biology technologies, people can now identify how diseases are caused by specific molecules involved in transcription and translation. It is becoming increasingly common to use these molecules as diagnostic markers for diseases related to cancer and other diseases. Nevertheless, the development of imaging and molecular marker-related technologies is considered the most effective method of improving the resolution and sensitivity of detection. Since AuNPs enable image transduction and are non-invasive, many studies have been conducted to label many key disease-related molecules identified via AuNPs, thereby improving the sensitivity of diagnosis (Figure 5).

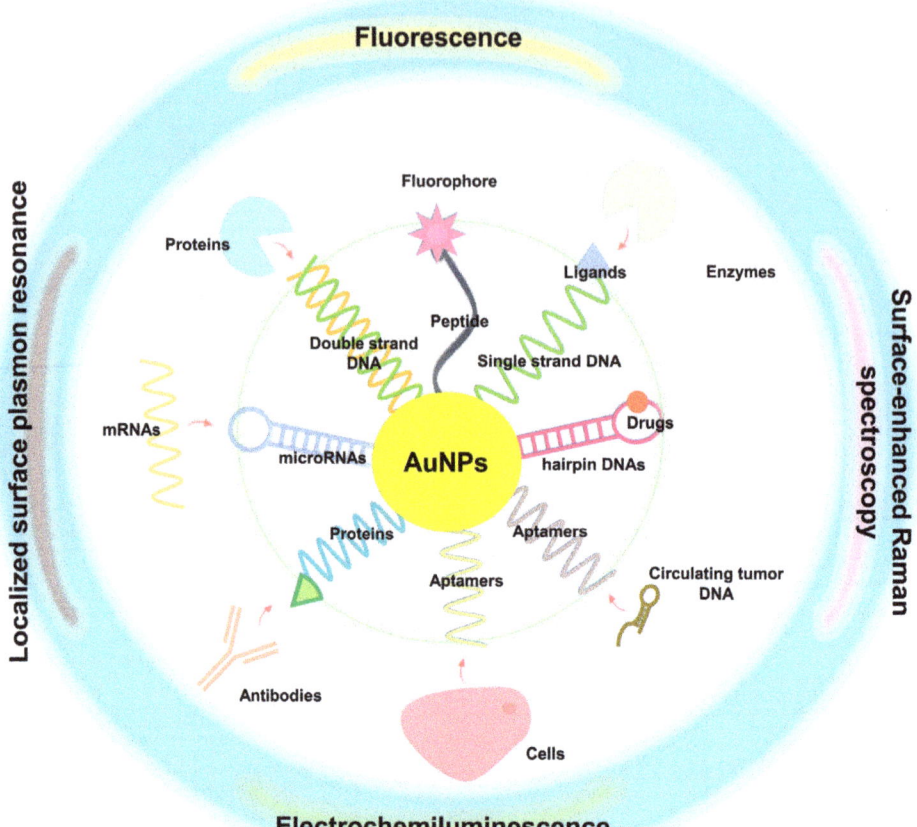

Figure 5. The development of cross-disciplinary nanotechnology has been vigorous, and one critical application is in biosensors produced by combining the semiconductor industry and biotechnology. AuNPs induce more biomolecular bonds, such as those of DNA, RNA, protein, and drugs to achieve the benefit of expanding the output signal. In addition, immobilization techniques can also be applied to a novel metal–semiconductor–metal material wafer. Using the chemiluminescence reaction in the system, the photons are converted into electronic signals to measure target biomolecules.

4.1. DNA

DNA is one of the most critical substances cells used to store genetic information. Most commonly, it is found in the nucleus as a chromosome. Chromatin stability can be regulated by histone modifications, DNA methylation and DNA-histone interactions, the results of which is only gene expression, but not the intrinsic sequence of genes, known as epigenetic regulation [51]. Most eukaryotic cells, including mammals, regulate DNA methylation in GC-rich regions or CpG islands, primarily through DNA methyltransferases. Such modification leads to dysregulation and abnormal expression of specific genes. These expression imbalances have been linked to the development of several diseases [52]. It was therefore necessary to develop these specific variants of methyltransferases as a means of detection.

The detection of DNA methylation has been carried out using several techniques [53], including genome-wide methylation extent analysis, gene-specific methylation analysis, and screening for new methylated sites. Assays are restricted by restriction sites [54], including enzymatic hydrolysis-based traditional methods, such as high-performance liquid chromatography (HPLC), high-performance capillary electrophoresis (HPCE), restriction landmark genome scanning (RLGS), and methylation-sensitive arbitrary primed polymerase chain reaction (MS-AP-PCR), as well as restriction enzyme-free methods, such as methylation-specific PCR (MSP) and real-time methylation-specific PCR. However, many technologies may generate false-positive results and lack the sensitivity to detect them. Nevertheless, there are methods that can be used to improve this insufficiency [54], such as combined bisulfite restriction analysis (COBRA), a methylation-sensitive single nucleotide primer extension (Ms-SNuPE) assay, Ms-SNuPE assays, matrix-assisted laser desorption/ionization mass spectrometry (MALDI-MS), double-labeled probes, and quantitative PCR (qPCR), but these methods are not cost-effective.

The introduction of AuNPs is considered to be an effective method to improve these shortcomings. Upon exposure to methyltransferases, AuNPs modified by a double-strand DNA probe will change from a well-dispersed state to an aggregated state. By measuring the A620/A520 difference, a colorimetric assay based on AuNPs can be developed to detect the presence of methyltransferases in the environment [55]. It is also possible to apply the same logic by using surface-enhanced Raman spectroscopy (SERS) to detect single-nucleotide polymorphisms. Detecting changes in SERS was found to be enhanced by comparing the conversion of cytosine and 5-methylcytosine into sodium bisulfite with the binding difference between AuNP-dGTP-cy5-probe and sodium bisulfite [55]. According to Liu's research, a polyadenine-DNA hairpin probe that is methylene blue-labeled and can be methylated by methyltransferases can be used as a restriction enzyme recognition tool for determining the differences in electrochemical response and DNA methylation [56]. Additionally, a specific probe designed to detect methylation combined with a fiber optic nanogold-linked sorbent assay can directly detect the methylation state of the tumor suppressor SOCS-1 through transmitted light intensity [57].

Furthermore, biomarkers that identify somatic mutations are also used to diagnose potential genetic diseases as a preventative measure. For example, the BRCA1 mutation is one of breast and ovarian cancer most commonly occurring oncogenes. It was found that a DNA capture probe immobilized in AuNP can be used to detect these discovered prognosis biomarkers, with a detection accuracy reaching femtomolar levels [58]. Additionally, an electrochemical DNA sensor based on a BRCA1 DNA sequence-based tetrahedral-structured probe was fabricated with AuNPs to precisely detect BRCA1 [59]. It has also been found that using Bi_2Se_3-AuNPs as a load signal probe to form a sandwich with BRCA1-immobilized silicon improves diagnostic accuracy [60]. It is also possible to use boron nitride quantum dot AuNPs to enhance the detection of BRCA1/2 [61]. An electrochemical method of prostate cancer-specific DNA sequences (PCA3) using chondroitin sulfate-AuNPs has been demonstrated to treat prostate cancer [62].

The cell-systematic evolution of ligands by exponential enrichment (cell-SELEX) is the process of discovering small oligonucleotides through arithmetic. It has been found that these less than 100-mer aptamers bind with a specific affinity to the cell surface through a unique folding structure and have applications in the detection and diagnosis of diseases [63], as well as in the annotation of metals [64–66]. The identification of cancer cells was found to be effective when combining specific aptamers with AuNPs and simultaneously forming complexes with magnetic beads and magnetic forces [67]. Using this method, it is possible to separate a sample using complementary sequence DNA (capture DNA) and determine whether the sample contains cancer cells via electrochemiluminescence. Furthermore, AuNPs have a good solubility, high surface reactivity, and excellent bioactivity. Taking advantage of their high fluorescence efficiency and versatility in surface modification, concatemer quantum dots were further used as a component of nanocomposites with AuNPs. This study demonstrated that MWCNTs@PDA@AuNP nanocomposites can improve aptamer recognition specificity [68].

Similar oncogenes such as KRAS can also be used as biomarkers for cancer diagnosis. MWCNTs–PA6–PTH conjugates can recognize mismatched bases in KRAS single-strand DNA-AuNPs with over 50% efficiency when used with KRAS single-strand DNA-AuNPs [69]. This method can also be applied to gastric tumors using KRAS as the target. A hairpin-DNA that recognizes KRAS mutations and conjugates with Cy3-AuNPs can be highly effective in identifying gastric tumors. As a result, KRAS-specific hairpin-DNA can go against the sequence of the KRAS sequence to inhibit its biological functions, which include vascularization and metastasis, resulting in an improvement in the survival rate [70].

Apart from forming nanocomposite materials with different elements, aptamer-AuNPs can also be used to separate circulating tumor cells from blood using microfluidics, with a 39-fold increase in efficiency [71]. According to this method, when using aptamer-AuNP probes for detection, one can determine the presence or absence of binding based on the color difference between the two probes [72]. It is therefore possible to apply similar assays to identify different molecules in liquid biopsies, such as circulating tumor DNA (ctDNA) and PIK3CA. In conjunction with peptide nucleic acid-AuNPs, the anti-5-methylcytosine antibody conjugate apoferritin can be used to specifically identify ctDNA in patient blood by detecting differences in ions with an accuracy of 50–10,000 fM [73]. The use of aptamers is helpful for not only identifying specific cells as a diagnostic tool but also targeting therapies. A Pt@Au nanoring@DNA nanocomposite was found to be helpful as a diagnostic and therapeutic strategy due to its photothermal therapy abilities for tumor cells when exposed to near-infrared light [74]. It has also been shown that combining tetrahedral DNA frameworks and AuNPs can further enhance the capture efficiency of BRCA1 ctDNA from 1 aM to 1 pM [75].

In addition to the DNA of a tumor itself, different genotypes of tumor-associated viruses have been designed as probes in conjunction with polyaniline-AuNP to develop biosensors that can identify the specific DNA targets of human papillomavirus, as well as other viruses, to improve diagnosis efficiency [76]. Studies have also shown trichomoniasis infections cause several cancers. The abovementioned probes can also be used detect trichomoniasis by designing a sequence that targets trichomoniasis and forming a biosensor with AuNPs [77].

4.2. RNA/miRNA

RNA messengers, also known as messenger RNA (mRNA), are essential molecules that are transcribed from DNA, transmit information, and serve as templates for translation into proteins. The use of dysregulated mRNA is also considered to be one of the most important cancer detection methods. With an accuracy of 0.31 nM, polyadenine can be quantified with fluorescent spherical nucleic acid labeled at the end of the mRNA in AuNPs [78]. This is therefore considered a precision medicine strategy to predict diseases by quantifying associated mRNAs. Among the major enzymes involved in DNA replication and repair is topoisomerase 1/2, which is implicated in cancer development or resistance to chemotherapy [79]. As a result of the labeling of a probe capable of detecting topoisomerase 1/2 in AuNPs, the absorbance ratio in a gold-aggregating assay can be calculated to determine the amount of AuNPs [79]. Likewise, quantitative measurements of biomarkers associated with cancer, such as glypican-1, facilitate the diagnosis of pancreatic cancer [80]. In Li's work, glypican-1 was found to be pre-amplifiable when using the catalytic hairpin assembly and point-of-care-testing methods. Capturing glypican-1 mRNA through AuNPs is performed with paper-based strips with a 100 fM sensitivity [80].

In addition to regulating the transcription of DNA into RNA, a noncoding RNA can also regulate the translation of mRNA. The use of microRNA (miRNA) is one of the most common methods for regulating mRNA, so detecting the level of miRNA is considered a means of diagnosing diseases. It is possible to successfully capture miRNAs by designing corresponding DNA probe-AuNPs [81–84]. Various methods of detection have been used to quantify miRNAs in many studies. miRNA levels can be detected using electrochemical differential pulse voltammetry [85]. The results of Pothipor's study suggested that the detection limit of miRNA-21 can reach as low as 0.020 fM after optimizing the electron transfer reaction between graphene and polypyrrole in AuNPs [85]. The introduction of MXene–MoS2 heterostructure nanocomposites was also found to improve electrochemistry and to form a catalytic hairpin assembly that can amplify the signal and enhance data detection [86]. Consequently, changing DNA structure may improve the efficiency of miRNA recognition. It has been shown that using three-dimensional or tetrahedral DNA nanostructure probes enhances electrochemical signals [87,88]. Furthermore, this DNA structure can also be applied to detect magnetic nanocomposites based on AuNPs [89]. Magnet-based devices for miRNA detection, such as the SERS-based biosensor, can be applied to serum and urine samples [90,91]. It is important to note that when using specific DNA structures and SERS, anticancer drugs can also be loaded to target miR21 for precision therapy [92].

4.3. Protein

As the final product of mRNA translation, a protein is a larger molecule than DNA and RNA, and conjugates with AuNPs can also be used to detect proteins. In cancer cells, telomerase, an enzyme responsible for protecting DNA replication from cell division, is overexpressed. It was found that the use of (TTAGGG)n sequences and AuNPs as probes could be used to diagnose cancer cells or the telomerase level in a liquid biopsy as a method of cancer diagnosis [93–95]. As a similar approach, single-strand DNA-rhodamine 6G-AuNPs were designed to be cleaved by Flap endonuclease 1, another molecule related to DNA structure; the potential Flap was calculated by exposing the rhodamine 6G after cleavage at the Flap endonuclease 1 level [96,97]. In addition to being used to screen for specific cell types, the aptamer can also be used to identify specific proteins. The use of designed aptamer-AuNPs has been reported to be specific for diagnosing prostate cancer through a prostate-specific antigen [98]. In breast cancer, the aptamer-AuNP method can also detect HER2 [99], and surface plasmon resonance technology can improve this method's detection sensitivity [100].

5. Biosensing of Cancer Biomarkers

As most molecules can only be quantified through precision instruments, many studies have primarily focused on optical and enzyme-dependent biosensing to reduce detection time and cost. These two methods' numerous applications in detecting cancer biomarkers are discussed in this section.

5.1. Optical Biosensing

Many studies have identified and determined the identity of generated AuNPs through their spectral differences since AuNPs require different media for ultraviolet–visible spectroscopy. Most studies have used electrochemical responses to assess whether targets are captured [101]. However, it should be noted that optical property can be used to facilitate detection and interpretation, thus reducing the requirements for equipment level or response time. Based on the surface plasmon resonance of AuNPs formed by incident photon frequency and free electrons, state changes can occur between dispersion and aggregation and are accompanied by color changes [102]. For this reason, colorimetric assays are used to interpret data. According to Miti's study, localized surface plasmon resonance can be used to detect miR-17 levels in cancer cells [103]. It is also possible to combine it with the upconversion emission nanoparticle $NaYF_4$ to amplify the signal for detecting miR-21-AuNPs [104]. This application can be used to detect not only miRNA sequences but also target miRNAs for the delivery of drugs. Introducing nanocomposite elements with AuNPs can not only enable the detection of miR-21 but also allow doxorubicin to be released through localized surface plasmon resonance to achieve therapeutic effects. miR-21 levels and cell death incidence can also be determined via color conversion. A nanocomposite that undergoes electron transfer can also be detected via electrochemiluminescence [105]. In Zhang's study, Ru(bpy)32+ and boron nitride quantum dots were used to detect miR-21-AuNPs, and the resonance energy transfer was quantified through electrochemiluminescence [106,107].

Fluorescence labeling has also been employed as a further method of identifying AuNP probe signals. In Li's research, miR-21 and miR-141 were simultaneously labeled with two fluorescent dyes on AuNPs to hybridize them with intracellular microRNA. The fluorescent signal changed when the target RNA was bound to the probe. Due to how flares are made, diagnostic sensitivity can be achieved [108]. Labeling DNA sequences with fluorescent signals can also be used to detect proteins. There is evidence that labeling Cy5 with (TTAGGG)n sequence-AuNPs can be combined with surface-enhanced Raman spectroscopy (SERS) to identify the telomerase level in tumor cells [109]. Moreover, SERS and AuNPs, in combination with doxorubicin in DNA sequence, were shown to detect miR-21 and drug release [92]. With fluorescence resonance energy transfer, specific proteins from a mixture can be detected using N-doped carbon dots if there is an interaction between the target protein and the carbon dots [110]. Using antibodies to recognize specific DNA sequences can also increase the signal. A prostate-specific antigen, kallikrein-3, is considered to be useful for the diagnosis of prostate cancer. By simultaneously labeling aptamer-AuNPs with streptavidin and anti-kallikrein-3-AuNPs, kallikrein-3 can be effectively detected. It is noteworthy that the detected signal can also be embedded in cellulose paper and read with a smartphone [111].

5.2. Enzyme-Dependent Biosensing

The enzyme-dependent biosensing method uses a catalytic reaction or formation of a structure when the target and probe are combined, resulting in the cleavage of a specific enzyme. The methylation of DNA is the most common detection method. Various restriction enzymes may cleave a methylated DNA sequence to reveal signal differences. It is also possible to apply a similar approach to the incidence of miRNA detection. Upon interaction between the target miRNA and the probe, a structure can be formed that can be recognized by a cleaved, duplex-specific nuclease. A specific antibody with a label can identify the cleaved structure and interpret the structure using methods such as electrochemistry [112]. It has been shown that this duplex-specific nuclease approach can be used in conjunction with fluorescence [113,114], surface plasmon resonance [115], or SERS [116] for the identification of signals in other ways.

The development of methods using optical and enzyme-dependent biosensing has also been gradual. According to Zhang's research, noncoding RNAs can be bound and cleaved by exonuclease III, and the cleaved products can release signal DNA and be detected by fluorescence resonance energy transfer [117]. A prostate-specific antigen was used in Wang's experiment to design a unique sequence that would be vulnerable to CRISPR-Cas12a cleavage. This method produces a product that can be quantified via the colorimetric method and used to diagnose prostate cancer [118].

6. Clinical Applications of Cancer Based AuNP Biomarkers

Based on the above discussion, AuNPs can distinguish between single bases and therefore be applied to detect DNA methylation. Accordingly, this section discusses single-point mutations and single-nucleotide polymorphisms as related diagnostic cases.

6.1. Single-Point Mutations or Single-Nucleotide Polymorphism (SNP)

AuNPs have also been used for the detection of single-nucleotide polymorphisms [119]. Analyses are primarily conducted by hybridizing a DNA microarray with single-strand DNA-AuNPs, and interpretation is obtained through analysis of various signals, including resonance light scattering [120] and SERS [121].

A single-point mutation is also a type of single-nucleotide polymorphism that can lead to the continuous activation of specific genes, ultimately resulting in a mutation. As a result of Park's work, it was found that the mismatch recognition protein, the MutS protein, can recognize mutations of the KRAS gene. Combining MutS-AuNPs with sequence hybridization can change the resonance frequency detected by a microcantilever resonator [122]. It can also be applied to liquid biopsy-related incidents, including the detection of circulating tumor cells and the detection of ctDNA for the diagnosis of cancer [123–125].

6.2. Exon or Gene Copy-Number Changes Detection

There is an association between cancer progression and mutations in the EGFR and the development of drug resistance. Mainly, mutations in exon 19/21 are often detected in tissues and cells related to lung cancer. Via the hybridization of specific probe-AuNPs with a target sequence, AuNPs can be changed between dispersed and aggregated states that colorimetric assays can detect and applied to different tissues, including tumor cells and liquid biopsies [126,127].

6.3. Protein Structural Modifications Detection

Proteins such as transcription factors can be detected using AuNPs. This is accomplished by designing a suitable ligand or aptamer. It has been demonstrated that a protein–captured target interaction product can change signaling through fluorescence resonance energy transfer [110]. Additionally, a proper ligand can also be utilized in high-throughput screening to identify special drug candidates for alternative treatments [128]. Interestingly, a similar detection method was recently used to identify differences in proteins subjected to glycosylation [129].

7. Conclusions

The use of nanoparticles for bio-detection, drug delivery, and drug screening has become a new research direction for scientists. In addition, the combination of semiconductor etching technology and biomaterials can be used for biological disease detection or drug screening. Due to the unique spectroscopic properties of AuNPs, more detection methods can be established to perform a variety of analytical experiments on a single biochip. In pharmaceutical science and biotechnology, the rapid detection and reading of different DNA sequences is a more compelling development target. Combining DNA sequences with fluorescent probes, gold nanoparticles, and chemical luminescence has partially replaced old radio-isotope detection methods. The development of nanotechnology will bring unlimited opportunities for the development of life sciences, mainly regarding: (1) the study of the structure and function of various intracellular organelles (e.g., granulosa and nucleus) at the nanoscale in terms of the exchange of materials, energy, and information between cells and organisms; (2) biological response mechanisms in repair, replication, and regulation, as well as the development of molecular engineering (including the use of AuNP biomolecular robots), based on biological principles; (3) nanoscale imaging techniques, such as optical coherence chromatography (OCT), which is known by scientists as the "molecular radar" and has a resolution of 1 micron (which is thousands of times higher than the precision of available chromatography and nuclear magnetic resonance techniques), can conduct the dynamic imaging of living cells in living organisms 2000 times per second, and observe the dynamics of living cells without killing them as with X-ray, general chromatography, and MRI; (4) laser single-atom molecular detection, which is also ultra-sensitive and can accurately acquire one of the 100 billion atoms or molecules in gaseous materials at the single-atom molecular level; and (5) microprobe technology, which can be implanted into different parts of human body according to different research purposes and can also run with the blood in the body so that various types of biological information can be delivered to the external recording device at any time. Microprobe technology has the potential to become a standard tool for life science research in the 21st century. The development of nanotechnology will bring infinite hope to the development of biotechnology as it can be used to investigate the activities of biomolecules in the body to investigate issues affecting human health.

Author Contributions: Conceptualization, C.-H.L., M.-H.C. and M.H.; writing—original draft preparation, C.-H.L., M.-H.C. and Y.-C.C.; reviewing and editing, C.-H.L., M.-H.C., Y.-C.C. and M.H.; supervision, M.H.; project administration, M.H. All authors have read and agreed to the published version of the manuscript.

Funding: This research received no external funding.

Institutional Review Board Statement: Not applicable.

Informed Consent Statement: Not applicable.

Data Availability Statement: Not applicable.

Acknowledgments: The authors would like to express their gratitude to the Genomics Research Center, Academia Sinica for supporting Michael Hsiao. Ming-Hsien Chan is grateful for the support of the Academia Sinica Outstanding Postdoctoral Fellowship.

Conflicts of Interest: The authors declare no conflict of interest.

References

1. Daniel, M.C.; Astruc, D. Gold nanoparticles: Assembly, supramolecular chemistry, quantum-size-related properties, and applications toward biology, catalysis, and nanotechnology. *Chem. Rev.* **2004**, *104*, 293–346. [CrossRef]
2. Baranova, N.N.; Zotov, A.V.; Bannykh, L.N.; Darina, T.G.; Savelev, B.V. Experimental-Study of the Solubility of Gold in Water under 450-Degrees-C and 500-Atm in Relation to Redox Conditions. *Geokhimiya* **1983**, *8*, 1133–1138.
3. Nerle, U.; Hodlur, R.M.; Rabinal, M.K. A sharp and visible range plasmonic in heavily doped metal oxide films. *Mater. Res. Express* **2014**, *1*, 015910. [CrossRef]
4. Sun, M.Z.; Xu, L.G.; Ma, W.; Wu, X.L.; Kuang, H.; Wang, L.B.; Xu, C.L. Hierarchical Plasmonic Nanorods and Upconversion Core-Satellite Nanoassemblies for Multimodal Imaging-Guided Combination Phototherapy. *Adv. Mater.* **2016**, *28*, 898–904. [CrossRef]
5. Martin, C.R. Nanomaterials: A Membrane-Based Synthetic Approach. *Science* **1994**, *266*, 1961–1966. [CrossRef]
6. Billot, L.; de la Chapelle, M.L.; Grimault, A.-S.; Vial, A.; Barchiesi, D.; Bijeon, J.-L.; Adam, P.-M.; Royer, P. Surface enhanced Raman scattering on gold nanowire arrays: Evidence of strong multipolar surface plasmon resonance enhancement. *Chem. Phys. Lett.* **2006**, *422*, 303–307. [CrossRef]
7. Taub, N.; Krichevski, A.O.; Markovich, G. Growth of Gold Nanorods on Surfaces. *J. Phys. Chem. B* **2003**, *107*, 11579–11582. [CrossRef]
8. Reetz, M.T.; Helbig, W. Size-Selective Synthesis of Nanostructured Transition Metal Clusters. *J. Am. Chem. Soc.* **1994**, *116*, 7401–7402. [CrossRef]
9. Jana, N.R. Gram-Scale Synthesis of Soluble, Near-Monodisperse Gold Nanorods and Other Anisotropic Nanoparticles. *Small* **2005**, *1*, 875–882. [CrossRef]
10. Jana, N.R.; Gearheart, L.; Murphy, C.J. Evidence for Seed-Mediated Nucleation in the Chemical Reduction of Gold Salts to Gold Nanoparticles. *Chem. Mater.* **2001**, *13*, 2313–2322. [CrossRef]
11. Wiesner, J.; Wokaun, A. Anisometric Gold Colloids—Preparation, Characterization, and Optical-Properties. *Chem. Phys. Lett.* **1989**, *157*, 569–575. [CrossRef]
12. Nikoobakht, B.; El-Sayed, M.A. Preparation and Growth Mechanism of Gold Nanorods (NRs) Using Seed-Mediated Growth Method. *Chem. Mater.* **2003**, *15*, 1957–1962. [CrossRef]
13. Johnson, C.J.; Dujardin, E.; Davis, S.A.; Murphy, C.J.; Mann, S. Growth and form of gold nanorods prepared by seed-mediated, surfactant-directed synthesis. *J. Mater. Chem.* **2002**, *12*, 1765–1770. [CrossRef]
14. Murphy, C.J.; San, T.K.; Gole, A.M.; Orendorff, C.J.; Gao, J.X.; Gou, L.; Hunyadi, S.E.; Li, T. Anisotropic metal nanoparticles: Synthesis, assembly, and optical applications. *J. Phys. Chem. B* **2005**, *109*, 13857–13870. [CrossRef]
15. Pérez-Juste, J.; Liz-Marzán, L.M.; Carnie, S.; Chan, D.Y.C.; Mulvaney, P. Electric-Field-Directed Growth of Gold Nanorods in Aqueous Surfactant Solutions. *Adv. Funct. Mater.* **2004**, *14*, 571–579. [CrossRef]
16. Alkilany, A.M.; Thompson, L.B.; Boulos, S.P.; Sisco, P.N.; Murphy, C.J. Gold nanorods: Their potential for photothermal therapeutics and drug delivery, tempered by the complexity of their biological interactions. *Adv. Drug Deliv. Rev.* **2012**, *64*, 190–199. [CrossRef]
17. Wood, R.W. On a Remarkable Case of Uneven Distribution of Light in a Diffraction Grating Spectrum. *PPSL* **1902**, *18*, 269. [CrossRef]
18. Fano, U. The Theory of Anomalous Diffraction Gratings and of Quasi-Stationary Waves on Metallic Surfaces (Sommerfeld's Waves). *J. Opt. Soc. Am.* **1941**, *31*, 213–222. [CrossRef]
19. Kelly, K.L.; Coronado, E.; Zhao, L.L.; Schatz, G.C. The Optical Properties of Metal Nanoparticles: The Influence of Size, Shape, and Dielectric Environment. *J. Phys. Chem. B* **2003**, *107*, 668–677. [CrossRef]
20. Weissleder, R. A clearer vision for in vivo imaging. *Nat. Biotechnol.* **2001**, *19*, 316–317. [CrossRef]

21. Jain, P.K.; Huang, X.H.; El-Sayed, I.H.; El-Sayed, M.A. Noble Metals on the Nanoscale: Optical and Photothermal Properties and Some Applications in Imaging, Sensing, Biology, and Medicine. *Accounts Chem. Res.* **2008**, *41*, 1578–1586. [CrossRef]
22. Ye, X.; Jin, L.; Caglayan, H.; Chen, J.; Xing, G.; Zheng, C.; Doan-Nguyen, V.; Kang, Y.; Engheta, N.; Kagan, C.R.; et al. Improved Size-Tunable Synthesis of Monodisperse Gold Nanorods through the Use of Aromatic Additives. *ACS Nano* **2012**, *6*, 2804–2817. [CrossRef]
23. Wijaya, A.; Schaffer, S.B.; Pallares, I.G.; Hamad-Schifferli, K. Selective Release of Multiple DNA Oligonucleotides from Gold Nanorods. *ACS Nano* **2008**, *3*, 80–86. [CrossRef]
24. Chen, C.-C.; Lin, Y.-P.; Wang, C.-W.; Tzeng, H.-C.; Wu, C.-H.; Chen, Y.-C.; Chen, C.-P.; Chen, L.-C.; Wu, Y.-C. DNA−Gold Nanorod Conjugates for Remote Control of Localized Gene Expression by near Infrared Irradiation. *J. Am. Chem. Soc.* **2006**, *128*, 3709–3715. [CrossRef]
25. Lee, S.E.; Liu, G.L.; Kim, F.; Lee, L.P. Remote Optical Switch for Localized and Selective Control of Gene Interference. *Nano Lett.* **2009**, *9*, 562–570. [CrossRef]
26. Dujardin, E.; Mann, S.; Hsin, L.-B.; Wang, C.R.C. DNA-driven self-assembly of gold nanorods. *Chem. Commun.* **2001**, 1264–1265. [CrossRef]
27. Huff, T.B.; Tong, L.; Zhao, Y.; Hansen, M.N.; Cheng, J.-X.; Wei, A. Hyperthermic effects of gold nanorods on tumor cells. *Nanomedicine* **2007**, *2*, 125–132. [CrossRef]
28. von Maltzahn, G.; Park, J.-H.; Agrawal, A.; Bandaru, N.K.; Das, S.K.; Sailor, M.J.; Bhatia, S.N. Computationally Guided Photothermal Tumor Therapy Using Long-Circulating Gold Nanorod Antennas. *Cancer Res* **2009**, *69*, 3892–3900. [CrossRef]
29. Yu, C.; Varghese, L.; Irudayaraj, J. Surface Modification of Cetyltrimethylammonium Bromide-Capped Gold Nanorods to Make Molecular Probes. *Langmuir* **2007**, *23*, 9114–9119. [CrossRef]
30. Pissuwan, D.; Valenzuela, S.M.; Killingsworth, M.C.; Xu, X.; Cortie, M.B. Targeted destruction of murine macrophage cells with bioconjugated gold nanorods. *J. Nanoparticle Res.* **2007**, *9*, 1109–1124. [CrossRef]
31. Hauck, T.S.; Ghazani, A.A.; Chan, W.C.W. Assessing the Effect of Surface Chemistry on Gold Nanorod Uptake, Toxicity, and Gene Expression in Mammalian Cells. *Small* **2008**, *4*, 153–159. [CrossRef] [PubMed]
32. Gole, A.; Murphy, C. Polyelectrolyte-Coated Gold Nanorods: Synthesis, Characterization and Immobilization. *Chem. Mater.* **2005**, *17*, 1325–1330. [CrossRef]
33. Alkilany, A.M.; Thompson, L.B.; Murphy, C.J. Polyelectrolyte Coating Provides a Facile Route to Suspend Gold Nanorods in Polar Organic Solvents and Hydrophobic Polymers. *ACS Appl. Mater. Interfaces* **2010**, *2*, 3417–3421. [CrossRef]
34. Sendroiu, I.E.; Warner, M.E.; Corn, R.M. Fabrication of Silica-Coated Gold Nanorods Functionalized with DNA for Enhanced Surface Plasmon Resonance Imaging Biosensing Applications. *Langmuir* **2009**, *25*, 11282–11284. [CrossRef]
35. Zhang, Z.J.; Wang, L.M.; Wang, J.; Jiang, X.M.; Li, X.H.; Hu, Z.J.; Ji, Y.H.; Wu, X.C.; Chen, C.Y. Mesoporous Silica-Coated Gold Nanorods as a Light-Mediated Multifunctional Theranostic Platform for Cancer Treatment. *Adv. Mater.* **2012**, *24*, 1418–1423. [CrossRef]
36. Malsch, I. Nanotechnology in Europe: Scientific trends and organizational dynamics. *Nanotechnology* **1999**, *10*, 1–7. [CrossRef]
37. Ojea-Jiménez, I.; Tort, O.; Lorenzo, J.; Puntes, V. Engineered nonviral nanocarriers for intracellular gene delivery applications. *Biomed. Mater.* **2012**, *7*, 54106. [CrossRef] [PubMed]
38. Roduner, E. Size matters: Why nanomaterials are different. *Chem. Soc. Rev.* **2006**, *35*, 583–592. [CrossRef]
39. Li, Y.; Boone, E.; El-Sayed, M.A. Size Effects of PVP−Pd Nanoparticles on the Catalytic Suzuki Reactions in Aqueous Solution. *Langmuir* **2002**, *18*, 4921–4925. [CrossRef]
40. Zhan, Q.; Zhang, X.; Zhao, Y.; Liu, J.; He, S. Tens of thousands-fold upconversion luminescence enhancement induced by a single gold nanorod. *Laser Photon.- Rev.* **2015**, *9*, 479–487. [CrossRef]
41. Dadmehr, M.; Shahi, S.C.; Malekkiani, M.; Korouzhdehi, B.; Tavassoli, A. A stem-loop like aptasensor for sensitive detection of aflatoxin based on graphene oxide/AuNPs nanocomposite platform. *Food Chem.* **2023**, *402*, 134212. [CrossRef]
42. Hu, A.; Chen, G.; Yang, T.; Ma, C.; Li, L.; Gao, H.; Gu, J.; Zhu, C.; Wu, Y.; Li, X.; et al. A fluorescent probe based on FRET effect between carbon nanodots and gold nanoparticles for sensitive detection of thiourea. *Spectrochim. Acta Part A: Mol. Biomol. Spectrosc.* **2022**, *281*, 121582. [CrossRef]
43. Wu, D.M.; García-Etxarri, A.; Salleo, A.; Dionne, J.A. Plasmon-Enhanced Upconversion. *J. Phys. Chem. Lett.* **2014**, *5*, 4020–4031. [CrossRef]
44. Li, Y.; Wen, T.; Zhao, R.; Liu, X.; Ji, T.; Wang, H.; Shi, X.; Shi, J.; Wei, J.; Zhao, Y.; et al. Localized Electric Field of Plasmonic Nanoplatform Enhanced Photodynamic Tumor Therapy. *ACS Nano* **2014**, *8*, 11529–11542. [CrossRef]
45. Huang, K.-W.; Yu, C.-J.; Tseng, W.-L. Sensitivity enhancement in the colorimetric detection of lead(II) ion using gallic acid–capped gold nanoparticles: Improving size distribution and minimizing interparticle repulsion. *Biosens. Bioelectron.* **2010**, *25*, 984–989. [CrossRef]
46. Wang, C.-I.; Huang, C.-C.; Lin, Y.-W.; Chen, W.-T.; Chang, H.-T. Catalytic gold nanoparticles for fluorescent detection of mercury(II) and lead(II) ions. *Anal. Chim. Acta* **2012**, *745*, 124–130. [CrossRef]
47. Li, Y.; Tang, L.; Zhu, C.; Liu, X.; Wang, X.; Liu, Y. Fluorescent and colorimetric assay for determination of Cu(II) and Hg(II) using AuNPs reduced and wrapped by carbon dots. *Mikrochim. Acta* **2021**, *189*, 1–11. [CrossRef] [PubMed]
48. Dadmehr, M.; Mortezaei, M.; Korouzhdehi, B. Dual mode fluorometric and colorimetric detection of matrix metalloproteinase MMP-9 as a cancer biomarker based on AuNPs@gelatin/ AuNCs nanocomposite. *Biosens. Bioelectron.* **2023**, *220*. [CrossRef]

49. Shahi, S.C.; Dadmehr, M.; Korouzhdehi, B.; Tavassoli, A. A novel colorimetric biosensor for sensitive detection of aflatoxin mediated by bacterial enzymatic reaction in saffron samples. *Nanotechnology* **2021**, *32*, 505503. [CrossRef]
50. Strasser, B.J. A world in one dimension: Linus Pauling, Francis Crick and the central dogma of molecular biology. *Hist. Philos. Life Sci.* **2006**, *28*, 491–512.
51. Zhou, V.W.; Goren, A.; Bernstein, B.E. Charting histone modifications and the functional organization of mammalian genomes. *Nat. Rev. Genet.* **2010**, *12*, 7–18. [CrossRef] [PubMed]
52. Tsankova, N.; Renthal, W.; Kumar, A.; Nestler, E.J. Epigenetic regulation in psychiatric disorders. *Nat. Rev. Neurosci.* **2007**, *8*, 355–367. [CrossRef] [PubMed]
53. Laird, P.W. Principles and challenges of genome-wide DNA methylation analysis. *Nat. Rev. Genet.* **2010**, *11*, 191–203. [CrossRef]
54. Olkhov-Mitsel, E.; Bapat, B. Strategies for discovery and validation of methylated and hydroxymethylated DNA biomarkers. *Cancer Med.* **2012**, *1*, 237–260. [CrossRef]
55. Hu, J.; Zhang, C.-Y. Single base extension reaction-based surface enhanced Raman spectroscopy for DNA methylation assay. *Biosens. Bioelectron.* **2012**, *31*, 451–457. [CrossRef]
56. Liu, P.; Wang, D.; Zhou, Y.; Wang, H.; Yin, H.; Ai, S. DNA methyltransferase detection based on digestion triggering the combination of poly adenine DNA with gold nanoparticles. *Biosens. Bioelectron.* **2016**, *80*, 74–78. [CrossRef]
57. Guthula, L.S.; Yeh, K.-T.; Huang, W.-L.; Chen, C.-H.; Chen, Y.-L.; Huang, C.-J.; Chau, L.-K.; Chan, M.W.; Lin, S.-H. Quantitative and amplification-free detection of SOCS-1 CpG methylation percentage analyses in gastric cancer by fiber optic nanoplasmonic biosensor. *Biosens. Bioelectron.* **2022**, *214*, 114540. [CrossRef]
58. Rasheed, P.A.; Sandhyarani, N. Femtomolar level detection of BRCA1 gene using a gold nanoparticle labeled sandwich type DNA sensor. *Colloids Surfaces B: Biointerfaces* **2014**, *117*, 7–13. [CrossRef]
59. Feng, D.; Su, J.; He, G.; Xu, Y.; Wang, C.; Zheng, M.; Qian, Q.; Mi, X. Electrochemical DNA Sensor for Sensitive BRCA1 Detection Based on DNA Tetrahedral-Structured Probe and Poly-Adenine Mediated Gold Nanoparticles. *Biosensors* **2020**, *10*, 78. [CrossRef]
60. Bai, Y.; Li, H.; Xu, J.; Huang, Y.; Zhang, X.; Weng, J.; Li, Z.; Sun, L. Ultrasensitive colorimetric biosensor for BRCA1 mutation based on multiple signal amplification strategy. *Biosens. Bioelectron.* **2020**, *166*, 112424. [CrossRef]
61. Liang, Z.; Nie, Y.; Zhang, X.; Wang, P.; Ma, Q. Multiplex Electrochemiluminescence Polarization Assay Based on the Surface Plasmon Coupling Effect of Au NPs and Ag@Au NPs. *Anal. Chem.* **2021**, *93*, 7491–7498. [CrossRef] [PubMed]
62. Rodrigues, V.C.; Soares, J.C.; Soares, A.C.; Braz, D.C.; Melendez, M.E.; Ribas, L.C.; Scabini, L.F.; Bruno, O.M.; Carvalho, A.L.; Reis, R.M.; et al. Electrochemical and optical detection and machine learning applied to images of genosensors for diagnosis of prostate cancer with the biomarker PCA3. *Talanta* **2020**, *222*, 121444. [CrossRef] [PubMed]
63. Sefah, K.; Shangguan, D.; Xiong, X.; O'Donoghue, M.B.; Tan, W. Development of DNA aptamers using Cell-SELEX. *Nat. Protoc.* **2010**, *5*, 1169–1185. [CrossRef]
64. Ciesiolka, J.; Yarus, M. Small RNA-divalent domains. *RNA* **1996**, *2*, 785–793.
65. Hofmann, H.P.; Limmer, S.; Hornung, V.; Sprinzl, M. Ni2+-binding RNA motifs with an asymmetric purine-rich internal loop and a G-A base pair. *RNA* **1997**, *3*, 1289–1300.
66. Rajendran, M.; Ellington, A.D. Selection of fluorescent aptamer beacons that light up in the presence of zinc. *Anal. Bioanal. Chem.* **2007**, *390*, 1067–1075. [CrossRef]
67. Ding, C.; Wei, S.; Liu, H. Electrochemiluminescent Determination of Cancer Cells Based on Aptamers, Nanoparticles, and Magnetic Beads. *Chem. – A Eur. J.* **2012**, *18*, 7263–7268. [CrossRef]
68. Liu, H.; Xu, S.; He, Z.; Deng, A.; Zhu, J.-J. Supersandwich Cytosensor for Selective and Ultrasensitive Detection of Cancer Cells Using Aptamer-DNA Concatamer-Quantum Dots Probes. *Anal. Chem.* **2013**, *85*, 3385–3392. [CrossRef]
69. Wang, X.; Shu, G.; Gao, C.; Yang, Y.; Xu, Q.; Tang, M. Electrochemical biosensor based on functional composite nanofibers for detection of K-ras gene via multiple signal amplification strategy. *Anal. Biochem.* **2014**, *466*, 51–58. [CrossRef]
70. Bao, C.; Conde, J.; Curtin, J.; Artzi, N.; Tian, F.; Cui, D. Bioresponsive antisense DNA gold nanobeacons as a hybrid in vivo theranostics platform for the inhibition of cancer cells and metastasis. *Sci. Rep.* **2015**, *5*, 12297. [CrossRef]
71. Sheng, W.; Chen, T.; Tan, W.; Fan, Z.H. Multivalent DNA Nanospheres for Enhanced Capture of Cancer Cells in Microfluidic Devices. *ACS Nano* **2013**, *7*, 7067–7076. [CrossRef] [PubMed]
72. Borghei, Y.-S.; Hosseini, M.; Dadmehr, M.; Hosseinkhani, S.; Ganjali, M.R.; Sheikhnejad, R. Visual detection of cancer cells by colorimetric aptasensor based on aggregation of gold nanoparticles induced by DNA hybridization. *Anal. Chim. Acta* **2016**, *904*, 92–97. [CrossRef] [PubMed]
73. Cai, J.; Guo, Z.; Cao, Y.; Zhang, W.; Chen, Y. A dual biomarker detection platform for quantitating circulating tumor DNA (ctDNA). *Nanotheranostics* **2018**, *2*, 12–20. [CrossRef]
74. Zhang, H.; Wang, Y.; Zhong, H.; Li, J.; Ding, C. Near-Infrared Light-Activated Pt@Au Nanorings-Based Probe for Fluorescence Imaging and Targeted Photothermal Therapy of Cancer Cells. *ACS Appl. Bio Mater.* **2019**, *2*, 5012–5020. [CrossRef]
75. Wang, C.; Wang, W.; Xu, Y.; Zhao, X.; Li, S.; Qian, Q.; Mi, X. Tetrahedral DNA Framework-Programmed Electrochemical Biosensors with Gold Nanoparticles for Ultrasensitive Cell-Free DNA Detection. *Nanomaterials* **2022**, *12*, 666. [CrossRef]
76. Avelino, K.Y.; Oliveira, L.S.; Lucena-Silva, N.; de Melo, C.P.; de Andrade, C.A.S.; Oliveira, M.D. Metal-polymer hybrid nanomaterial for impedimetric detection of human papillomavirus in cervical specimens. *J. Pharm. Biomed. Anal.* **2020**, *185*, 113249. [CrossRef]

87. Ilbeigi, S.; Vais, R.D.; Sattarahmady, N. Photo-genosensor for Trichomonas vaginalis based on gold nanoparticles-genomic DNA. *Photodiagnosis Photodyn. Ther.* **2021**, *34*, 102290. [CrossRef]
88. Zhu, D.; Zhao, D.; Huang, J.; Zhu, Y.; Chao, J.; Su, S.; Li, J.; Wang, L.; Shi, J.; Zuo, X.; et al. Poly-adenine-mediated fluorescent spherical nucleic acid probes for live-cell imaging of endogenous tumor-related mRNA. *Nanomedicine: Nanotechnology, Biol. Med.* **2018**, *14*, 1797–1807. [CrossRef]
89. Shawky, S.M.; Awad, A.M.; Abugable, A.A.; El-Khamisy, S.F. Gold nanoparticles—An optical biosensor for RNA quantification for cancer and neurologic disorders diagnosis. *Int. J. Nanomed.* **2018**, *13*, 8137–8151. [CrossRef]
90. Li, H.; Warden, A.R.; Su, W.; He, J.; Zhi, X.; Wang, K.; Zhu, L.; Shen, G.; Ding, X. Highly sensitive and portable mRNA detection platform for early cancer detection. *J. Nanobiotechnology* **2021**, *19*, 1–10. [CrossRef]
91. Xia, N.; Zhang, L.; Wang, G.; Feng, Q.; Liu, L. Label-free and sensitive strategy for microRNAs detection based on the formation of boronate ester bonds and the dual-amplification of gold nanoparticles. *Biosens. Bioelectron.* **2013**, *47*, 461–466. [CrossRef] [PubMed]
92. Liu, L.; Xia, N.; Liu, H.; Kang, X.; Liu, X.; Xue, C.; He, X. Highly sensitive and label-free electrochemical detection of microRNAs based on triple signal amplification of multifunctional gold nanoparticles, enzymes and redox-cycling reaction. *Biosens. Bioelectron.* **2014**, *53*, 399–405. [CrossRef] [PubMed]
93. Borghei, Y.-S.; Hosseini, M. A New Eye Dual-readout Method for MiRNA Detection based on Dissolution of Gold nanoparticles via LSPR by CdTe QDs Photoinduction. *Sci. Rep.* **2019**, *9*, 5453. [CrossRef] [PubMed]
94. Qian, Q.; He, G.; Wang, C.; Li, S.; Zhao, X.; Xu, Y.; Mi, X. Poly-adenine-mediated spherical nucleic acid probes for live cell fluorescence imaging of tumor-related microRNAs. *Mol. Biol. Rep.* **2022**, *49*, 3705–3712. [CrossRef]
95. Pothipor, C.; Aroonyadet, N.; Bamrungsap, S.; Jakmunee, J.; Ounnunkad, K. A highly sensitive electrochemical microRNA-21 biosensor based on intercalating methylene blue signal amplification and a highly dispersed gold nanoparticles/graphene/polypyrrole composite. *Analyst* **2021**, *146*, 2679–2688. [CrossRef]
96. Zhao, J.; He, C.; Wu, W.; Yang, H.; Dong, J.; Wen, L.; Hu, Z.; Yang, M.; Hou, C.; Huo, D. MXene-MoS2 heterostructure collaborated with catalyzed hairpin assembly for label-free electrochemical detection of microRNA-21. *Talanta* **2021**, *237*, 122927. [CrossRef]
97. Liu, S.; Su, W.; Li, Z.; Ding, X. Electrochemical detection of lung cancer specific microRNAs using 3D DNA origami nanostructures. *Biosens. Bioelectron.* **2015**, *71*, 57–61. [CrossRef]
98. Shen, D.; Hu, W.; He, Q.; Yang, H.; Cui, X.; Zhao, S. A highly sensitive electrochemical biosensor for microRNA122 detection based on a target-induced DNA nanostructure. *Anal. Methods* **2021**, *13*, 2823–2829. [CrossRef]
99. Lu, H.; Hailin, T.; Yi, X.; Wang, J. Three-Dimensional DNA Nanomachine Combined with Toehold-Mediated Strand Displacement Reaction for Sensitive Electrochemical Detection of MiRNA. *Langmuir* **2020**, *36*, 10708–10714. [CrossRef]
100. Wu, J.; Zhou, X.; Li, P.; Lin, X.; Wang, J.; Hu, Z.; Zhang, P.; Chen, D.; Cai, H.; Niessner, R.; et al. Ultrasensitive and Simultaneous SERS Detection of Multiplex MicroRNA Using Fractal Gold Nanotags for Early Diagnosis and Prognosis of Hepatocellular Carcinoma. *Anal. Chem.* **2021**, *93*, 8799–8809. [CrossRef]
101. Kim, W.H.; Lee, J.U.; Jeon, M.J.; Park, K.H.; Sim, S.J. Three-dimensional hierarchical plasmonic nano-architecture based label-free surface-enhanced Raman spectroscopy detection of urinary exosomal miRNA for clinical diagnosis of prostate cancer. *Biosens. Bioelectron.* **2022**, *205*, 114116. [CrossRef] [PubMed]
102. Liu, X.; Wang, X.; Ye, S.; Li, R.; Li, H. A One–Two–Three Multifunctional System for Enhanced Imaging and Detection of Intracellular MicroRNA and Chemogene Therapy. *ACS Appl. Mater. Interfaces* **2021**, *13*, 27825–27835. [CrossRef] [PubMed]
103. Wang, J.; Zhang, J.; Li, T.; Shen, R.; Li, G.; Ling, L. Strand displacement amplification-coupled dynamic light scattering method to detect urinary telomerase for non-invasive detection of bladder cancer. *Biosens. Bioelectron.* **2019**, *131*, 143–148. [CrossRef] [PubMed]
104. Pu, F.; Ren, J.; Qu, X. Primer-Modified G-Quadruplex-Au Nanoparticles for Colorimetric Assay of Human Telomerase Activity and Initial Screening of Telomerase Inhibitors. *Methods Mol. Biol.* **2019**, *2035*, 347–356. [CrossRef]
105. Liu, L.; Chang, Y.; Ji, X.; Chen, J.; Zhang, M.; Yang, S. Surface-tethered electrochemical biosensor for telomerase detection by integration of homogeneous extension and hybridization reactions. *Talanta* **2023**, *253*, 123597. [CrossRef]
106. Tang, Y.; Zhang, D.; Lu, Y.; Liu, S.; Zhang, J.; Pu, Y.; Wei, W. Fluorescence imaging of FEN1 activity in living cells based on controlled-release of fluorescence probe from mesoporous silica nanoparticles. *Biosens. Bioelectron.* **2022**, *214*, 114529. [CrossRef]
107. Aayanifard, Z.; Alebrahim, T.; Pourmadadi, M.; Yazdian, F.; Dinani, H.S.; Rashedi, H.; Omidi, M. Ultra pH-sensitive detection of total and free prostate-specific antigen using electrochemical aptasensor based on reduced graphene oxide/gold nanoparticles emphasis on TiO_2/carbon quantum dots as a redox probe. *Eng. Life Sci.* **2021**, *21*, 739–752. [CrossRef]
108. Wei, B.; Mao, K.; Liu, N.; Zhang, M.; Yang, Z. Graphene nanocomposites modified electrochemical aptamer sensor for rapid and highly sensitive detection of prostate specific antigen. *Biosens. Bioelectron.* **2018**, *121*, 41–46. [CrossRef]
109. Poturnayová, A.; Dzubinová, L.; Buríková, M.; Bízik, J.; Hianik, T. Detection of Breast Cancer Cells Using Acoustics Aptasensor Specific to HER2 Receptors. *Biosensors* **2019**, *9*, 72. [CrossRef]
110. Chen, W.; Li, Z.; Wu, T.; Li, J.; Li, X.; Liu, L.; Bai, H.; Ding, S.; Li, X.; Yu, X. Surface plasmon resonance biosensor for exosome detection based on reformative tyramine signal amplification activated by molecular aptamer beacon. *J. Nanobiotechnology* **2021**, *19*, 1–10. [CrossRef]
111. Gundagatti, S.; Srivastava, S. Development of Electrochemical Biosensor for miR204-Based Cancer Diagnosis. *Interdiscip. Sci. Comput. Life Sci.* **2022**, *14*, 596–606. [CrossRef] [PubMed]

102. Cai, J.; Ding, L.; Gong, P.; Huang, J. A colorimetric detection of microRNA-148a in gastric cancer by gold nanoparticle–RNA conjugates. *Nanotechnology* **2019**, *31*, 095501. [CrossRef] [PubMed]
103. Miti, A.; Thamm, S.; Müller, P.; Csáki, A.; Fritzsche, W.; Zuccheri, G. A miRNA biosensor based on localized surface plasmon resonance enhanced by surface-bound hybridization chain reaction. *Biosens. Bioelectron.* **2020**, *167*, 112465. [CrossRef] [PubMed]
104. Zhang, K.Y.; Song, S.T.; Huang, S.; Yang, L.; Min, Q.H.; Wu, X.C.; Lu, F.; Zhu, J.J. Lighting Up MicroRNA in Living Cells by the Disassembly of Lock-Like DNA-Programmed UCNPs-AuNPs through the Target Cycling Amplification Strategy. *Small* **2018**, *14*, 1802292. [CrossRef] [PubMed]
105. Zhang, D.; Wang, K.; Wei, W.; Liu, Y.; Liu, S. Multifunctional Plasmonic Core-Satellites Nanoprobe for Cancer Diagnosis and Therapy Based on a Cascade Reaction Induced by MicroRNA. *Anal. Chem.* **2021**, *93*, 9521–9530. [CrossRef]
106. Zhang, Y.; Chai, Y.; Wang, H.; Yuan, R. Target-Induced 3D DNA Network Structure as a Novel Signal Amplifier for Ultrasensitive Electrochemiluminescence Detection of MicroRNAs. *Anal. Chem.* **2019**, *91*, 14368–14374. [CrossRef]
107. Cui, A.; Zhang, J.; Bai, W.; Sun, H.; Bao, L.; Ma, F.; Li, Y. Signal-on electrogenerated chemiluminescence biosensor for ultrasensitive detection of microRNA-21 based on isothermal strand-displacement polymerase reaction and bridge DNA-gold nanoparticles. *Biosens. Bioelectron.* **2019**, *144*, 111664. [CrossRef]
108. Li, J.; Huang, J.; Yang, X.; Yang, Y.; Quan, K.; Xie, N.; Wu, Y.; Ma, C.; Wang, K. Two-Color-Based Nanoflares for Multiplexed MicroRNAs Imaging in Live Cells. *Nanotheranostics* **2018**, *2*, 96–105. [CrossRef]
109. Qi, G.; Yi, X.; Wang, M.; Sun, D.; Zhu, H. SERS and fluorescence dual-channel microfluidic droplet platform for exploring telomerase activity at single-cell level. *Anal.* **2022**, *147*, 5062–5067. [CrossRef]
110. Mahani, M.; Taheri, M.; Divsar, F.; Khakbaz, F.; Nomani, A.; Ju, H. Label-free triplex DNA-based biosensing of transcription factor using fluorescence resonance energy transfer between N-doped carbon dot and gold nanoparticle. *Anal. Chim. Acta* **2021**, *1181*, 338919. [CrossRef]
111. Huang, J.-Y.; Lin, H.-T.; Chen, T.-H.; Chen, C.-A.; Chang, H.-T.; Chen, C.-F. Signal Amplified Gold Nanoparticles for Cancer Diagnosis on Paper-Based Analytical Devices. *ACS Sens.* **2018**, *3*, 174–182. [CrossRef] [PubMed]
112. Bo, B.; Zhang, T.; Jiang, Y.; Cui, H.; Miao, P. Triple Signal Amplification Strategy for Ultrasensitive Determination of miRNA Based on Duplex Specific Nuclease and Bridge DNA–Gold Nanoparticles. *Anal. Chem.* **2018**, *90*, 2395–2400. [CrossRef] [PubMed]
113. Huang, J.; Shangguan, J.; Guo, Q.; Ma, W.; Wang, H.; Jia, R.; Ye, Z.; He, X.; Wang, K. Colorimetric and fluorescent dual-mode detection of microRNA based on duplex-specific nuclease assisted gold nanoparticle amplification. *Anal.* **2019**, *144*, 4917–4924. [CrossRef]
114. Sun, Z.; Li, J.; Yang, Y.; Tong, Y.; Li, H.; Wang, C.; Du, L.; Jiang, Y. Ratiometric Fluorescent Biosensor Based on Self-Assembled Fluorescent Gold Nanoparticles and Duplex-Specific Nuclease-Assisted Signal Amplification for Sensitive Detection of Exosomal miRNA. *Bioconjugate Chem.* **2022**, *33*, 1698–1706. [CrossRef] [PubMed]
115. Ki, J.S.; Lee, H.Y.; Son, H.Y.; Huh, Y.-M.; Haam, S. Sensitive Plasmonic Detection of miR-10b in Biological Samples Using Enzyme-Assisted Target Recycling and Developed LSPR Probe. *ACS Appl. Mater. Interfaces* **2019**, *11*, 18923–18929. [CrossRef]
116. Ma, D.; Huang, C.; Zheng, J.; Tang, J.; Li, J.; Yang, J.; Yang, R. Quantitative detection of exosomal microRNA extracted from human blood based on surface-enhanced Raman scattering. *Biosens. Bioelectron.* **2018**, *101*, 167–173. [CrossRef] [PubMed]
117. Zhang, K.Y.; Yang, L.; Lu, F.; Wu, X.C.; Zhu, J.J. A Universal Upconversion Sensing Platform for the Sensitive Detection of Tumour-Related ncRNA through an Exo III-Assisted Cycling Amplification Strategy. *Small* **2018**, *14*, 1703858. [CrossRef]
118. Wang, W.; Liu, J.; Wu, L.-A.; Ko, C.-N.; Wang, X.; Lin, C.; Liu, J.; Ling, L.; Wang, J. Nicking enzyme-free strand displacement amplification-assisted CRISPR-Cas-based colorimetric detection of prostate-specific antigen in serum samples. *Anal. Chim. Acta* **2022**, *1195*, 339479. [CrossRef]
119. Li, Y.; Wark, A.W.; Lee, H.J.; Corn, R.M. Single-Nucleotide Polymorphism Genotyping by Nanoparticle-Enhanced Surface Plasmon Resonance Imaging Measurements of Surface Ligation Reactions. *Anal. Chem.* **2006**, *78*, 3158–3164. [CrossRef]
120. Gao, J.; Ma, L.; Lei, Z.; Wang, Z. Multiple detection of single nucleotide polymorphism by microarray-based resonance light scattering assay with enlarged gold nanoparticle probes. *Analyst* **2016**, *141*, 1772–1778. [CrossRef]
121. Lyu, N.; Rajendran, V.K.; Li, J.; Engel, A.; Molloy, M.P.; Wang, Y. Highly specific detection of KRAS single nucleotide polymorphism by asymmetric PCR/SERS assay. *Analyst* **2021**, *146*, 5714–5721. [CrossRef] [PubMed]
122. Park, C.; Kang, J.; Baek, I.; You, J.; Jang, K.; Na, S. Highly sensitive and selective detection of single-nucleotide polymorphisms using gold nanoparticle MutS enzymes and a micro cantilever resonator. *Talanta* **2019**, *205*, 120154. [CrossRef] [PubMed]
123. Kalligosfyri, P.; Nikou, S.; Bravou, V.; Kalogianni, D.P. Liquid biopsy genotyping by a simple lateral flow strip assay with visual detection. *Anal. Chim. Acta* **2021**, *1163*, 338470. [CrossRef] [PubMed]
124. Kalligosfyri, P.M.; Nikou, S.; Karteri, S.; Kalofonos, H.P.; Bravou, V.; Kalogianni, D.P. Rapid Multiplex Strip Test for the Detection of Circulating Tumor DNA Mutations for Liquid Biopsy Applications. *Biosensors* **2022**, *12*, 97. [CrossRef]
125. Wang, Y.; Kong, S.L.; Di Su, X. A centrifugation-assisted visual detection of SNP in circulating tumor DNA using gold nanoparticles coupled with isothermal amplification. *RSC Adv.* **2020**, *10*, 1476–1483. [CrossRef]
126. Lee, H.; Kang, T.; Yoon, K.-A.; Lee, S.Y.; Joo, S.-W.; Lee, K. Colorimetric detection of mutations in epidermal growth factor receptor using gold nanoparticle aggregation. *Biosens. Bioelectron.* **2010**, *25*, 1669–1674. [CrossRef]
127. You, J.; Park, C.; Jang, K.; Park, J.; Na, S. Novel Detection Method for Circulating EGFR Tumor DNA Using Gravitationally Condensed Gold Nanoparticles and Catalytic Walker DNA. *Materials* **2022**, *15*, 3301. [CrossRef]

28. New, S.Y.; Aung, K.M.M.; Lim, G.L.; Hong, S.; Tan, S.K.; Lu, Y.; Cheung, E.; Su, X. Fast Screening of Ligand-Protein Interactions based on Ligand-Induced Protein Stabilization of Gold Nanoparticles. *Anal. Chem.* **2014**, *86*, 2361–2370. [CrossRef]
29. Liu, Z.; Liang, Y.; Cao, W.; Gao, W.; Tang, B. Proximity-Induced Hybridization Chain Reaction-Based Photoacoustic Imaging System for Amplified Visualization Protein-Specific Glycosylation in Mice. *Anal. Chem.* **2021**, *93*, 8915–8922. [CrossRef]

Disclaimer/Publisher's Note: The statements, opinions and data contained in all publications are solely those of the individual author(s) and contributor(s) and not of MDPI and/or the editor(s). MDPI and/or the editor(s) disclaim responsibility for any injury to people or property resulting from any ideas, methods, instructions or products referred to in the content.

Review

Application and Research Status of Long-Wavelength Fluorescent Carbon Dots

Yujia Cheng and Guang Yu *

Mechanical and Electrical Engineering Institute, Zhongshan Institute, University of Electronic Science and Technology of China, Zhongshan 528400, China; chengyujia@zsc.edu.cn
* Correspondence: yuguang@zsc.edu.cn; Tel.: +86-0760-8826-9835

Abstract: This article discusses the application and research status of long-wavelength fluorescent carbon dots. Currently, there are two main methods for synthesising carbon dots (CDs), either from top to bottom, according to the bulk material, or from bottom to top, according to the small molecules. In previous research, mainly graphite and carbon fibres were used as raw materials with which to prepare CDs, using methods such as arc discharge, laser corrosion, and electrochemistry. These preparation methods have low quantum efficiencies and afford CDs that are limited to blue short-wavelength light emissions. With advancing research, the raw materials used for CD preparation have expanded from graphite to biomaterials, such as strawberry, lime juice, and silkworm chrysalis, and carbon-based molecules, such as citric acid, urea, and ethylenediamine (EDA). The preparation of CDs using carbon-based materials is more rapid and convenient because it involves the use of microwaves, ultrasonication, and hydrothermal techniques. Research on developing methods through which to prepare CDs has made great progress. The current research in this regard is focused on the synthesis of CDs, including long-wavelength fluorescent CDs, with a broader range of applications.

Keywords: carbon dots; fluorescent nanomaterial; versatile applications

1. Introduction

Carbon dots (CDs) are fluorescent nanomaterials that are smaller than 10 nm. The origin of these types of materials can be traced back to 2004, when Xu et al. [1] purified a single-walled carbon nanotube fragment using electrophoresis and isolated nanoparticles that exhibited fluorescence properties. In 2006, Sun et al. [2] used graphite and black charcoal as raw materials with which to synthesise nanoparticles with improved fluorescence properties. This led to the nomination of CDs as the first official carbon-based fluorescent nanoparticles. Compared with traditional fluorescent materials, such as organic dyestuff and quantum dots, CDs are simple to synthesise, have low consumption, and possess superior luminescence properties and biocompatibility [3,4]. In addition, the surfaces of CDs are enriched in water-soluble groups. CDs readily combine with inorganic molecules, organic molecules, and biomolecules, allowing them to exhibit multifunctionality. CDs are widely used in biological imaging, heavy metal detection, catalysis, and drug loading, making them a current research hotspot [5–8]. The research targets are divided into two aspects: improving the quantum yield (QY) and exploring the versatile applications of CDs [9,10]. Exploring the possible applications of CDs, Tian [11] developed full-colour CDs for use in white-light-emitting diodes (WLEDs). In 2017, Zhong [12] synthesised green CDs for Ag$^+$ detection, using 1,2-phenylenediamine and formaldehyde as raw materials via hydrothermal synthesis. Additionally, CDs are widely used in catalytic reactions and electro-optical devices. For example, Wang synthesised silane-functionalised CDs, G-SiCDs, and R-SiCDs with green and red emissions, using a one-step solvothermal route. Initially, a red light-emitting diode (LED) was prepared by combining blue-fluorescence LED chips with R-SiCDs. Subsequently, the ratios of G-SiCDs and R-SiCDs were adjusted, and a

Citation: Cheng, Y.; Yu, G. Application and Research Status of Long-Wavelength Fluorescent Carbon Dots. *Molecules* **2023**, *28*, 7473. https://doi.org/10.3390/molecules28227473

Academic Editors: Sudeshna Chandra and Heinrich Lang

Received: 16 June 2023
Revised: 15 July 2023
Accepted: 18 July 2023
Published: 8 November 2023

Copyright: © 2023 by the authors. Licensee MDPI, Basel, Switzerland. This article is an open access article distributed under the terms and conditions of the Creative Commons Attribution (CC BY) license (https://creativecommons.org/licenses/by/4.0/).

WLED with a chrominance value of 88 was synthesised based on trichromatic fluorescence. The chrominance of the WLED was higher than that of the double-colour-based LED prepared using G-SiCDs by a value of 30. This led to an improvement in the ability of the WLED to present the true colour of the irradiated object. Ge [13] used polythiophene phenylpropionic acid as a raw material with which to prepare new-type red fluorescent CDs. The range of these CDs is 400–750 nm, and their maximum emission wavelength is 640 nm. The in vivo fluorescence imaging, which was conducted on mouse models, is shown in Figure 1. Under a near-infrared laser, the photoacoustic response was very strong, and photothermal conversion efficiency was high (η = 38.5%), proving that these CDs can be applied to the areas of multichannel fluorescence imaging, photoacoustic imaging, and photothermal therapy. Moreover, it is hopeful that these CDs have the potential to be used for cancer diagnosis and treatment. Zhong [14] used 1,2-paraphenylenediamine and carbofural as precursors. Nitrogen-doped orange florescence CDs (QE = 14.3%) were composed via 180 °C hydrothermal reaction for 3 h, and can be applied to Ag^+ detection in water and to A549 human lung cancer cell imaging.

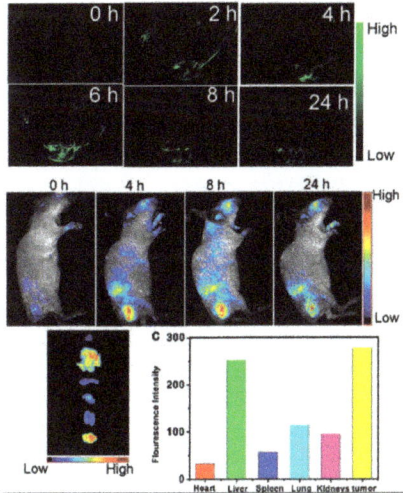

Figure 1. Carbon dots utilised for multimodal imaging in mice [13].

2. Summary of Previously Reported Synthetic Routes to CDs

The incomplete carbonisation of CD surfaces is advantageous because the existing interstitial structure makes the CD surface easy to modify. Research has shown that atom doping can improve the fluorescence properties of CDs [15–18].

2.1. Non-Metallic Element Doping

Early experiments suggested that nitrogen atom doping is the most effective, due to the similarity in radius between nitrogen and carbon atoms and the high electronegativity of nitrogen. Nitrogen atoms can provide electrons to carbon atoms, thus altering the internal structure and improving the luminescent properties of CDs. Qian et al. [19] used 1,2-ethylenediamine (EDA), 1,3-propylenediamine (PDA), 1,4-tetramethylenediamine, and ethylene glycol as dopants, and the different nitrogen contents in these materials had different effects on the luminescent properties of CDs. The QYs of these nitrogen-doped CDs were between 20.4% and 36.3%, far surpassing those of undoped CDs. Reckmeier et al. [20] utilised a hydrothermal method through which to prepare nitrogen-doped carbon quantum dots (CQDs), which exhibited a fluorescence quantum efficiency of nearly 80%.

Furthermore, the homology of carbon atoms and nearby elements, like Si, P, and S, can have an effect on doping. Guo et al. [21] utilised sodium citrate and L-cysteine as precursors in order to synthesise CDs with a quantum efficiency up to 68%. He et al. [22] used triethoxysilane and EDA as raw materials with which to prepare blue CDs with a quantum efficiency of 29.7%. These CDs were successfully utilised for the detection of ferric ions, demonstrating that doping with Si and N elements can enhance the fluorescence properties of CDs. Wang et al. [23] used maltose, hydrochloric acid, and phosphoric acid as raw materials, increasing the quantum efficiency of the tested CDs from 9.3% to 15%. The CDs also showed a linear response to ferric ions, making them a promising material for sensing applications. Two types of nitrogen–boron-doped CDs were prepared by Ye et al. [24] in order to detect 2,4,6-trinitrophenol in cells and Hg^{2+} in water. Yang et al. [25] used phytic and citric acids, which are rich in phosphorus, to prepare P-doped CDs, which were effective in the sensitive detection of copper ions.

2.2. Metallic Element Doping

Doping CDs with metal ions has been reported to significantly impact their properties and potential applications. Near-infrared (NIR) laser induction therapy is a unique, minimally invasive or non-invasive cancer treatment that has attracted considerable attention. Guo et al. [26] prepared excitation-dependent blue-fluorescent Cu-N-doped CDs using ethylenediaminetetraacetic acid (EDTA-2Na) and $CuCl_2$ via one-step hydrothermal synthesis. Cooperative photothermal/photodynamic therapy has been found to substantially inhibit cancer. Moreover, Cu-N-CDs can function as a type of fluorescent probe and infrared thermal developer. In addition, the treatment process using Cu-N-CDs is non-invasive, making it a promising approach for application in biomedical imaging and therapy. Pakkath [27] employed citric acid as the carbon source and rapidly synthesised (within 6 min) CDs doped with ethylenediamine-functionalised transition-metal ions (Mn^{2+}, Fe^{2+}, Co^{2+}, and Ni^{2+}) via the microwave method. These CDs had a quantum efficiency of up to 50.84% and exhibited biocompatibility when used for fluorescence biological imaging of human colon cancer cells (SW480). He et al. [28] developed Gd-doped CDs with excellent magnetic resonance and an excellent cell imaging capability, as well as a high gene transfer efficiency and potent antiserum capacity. Even in an environment containing 10% serum, these CDs exhibited a gene transfer efficiency 74 times higher than that of polyethyleneimine (PEI, 25 kDa). Furthermore, they produced high permeability and carryover effects in solid tumours, resulting in enhanced imaging. Thus, metal ion-doped CDs are promising for medical applications—particularly as contrast media and targeting agents—owing to their unique properties.

CD preparation methods can be divided into two categories according to the mechanism used: top-to-bottom (or 'top-down') and bottom-to-top (or 'bottom-up') (Figure 2a). The quintessence of the top-down method is physical shearing or chemical stripping of carbon-rich fragments, such as graphene and carbon nanotubes (CNTs), into nanoscale CDs (graphene is shown as green, carbon nanotubes are shown in yellow). For instance, in a previous study, graphene was used as the raw material. A two-step shearing process was employed to produce monodispersed CDs from HNO_3-octadecene amine hydrazine, and these CDs were then used for the preparation of white light-emitting diode (WLED) components. Lu used an electrochemical method to strip graphite in an ionic solution and produce blue-fluorescence CDs. The structure and fluorescence properties of these CDs were subsequently characterised. In the bottom-up method, carbon-based small molecules and carbon-rich fragments are used as raw materials. In the hydrothermal, microwave, and ultrasonic methods, polymerisation and carbonisation of the carbon source result in the formation of CDs such as those shown in Figure 2b. Compared with the top-down method, the bottom-up method involves materials that can be obtained from a wider variety of sources. Further, it involves simpler synthesis routes and incurs lower equipment costs; therefore, it is preferred for CD preparation.

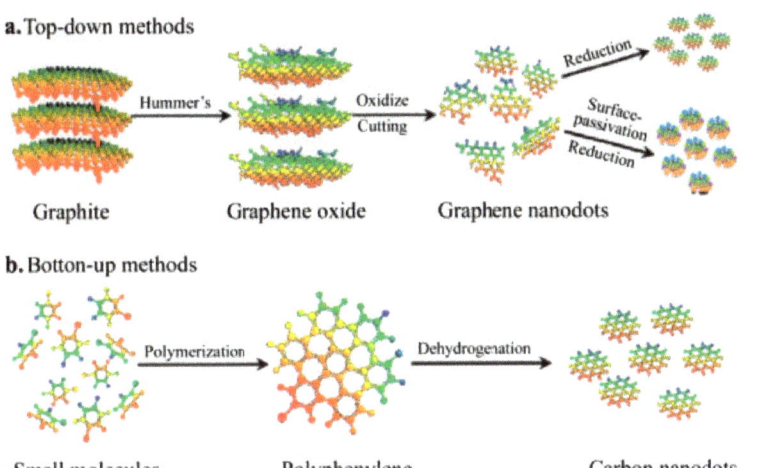

Figure 2. Schematic of the synthesis methods for carbon dots [28]: (**a**) 'top-down' method, (**b**) 'bottom-up' method.

2.3. Arc Discharge Method

In 2004, Xu [29] inadvertently obtained fluorescent substances during the preparation of pure single-walled CNTs by using a 3 M HNO_3 solution as the oxidant to achieve oxidation arc discharge. Under alkaline conditions (NaOH, pH = 8),e HNO_3 was neutralised, yielding a suspension containing single-walled CNTs and other impurities. With subsequent separation, a fluorescent substance was obtained. Gel electrophoresis was used to isolate the purified fluorescent substance, from which electrophoretic bands corresponding to blue-green, yellow, and orange lasers were obtained. It was found that three types of fluorescent substances could be separated via arc discharge. However, this approach tends to produce separated substances that contain significant amounts of impurities that are difficult to separate, leading to low yields.

2.4. Laser Erosion Method

Graphite powder and cement have been used to obtain carbon materials via heat treatment [30]. In the presence of a water vapour current, these carbon materials were transformed into carbon particles via laser erosion. The resulting carbon particles were then compounded to form fluorescent particles of various sizes. Subsequently, the fluorescent particles were placed in a 2.6 M HNO_3 solution for 12 h under backflow conditions, and the resulting uniformly sized carbon particles were decorated with PEG1500 to obtain particles with an average diameter of 5 nm. At an excitation wavelength of 400 nm, the yield of the prepared carbon quantum dots was found to increase from 4% to 10%. Further, Hu [31] used a pulsed laser to illuminate graphite sheets in a polymer solution. The pulse duration was found to affect the resulting particle size and the fluorescence quantum yield (QY). The laser erosion method is advantageous; however, large amounts of the carbon source are required, and the preparation of heterogeneously sized carbon nanoparticles significantly restricts the application of the resulting CDs.

2.5. Microwave Method

The microwave method is simple, rapid, and highly sensitive; therefore, it has been widely employed for the preparation of CDs. For example, Qu [32] used citric acid and urea as raw materials to prepare water-soluble CDs. This simple route involves mixing of the reactants in the microwave digestion system and heating at 750 W for 4–5 min to produce a brown CD solution. The desired CDs are then isolated via centrifugal purification. In addition, Zhang [33] used the microwave method to prepare green CDs for the fluorescence probe detection of folate receptor-positive cancer cells. Additionally, He prepared CDs via the carbonisation of active dry yeast through the microwave approach. The average particle size of the resulting CDs was determined to be 3.4 nm, and their maximum emission peak was observed at ~516 nm. Furthermore, Liu [34] used the microwave method (400 W, 30 min) to prepare orange, yellow, and blue CDs that can be used in colour-based cell imaging and illumination, thereby expanding the application scope of CDs. It was also possible to use these CDs for the detection of cell pH levels. Compared with other methods, the synthesis time required by the microwave method is short, the equipment is simple, and the raw materials are inexpensive and widely available. Consequently, this is one of the most popular approaches for preparing CDs.

2.6. Hydrothermal Method

The hydrothermal method has numerous advantages, including a simple apparatus, facile operation, low cost, and eco-friendliness. Zhou used ammonium citrate as the carbon source, and, through a hydrothermal reaction at 180 °C for 30 min, obtained the desired CDs possessing applicability in the rapid detection of ochratoxin A in flour and beers. Chandra [35] used citric acid and ammonium dihydrogen phosphate as the raw materials for their hydrothermal reaction. After 4 h of reaction at 180 °C and subsequent centrifugation, the purified CDs were obtained via dialysis. The prepared CDs were then used to measure the Fe^{3+} contents of cancer cells. Further, Long used the hydrothermal method to prepare CDs containing F and N. When these CDs were combined with room-temperature phosphors, they were suitable for use as protective materials for data security. Furthermore, the hydrothermal method has been used to prepare organic-soluble CDs and multicolour fluorescence emission-adjustable CDs. Moreover, Yuan [36] used citric acid and either 2,3-diaminonaphthalene (2,3-DAN) or 1,5-DAN as the raw materials under different reaction times and solvent contents to control the carbonisation process. They found that the emission peaks of the resulting fluorescent substances exhibited a red shift. Fluorescent CDs exhibiting blue, green, yellow, orange, and red colours were also prepared. It has been reported that the luminescence mechanism of multicolour fluorescence emission-adjustable CDs is affected by various factors, e.g., changes in the CD bandgap, the charge transfer on the CD surface, and the sizes of the CD molecules. Overall, the CDs prepared using the hydrothermal approach are widely applicable, and the carbon sources tend to be cheap and readily available. However, the CD luminescence mechanism remains to be elucidated. Further, novel methods are needed to increase the CD QYs.

3. Fluorescence Properties of CDs

3.1. Synthesis of High-Efficiency N,S-Doped Blue CDs

At present, the quantum efficiency of CDs is low. Thus, increasing the quantum efficiency is the main priority in the CD research field. High-efficiency CDs are prerequisites for composite applications. For the experiment performed in this study, the convenient hydrothermal method was selected. Citrate sodium and cysteine were used as raw materials. Ethylenediamine was used as a passivation agent. High-efficiency N,S-doped blue CDs were synthesised. Quinine sulphate (0.1 M H_2SO_4, QY = 56%) was used as a standard reference solution. The quantum efficiency of the CDs was calculated as 68%. The fluorescence of the CDs was examined, along with the effects of the NaCl solution, pH, ultraviolet (UV) light, temperature, and metal ions on the CD fluorescence intensity. All the reagents used in the experiment are presented in Table 1.

The experiment instruments are shown in Table 2.

3.2. Synthesis of N,S-Doped Blue CDs

The blue CDs were synthesised using a one-step hydrothermal method via the following steps. First, 0.2 g of L-cysteine and 0.2 g of citrate sodium were weighed and added to a 25-mL beaker. Subsequently, 10 mL of ultrapure water and 500 μL of ethylenediamine (EDA) were added via pipette for 5 min until complete dissolution. This transparent liquid was shifted quickly to a 50-mL hydrothermal reactor. Then, the reactor was kept in an oven at 200 °C for 4 h. After the liquid cooled to room temperature, dialysis was performed for 24 h by using a dialysis bag (Mw = 1000). Thus, purified N,S-doped CDs with an average size of 5 nm were obtained.

Table 1. Chemical reagents used in the experiments.

Reagent	Purity	Manufacturer
Citrate sodium	Analytical pure	Aladdin Reagent Co., Ltd., Shanghai, China
L-Cysteine	Analytical pure	GHTECH Co., Ltd., Guangzhou, China
EDA	Analytical pure	Aladdin Reagent Co., Ltd., Shanghai, China
Citric acid	Analytical pure	DaMao chemical reagent factory, Tianjin, China
Citrate sodium	Analytical pure	Aladdin Reagent Co., Ltd., Shanghai, China

Table 2. Experimental instruments.

Instruments	Model	Manufacturer
Steady/transient state X-ray fluorescence (XRF) spectrometer	FLS980	Edinburgh Instruments company, Edinburgh, Britain
Fourier-transform infrared (FT-IR) spectroscope	Spectrum Two FT-IR	Thermo Fisher Scientific technology company, Waltham, MA, USA
UV-visible-near infrared light spectrophotometer	UV-5500PC	Shimadzu corporation, Kyoto, Japan
Transmission electron microscope	Tecnai G2 F20	Oxford instrument technology Co., Ltd., Shanghai, China
Constant magnetic stirring	85-2	Thermo Fisher Scientific technology company, Waltham, MA, USA
Table-top high-speed centrifuge	TG16-WS	Xiangyi laboratory Instrument Development Co., Ltd., Xiangtan, China
Collector type constant-temperature heating magnetic stirrer	DF-101S	Yuhua instrument Co., Ltd., Gongyi, China
Electronic analytical balance	AX124 ZH/E	OHAUS instrument Co., Ltd., Newark, NI, USA
Camera obscura UV analyser	ZF-20D	Yuhua instrument Co., Ltd., Gongyi, China
Electric blast drying oven	DHG-9145A	Yiheng scientific instrument Co., Ltd., Shanghai, China

3.3. Fluorescence Spectra

The excitation and emission spectra, temperature-dependence emission spectra, and fluorescence lifetimes of the samples were measured using X-ray fluorescence (FLS980). The illuminant in the emission and photostability tests was an Xe lamp, and that in the fluorescence lifetime test was a 320 nm laser. Next, 3 mL of the CD solution was added to a cuvette. The slit was set as 1.5 nm. The residence time was 0.05 s. The excitation wavelength was 350 nm. The emission spectra were acquired within the range of 360–750 nm. Additionally, the CD influence factors (pH, metal ions, and NaCl) were tested via the following steps. First, 3 mL of buffer solution with different pH values was prepared. Next, 100 μL of CD solution was added, and the mixture was shaken. The fluorescence intensity test (λ_{ex} = 350 nm) was then conducted. To examine the effect of the concentration of the NaCl solution on the CDs, NaCl solutions with concentrations of 0.2, 0.4, 0.6, 0.8, 1.0, and 1.2 M were used; 100 μL of the CD solution was added, followed by mixing for 3 min. Finally, the emission spectra were acquired.

The photoluminescence excitation (PLE) spectrum and emission spectrum (PL) are important for measuring the fluorescence properties of CDs. In this study, a steady/transient-state fluorescence analyser (FLS980) was used to analyse the PLE and PL of the samples (Figure 3a). Under natural light, the CD solution was faint yellow and transparent; however, under UV irradiation, the light emitted by the CDs was blue. The optimum excitation wavelength (λ_{ex}) and strongest emission peak of the CDs were 350 and 450 nm, respectively. Calculations using quinine sulphate (QY = 56%) as a standard reference solution indicated that the QY of the CDs increased to 68%; this QY exceeded those of other CDs, such as S/N-doped CDs (13.71%), full-colour electroluminescent CDs (35%), and BNS CDs (5.44%). The emission behaviour of the samples is also important, as differences in the excitation wavelength can result in a redshift or blueshift of the emission spectrum. The emission behaviour can be divided into excitation wavelength-dependent behaviour and non-excitation wavelength-dependent behaviour. An excitation wavelength ranging from 300 to 360 nm was used to assess the emission behaviour of the CDs. The emission spectrum was analysed at 10 nm intervals within the testing range of the excitation wavelength. However, only one emission peak at 450 nm was observed for each spectrum (Figure 3b). Thus, the emission spectrum did not exhibit a redshift or blue shift across different excitation wavelengths, indicating that the emission behaviour of the CDs did not depend on the excitation wavelength. This is because the size of the aromatic ring on the surface of the CDs was relatively uniform, and the emission locus was close.

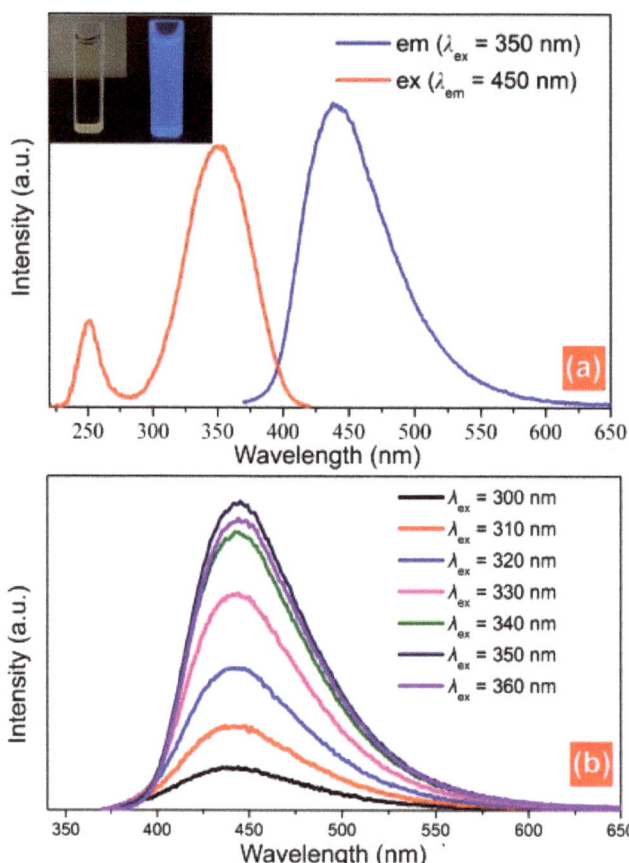

Figure 3. (a) Excitation and emission spectra of CDs and (b) effect of different excitation wavelengths on the fluorescence intensity of CDs.

The fluorescence lifetime test was performed to analyse the fluorescence decay characteristics of the CDs. The double-exponential decay dynamics function ($\chi^2 = 1.093$) was successfully fitted. At room temperature, the fluorescence lifetimes were 2.48 ns (3.48%) and 14.87 ns (96.52%) (Figure 4a), and the average lifetime of the CDs was 14.44 ns; this short lifespan was consistent with the blue fluorescence decay characteristics. Chen previously reported an average lifetime (12 ns) that was similar to the results of this study. The fluorescence lifetime was also affected by the temperature; an increase in the temperature from 283 to 343 K reduced the fluorescence lifetime from 15.03 to 11.75 ns (Figure 4b). This occurred because of nonradiative relaxation. The blue arrow meas the Non radiative relaxation process.

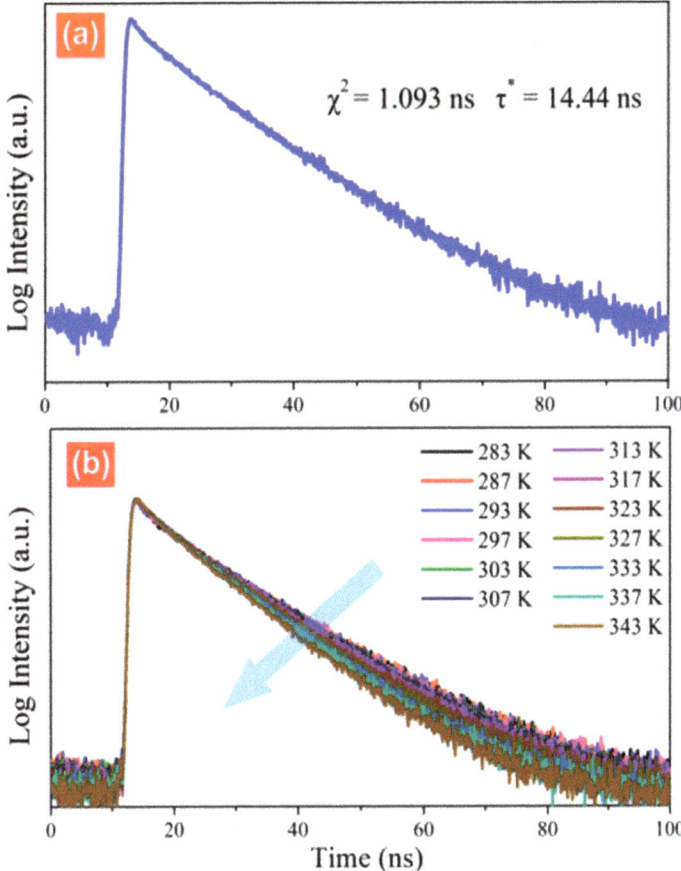

Figure 4. Fitted curve for the fluorescence decay of CDs. (**a**) Everage life of CD (**b**) Temperature-life dependence graph of CD.

3.4. Analysis of CD Morphology

The CD size and morphology were examined via TEM, as shown in Figure 5a. The CDs synthesised via the hydrothermal method were spherical and well-dispersed. The average particle size was 3.6 nm. As shown in Figure 5b, these findings were supported by atomic force microscopy (AFM) test results (The green arrow point to the CDs). To measure the electric charges in the CDs, a potential test was performed using a Zetasizer Nano potentiostat. After parallel testing was performed three times, the charge in the CDs was −38 mV. The groups on the CD surface were ionised easily, yielding anions; thus, the negative electricity was strong. These CDs can be used for the detection of positively charged species. Additionally, they can combine with positively charged species, forming composites.

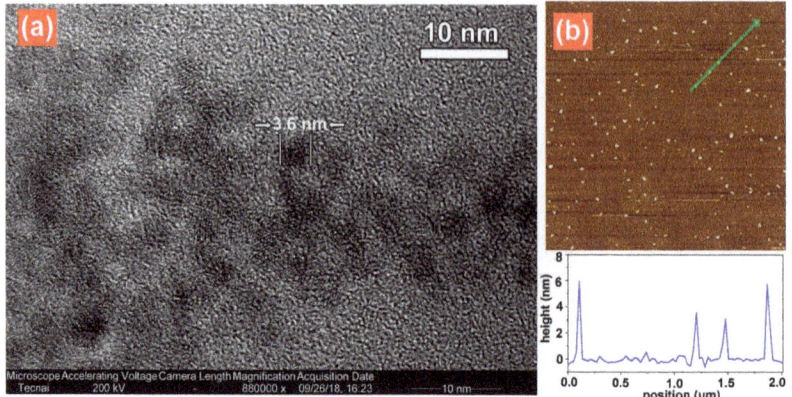

Figure 5. TEM and AFM images of CDs. (**a**) TEM image of CDs (**b**) AFM image of CDs.

4. Carbon-Dot Applications

Given their exceptional fluorescence and biocompatibility, CDs are commonly used in various applications, including drug testing, cell imaging, chemical sensing, and information forgery-proofing. In this section, we explore the use of CDs in four key areas—electro-optical devices, detection of drug ions, biological imaging, and information forgery-proofing.

4.1. Electro-Optical Devices

Typically, CDs are prepared in aqueous solutions. Even when prepared using the microwave method, CDs only exhibit fluorescence when dissolved in water. However, CDs must be in the solid state for application to electro-optical devices, and severe aggregation-induced quenching remains a significant challenge in this regard. Tian [37] synthesised controllable-luminescence CDs that emitted light ranging from blue to red. To overcome the issue of aggregation-induced quenching, a microwave-assisted rapid heating method was used to solidify the CDs in a sodium silicate aqueous solution. The solid-state CDs maintained their fluorescence, which allowed the preparation of multicolour light-emitting diode (LED) and WLED lighting devices based on CDs. Zhai [38] synthesised red-, green-, and yellow-fluorescent CDs, which were combined with blue chips to construct WLEDs with adjustable colour temperatures. Wang [39] synthesised red CDs with a quantum efficiency of 53%, which were combined with blue- and green-fluorescent powders to prepare warm white LEDs. These LEDs exhibited excellent overall performance, including a correlation colour temperature of 2875 K, colour rendering index of 97, and luminous efficacy of 31.3 lm/W, representing a significant contribution to the development and application of high-efficiency composite LEDs based on CDs.

4.2. Drug and Ion Detection

CDs exhibit excellent fluorescent properties; however, their fluorescence is quenched when they are combined with drugs or ions. This property makes CDs useful for the quantitative detection of drugs and ions.

Tetracycline is a broad-spectrum antibiotic that inhibits the growth of multiple Gram-positive and Gram-negative bacteria. It is also effective against rickettsia and trachoma virus. Guo [40] developed a method to produce blue-fluorescent CDs by thermally cracking crab waste material. These CDs have been successfully utilised to detect tetracycline with good detection stability and low detection limits. They can be applied to quantitative detection of tetracycline in most acidic solutions. Furthermore, this method has been shown to be useful for sewage water analysis. Ehtesabi [41] synthesised blue-fluorescent CDs via pyrolysis and characterised them. Upon exposure to tetracycline, the fluorescence emission intensity of the CDs was significantly reduced. To develop a user-friendly tetracycline detection method, the synthesised CDs were encapsulated in a chondrocyte-sodium hydrogel. Figure 6a shows the fluorescence quenching intensity of CDs encapsulated in the hydrogel structure measured in the presence of tetracycline. The encapsulation of CDs in hydrogels enables their use as a tetracycline sensor and as an adsorbent for environmental tetracycline pollutants. The new method is faster and more efficient than traditional tetracycline detection methods. Folic acid is a pteridine derivative that was initially separated from heparin and is abundant in green plants. Tu [42] used citric acid, EDTA, and mesoporous silica (MCM-41) to prepare blue-fluorescent CDs. Their research indicated that folic acid can selectively quench CD fluorescence. Zhang [43] used these CDs to analyse folic acid in human urine samples, and the recovery rates for the samples were between 82.0% and 113.1%. The results indicated the effectiveness of this method for folic-acid analysis.

Heavy-metal ions can cause severe environmental pollution, as they cannot be easily dispersed and tend to bioaccumulate in the food chain and ultimately in the human body, resulting in metal poisoning. Therefore, detecting their presence in the environment and food is crucial. Qin [44] used microwave irradiation to prepare CDs that served as effective fluorescence sensing probes that enable sensitive and selective detection of Hg^{2+} with a detection limit as low as 0.5 nM. Through a hydrothermal method with coconut juice as the raw material, Murugesan [45] prepared fluorescent CDs that could detect Ag ions in aqueous solutions with a detection limit of 0.26 nM. These CDs can be used as a material for fluorescent probes for environmental protection against metal ions. Figure 6 shows the fluorescence of CDs prepared by Chaudhary [46]; these CDs could measure the concentration of Cu ions with a detection range of 1–800 µg/mL and a detection limit of 0.3 µg/mL.

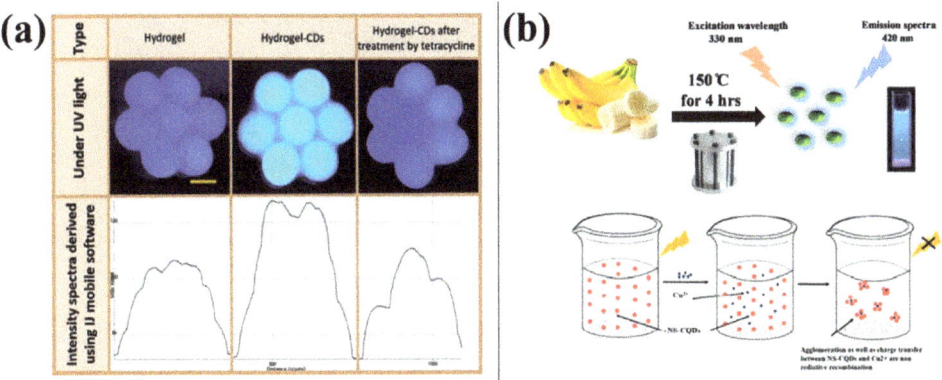

Figure 6. Carbon dots (CDs) for drug and ion detection: (**a**) CD composite hydrogels for tetracycline detection and absorbance [41], (**b**) CDs for copper ion detection [46].

4.3. Biological Imaging

Compared with traditional fluorescent materials, CDs exhibit low toxicity and high biocompatibility, thereby reducing the risk of side effects when they enter the body [47,48]. Additionally, most CDs are water-soluble, which simplifies sample preparation by avoiding the need for complex pre-treatment procedures. Moreover, the fluorescence signal of CDs is stable and is less likely to be affected by the complex biological environment. Owing to these significant advantages, CDs are increasingly being used in biological imaging.

CDs with efficient excitation and ionisation emissions in the deep-red/NIR spectral range are important for bioimaging applications. Liu developed a simple and effective method to significantly enhance the absorption and emission of CDs in the deep-red/NIR spectral range by suppressing nonradiative charge recombination via deprotonation of the CD surface. Owing to an enhanced deprotonation ability and increased viscosity, the NIR emissions of CDs in N,N-dimethylformamide and glycerol at −20 °C exhibited

50- and 70-fold increases, respectively, compared with those of CDs in aqueous solutions at room temperature. Given the adjustable NIR fluorescence intensity of CDs, multilevel data encryption in the NIR region was achieved by controlling the humidity and temperature of CD-ink-stamped paper.

Gong et al. [49] used amylacea, EDA, and strong phosphoric acid (SPA) to prepare a cavity CD with P and N double doping. The CD prepared by amylacea, EDA, and strong phosphoric acid are shown in Figure 7a(i), (ii) and (iii) respectively. As shown in Figure 7a, these CDs can serve as nanocarriers for anticancer drugs such as doxorubicin, inhibiting tumour growth while allowing fluorescence imaging. Chen [50] synthesised a series of hydrophobically modified CDs with low cytotoxicity via a ring-opening reaction. These CDs exhibited reduced cytotoxicity and improved serum survivability. Moreover, they enable dual-channel imaging and can be used to track the transfer of DNA in cells. Cell-uptake experiments confirmed that these CDs have excellent serum survivability and a strong structure–activity relationship. Yang [51] prepared F-doped CDs using a simple and eco-friendly one-step microwave-assisted carbonation method. As shown in Figure 7b, these CDs emitted red fluorescence, and the researchers suggested a possible mechanism for the emission redshift caused by F doping. These CDs can be used as optical nanoprobes for biological imaging within the human body. The results indicate that F-doped CDs have potential for tumour biological imaging and diagnosis. In summary, research on CDs for biological imaging is constantly advancing, and CDs are expected to be increasingly used as internal fluorescence probes.

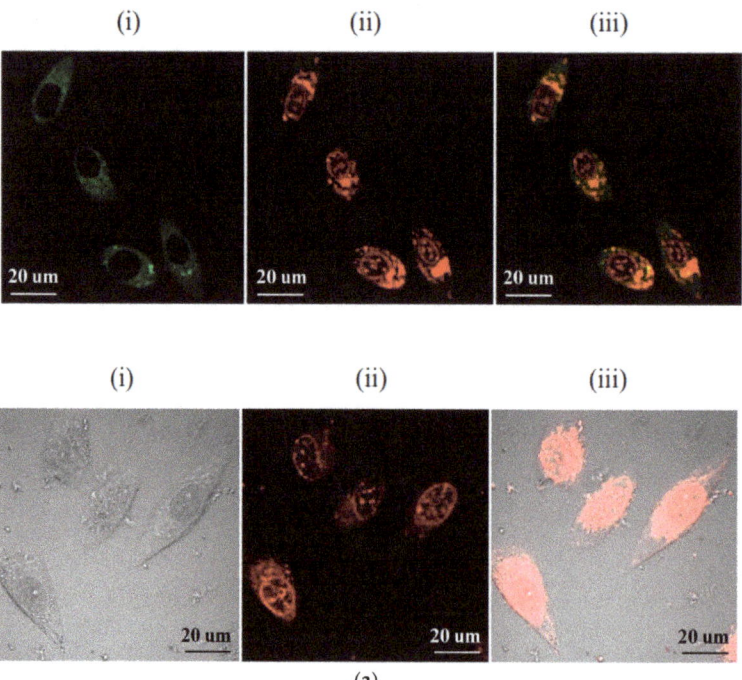

(a)

Figure 7. Cont.

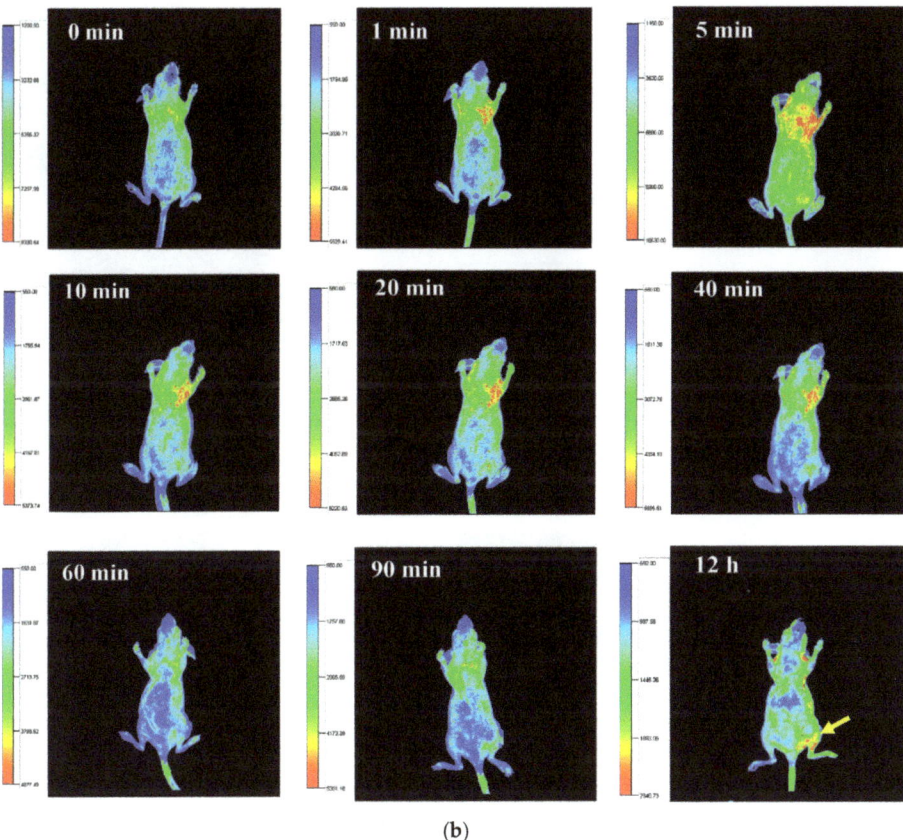

Figure 7. Carbon dots (CDs) for biological imaging: (**a**) CDs for cell imaging [49], (**b**) CDs for in vivo imaging of mice [51].

4.4. Information Anti-Counterfeiting and Encryption

Fluorescent anti-counterfeiting refers to security patterns or characters printed using fluorescent ink that are invisible under normal lighting conditions. However, when exposed to UV radiation, these patterns become visible. The recoverable fluorescence characteristics of CDs can be used for information anti-counterfeiting and encryption.

Guo [52] prepared bright green-yellow-emitting CDs via a hydrothermal method using 2-hydroxyphenyl boric acid and EDTA as a solvent. These CDs can be used to measure the concentration of $Cr_2O_7^{2-}$ in water. Furthermore, they can be used as fluorescent ink, emitting bright green light under UV radiation. Yuan [53] synthesised red-fluorescent CDs that were sensitive to acid and attached them to a polyvinyl alcohol (PVA) membrane. As shown in Figure 8a, the CDs without hydrochloric acid treatment exhibited red fluorescence under UV irradiation. After treatment with hydrochloric acid, the fluorescence was blue, facilitating a dual-mode information anti-counterfeiting application. Zhao [54] synthesised three types of CDs with red, green, and blue fluorescence. The blue-fluorescent CDs were used to prepare a fluorescent ink, as shown in Figure 8b. When this ink was applied to paper, the blue fluorescence was visible under UV radiation. However, it disappeared when the ink was sprayed with a Cu^{2+} solution, providing encryption. The fluorescence can be recovered upon spraying with a cysteine solution, providing decryption. This method based on fluorescent CDs is a promising new approach to anti-counterfeiting and encryption.

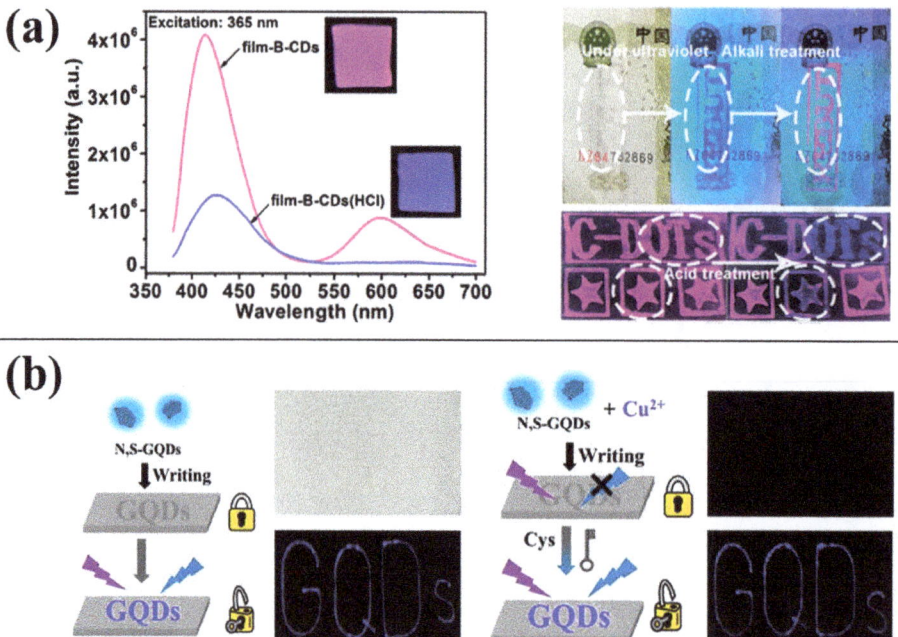

Figure 8. Carbon dots (CDs) used for information security: (**a**) double-mode information security effect of CDs [53], (**b**) encryption and decryption function of CDs [54].

4.5. Introduction to Danio rerio Model

The zebrafish (*Danio rerio*) is a model organism with several advantages over tradition animal models, for example, small size, highly transparent bodies, high reproduction rates, and high hatchability. The sequencing of the *Danio rerio* genome has revealed considerable homology with the genomes of mammals. Of the 26000 genes of the zebrafish, 70% are similar to those of humans, including genes encoding cytokines and histocompatibility system molecules, which are immune-reaction regulatory factors. Additionally, 84% of known human disease-causing genes correspond to those of the zebrafish [55,56]. Owing to these similarities, the related mechanisms of zebrafish and the pathogenic hosts are similar to those of humans [57,58]. The *Danio rerio* embryo is considered a type of in vitro animal model; accordingly, simple cell or tissue culture experiments can be used to validate the results of animal (e.g., rodent) experiments.

Because of their transparency and short reproductive cycle, zebrafish embryos are particularly well-suited for fluorescence imaging. Wei [59] synthesised a type of blue CDs, and the results showed that different concentrations of CDs had low toxicity to zebrafish embryo development. The nervous and circulatory systems of zebrafish embryos developed normally in the presence of CDs. Figure 9a,b show that the CDs possess great florescent stability and biocompatibility, making them suitable for biological imaging in zebrafish. The CDs can enter the zebrafish embryo through the chorion or oral cavity. Wang [60] prepared red CDs as an unmarked nanoprobe, and the addition of formaldehyde increased the fluorescence of the CDs in zebrafish, allowing for selective testing of formaldehyde in living cells and zebrafish, which is shown in Figure 9c (The different addition of formaldehyde in A–K is 10, 20, 30, 40, 50, 60, 70, 80, 90, 100, 110 mol/L).

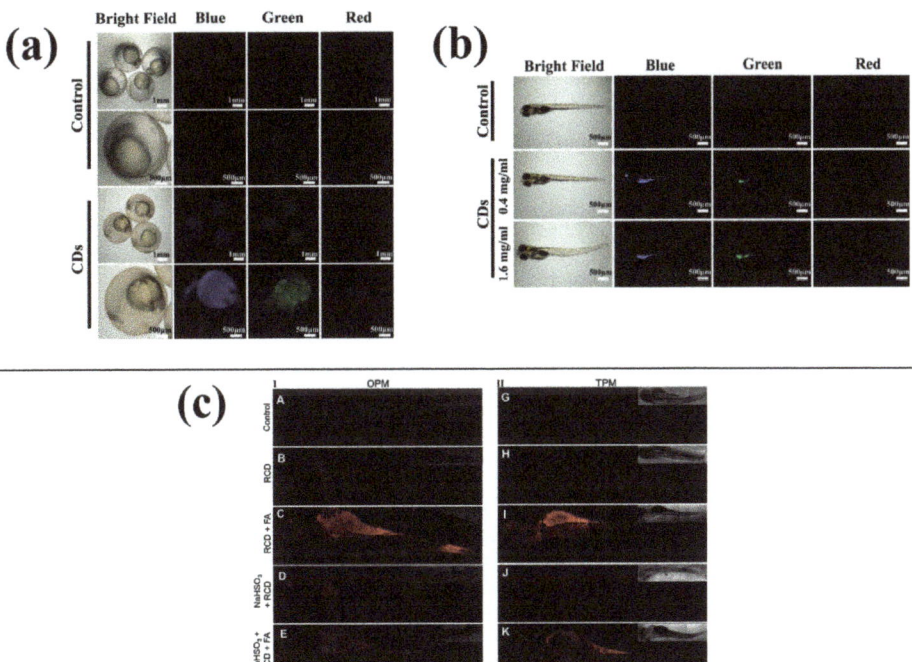

Figure 9. Imaging of zebrafish: (**a**) CDs for imaging zebrafish embryos [59], (**b**) CDs for imaging adult zebrafish [59], (**c**) CDs for imaging formaldehyde in zebrafish [60].

5. Conclusions

To summarise, CDs are a type of florescent nanomaterial that has garnered a lot of interest due to their excellent fluorescence properties and biocompatibility. Despite the fact that the structure, luminescence mechanism and quenching mechanism of CDs are still not completely understood, research on CDs continues. The research on CD composite methods is relatively mature, and breakthroughs have been made in their application in drug and ion detection, electro-optical devices and biological imaging. The synthesis and application of long-wavelength fluorescent CDs are also widely researched. Additionally, CDs are a new type of florescent nanomaterial that possess many advantages, such as low toxicity, good biocompatibility, unique optical properties and non-bleaching fluorescence, which overcome some of the disadvantages of traditional florescent materials. With the development of society, the demand for functionalised composites is increasing, and CDs are playing an important role. Synthesis methods for CDs have increased along with the development of CDs. The physicochemical properties of CDs, such as particle size, fluorescence intensity, stability, and water-solubility, can be significantly affected by the synthesis method used, as well as reaction times, temperatures, and pH values. Therefore, controlling these parameters during the synthesis process is crucial for obtaining CDs with desired properties for specific applications.

Author Contributions: Conceptualization, Y.C. and G.Y.; methodology, G.Y.; software, G.Y.; validation, G.Y.; formal analysis, G.Y.; investigation, Y.C.; resources, G.Y.; data curation, G.Y.; writing—original draft preparation, G.Y.; writing—review and editing, G.Y.; visualization, Y.C.; supervision, Y.C.; project administration, Y.C.; funding acquisition, G.Y. All authors have read and agreed to the published version of the manuscript.

Funding: This research was aided by Colleges Feature Innovation Project of Guangdong Province, China, Grant Number 2022KTSCX194; the Science and Technology Foundation of Guangdong Province under Grant 2021A0101180005.

Conflicts of Interest: The authors declare no conflict of interest.

References

1. Yin, P.; Niu, Q.; Yang, Q.; Lan, L.; Li, T. A new "naked-eye" colorimetric and ratiometric fluorescent sensor for imaging Hg^{2+} in living cells. *Tetrahedron* **2019**, *75*, 130687. [CrossRef]
2. Gao, G.; Busko, D.; Kauffmann-Weiss, S.; Turshatov, A.; Howard, I.A.; Richards, B.S. Wide-range non-contact fluorescence intensity ratio thermometer based on Yb^{3+}/Nd^{3+} co-doped La_2O_3 microcrystals operating from 290 to 1230 K. *J. Mater. Chem. C* **2018**, *6*, 4163–4170. [CrossRef]
3. Chang, C.-Y.; Venkatesan, S.; Herman, A.; Wang, C.-L.; Teng, H.; Lee, Y.-L. Carbon quantum dots with high quantum yield prepared by heterogeneous nucleation processes. *J. Alloys Compd.* **2023**, *938*, 168654. [CrossRef]
4. Dang, Y.; Tian, J.; Wang, W.; Ma, B. Insight into the whole characteristics of (Pd/WP)/CdS for photocatalytic hydrogen evolution. *J. Colloid Interface Sci.* **2023**, *633*, 649–656. [CrossRef]
5. Yu, H.; Liang, H.; Bai, J.; Li, C. Sulfur vacancy and CdS phase transition synergistically boosting one-dimensional $CdS/Cu_2S/SiO_2$ hollow tube for photocatalytic hydrogen evolution. *Int. J. Hydrogen Energy* **2023**, *48*, 15908–15920. [CrossRef]
6. Liu, C.; Zhang, Y.; Shi, T.; Liang, Q.; Chen, Z. Hierarchical hollow-microsphere cadmium sulfide-carbon dots composites with enhancing charge transfer efficiency for hotocatalytic CO_2 reduction. *J. Alloys Compd.* **2023**, *936*, 168286. [CrossRef]
7. Zhong, J.; Li, Y.; Zhang, H.; Zhang, Z.; Qi, K.; Zhang, H.; Gao, C.; Li, Y.; Wang, L.; Sun, Z.; et al. Highly efficient charge transfer from small-sized cadmium sulfide nanosheets to large-scale nitrogen-doped carbon for visible-light dominated hydrogen evolution. *J. Colloid Interface Sci.* **2023**, *630*, 260–268. [CrossRef] [PubMed]
8. Yang, Y.; Wang, X.; Xia, Y.; Dong, M.; Zhou, Z.; Zhang, G.; Li, L.; Hu, Q.; Zhu, X.; Yi, J. The role of facet engineered surface and interface in CdS nanostructures toward solar driven hydrogen evolution. *Appl. Surf. Sci.* **2023**, *615*, 156402. [CrossRef]
9. Liu, B.; Guo, C.; Ke, C.; Chen, K.; Dang, Z. Colloidal stability and aggregation behavior of CdS colloids in aquatic systems: Effects of macromolecules, cations, and pH. *Sci. Total Environ.* **2023**, *869*, 161814. [CrossRef]
10. Song, X.; Zhao, S.; Xu, Y.; Chen, X.; Wang, S.; Zhao, P.; Pu, Y.; Ragauskas, A.J. Preparation, Properties, and Application of Lignocellulosic-Based Fluorescent Carbon Dots. *ChemSusChem* **2022**, *15*, e202102486. [CrossRef]
11. Liu, X.; Liu, J.; Zhou, H.; Yan, M.; Liu, C.; Guo, X.; Xie, J.; Li, S.; Yang, G. Ratiometric dual fluorescence tridurylboron thermometers with tunable measurement ranges and colors. *Talanta* **2020**, *210*, 120630. [CrossRef] [PubMed]
12. Zhong, Y.; Li, J.; Jiao, Y.; Zuo, G.; Pan, X.; Su, T.; Dong, W. One-step synthesis of orange luminescent carbon dots for Ag^+ sensing and cell imaging. *J. Lumin.* **2017**, *190*, 188–193. [CrossRef]
13. Sun, C.; Zhang, Y.; Sun, K.; Reckmeier, C.; Zhang, T.; Zhang, X.; Zhao, J.; Wu, C.; Yu, W.W.; Rogach, A.L. Combination of carbon dot and polymer dot phosphors for white light-emitting diodes. *Nanoscale* **2015**, *7*, 12045–12050. [CrossRef] [PubMed]
14. Macairan, J.R.; Jaunky, D.B.; Piekny, A.; Naccache, R. Intracellular ratiometric temperature sensing using fluorescent carbon dots. *Nanoscale Adv.* **2019**, *1*, 105–113. [CrossRef]
15. Fang, W.-K.; Zhou, S.-H.; Liu, D.; Liu, L.; Zhang, L.-L.; Xu, D.-D.; Li, Y.-Y.; Liu, M.-H.; Tang, H.-W. Tunable emissive carbon polymer dots with solvatochromic behaviors for trace water detection and cell imaging. *New J. Chem.* **2023**, *47*, 1985–1992. [CrossRef]
16. Ghosh, T.; Nandi, S.; Bhattacharyya, S.K.; Ghosh, S.K.; Mandal, M.; Banerji, P.; Das, N.C. Nitrogen and sulphur doped carbon dot: An excellent biocompatible candidate for in-vitro cancer cell imaging and beyond. *Environ. Res.* **2023**, *217*, 114922. [CrossRef]
17. de Boëver, R.; Town, J.R.; Li, X.; Claverie, J.P. Carbon Dots for Carbon Dummies: The Quantum and The Molecular Questions Among Some Others. *Chem. Eur. J.* **2022**, *28*, e202200748. [CrossRef]
18. Li, Y.; Yang, H.-P.; Chen, S.; Wu, X.-J.; Long, Y.-F. Simple Preparation of Carbon Dots and Application in Cephalosporin Detection. *J. Nanosci. Nanotechnol.* **2021**, *21*, 6024–6034. [CrossRef]
19. Qian, Z.; Ma, J.; Shan, X.; Feng, H.; Shao, L.; Chen, J. Highly Luminescent N-Doped Carbon Quantum Dots as an Effective Multifunctional Fluorescence Sensing Platform. *Chem. Eur. J.* **2014**, *20*, 2254–2263. [CrossRef]
20. Reckmeier, C.J.; Wang, Y.; Zboril, R.; Rogach, A.L. Influence of Doping and Temperature on Solvatochromic Shifts in Optical Spectra of Carbon Dots. *J. Phys. Chem. C* **2016**, *120*, 10591–10604. [CrossRef]
21. Guo, Z.; Luo, J.; Zhu, Z.; Sun, Z.; Zhang, X.; Wu, Z.-C.; Mo, F.; Guan, A. A facile synthesis of high-efficient N,S co-doped carbon dots for temperature sensing application. *Dye. Pigment.* **2020**, *173*, 107952. [CrossRef]
22. He, S.; Qi, S.; Sun, Z.; Zhu, G.; Zhang, K.; Chen, W. Si, N-codoped carbon dots: Preparation and application in iron overload diagnosis. *J. Mater. Sci.* **2018**, *54*, 4297–4305. [CrossRef]
23. Wang, W.; Peng, J.; Li, F.; Su, B.; Chen, X.; Chen, X. Phosphorus and chlorine co-doped carbon dots with strong photoluminescence as a fluorescent probe for ferric ions. *Microchim. Acta* **2018**, *186*, 32. [CrossRef] [PubMed]
24. Ye, Q.; Yan, F.; Shi, D.; Zheng, T.; Wang, Y.; Zhou, X.; Chen, L. N, B-doped carbon dots as a sensitive fluorescence probe for Hg^{2+} ions and 2,4,6-trinitrophenol detection for bioimaging. *J. Photochem. Photobiol. B Biol.* **2016**, *162*, 1–13. [CrossRef] [PubMed]

25. Yang, F.; He, X.; Wang, C.; Cao, Y.; Li, Y.; Yan, L.; Liu, M.; Lv, M.; Yang, Y.; Zhao, X.; et al. Controllable and eco-friendly synthesis of P-riched carbon quantum dots and its application for copper (II) ion sensing. *Appl. Surf. Sci.* **2018**, *448*, 589–598. [CrossRef]
26. Guo, X.-L.; Ding, Z.-Y.; Deng, S.-M.; Wen, C.-C.; Shen, X.-C.; Jiang, B.-P.; Liang, H. A novel strategy of transition-metal doping to engineer absorption of carbon dots for near-infrared photothermal/photodynamic therapies. *Carbon* **2018**, *134*, 519–530. [CrossRef]
27. Pakkath, S.A.R.; Chetty, S.S.; Selvarasu, P.; Murugan, A.V.; Kumar, Y.; Periyasamy, L.; Santhakumar, M.; Sadras, S.R.; Santhakumar, K. Transition Metal Ion ($Mn^{(2+)}$, $Fe^{(2+)}$, $Co^{(2+)}$, and $Ni^{(2+)}$)-Doped Carbon Dots Synthesized via Microwave-Assisted Pyrolysis: A Potential Nanoprobe for Magneto-fluorescent Dual-Modality Bioimaging. *ACS Biomater. Sci. Eng.* **2018**, *4*, 2582–2596. [CrossRef] [PubMed]
28. He, X.; Luo, Q.; Zhang, J.; Chen, P.; Wang, H.-J.; Luo, K.; Yu, X.-Q. Gadolinium-doped carbon dots as nano-theranostic agents for MR/FL diagnosis and gene delivery. *Nanoscale* **2019**, *11*, 12973–12982. [CrossRef]
29. Xu, X.; Ray, R.; Gu, Y.; Ploehn, H.J.; Gearheart, L.; Raker, K.; Scrivens, W.A. Electrophoretic Analysis and Purification of Fluorescent Single-Walled Carbon Nanotube Fragments. *J. Am. Chem. Soc.* **2004**, *126*, 12736–12737. [CrossRef]
30. Sun, Y.P.; Zhou, B.; Lin, Y.; Wang, W.; Fernando, K.S.; Pathak, P.; Meziani, M.J.; Harruff, B.A.; Wang, X.; Wang, H.; et al. Quantum-sized carbon dots for bright and colorful photoluminescence. *J. Am. Chem. Soc.* **2006**, *128*, 7756–7757. [CrossRef]
31. Hu, B. Laser synthesis and size tailor of carbon quantum dots. *J. Nanoparticle Res.* **2011**, *13*, 7247–7252. [CrossRef]
32. Qu, S.; Wang, X.; Lu, Q.; Liu, X.; Wang, L. A Biocompatible Fluorescent Ink Based on Water-Soluble Luminescent Carbon Nanodots. *Angew. Chem. Int. Ed.* **2012**, *51*, 12215–12218. [CrossRef] [PubMed]
33. Zhang, J.; Zhao, X.; Xian, M.; Dong, C.; Shuang, S. Folic acid-conjugated green luminescent carbon dots as a nanoprobe for identifying folate receptor-positive cancer cells. *Talanta* **2018**, *183*, 39–47. [CrossRef]
34. Liu, C.; Wang, R.; Wang, B.; Deng, Z.; Jin, Y.; Kang, Y.; Chen, J. Orange, yellow and blue luminescent carbon dots controlled by surface state for multicolor cellular imaging, light emission and illumination. *Microchim. Acta* **2018**, *185*, 539. [CrossRef] [PubMed]
35. Chandra, S.; Laha, D.; Pramanik, A.; Chowdhuri, A.R.; Karmakar, P.; Sahu, S.K. Synthesis of highly fluorescent nitrogen and phosphorus doped carbon dots for the detection of Fe^{3+} ions in cancer cells. *Luminescence* **2016**, *31*, 81–87. [CrossRef]
36. Yuan, F.; Wang, Z.; Li, X.; Li, Y.; Tan, Z.A.; Fan, L.; Yang, S. Bright Multicolor Bandgap Fluorescent Carbon Quantum Dots for Electroluminescent Light-Emitting Diodes. *Adv. Mater.* **2017**, *29*, 1604436. [CrossRef]
37. Tian, Z.; Zhang, X.; Li, D.; Zhou, D.; Jing, P.; Shen, D.; Qu, S.; Zboril, R.; Rogach, A.L. Full-Color Inorganic Carbon Dot Phosphors for White-Light-Emitting Diodes. *Adv. Opt. Mater.* **2017**, *5*, 1700416. [CrossRef]
38. Zhai, Y.; Wang, Y.; Li, D.; Zhou, D.; Jing, P.; Shen, D.; Qu, S. Red carbon dots-based phosphors for white light-emitting diodes with color rendering index of 92. *J. Colloid Interface Sci.* **2018**, *528*, 281–288. [CrossRef]
39. Wang, Z.; Yuan, F.; Li, X.; Li, Y.; Zhong, H.; Fan, L.; Yang, S. 53% Efficient Red Emissive Carbon Quantum Dots for High Color Rendering and Stable Warm White-Light-Emitting Diodes. *Adv. Mater.* **2017**, *29*, 1702910. [CrossRef]
40. Guo, F.; Zhu, Z.; Zheng, Z.; Jin, Y.; Di, X.; Xu, Z.; Guan, H. Facile synthesis of highly efficient fluorescent carbon dots for tetracycline detection. *Environ. Sci. Pollut. Res.* **2020**, *27*, 4520–4527. [CrossRef]
41. Ehtesabi, H.; Roshani, S.; Bagheri, Z.; Yaghoubi-Avini, M. Carbon dots—Sodium alginate hydrogel: A novel tetracycline fluorescent sensor and adsorber. *J. Environ. Chem. Eng.* **2019**, *7*, 103419. [CrossRef]
42. Tu, Y.; Chen, X.; Xiang, Y.; Yuan, X.; Qin, K.; Wei, Y.; Xu, Z.; Zhang, Q.; Ji, X. Hydrothermal Synthesis of a Novel Mesoporous Silica Fluorescence Carbon Dots and Application in Cr(VI) and Folic Acid Detection. *Nano* **2020**, *15*, 2050090. [CrossRef]
43. Zhang, W.; Wu, B.; Li, Z.; Wang, Y.; Zhou, J.; Li, Y. Carbon quantum dots as fluorescence sensors for label-free detection of folic acid in biological samples. *Spectrochim. Acta Part A Mol. Biomol. Spectrosc.* **2020**, *229*, 117931. [CrossRef] [PubMed]
44. Qin, X.; Lu, W.; Asiri, A.M.; Al-Youbi, A.O.; Sun, X. Microwave-assisted rapid green synthesis of photoluminescent carbon nanodots from flour and their applications for sensitive and selective detection of mercury(II) ions. *Sens. Actuators B Chem.* **2013**, *184*, 156–162. [CrossRef]
45. Murugesan, P.; Moses, J.A.; Anandharamakrishnan, C. One step synthesis of fluorescent carbon dots from neera for the detection of silver ions. *Spectrosc. Lett.* **2020**, *53*, 407–415. [CrossRef]
46. Chaudhary, N.; Gupta, P.K.; Eremin, S.; Solanki, P.R. One-step green approach to synthesize highly fluorescent carbon quantum dots from banana juice for selective detection of copper ions. *J. Environ. Chem. Eng.* **2020**, *8*, 103720. [CrossRef]
47. Mate, N.; Pranav; Nabeela, K.; Kaur, N.; Mobin, S.M. Insight into the Modulation of Carbon-Dot Optical Sensing Attributes through a Reduction Pathway. *ACS Omega* **2022**, *7*, 43759–43769. [CrossRef]
48. Arul, V.; Chandrasekaran, P.; Sivaraman, G.; Sethuraman, M.G. Biogenic preparation of undoped and heteroatoms doped carbon dots: Effect of heteroatoms doping in fluorescence, catalytic ability and multicolour in-vitro bio-imaging applications—A comparative study. *Mater. Res. Bull.* **2022**, *162*, 112204. [CrossRef]
49. Gong, X.; Zhang, Q.; Gao, Y.; Shuang, S.; Choi, M.M.F.; Dong, C. Phosphorus and Nitrogen Dual-Doped Hollow Carbon Dot as a Nanocarrier for Doxorubicin Delivery and Biological Imaging. *ACS Appl. Mater. Interfaces* **2016**, *8*, 11288–11297. [CrossRef]
50. Chen, P.; Zhang, J.; He, X.; Liu, Y.-H.; Yu, X.-Q. Hydrophobically modified carbon dots as a multifunctional platform for serum-resistant gene delivery and cell imaging. *Biomater. Sci.* **2020**, *8*, 3730–3740. [CrossRef]
51. Yang, W.; Zhang, H.; Lai, J.; Peng, X.; Hu, Y.; Gu, W.; Ye, L. Carbon dots with red-shifted photoluminescence by fluorine doping for optical bio-imaging. *Carbon* **2017**, *128*, 78–85. [CrossRef]

52. Guo, Y.; Chen, Y.; Cao, F.; Wang, L.; Wang, Z.; Leng, Y. Hydrothermal synthesis of nitrogen and boron doped carbon quantum dots with yellow-green emission for sensing Cr(vi), anti-counterfeiting and cell imaging. *RSC Adv.* **2017**, *7*, 48386–48393. [CrossRef]
53. Yuan, K.; Zhang, X.; Li, X.; Qin, R.; Cheng, Y.; Li, L.; Yang, X.; Yu, X.; Lu, Z.; Liu, H. Great enhancement of red emitting carbon dots with B/Al/Ga doping for dual mode anti-counterfeiting. *Chem. Eng. J.* **2020**, *397*, 125487. [CrossRef]
54. Zhao, J.; Zheng, Y.; Pang, Y.; Chen, J.; Zhang, Z.; Xi, F.; Chen, P. Graphene quantum dots as full-color and stimulus responsive fluorescence ink for information encryption. *J. Colloid Interface Sci.* **2020**, *579*, 307–314. [CrossRef]
55. Lieschke, G.J.; Currie, P.D. Animal models of human disease: Zebrafish swim into view. *Nat. Rev. Genet.* **2007**, *8*, 353–367. [CrossRef]
56. Guo, T.; Wang, X.; Hong, X.; Xu, W.; Shu, Y.; Wang, J. Modulation of the binding ability to biomacromolecule, cytotoxicity and cellular imaging property for ionic liquid mediated carbon dots. *Colloids Surfaces B Biointerfaces* **2022**, *216*, 112552. [CrossRef] [PubMed]
57. Deng, L.; Fang, N.; Wu, S.; Shu, S.; Chu, Y.; Guo, J.; Cen, W. Uniform H-CdS@Nicop core-shell nanosphere for highly efficient visible-driven photocatalytic H_2 evolution. *J. Colloid Interface Sci.* **2022**, *608*, 2730–2739. [CrossRef] [PubMed]
58. Yang, X.; Li, X.; Wang, B.; Ai, L.; Li, G.; Yang, B.; Lu, S. Advances, opportunities, and challenge for full-color emissive carbon dots. *Chin. Chem. Lett.* **2022**, *33*, 613–625. [CrossRef]
59. Wei, X.; Li, L.; Liu, J.; Yu, L.; Li, H.; Cheng, F.; Yi, X.; He, J.; Li, B. Green Synthesis of Fluorescent Carbon Dots from Gynostemma for Bioimaging and Antioxidant in Zebrafish. *ACS Appl. Mater. Interfaces* **2019**, *11*, 9832–9840. [CrossRef]
60. Wang, H.; Wei, J.; Zhang, C.; Zhang, Y.; Zhang, Y.; Li, L.; Yu, C.; Zhang, P.; Chen, J. Red carbon dots as label-free two-photon fluorescent nanoprobes for imaging of formaldehyde in living cells and zebrafishes. *Chin. Chem. Lett.* **2020**, *31*, 759–763. [CrossRef]

Disclaimer/Publisher's Note: The statements, opinions and data contained in all publications are solely those of the individual author(s) and contributor(s) and not of MDPI and/or the editor(s). MDPI and/or the editor(s) disclaim responsibility for any injury to people or property resulting from any ideas, methods, instructions or products referred to in the content.

Article

Cannabidiol-Loaded Lipid-Stabilized Nanoparticles Alleviate Psoriasis Severity in Mice: A New Approach for Improved Topical Drug Delivery

Mark Zamansky [1,2,†], Doron Yariv [2], Valeria Feinshtein [3], Shimon Ben-Shabat [3,*] and Amnon C. Sintov [1,2,*]

1. Department of Biomedical Engineering, Ben Gurion University of the Negev, Be'er Sheva 84105, Israel; markzam@post.bgu.ac.il
2. Laboratory for Biopharmaceutics, E.D. Bergmann Campus, Ben-Gurion University of the Negev, Be'er Sheva 84105, Israel; yarivdo@post.bgu.ac.il
3. Department of Biochemistry and Pharmacology, Ben Gurion University of the Negev, Be'er Sheva 84105, Israel; shteiman@bgu.ac.il
* Correspondence: sbs@bgu.ac.il (S.B.-S.); asintov@bgu.ac.il (A.C.S.)
† This work was performed in partial fulfillment of the requirements for the Ph.D. degree of M.Z.

Abstract: Cannabidiol (CBD) is a promising natural agent for treating psoriasis. CBD activity is attributed to inhibition of NF-kB, IL-1β, IL-6, and IL-17A. The present study evaluated the antipsoriatic effect of cannabidiol in lipid-stabilized nanoparticles (LSNs) using an imiquimod (IMQ)-induced psoriasis model in mice. CBD-loaded LSNs were stabilized with three types of lipids, Cetyl alcohol (CA), Lauric acid (LA), and stearic-lauric acids (SALA), and were examined in-vitro using rat skin and in-vivo using the IMQ-model. LSNs loaded with coumarin-6 showed a localized penetration depth of about 100 μm into rat skin. The LSNs were assessed by the IMQ model accompanied by visual (psoriasis area severity index; PASI), histological, and pro-psoriatic IL-17A evaluations. Groups treated with CBD-loaded LSNs were compared to groups treated with CBD-containing emulsion, unloaded LSNs, and clobetasol propionate, and to an untreated group. CBD-loaded LSNs significantly reduced PASI scoring compared to the CBD emulsion, the unloaded LSNs, and the untreated group (negative controls). In addition, SALA- and CA-containing nanoparticles significantly inhibited IL-17A release, showing a differential response: SALA > CA > LA. The data confirms the effectiveness of CBD in psoriasis therapy and underscores LSNs as a promising platform for delivering CBD to the skin.

Keywords: cannabidiol-loaded nanoparticles; lipid-stabilized nanoparticles; skin permeability; IMQ induced psoriasis; interleukin release

1. Introduction

There has been growing evidence related to the anti-inflammatory activity of cannabidiol (CBD). CBD, one of the main components of the *Cannabis sativa* extract, is a non-psychoactive phytocannabinoid shown to have therapeutic potential for various disease states [1]. Whereas CBD acts as an antagonist or partial agonist via allosteric binding to CB1 and CB2 receptors [2–4], its anti-inflammatory activity is attributed to its effect exerted on the adenosine A_2A receptor [5]. This mechanism was demonstrated in LPS-induced inflammation in a mouse model, where CBD reduced the production of pro-inflammatory cytokines (TNF-α and IL-6) and chemokines (MCP-1 and MIP-2). A study on Toll-like receptor (TLR)-activated human monocytes showed that CBD modulated the production of TNF-α, IL-1β, and IL-6 [6]. In addition, CBD is effective in treating psoriasis [7,8]. Such activity of CBD is explained by its ability to inhibit TNF-α-induced NF-kB transcription in a dose-dependent manner in HaCaT cells [9]. The potential applicability of CBD to psoriasis treatment is well established, however, its cutaneous delivery and retention in the skin remain to be addressed. The retention and prolonged mode of action are significant in chronic dermatological conditions such as psoriasis due to increasing efficacy, patient compliance, and chances of successful treatment. Formulation of nanoparticles (NPs) is a technology platform that enables active substances to penetrate the skin, retain them in the skin, and control the release of these active compounds in skin layers. The mechanisms of skin permeation and retention of nanoparticles include an entry of the applied NPs through the hair follicles [10–12] and penetration of the nanoparticles through the stratum corneum between corneocytes [13,14]. Published reports on the topical application of CBD in a nanoparticulate system are relatively limited. Lodzki et al. reported successful transdermal delivery of CBD using an ethosome-based formulation with 3% w/w CBD and 40% w/w ethanol combined with a phospholipid carbomer gel [15]. Another study reports successful trans-corneal delivery of CBD loaded in mixed polymeric micelles of chitosan/polyvinyl alcohol and polymethyl methacrylate under air-liquid and liquid-liquid conditions [16]. In recent publications by our group, we have shown that ethyl cellulose NPs stabilized by various lipids (Figure 1) could deliver CBD into and through rat skin [17]. We also showed that the stabilizing lipids affected the in vitro release of CBD and its ex-vivo permeation through rat skin. Particularly, the incorporation of the relatively high melting point (m.p. 69.5 °C) of stearic acid (SA) reduced the permeation, whereas the incorporation of a eutectic mixture of lauric acid (LA) (m.p. 46.4 °C) and SA (SALA) (m.p. ~36 °C) increased skin permeation. The lipid-stabilized NPs (LSNs) showed a significant anti-inflammatory activity by reduction in IL-6 and IL-8 release in TNF-α induced HaCaT cells [18].

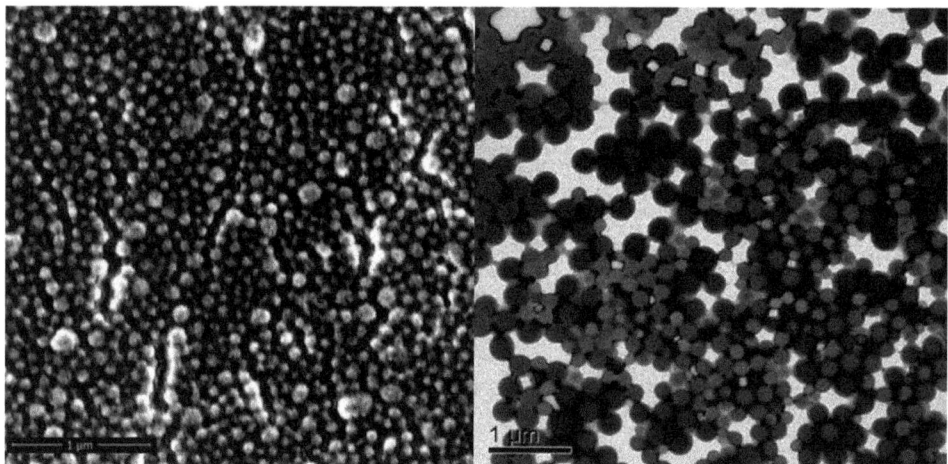

Figure 1. A representative scanning electron micrograph (SEM) (**left**) and transmission electron micrograph (TEM) (**right**) of cetyl alcohol (CA)-stabilized LSNs.

The aim of the current research was an in-vivo efficacy evaluation of these CBD-loaded LSNs as a new potential platform for the treatment of psoriasis in humans. This evaluation was performed using the imiquimod (IMQ)-induced psoriasis model in mice. We evaluated LSNs stabilized with different lipids in terms of their effectiveness in reducing psoriasis manifestations measured by psoriasis area and severity index (PASI) score and histological and cytokine profiles. IMQ-induced psoriasis in mice is a widely used model for assessing potential psoriasis treatments with a good translation to humans [19] when using the C57BL/6 mice strain. IMQ is a Toll-like receptor (TLR7/8) agonist that can be applied to mouse skin to elicit erythema, scaling, and keratinocyte proliferation. At the same time, the phenotype involves the induction of the IL-17/IL-23 axis cytokines [20]. The ability of CBD to reduce the production of pro-inflammatory cytokines through the A_2A and the TLR pathways, suggests the applicability of the IMQ-induced psoriasis model for its evaluation.

2. Results and Discussion

2.1. Formulation Development

The various LSNs were first characterized for particle size, particle concentration, loaded CBD concentration, and Entrapment Efficiency (EE) (Table 1). Prior to the preclinical studies, CBD-loaded LSNs were compounded into a conveniently used topical dosage form. Various delivery forms have been previously used for nanoparticles, including directly applying the NPs dispersed in water, saline, or PBS [21] and semi-solid formulations. There are several examples of NPs formulated in an acrylate gel based on Carbopol® ETD 2020 [14] or Carbopol® 934 [22], hydroxypropyl methylcellulose (HPMC) [23,24], Xanthan gum [25], Poloxamer 407 [26], and colloidal silica [27]. For the LSN systems, the use of polysaccharide thickeners such as xanthan gum or modified celluloses (e.g., hydroxypropyl cellulose, hydroxyethyl cellulose, or carboxymethyl cellulose) were not utilized since these linear fiber-based polymers may hinder permeation of the nanoparticles. Thus, inorganic thickeners such as colloidal silica gel and magnesium aluminum silicates (MAS) were used. In addition, Poloxamer 407 was also considered as a candidate thickener due to its thermogelling properties.

Table 1. Nanoparticles and their parameters.

LSN-Type	Size [nm]	PDI	NP Conc. [NPs/mL]	CBD Conc. [mg/g]	EE [%]
SALA	238.0	0.028	2.06×10^{13}	13.8	86.3
CA	220.8	0.102	1.52×10^{13}	12.5	69.2
LA	245.0	0.182	1.78×10^{13}	20.0	68.0
Unloaded	226.2	0.114	1.60×10^{13}	NA	NA

2.2. In Vitro Skin Permeability of Semi-Solid LSN Formulation Development

The skin permeability of CBD from the LSN semi-solid formulations was analyzed using fresh rat skin mounted on a Franz diffusion cell system. No significant differences were observed between the skin permeation and retention of the various LSNs. However, since SALA-stabilized nanoparticles showed a relatively high CBD release in our previous study [18], they were used in skin penetration experiments as a representative LSN system. The LSN dispersion was thickened with fumed silica (CAB-O-SIL® 530), MAS (Veegum HV), and Poloxamer 407 thermogel, while LSNs dispersed in water served as a non-gelled control. Volumes of 0.2 mL (about 0.2 g) from the various LSN formulations, each containing CBD at a concentration of 500 μg/g, were applied on the excised skin, providing 100 μg dose of CBD over 1.77 cm² of skin surface area. (i.e., 100 μg CBD/0.2 g quantity and 56.5 μg/cm² skin surface area). The unstirred volume of these LSN products (about 0.2 mL) was relatively high for typical dermal application, and only a tiny portion of CBD in the SLNs came in direct contact with the skin. Typically, semi-solid formulations are applied to the skin in much lower volumes per cm², constituting about 1 μL/cm² [28,29] and accompanied by rubbing. Thus, considerably higher availability of LSNs would be expected in clinical use. Considering the hindrance caused by the gel structures, the total permeation from the liquid aqueous dispersion was expected to be the highest.

As seen in Figure 2, there was a quantitative penetration of CBD into rat skin. The extent of CBD penetration depended on the type of gelling agent used for formulating the LSNs. Skin penetration of CBD from the liquid dispersion of the LSNs was relatively high, as expected (100% ± 28%, with actual values of 2.0 ± 0.8 μg/cm²), compared to the penetration of CBD from the poloxamer thermogel, which was the lowest (15.4% ± 14.6%, with actual values of 0.4 ± 0.3 μg/cm²; $p < 0.001$, ANOVA test). The relatively low skin penetration of LSNs from the poloxamer gel can be explained by the micelle density formed by poloxamers. Liu et al. [30] found that the micellar face-centered cubic lattice length is 29.5 nm for 15–45% w/w poloxamer 407 in an aqueous solution. Thus, with such small distances between the poloxamer micelles, the mobility of the 200 nm diameter LSNs through the Poloxamer gel network should have been significantly obstructed at rest (without external mechanical shear, i.e., rubbing). MAS particles are extremely thin, negatively charged plates approximately 2 nm thick and about 200–500 nm long, forming a three-dimensional network after water dispersion [31]. CAB-O-SIL® 530 fumed silica has an average aggregate size of 200–300 nm, though the distance between the aggregates may be even greater. According to localization and hopping theory [32,33], the permeation of nanoparticles through an entangled polymer network depends on the confinement 'parameter C' described as the ratio of effective particle diameter to the effective tube diameter, which roughly corresponds to the distance between the polymer or mesh crosslinks. Particles with the confinement parameter $C < 1$ permeate relatively free, whereas the movement in the mesh of particles with $1 < C < 3$ is due to hopping, given a sufficient activation energy. According to this theory, the diffusion coefficient in the mesh quickly drops proportionally to the exponent of $(-C^2)$. Thus, CAB-O-SIL® 530 gel with the largest distances between the gel crosslinks would have the highest diffusion coefficient compared to the other tested gels, explaining the differences observed in Figure 2. It was also noted that the skin penetration of CBD was dependent on the silica concentration. Compared to 5% w/w gel, penetration extent decreased to $63.5 \pm 17.5\%$ for 4% w/w gel and further to $33.5\% \pm 7.4\%$ for 3% w/w gel. This result is counter-intuitive to the expectation that reducing the gelling agent's content would increase the molecules' or NPs' permeation through its polymeric network. It has been shown by Binder et al. [34] that an increase in the concentration of hydroxypropyl methylcellulose (HPMC) and hydroxyethyl cellulose (HEC) reduced the permeation of sulphadiazine sodium through the skin, which was explained by increased viscosity and entanglement of these cellulosic gels. In contrast to these polymers, the higher concentration of CAB-O-SIL® 530 silica gel increased skin permeation, probably due to the hydrophobicity of the silica particles. Additional silica particles tend to form aggregates, building a mesh with thicker strands-fibers and larger openings. These results established CAB-O-SIL® 530 5% gel as a more suitable semi-solid vehicle for LSNs and NPs in general.

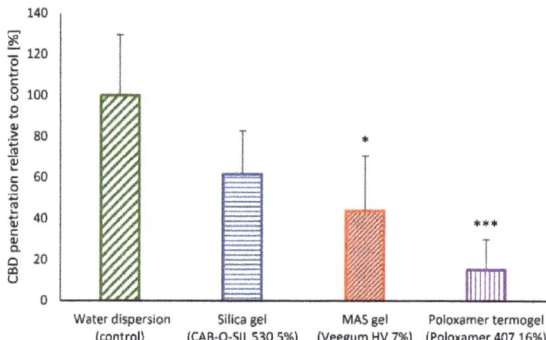

Figure 2. CBD penetration following the application of NP formulated with different gelling agents recovered from the skin. The values are presented relative to NP dispersion in water (100%), * $p < 0.05$, *** $p < 0.001$ ANOVA followed by Tuckey post-hoc analysis. Data are presented as mean $\pm$ SD.

2.3. In-Vivo Skin Penetration and Retention

Further assessment of the skin penetrability of LSNs from the topical silica gel vehicle was performed in an in-vivo study using SALA-stabilized NPs loaded with a fluorescent marker—coumarin 6 (C6) [35]. The LSN dispersion was combined with CAB-O-SIL® 530 to obtain a 5% silica gel with 7.5×10^{11} NPs/g. Then, the gel was applied for 2 h onto the abdomen of an anesthetized rat, which had previously been shaved and depilated as described in the experimental section. As seen in Figure 3, C6 penetrated to about 100 μm skin depth, roughly corresponding to or slightly exceeding the thickness of the epidermis layer of rat skin [36,37].

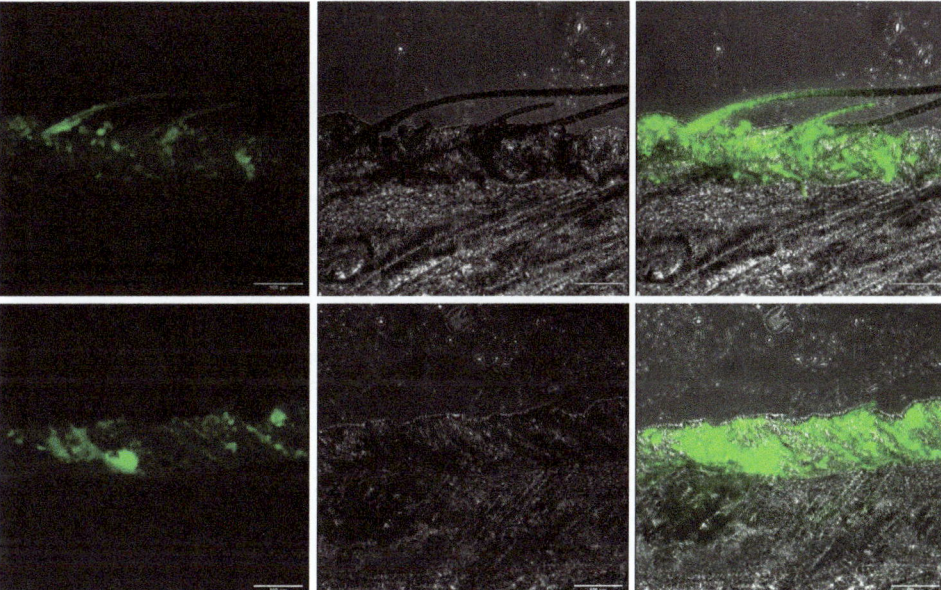

Figure 3. Microphotographs of excised rat skin after application of 5% CAB-O-SIL® 530 gel with coumarin-6 loaded LSNs on rat skin, unshaved (**top**) and shaved (**bottom**). Left: green, fluorescent light; Middle: bright field; Right: merged pictures. Scale bar: 100 μm.

2.4. Imiquimod-Induced Psoriasis Model

IMQ-induced psoriasis in mice is a widely accepted psoriasis-like model for testing potential treatments with a good translation power to humans [38], especially when performed on the commonly used C57BL/6 mice strain [19]. IMQ-induced psoriasis study conducted on Cannabinoid 2 Receptor (CB2R) knockout mice showed that CB2R deficiency exacerbated psoriasis disease [39]. These results further suggest the IMQ model's applicability for the evaluation of CBD as a treatment for psoriasis. This report presents for the first time an evaluation of CBD in IMQ-induced psoriasis model in mice, particularly the evaluation of CBD delivered by LSNs. We have previously shown that incorporating different stabilizing lipids in ethyl cellulose NPs could influence the rate and extent of both release and dermal permeability of the loaded CBD [18]. By using the IMQ-induced psoriasis model, we compared the effectiveness of three types of LSNs stabilized with either CA, LA, or SALA to the efficacy of emulsified CBD solution ('free CBD' or f-CBD), clobetasol (CLO, positive control), and unloaded nanoparticles (Unloaded NP, negative control).

2.4.1. Erythema and Scaling

The first appearance of IMQ-related symptoms was on the second day after induction. If not treated, the symptoms increased continuously during the following days. As shown in Figure 4d (cumulative PASI score), a significant improvement was observed in the treatment groups, CBD-loaded CA-NPs, LA-NPs, and SALA-NPs, as well as CLO (positive treatment groups—G+), compared to the apparent progression of the disease in the non-treatment (NT) group, the treatment with unloaded LSNs group, and the treatment with CBD solution (f-CBD) group (negative treatment groups—G−). On study day 4, the NT and treatment with f-CBD were significantly less effective than all G+ treatments ($p < 0.001$ for CBD-loaded LA-NPs and CLO, and $p < 0.01$ for CBD-loaded CA-NPs and CBD-loaded SALA-NPs, ANOVA). In addition, treatment with CBD-unloaded LSNs had lower effectiveness than treatments with CBD-loaded LA-NPs and CLO ($p < 0.05$, ANOVA).

On the last day of the study (day 5), the differences between each treatment of G+ groups and each treatment of G- groups were highly statistically significant ($p < 0.001$, ANOVA). Similarly, no significant differences in erythema symptoms were observed between any of the treatments during the first three days. Only on day 4, a significant difference was observed between the f-CBD treatment group and each of the treatments in G+ groups ($p < 0.001$ for CLO and CBD-loaded LA-NPs, $p < 0.01$ for CBD-loaded CA-NPs and CBD-loaded SALA-NPs, Figure 4a). On day 5, all treatments in the positive G+ treatment groups were significantly more effective in the prevention of erythema than each of the treatments in the negative groups ($p < 0.001$ for NT and f-CBD groups compared to CBD-loaded CA-NPs, CBD-loaded LA-NPs, CBD-loaded SALA-NPs, and CLO; $p < 0.001$ for treatment with unloaded-LSNs compared to CBD-loaded SALA-NPs and CLO, $p < 0.01$ for treatment with unloaded-LSNs compared to CBD-loaded CA-NPs, and $p < 0.05$ for treatment with unloaded-LSNs compared to CBD-loaded LA-NP, ANOVA). As seen in Figure 4c, skin scaling was significantly minimized by CBD-loaded CA-NPs and CBD-loaded LA-NPs compared to the untreated group. In contrast, scaling increased on treatment with CBD-loaded SALA-NPs on days 2 and 3, but on day 4, these nanoparticles hindered the scaling progress. Thus, on day 4 the NT group and the f-CBD group were significantly less effective than any treatment in the positive G+ groups ($p < 0.001$ for CLO and CBD-loaded LA-NPs; $p < 0.01$ for CBD-loaded CA-NPs and CBD-loaded SALA-NPs, ANOVA). Unloaded LSNs were less effective than treatments with CLO and CBD-loaded LA-NPs ($p < 0.05$, ANOVA). On day 5, all treatments in the G+ groups were significantly more effective than those in the negative G- groups ($p < 0.001$, ANOVA).

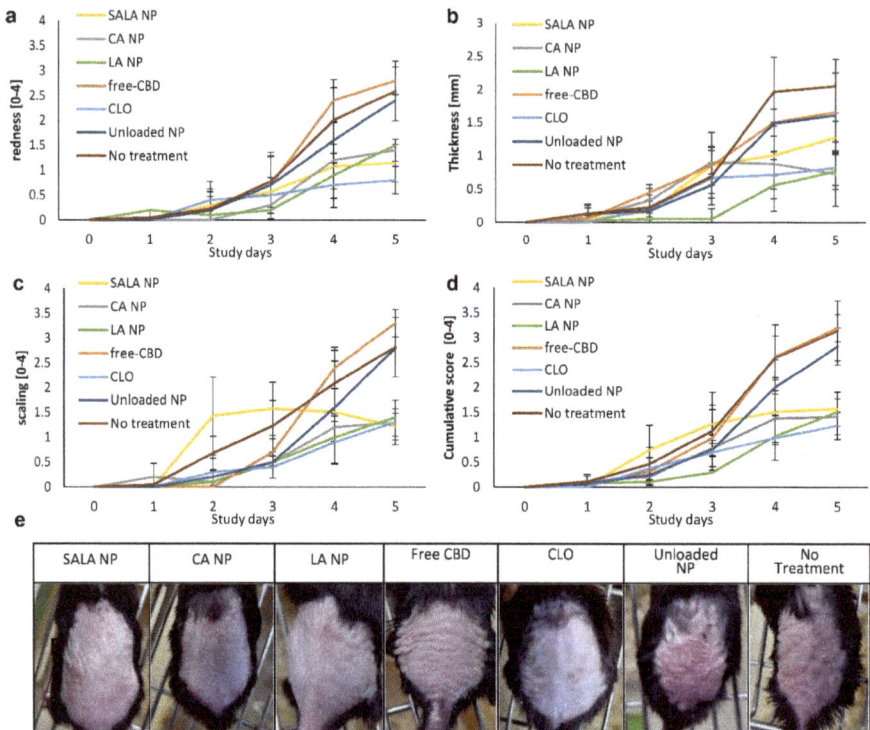

Figure 4. Evaluation of dorsal skin from Day 0 (D0) to Day 5 (D5) according to PASI score during IMQ induction of psoriasis-like dermatitis. Observed criteria are skin redness (**a**), thickness (**b**), scaling (**c**), and total cumulative PASI score (**d**). Representative images of mice in each group on Day 5 (D5) (**e**). Data are presented as mean ± SD.

2.4.2. Skin Thickness

A gradual increase in skin thickness, another parameter of inflamed area, was observed in the NT group and the unloaded LSNs treatment group (Figure 4b). However, significant inhibition was noted by CBD-loaded LA-NPs and CBD-loaded CA-NPs (days 4 and 5). Clobetasol cream also inhibited skin thickening on days 4 and 5. All treatments in the G+ groups were significantly more effective on day 4 than the NT group ($p < 0.001$, ANOVA). Treatments with f-CBD and unloaded LSNs were significantly less effective in reducing skin thickening when compared to treatment with CBD-loaded LA-NPs ($p < 0.001$, ANOVA). On day 5, NT group showed a significant skin thickening compared to each treatment in G+ ($p < 0.001$ for CA-NP, LA-NP, and CLO and $p < 0.01$ for SALA-NP, ANOVA), f-CBD treatment was also less effective compared to treatments with CBD-loaded CA-NPs, CBD-loaded LA-NPs, and CLO ($p < 0.05$, ANOVA). At the same time, unloaded LSNs were significantly less effective than treatments with CBD-loaded CA-NPs and CBD-loaded LA-NPs only ($p < 0.05$, ANOVA).

2.4.3. Cumulative Day-to-Day Scoring

To perform an overall comparison between the treatments throughout the study rather than on a day-to-day basis, we summed up the daily cumulative scores from day 0 to day 5 (Figure 5a). As shown in Figure 5a, All the G+ treatment groups were significantly more effective than the untreated group in reducing IMQ-induced psoriasis symptoms ($p < 0.001$ for CBD-loaded CA-NPs, CBD-loaded LA-NPs, and CLO, and $p < 0.01$ for CBD-loaded SALA-NPs, ANOVA). Unloaded LSNs were the most effective treatment in the G- group, being significantly less effective than treatments with CBD-loaded LA-NPs ($p < 0.01$) and CLO ($p < 0.05$). Among the G+ treatment groups, treatment with CBD-loaded SALA-NPs was the least effective, although no significant difference ($p > 0.05$) was noted compared to the other G+ groups. The differences between the LSNs stabilized with CA, LA, or SALA could be attributed to differences in the release rates of CBD from these NPs. NPs, such as those stabilized with LA, released CBD at a higher rate and, therefore, had a lower day-to-day score. In contrast, NPs stabilized with SALA released their CBD at a slower rate but allowed its accumulation, thus providing a similar effect on day 5.

2.4.4. Weight Loss

Examination of body weight changes during the study and the spleen-to-body weight ratio did not show significant differences between the G+ and the G- groups (Figure 5b). Figure 5b shows that treatments with CBD-loaded SALA-NPs and CBD-loaded CA-NPs were the only treatments that significantly kept the body weight of the diseased mice almost constant compared to the untreated group. These treatments prevented the decrease in body weight, which was significantly different from all other treatments except for the f-CBD treatment group. Body weight is a common measure of an animal's health condition, while its loss is usually associated with the IMQ model. Numerous studies have shown that the change in body weight in mice treated with topical imiquimod is related to reduced consumption of food and water [40,41]. Therefore, the extent of the weight loss reversal might be a measure of treatment effectiveness. Data available from previous studies regarding the influence of CBD on body weight have indicated that CBD induces weight loss by appetite depression through CB2 receptors in rats [42], as well as in humans [43]. Nevertheless, the effect of oral CBD on body weight in mice was significant only for the highest daily dose of 615 mg/kg/day, while no weight change occurred after lower doses of CBD [44]. For comparison, the total CBD dose applied on the psoriasis-like skin in the present study was about 50–70 mg/kg. It is interesting to note that topical application of clobetasol propionate used as a positive control in the IMQ-induced psoriasis model resulted in a marked decrease in body weight [45,46]. Thus, the significantly lower body weight loss in the groups treated with CBD-loaded SALA-NPs and CBD-loaded CA-NPs could be explained by their anti-inflammatory action that reversed the influence of IMQ.

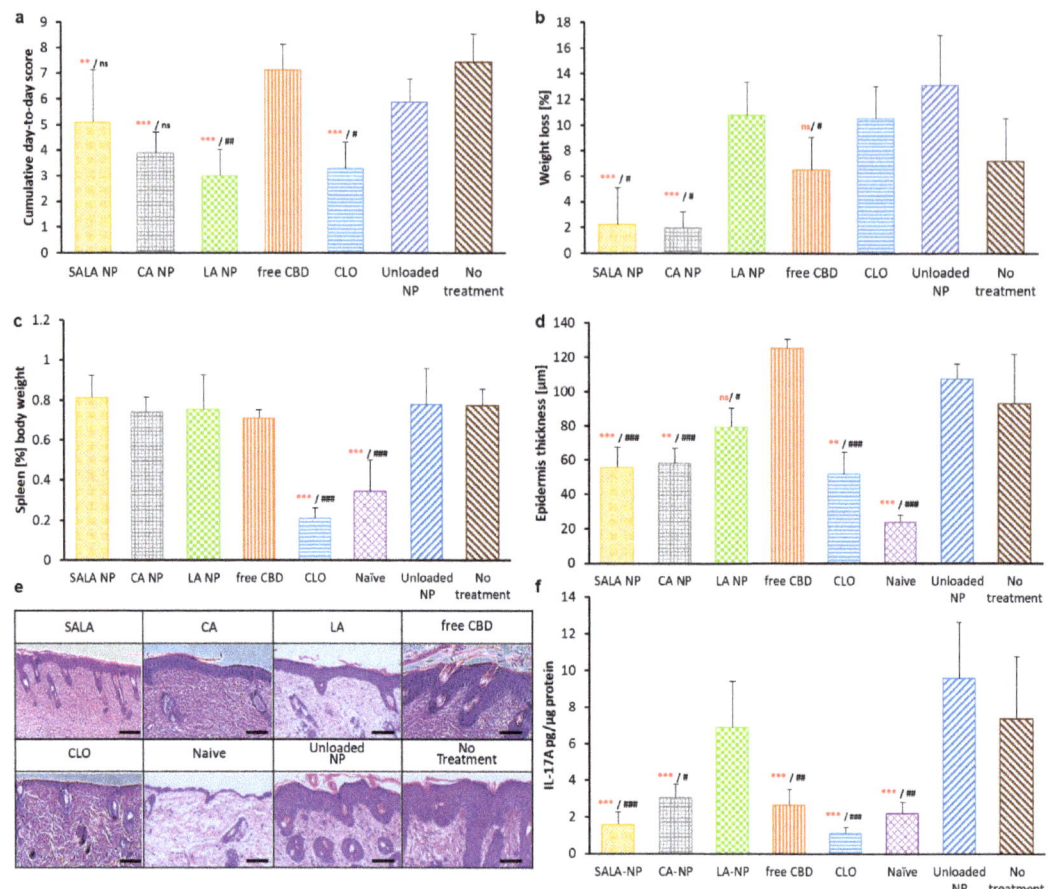

Figure 5. Comparison of the applied treatments with respect to the sum of their day-to-day cumulative PASI scores (**a**); Evaluation the mice body weight change from day 0 (D0) to day 5 (D5) (**b**); Evaluation of the spleen to body weight ratio [%] as an indication for systemic exposure of the treatments (**c**); Evaluation of the IMQ induced acanthosis measured as the epidermis thickness (**d**); Histological examination by H&E staining of the mice back skin samples (**e**); Evaluation of the IL-17A release in the mice back skin samples taken on D5 (**f**). ns/**/***—not significant/$p < 0.01$/$p < 0.001$ compared to the No treatment group, red symbols ns/#/##/###—not significant/$p < 0.05$/$p < 0.01$/$p < 0.001$ compared to the Unloaded NP treatment group, ANOVA followed by post-hoc Tukey test. Data are presented as mean ± SD. Scale bar: 100 μm.

2.4.5. Spleen

The spleen is the second major immune organ besides the lymph nodes. The observed splenomegaly and the subsequent increase in spleen-to-body weight ratio indicates a systemic effect exerted by the treatment on the immune system. It is proposed that IMQ-related splenomegaly is caused by inflammation [47], whereas the hyposplenism induced by clobetasol (CLO) was associated with the depletion of splenic dendritic cells [48]. According to the observed results (Figure 5c), neither of the CBD-including treatments prevented IMQ-induced splenomegaly as CLO, possibly indicating localization of the NPs in the skin, preventing systemic exposure to CBD.

2.4.6. Acanthosis Evaluation

The results obtained for the measured acanthosis values, as seen from Figure 5d,e, are well in line with the day-to-day cumulative scores, although without the distinct differences between the groups treated with the CBD-loaded LSNs. The most visually pronounced effect of CLO can be attributed to its anti-inflammatory activity as well as skin tissue atrophy generally associated with corticosteroid treatment [49].

2.4.7. Anti-Inflammatory Action of CBD-Loaded LSNs as Evaluated by Reduction in IL-17A Secretion

Pro-inflammatory cytokines play a significant role in psoriasis disease manifestation. It was shown that the IMQ-induced psoriasis model is mediated via the IL-23/IL-17 axis [20]. Other publications showed the effectiveness of IL-17A antagonists and anti-IL-17A ssDNA aptamers in reversing the action of IMQ in mice [50,51]. Although CBD has not been tested previously in the IMQ-induced psoriasis model in mice, several other mice and human models have demonstrated its effectiveness in reducing IL-17A secretion [52,53]. In the present study, we have selected to evaluate the influence of various CBD-loaded LSNs on the secretion of IL-17A as a supplementary measure of their effectiveness.

The results showed that treatment with CBD-loaded SALA-NPs was more effective in the reduction in IL-17A secretion compared to treatment with CBD-loaded LA-NPs ($p < 0.01$, ANOVA) (Figure 5f). While CBD-loaded LA-NPs had no effect on IL-17A release, both treatments with CBD-loaded SALA-NPs and CBD-loaded CA-NPs resulted in a significant reduction in IL-17A levels compared to the negative control groups, the untreated group and the group treated with unloaded LSNs. The finding that treatment with CBD-loaded LA-NPs was less effective in inhibiting IL-17A compared to other CBD-loaded LSNs may be explained by an inherent pro-inflammatory activity of LA. Such activity was previously shown by the ability of LA to induce the release of IL-12 from the RAW264.7 cells (BALB/c mouse macrophages) [54] and the release of IP-10 chemokine from human U937 macrophages [55]. Considering IL-17A, LA was shown to promote differentiation of Th17 cells, resulting in increased levels of IL-17A [56]. The difference in the effects between CBD-loaded NPs containing LA and those containing SALA eutectic mixture can be attributed to the lower solubility and release of free LA molecules from the eutectic mixture. For example, the individual solubility of each lidocaine and prilocaine is significantly reduced when both components form a eutectic mixture [57]. A similar effect of toxicity reduction in eutectic mixture components was shown for the reduction in menthol toxicity on HaCaT cells when applied as a component of a eutectic mixture with either lauric, stearic, or myristic acids [58].

3. Materials and Methods

3.1. Nano-Particles Preparation

Ethyl cellulose (EC) up to about 0.05% w/w, one of the stabilizing lipids: CA or SALA (24:76)—0.025% w/w, or LA—0.05% w/w, Triethyl citrate (TEC)—0.05% w/w, and CBD—0.015% w/w, all were dissolved in absolute ethanol. The solution was constantly stirred on a magnetic plate at about 700 RPM. The ratio of magnetic stirrer length to beaker diameter was at least 1:3. Deionized water was added by dripping at a constant rate of about 22–25 mL/min with a syringe pump NE-300, New Era Pump Systems (Farmingdale, NY, USA) to the final content of about 60% w/w of the final dispersion mass. The obtained NP dispersion was evaporated with R-205 Rotavapor (Buchi Labortechnik AG, Flawil, Switzerland) until about four times volume reduction. To obtain concentrated NP dispersion, several consecutive centrifugation steps were performed. The CBD content in the obtained nanoparticles was determined by High Pressure Liquid Chromatography HPLC, and the entrapment efficiency percentage (EE%) was calculated according to the following Equation (1):

$$EE\% = \frac{\text{Mass of CBD in formulation}}{\text{Total mass of CBD used for formulation}} \times 100 \quad (1)$$

3.2. Size and Microscopic Analysis

3.2.1. Dynamic Light Scattering (DLS)

The hydrodynamic diameter spectrum of the NPs was collected using a CGS-3 Compact Goniometer System (ALV GmbH, Langen, Germany). The laser power was 20 mW at the HeNe laser line (632.8 nm). Correlograms were calculated by ALV/LSE 5003 correlator, which were collected at 90°, for 20 s for 10 times, at 25 °C. The NP size was calculated using the Stokes–Einstein relationship, and the analysis was based on the regularization method as described by Provencher [59].

3.2.2. Nanoparticle Tracking Analysis (NTA)

Measurements were performed using a NanoSight NS300 instrument (Malvern Instruments Ltd., Worcestershire, UK), equipped with a 632 nm laser module and 450 nm long-pass filter, and a camera operating at 25 frames per second, capturing a video file of the particles moving under Brownian motion. The software for capturing and analyzing the data (NTA 3.4, Build 3.4.4) calculated the hydrodynamic diameters of the particles by using the Stokes–Einstein equation.

3.3. Determination of CBD in NP Dispersion

To quantify CBD (within nano-sized particles), to about 25 µL aliquots from each particle sample that was carefully weighed, 975 µL MeOH was added and stirred. After at least 10 min, the samples were further diluted 1:10, and then the liquid was injected into an HPLC system (Shimadzu VP series, Shimadzu Corp., Tokyo, Japan), equipped with a prepacked column (ReproSil-Pur 300 ODS-3, 5 µm, 250 mm 4.6 mm, Dr. Maisch, Ammerbuch, Germany), which was constantly maintained at 30 °C. The samples were chromatographed using a mobile phase of acetonitrile-35 mM acetic acid (75:25) at a 1 mL/min flow rate. A calibration curve, peak area measured at 208 nm versus CBD concentration, was constructed by running standard drug solutions in MeOH for each series of chromatographed samples.

3.4. In-Vitro Skin Penetration Study

3.4.1. In-Vitro Skin Penetration

The penetration of CBD from CBD-loaded NP formulated in various gel formulations into the skin was determined in vitro using a Franz diffusion cell system (Permegear, Inc., Bethlehem, PA, USA). The diffusion area was 1.767 cm^2 (15 mm diameter orifice), and the receptor compartment volume was 12 mL. The solutions in the water-jacketed cells were constantly set at 37 °C and stirred by externally driven, Teflon-coated magnetic bars. Each set of experiments was performed with twelve diffusion cells, each containing abdominal rat skin. The animal treatments were performed in accordance with a protocol reviewed and approved by the Institutional Committee for the Ethical Care and Use of Animals in Experiments, Ben-Gurion University of the Negev, which complies with the Israeli Law of Human Care and Use of Laboratory Animals". Authorization number: IL-30-06-2020(C). Sprague–Dawley rats were euthanized by aspiration of CO_2. Abdominal hair was carefully clipped, and sections of full-thickness skin were excised from the fresh carcasses of animals and used immediately. All skin sections were measured for transepidermal water loss (TEWL), and only those pieces in which TEWL levels were less than 10 g/m^2/h were used. TEWL testing was performed on skin pieces using the Dermalab Cortex Technology instrument (Hadsund, Denmark). The skin was placed on the receiver chambers with the stratum corneum facing upwards, and the donor chambers were then clamped in place. The receiver chamber, defined as the side facing the dermis, was filled with phosphate buffer (pH 7.4)—ethanol 50:50 solution [60] to allow sink conditions. Formulated NPs (0.2 mL or approx. 200 mg) containing 100 µg (about 0.05% w/w) of entrapped CBD was applied on the skin at time = 0. After a 6-h experimental period, each exposed skin tissue was removed, washed with plenty of water, wiped carefully, and tape-stripped (×15) to remove CBD adsorbed in the stratum corneum. Penetrated levels in the skin tissues were determined after overnight methanol extraction by HPLC (see Section 3.3).

3.4.2. Preparation of NPs for Formulation Carrier Selection and Optimization

The 16% w/w Poloxamer 407 gel NP dispersion was prepared by mixing previously prepared Poloxamer 407 (Kolliphor P 407™, BASF, Florham Park, NJ, USA) 20% w/w solution in deionized water stored at about 5 °C with NP dispersion and deionized water. The 7% MAS NP dispersion was prepared by mixing 10% w/w MAS (Veegum HV™, Vanderbilt, Norwalk, CT, USA) with NP dispersion and deionized water. The 3%, 4%, and 5% w/w colloidal silica gel NP dispersions were prepared by mixing previously prepared 10–11% w/w silica (CAB-O-SIL 530®, Cabot, Boston, MA, USA) gel with NP dispersion and deionized water. SALA-stabilized NPs were used for all comparative permeation studies intended for carrier selection and optimization.

3.5. In-Vivo Skin Penetration Study and Image Analysis

The in-vivo penetration study was performed on Sprague–Dawley rats. On the day prior to the experiment, the rats were anesthetized, and their abdominal hair was carefully clipped and depilated (Veet cream, Reckitt Benckiser, Chartres, France). On the day of the experiment, the rats were anesthetized, and Coumarin 6 (TCI, Tokyo, Japan) loaded NPs formulated in a 5% silica gel were applied at about 50 mg/cm^2. After two hours, the rats were euthanized by aspiration of CO_2. The abdominal skin was washed with plenty of water and removed from the carcasses. The animal treatments were performed in accordance with protocol authorization number: IL-30-06-2020(C). SALA stabilized, Coumarin 6 (C6) loaded NPs were prepared similarly to CBD loaded NPs (see Section 3.1), with initial $5 \times 10^{-4}\%$ w/w C6 content in the ethanol solution. The C6-loaded NP dispersion in 5% silica gel was prepared as described in Section 3.4.2 with a final concentration of about 0.002% w/w C6. The excised skin was snap-frozen in liquid nitrogen and sectioned with a cryotome using 100μm thickness for further confocal microscopy analysis (Spinning disc confocal microscope, 3i, Denver, CO, USA). The micrographs were collected with 1.5 μm depth steps. Further, they were processed with Fiji software (version 2.9.0/1.54f) [61] using the Z-project function.

3.6. Imiquimod-Induced Psoriasis in Mice
3.6.1. Animals

For the study, male C57BL/6 8 to 11-weeks-old mice were used. The animals were kept under standard conditions with free access to water and food. The animal treatments were performed in accordance with a protocol reviewed and approved by the Institutional Committee for the Ethical Care and Use of Animals in Experiments, Ben–Gurion University of the Negev, which complies with the Israeli Law of Human Care and Use of Laboratory Animals". Authorization number: IL-64-11-2021(C).

3.6.2. Preparation of the Formulated NP Dispersions and the CBD Emulsion

The formulated NP dispersions were prepared by mixing concentrated dispersions of CBD-loaded NP stabilized with either CA, LA, SALA or unloaded NP with 10% w/w silica gel (CAB-O-SIL 530®, see Section 3.4.2) aiming at a final concentration of 5% w/w silica. Since the lowest achieved CBD content for these dispersions was 1.25% w/w (12.5 mg/g for CA NP see Table 1), the content of CBD in the formulated dispersions was set to 0.6% w/w. The content of unloaded NP was determined by the highest achievable NP concentration, allowing to obtain 5% w/w silica gel resulting in about 8×10^{12} NP/g (based on initial 1.6×10^{13} NP/mL for Unloaded NP—see Table 1).

CBD emulsion was prepared by dissolving CBD (4.6% w/w) in C8-C10 triglycerides (Miglyol 810, Cremer Oleo, Hamburg, Germany) and emulsifying with Polysorbate-80 (J.T. Baker, Phillipsburg, NJ, USA) and deionized water. The prepared emulsion was mixed and dispersed with about 10% silica gel (CAB-O-SIL 530®, see Section 3.4.2) to obtain the following final contents (w/w): CBD—0.6%, Miglyol 810—13%, Polysorbate-80—2%, silica gel—5%, and deionized water ad 100%.

3.6.3. Imiquimod-Induced Psoriasis in Mice

On the first study day (D0), the back of the mice was shaved using an electric clipper and depilated (Veet cream, Reckitt Benckiser, Chartres, France), see Scheme 1. The mice with large black pigmentation areas on their skin—anagen areas, resulting in quick hair regrowth [62] and preventing the effective application of treatments, were excluded from the study. These animals were sacrificed on the first study day, n = 5. Samples obtained from these mice were designated "naive" and served as an IMQ non-treated control. Other experimental groups received a daily topical dose of 62.5 mg of commercially available IMQ cream (5%) (Aldara, 3M, Bracknell, UK) on the back for 5 consecutive days to establish a model of IMQ-induced psoriasis [20]. Negative control (No Treatment—NT) mice did not receive additional treatment on top of IMQ. Due to a large number of treatments, the experiments were performed in two series, each one including the NT group (n = 6 and n = 5) as an internal control. The presented data is for the combined result of both NT groups (n = 11). For other groups, the treatment was applied 1 h after the application of IMQ cream. The CA (n = 5), LA (n = 6), SALA (n = 7), and Unloaded (n = 5) groups received a daily dose of 50 mg of formulated CA, LA, and SALA-stabilized CBD-loaded NPs and formulated LA stabilized unloaded NPs, respectively. The Blank group served as a negative control for formulated CBD-loaded NPs. The formulated NPs' dispersions were prepared in 5% silica gel, as described in Section 3.6.2. The free CBD (f-CBD) group received a daily dose of 50 mg of formulated CBD emulsion (see Section 3.6.2). The CLO group received a daily dose of 100 mg 0.05% commercial clobetasol cream (Clobetasol 0.05%, Trima, Maabarot, Israel), serving as a positive control.

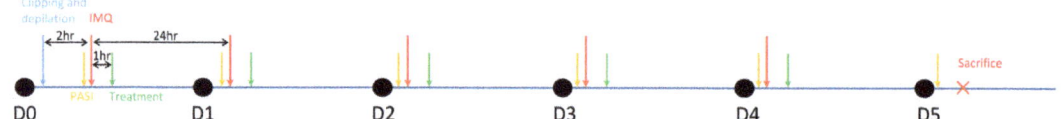

Scheme 1. IMQ-induced psoriasis in mice conduct.

3.6.4. Evaluation of Psoriasis Area Severity Index (PASI) Score and Spleen Index

PASI was used to assess the inflammatory condition in the mice on all study days (D0–D5). For this purpose, we visually examined each mouse's back skin erythema (redness) and scaling. The skin thickness was measured by a calibrated caliper (Mitutoyo, Japan) as the thickness of a back skin fold. The visually examined parameters were assigned a score between 0 and 4 (0-none, 1-slight, 2-moderate, 3-severe, 4-very severe) [63]. To accommodate the possible differences in evaluation between the series of experiments, the values obtained for each mouse were normalized by the maximum value obtained for all mice and multiplied by 4. To compare with the visually examined parameters, the skin thickness score was calculated as the thickness change from D0 for each mouse normalized by the maximum thickness change for all mice and then multiplied by 4 to bring the value to the 0–4 scale. The cumulative score (0 to 4), calculated as the average of the erythema, scaling, and thickness scores, indicates the severity of psoriasis inflammation (PASI score).

The animal weight was measured on the study's first (D0) and last (D5) days. Following the sacrifice, the spleens were weighed, and the spleen index was calculated as the percent of the animal weight on D5.

3.6.5. Histology

For histological analysis, following the sacrifice, the back skin samples were removed and fixed with 10% formaldehyde, embedded into paraffin blocks, cut, and finally stained with hematoxylin-eosin (H&E). The tissue sections on slides were then micrographed with a light microscope (Nikon Eclipse Ts2, Tokyo, Japan) with ×20 magnification. For each slide (corresponding to a specific animal), 12 micrographs were taken. The epidermal thickness was evaluated through area and length measurements [50] using Fiji software (version 2.9.0/1.54f) [61]. To avoid bias, the area of the viable epidermis containing the blue-purple stained nuclei was measured with the "wand" function. The epidermal thickness for each mouse was calculated as an average of 12 measurements.

3.6.6. Evaluation of IL-17A Levels in the Skin Tissue

For IL-17A evaluation, following the sacrifice, the back skin samples were snap-frozen and then dehydrated in a lyophilizer. The lyophilized samples were ground in a mortar with column sand (Sigma-Aldrich Inc., St. Louis, MO, USA) and then homogenized in a Tissue Extraction Reagent I (Thermo Fischer Scientific, Waltham, MA, USA) with Polytron homogenizer (Kinematica, Malters, Switzerland). Homogenates were centrifuged for 5 min at 10,000 RPM, and the supernatants were analyzed for protein and IL-17A content. The protein content was analyzed with Bradford assay using Protein Assay Dye Reagent Concentrate (Bio-Rad Laboratories, Hercules, CA, USA). The IL-17A content was evaluated by enzyme-linked immunosorbent assay (ELISA). The ELISA was performed according to the manufacturer's instructions for the kit (ELISA Max, Biolegend, San Diego, CA, USA). The results for each skin extract (corresponding to a specific animal) were obtained by normalization of the IL-17A content by the protein content of each sample.

3.7. Statistical Analysis

Data analysis was performed using the Graph-Pad Prism software (Version 5.01, San Diego, CA, USA). Data were expressed as mean ± standard deviation (SD) or original data represented. The one-way analysis of variance (ANOVA) followed by Tukey's post hoc test was used to compare groups. The p-value < 0.05 was considered statistically significant.

4. Conclusions

The present study demonstrates the potential of LSNs as a versatile platform for precisely delivering CBD to the skin. Formulation studies involving fumed silica, MAS, and Poloxamer thermogel underscored the critical role of vehicle selection and optimization for effective nanoparticle dispersion. In vitro skin permeation testing revealed that fumed silica was significantly more effective than other gelling agents. The in-vivo assessments using LSNs loaded with fluorescent marker C6 confirmed successful permeation and localization within the viable epidermis. Utilizing an IMQ-induced psoriasis model in mice provided additional evidence of CBD's efficacy in treating psoriasis. CBD-loaded LSNs significantly reduced the PASI score and acanthosis, as well as inhibited the IL-17A release compared to the control treatment groups, indicating a substantial improvement in psoriasis symptoms. Moreover, the difference in response between CBD-loaded LSNs and CBD emulsion suggests deeper skin penetration and localization due to the LSN formulation. Lastly, the variation in the anti-inflammatory response among LSNs stabilized with CA, LA, or SALA highlighted the significance of the stabilizing lipid selection, as LSNs stabilized with CA and SALA demonstrated greater anti-inflammatory effect compared to LA-stabilized LSNs. This has emphasized the importance of lipid selection in topical drug delivery.

Author Contributions: Conceptualization, M.Z., V.F., A.C.S. and S.B.-S.; methodology, M.Z., V.F. and D.Y.; software, M.Z.; validation, M.Z. and D.Y.; formal analysis, M.Z., V.F., A.C.S. and S.B.-S.; investigation, M.Z., V.F. and D.Y; resources, A.C.S. and S.B.-S.; data curation, M.Z., D.Y., A.C.S. and S.B.-S.; writing—original draft preparation, M.Z.; writing—review and editing, A.C.S. and S.B.-S.; visualization, M.Z., A.C.S. and S.B.-S.; supervision, A.C.S. and S.B.-S.; project administration, A.C.S. and S.B.-S.; funding acquisition, A.C.S. and S.B.-S. All authors have read and agreed to the published version of the manuscript.

Funding: This research did not receive any specific grant from funding agencies in the public, commercial, or not-for-profit sectors.

Institutional Review Board Statement: The animal treatments were performed in accordance with the protocol reviewed and approved by the Institutional Committee for the Ethical Care and Use of Animals in Experiments, Ben-Gurion University of the Negev, which complies with the Israeli Law of Human Care and Use of Laboratory Animals, authorization number: IL-30-06-2020 (C) and IL-64-11-2021(C).

Informed Consent Statement: Not applicable.

Data Availability Statement: Data is available upon request.

Acknowledgments: The authors are grateful for the professional assistance and technical support of the staff at Ilze Katz Institute for Nanoscale Science & Technology. The authors want to especially express their appreciation to Ilya Eydelman for his support and helpful assistance.

Conflicts of Interest: The authors declare no conflict of interest.

Sample Availability: Not applicable.

References

1. Peng, J.; Fan, M.; An, C.; Ni, F.; Huang, W.; Luo, J. A Narrative Review of Molecular Mechanism and Therapeutic Effect of Cannabidiol (CBD). *Basic Clin. Pharmacol. Toxicol.* **2022**, *130*, 439–456. [CrossRef] [PubMed]
2. Tham, M.; Yilmaz, O.; Alaverdashvili, M.; Kelly, M.E.M.; Denovan-Wright, E.M.; Laprairie, R.B. Allosteric and Orthosteric Pharmacology of Cannabidiol and Cannabidiol-Dimethylheptyl at the Type 1 and Type 2 Cannabinoid Receptors. *Br. J. Pharmacol.* **2019**, *176*, 1455–1469. [CrossRef] [PubMed]
3. Martínez-Pinilla, E.; Varani, K.; Reyes-Resina, I.; Angelats, E.; Vincenzi, F.; Ferreiro-Vera, C.; Oyarzabal, J.; Canela, E.I.; Lanciego, J.L.; Nadal, X.; et al. Binding and Signaling Studies Disclose a Potential Allosteric Site for Cannabidiol in Cannabinoid CB2 Receptors. *Front. Pharmacol.* **2017**, *8*, 744. [CrossRef]
4. Thomas, A.; Baillie, G.L.; Phillips, A.M.; Razdan, R.K.; Ross, R.A.; Pertwee, R.G. Cannabidiol Displays Unexpectedly High Potency as an Antagonist of CB1 and CB2 Receptor Agonists in Vitro. *Br. J. Pharmacol.* **2007**, *150*, 613–623. [CrossRef] [PubMed]
5. Ribeiro, A.; Ferraz-De-Paula, V.; Pinheiro, M.L.; Vitoretti, L.B.; Mariano-Souza, D.P.; Quinteiro-Filho, W.M.; Akamine, A.T.; Almeida, V.I.; Quevedo, J.; Dal-Pizzol, F.; et al. Cannabidiol, a Non-Psychotropic Plant-Derived Cannabinoid, Decreases Inflammation in a Murine Model of Acute Lung Injury: Role for the Adenosine A2A Receptor. *Eur. J. Pharmacol.* **2012**, *678*, 78–85. [CrossRef] [PubMed]
6. Sermet, S.; Li, J.; Bach, A.; Crawford, R.B.; Kaminski, N.E. Cannabidiol Selectively Modulates Interleukin (IL)-1β and IL-6 Production in Toll-like Receptor Activated Human Peripheral Blood Monocytes. *Toxicology* **2021**, *464*, 153016. [CrossRef] [PubMed]
7. Jarocka-Karpowicz, I.; Biernacki, M.; Wroński, A.; Gęgotek, A.; Skrzydlewska, E. Cannabidiol Effects on Phospholipid Metabolism in Keratinocytes from Patients with Psoriasis Vulgaris. *Biomolecules* **2020**, *10*, 367. [CrossRef]
8. Wilkinson, J.D.; Williamson, E.M. Cannabinoids Inhibit Human Keratinocyte Proliferation through a Non-CB1/CB2 Mechanism and Have a Potential Therapeutic Value in the Treatment of Psoriasis. *J. Dermatol. Sci.* **2007**, *45*, 87–92. [CrossRef]
9. Sangiovanni, E.; Fumagalli, M.; Pacchetti, B.; Piazza, S.; Magnavacca, A.; Khalilpour, S.; Melzi, G.; Martinelli, G.; Dell'Agli, M. Cannabis sativa L. Extract and Cannabidiol Inhibit in Vitro Mediators of Skin Inflammation and Wound Injury. *Phytother. Res.* **2019**, *33*, 2083–2093. [CrossRef]
10. Patzelt, A.; Lademann, J. Drug Delivery to Hair Follicles. *Expert Opin. Drug Deliv.* **2013**, *10*, 787–797. [CrossRef]
11. Blume-Peytavi, U.; Vogt, A. Human Hair Follicle: Reservoir Function and Selective Targeting. *Br. J. Dermatol.* **2011**, *165*, 13–17. [CrossRef] [PubMed]
12. Toll, R.; Jacobi, U.; Richter, H.; Lademann, J.; Schaefer, H.; Blume-Peytavi, U. Penetration Profile of Microspheres in Follicular Targeting of Terminal Hair Follicles. *J. Investig. Dermatol.* **2004**, *123*, 168–176. [CrossRef] [PubMed]
13. Palmer, B.C.; DeLouise, L.A. Nanoparticle-Enabled Transdermal Drug Delivery Systems for Enhanced Dose Control and Tissue Targeting. *Molecules* **2016**, *21*, 1719. [CrossRef] [PubMed]
14. Shah, K.A.; Date, A.A.; Joshi, M.D.; Patravale, V.B. Solid Lipid Nanoparticles (SLN) of Tretinoin: Potential in Topical Delivery. *Int. J. Pharm.* **2007**, *345*, 163–171. [CrossRef] [PubMed]
15. Lodzki, M.; Godin, B.; Rakou, L.; Mechoulam, R.; Gallily, R.; Touitou, E. Cannabidiol–Transdermal Delivery and Anti-Inflammatory Effect in a Murine Model. *J. Control. Release* **2003**, *93*, 377–387. [CrossRef] [PubMed]
16. Sosnik, A.; Shabo, R.B.; Halamish, H.M. Cannabidiol-Loaded Mixed Polymeric Micelles of Chitosan/Poly(Vinyl Alcohol) and Poly(Methyl Methacrylate) for Trans-Corneal Delivery. *Pharmaceutics* **2021**, *13*, 2142. [CrossRef] [PubMed]
17. Zamansky, M.; Zehavi, N.; Ben-Shabat, S.; Sintov, A.C. Characterization of Nanoparticles Made of Ethyl Cellulose and Stabilizing Lipids: Mode of Manufacturing, Size Modulation, and Study of Their Effect on Keratinocytes. *Int. J. Pharm.* **2021**, *607*, 121003. [CrossRef] [PubMed]

18. Zamansky, M.; Zehavi, N.; Sintov, A.C.; Ben-Shabat, S. The Fundamental Role of Lipids in Polymeric Nanoparticles: Dermal Delivery and Anti-Inflammatory Activity of Cannabidiol. *Molecules* **2023**, *28*, 1774. [CrossRef]
19. Swindell, W.R.; Michaels, K.A.; Sutter, A.J.; Diaconu, D.; Fritz, Y.; Xing, X.; Sarkar, M.K.; Liang, Y.; Tsoi, A.; Gudjonsson, J.E.; et al. Imiquimod Has Strain-Dependent Effects in Mice and Does Not Uniquely Model Human Psoriasis. *Genome Med.* **2017**, *9*, 24. [CrossRef]
20. van der Fits, L.; Mourits, S.; Voerman, J.S.A.; Kant, M.; Boon, L.; Laman, J.D.; Cornelissen, F.; Mus, A.-M.; Florencia, E.; Prens, E.P.; et al. Imiquimod-Induced Psoriasis-Like Skin Inflammation in Mice Is Mediated via the IL-23/IL-17 Axis. *J. Immunol.* **2009**, *182*, 5836–5845. [CrossRef]
21. Reis, C.P.; Martinho, N.; Rosado, C.; Fernandes, A.S.; Roberto, A. Design of Polymeric Nanoparticles and Its Applications as Drug Delivery Systems for Acne Treatment. *Drug Dev. Ind. Pharm.* **2014**, *40*, 409–417. [CrossRef] [PubMed]
22. Jain, A.K.; Jain, A.; Garg, N.K.; Agarwal, A.; Jain, A.; Jain, S.A.; Tyagi, R.K.; Jain, R.K.; Agrawal, H.; Agrawal, G.P. Adapalene Loaded Solid Lipid Nanoparticles Gel: An Effective Approach for Acne Treatment. *Colloids Surf. B Biointerfaces* **2014**, *121*, 222–229. [CrossRef] [PubMed]
23. Shi, Z.; Pan, S.; Wang, L.; Li, S. Topical Gel Based Nanoparticles for the Controlled Release of Oleanolic Acid: Design and in Vivo Characterization of a Cubic Liquid Crystalline Anti-Inflammatory Drug. *BMC Complement. Med. Ther.* **2021**, *21*, 224. [CrossRef] [PubMed]
24. Manna, S.; Lakshmi, U.S.; Racharla, M.; Sinha, P.; Kanthal, L.K.; Kumar, S.P.N. Bioadhesive HPMC Gel Containing Gelatin Nanoparticles for Intravaginal Delivery of Tenofovir. *J. Appl. Pharm. Sci.* **2016**, *6*, 22–29. [CrossRef]
25. Cai, X.J.; Mesquida, P.; Jones, S.A. Investigating the Ability of Nanoparticle-Loaded Hydroxypropyl Methylcellulose and Xanthan Gum Gels to Enhance Drug Penetration into the Skin. *Int. J. Pharm.* **2016**, *513*, 302–308. [CrossRef] [PubMed]
26. Al-Kassas, R.; Wen, J.; Cheng, A.E.M.; Kim, A.M.J.; Liu, S.S.M.; Yu, J. Transdermal Delivery of Propranolol Hydrochloride through Chitosan Nanoparticles Dispersed in Mucoadhesive Gel. *Carbohydr. Polym.* **2016**, *153*, 176–186. [CrossRef] [PubMed]
27. Rozman, B.; Gosenca, M.; Gasperlin, M.; Padois, K.; Falson, F. Dual Influence of Colloidal Silica on Skin Deposition of Vitamins C and e Simultaneously Incorporated in Topical Microemulsions. *Drug Dev. Ind. Pharm.* **2010**, *36*, 852–860. [CrossRef] [PubMed]
28. Schliemann, S.; Petri, M.; Elsner, P. How Much Skin Protection Cream Is Actually Applied in the Workplace? Determination of Dose per Skin Surface Area in Nurses. *Contact Dermat.* **2012**, *67*, 229–233. [CrossRef]
29. Stenberg, C.; Larkö, O. Sunscreen Application and Its Importance for the Sun Protection Factor. *Arch. Dermatol.* **1985**, *11*, 1400–1402. [CrossRef]
30. Liu, T.; Chu, B. Formation of Homogeneous Gel-like Phases by Mixed Triblock Copolymer Micelles in Aqueous Solution: FCC to BCC Phase Transition. *J. Appl. Cryst.* **2000**, *33*, 727–730. [CrossRef]
31. Brindley, G.W. Clays, Clay Minerals. In *Mineralogy*; Springer: Boston, MA, USA, 1981; pp. 69–80. [CrossRef]
32. Sorichetti, V.; Hugouvieux, V.; Kob, W. Dynamics of Nanoparticles in Polydisperse Polymer Networks: From Free Diffusion to Hopping. *Macromolecules* **2021**, *54*, 8575–8589. [CrossRef]
33. Dell, Z.E.; Schweizer, K.S. Theory of Localization and Activated Hopping of Nanoparticles in Cross-Linked Networks and Entangled Polymer Melts. *Macromolecules* **2014**, *47*, 405–414. [CrossRef]
34. Binder, L.; Mazál, J.; Petz, R.; Klang, V.; Valenta, C. The Role of Viscosity on Skin Penetration from Cellulose Ether-Based Hydrogels. *Skin Res. Technol.* **2019**, *25*, 725–734. [CrossRef] [PubMed]
35. Finke, J.H.; Richter, C.; Gothsch, T.; Kwade, A.; Büttgenbach, S.; Müller-Goymann, C.C. Coumarin 6 as a Fluorescent Model Drug: How to Identify Properties of Lipid Colloidal Drug Delivery Systems via Fluorescence Spectroscopy? *Eur. J. Lipid Sci. Technol.* **2014**, *116*, 1234–1246. [CrossRef]
36. Niczyporuk, M. Rat Skin as an Experimental Model in Medicine. *Prog. Health Sci.* **2018**, *8*, 223–228. [CrossRef]
37. Marquet, F.; Grandclaude, M.-C.; Ferrari, E.; Champmartin, C. Capacity of an in vitro rat skin model to predict human dermal absorption: Influences of aging and anatomical site. *Toxicol. In Vitro* **2019**, *61*, 104623. [CrossRef] [PubMed]
38. Dorjsembe, B.; Ham, J.Y.; Kim, J.C. The Imiquimod Induced Psoriatic Animal "Model: Scientific Implications. *Biomed. J. Sci. Tech. Res.* **2019**, *13*, 9722–9724. [CrossRef]
39. Li, L.; Liu, X.; Ge, W.; Chen, C.; Huang, Y.; Jin, Z.; Zhan, M.; Duan, X.; Liu, X.; Kong, Y.; et al. CB2R Deficiency Exacerbates Imiquimod-Induced Psoriasiform Dermatitis and Itch Through the Neuro-Immune Pathway. *Front. Pharmacol.* **2022**, *13*, 790712. [CrossRef]
40. Zhang, J.; Yang, X.; Qiu, H.; Chen, W. Weight Loss May Be Unrelated to Dietary Intake in the Imiquimod-Induced Plaque Psoriasis Mice Model. *Open Life Sci.* **2020**, *15*, 79–82. [CrossRef]
41. Alvarez, P.; Jensen, L.E. Imiquimod Treatment Causes Systemic Disease in Mice Resembling Generalized Pustular Psoriasis in an IL-1 and IL-36 Dependent Manner. *Mediat. Inflamm.* **2016**, *2016*, 6756138. [CrossRef]
42. Ignatowska-Jankowska, B.; Jankowski, M.M.; Swiergiel, A.H. Cannabidiol Decreases Body Weight Gain in Rats: Involvement of CB2 Receptors. *Neurosci. Lett.* **2011**, *490*, 82–84. [CrossRef] [PubMed]
43. Pinto, J.S.; Martel, F. Effects of Cannabidiol on Appetite and Body Weight: A Systematic Review. *Clin. Drug Investig.* **2022**, *42*, 909–919. [CrossRef] [PubMed]
44. Ewing, L.E.; Skinner, C.M.; Quick, C.M.; Kennon-McGill, S.; McGill, M.R.; Walker, L.A.; ElSohly, M.A.; Gurley, B.J.; Koturbash, I. Hepatotoxicity of a Cannabidiol-Rich Cannabis Extract in the Mouse Model. *Molecules* **2019**, *24*, 1694. [CrossRef] [PubMed]

45. D'Souza, L.; Badanthadka, M.; Salwa, F. Effect of Animal Strain on Model Stability to Imiquimod-Induced Psoriasis. *Indian J. Physiol. Pharmacol.* **2020**, *64*, 83–91.
46. Salwa, F.; Badanthadka, M.; D'Souza, L. Differential Psoriatic Effect of Imiquimod on Balb/c and Swiss Mice. *J. Health Allied Sci.* **2021**, *11*, 170–177. [CrossRef]
47. Shinno-Hashimoto, H.; Eguchi, A.; Sakamoto, A.; Wan, X.; Hashimoto, Y.; Fujita, Y.; Mori, C.; Hatano, M.; Matsue, H.; Hashimoto, K. Effects of Splenectomy on Skin Inflammation and Psoriasis-like Phenotype of Imiquimod-Treated Mice. *Sci. Rep.* **2022**, *12*, 14738. [CrossRef] [PubMed]
48. Krummen, M.B.W.; Varga, G.; Steinert, M.; Stuetz, A.; Luger, T.A.; Grabbe, S. Effect of Pimecrolimus vs. Corticosteroids on Murine Bone Marrow-Derived Dendritic Cell Differentiation, Maturation and Function. *Exp. Dermatol.* **2006**, *15*, 43–50. [CrossRef]
49. Schoepe, S.; Schäcke, H.; May, E.; Asadullah, K. Glucocorticoid Therapy-Induced Skin Atrophy. *Exp. Dermatol.* **2006**, *15*, 406–420. [CrossRef]
50. Shobeiri, S.S.; Rezaee, M.A.; Pordel, S.; Haghnnavaz, N.; Dashti, M.; Moghadam, M.; Sankian, M. Anti-IL-17A SsDNA Aptamer Ameliorated Psoriasis Skin Lesions in the Imiquimod-Induced Psoriasis Mouse Model. *Int. Immunopharmacol.* **2022**, *110*, 108963. [CrossRef]
51. Li, Q.; Liu, W.; Gao, S.; Mao, Y.; Xin, Y. Application of Imiquimod-Induced Murine Psoriasis Model in Evaluating Interleukin-17A Antagonist. *BMC Immunol.* **2021**, *22*, 11. [CrossRef]
52. Kozela, E.; Juknat, A.; Kaushansky, N.; Rimmerman, N.; Ben-Nun, A.; Vogel, Z. Cannabinoids Decrease the Th17 Inflammatory Autoimmune Phenotype. *J. Neuroimmune Pharmacol.* **2013**, *8*, 1265–1276. [CrossRef] [PubMed]
53. Hegde, V.L.; Nagarkatti, P.S.; Nagarkatti, M. Role of Myeloid-Derived Suppressor Cells in Amelioration of Experimental Autoimmune Hepatitis Following Activation of TRPV1 Receptors by Cannabidiol. *PLoS ONE* **2011**, *6*, e18281. [CrossRef] [PubMed]
54. Wang, J.; Wu, X.; Simonavicius, N.; Tian, H.; Ling, L. Medium-Chain Fatty Acids as Ligands for Orphan G Protein-Coupled Receptor GPR84. *J. Biol. Chem.* **2006**, *281*, 34457–34464. [CrossRef] [PubMed]
55. Laine, P.S.; Schwartz, E.A.; Wang, Y.; Zhang, W.Y.; Karnik, S.K.; Musi, N.; Reaven, P.D. Palmitic Acid Induces IP-10 Expression in Human Macrophages via NF-KB Activation. *Biochem. Biophys. Res. Commun.* **2007**, *358*, 150–155. [CrossRef]
56. Hammer, A.; Schliep, A.; Jörg, S.; Haghikia, A.; Gold, R.; Kleinewietfeld, M.; Müller, D.N.; Linker, R.A. Impact of Combined Sodium Chloride and Saturated Long-Chain Fatty Acid Challenge on the Differentiation of T Helper Cells in Neuroinflammation. *J. Neuroinflamm.* **2017**, *14*, 184. [CrossRef]
57. Brodin, A.; Nyqvist-Mayer, A.; Broberg, F.; Wadsten, T.; Forslund, B. Phase Diagram and Aqueous Solubility of the Lidocaine-prilocaine Binary System. *J. Pharm. Sci.* **1984**, *73*, 481–484. [CrossRef]
58. Silva, J.M.; Pereira, C.V.; Mano, F.; Silva, E.; Castro, V.I.B.; Sá-Nogueira, I.; Reis, R.L.; Paiva, A.; Matias, A.A.; Duarte, A.R.C. Therapeutic Role of Deep Eutectic Solvents Based on Menthol and Saturated Fatty Acids on Wound Healing. *ACS Appl. Bio Mater.* **2019**, *2*, 4346–4355. [CrossRef]
59. Provencher, S.W. A Constrained Regularization Method for Inverting Data Represented by Linear Algebraic or Integral Equations. *Comput. Phys. Commun.* **1982**, *27*, 213–227. [CrossRef]
60. Casiraghi, A.; Musazzi, U.M.; Centin, G.; Franzè, S.; Minghetti, P. Topical Administration of Cannabidiol: Influence of Vehicle-Related Aspects on Skin Permeation Process. *Pharmaceuticals* **2020**, *13*, 337. [CrossRef]
61. Schindelin, J.; Arganda-Carreras, I.; Frise, E.; Kaynig, V.; Longair, M.; Pietzsch, T.; Preibisch, S.; Rueden, C.; Saalfeld, S.; Schmid, B.; et al. Fiji: An Open-Source Platform for Biological-Image Analysis. *Nat. Methods* **2012**, *9*, 676–682. [CrossRef]
62. Sundberg, J.P.; Silva, K.A. What Color Is the Skin of a Mouse? *Vet. Pathol.* **2012**, *49*, 142–145. [CrossRef]
63. Jabeen, M.; Boisgard, A.S.; Danoy, A.; Kholti, N.E.; Salvi, J.P.; Boulieu, R.; Fromy, B.; Verrier, B.; Lamrayah, M. Advanced Characterization of Imiquimod-induced Psoriasis-like Mouse Model. *Pharmaceutics* **2020**, *12*, 789. [CrossRef]

Disclaimer/Publisher's Note: The statements, opinions and data contained in all publications are solely those of the individual author(s) and contributor(s) and not of MDPI and/or the editor(s). MDPI and/or the editor(s) disclaim responsibility for any injury to people or property resulting from any ideas, methods, instructions or products referred to in the content.

Review

Metal Sulfide Nanoparticles for Imaging and Phototherapeutic Applications

Aishwarya Shetty [1], Heinrich Lang [2,*] and Sudeshna Chandra [3,*]

1. Journal of Visualized Experiments 625, Massachusetts Avenue, Cambridge, MA 02139, USA
2. Chemnitz Research Group Organometallics, MAIN Research Center, Technische Universität, Rosenbergstr. 6, 09126 Chemnitz, Germany
3. Institute of Analytical Chemistry, University of Regensburg, 93040 Regensburg, Germany
* Correspondence: heinrich.lang@chemie.tu-chemnitz.de (H.L.); chandra.sudeshna@chemie.uni-regensburg.de (S.C.)

Abstract: The intriguing properties of metal sulfide nanoparticles (=MxSy-NPs), particularly transition metal dichalcogenides, are discussed for their use in diverse biological applications. Herein, recent advances in MxSy-NPs-based imaging (MRI, CT, optical and photoacoustic) and phototherapy (photothermal and photodynamic) are presented. Also, recent made progress in the use of immuno-phototherapy combinatorial approaches in vitro and in vivo are reported. Furthermore, challenges in nanomaterials-based therapies and future research directions by applying MxSy-NPs in combinatorial therapies are envisaged.

Keywords: metal sulfide nanoparticles; bioimaging; photothermal therapy; photodynamic therapy; immunotherapy

1. Introduction

In recent years, applications of nanotechnology have expanded into different branches of the biomedical field [1–3]. Efforts are continually being made towards the development of unique nanoparticles (=NPs) which can overcome limitations of traditional therapeutics and, hence, are able to improve management of diseases [4]. Large surface area-to-volume ratios of NPs provide a platform for easy chemical functionalization for excellent interaction with biological systems. Among the broad range of NPs studied for biomedical applications, metal sulfide nanoparticles (=M_xS_y-NPs) have been the focus of several studies in recent years [5–7]. In addition to properties found at the nanoscale, M_xS_y-NPs also exhibit favorable properties such as light conversion, Fenton catalysis, immune activation and radiation enhancement [8,9]. The lower electronegativity of sulfur in comparison to oxygen makes M_xS_y-NPs naturally versatile in comparison to highly exploited metal oxide ones [10]. The versatility of M_xS_y-NPs becomes evident by the fact that they can be successfully used for various applications including different types of imaging and therapy, often alone or in combination with other materials to enhance their intended application [11]. In addition, M_xS_y-NPs possess the ability to impart multiple functionalities as "stand-alone" systems without addition of other materials. For example, transition metal dichalcogenide-based molybdenum disulfide (MoS_2-) and tungsten disulfide (WS_2-) NPs are increasingly found in theranostic and biosensing applications [12,13]. Tunable bandgap and strong spin-orbit coupling make MoS_2-NPs particularly interesting for biomedical applications, whereas strong near-infrared (NIR) absorptions has led to the efficacious use of copper sulfide (CuS-) NPs as photothermal agents [14,15].

Hence, herein, various uses of M_xS_y-NPs towards the above-mentioned background will be discussed on selected examples.

2. Applications of Metal Sulfide Nanoparticles in Bioimaging

2.1. Magnetic Resonance Imaging

As a result of the use of non-ionizing radiation, high spatial resolution and non-invasive magnetic resonance imaging (=MRI) has become one of the most used imaging techniques in the medical field [16]. MRI makes use of pulsed magnetic waves to align protons present in water and images are produced by recording radio-waves released by these protons upon their relaxation to the ground state [17]. Contrast agents are applied to significantly improve resolution and work by reducing the longitudinal or transverse (i.e., T_1 or T_2) relaxation time of protons in water [18]. Studies on NPs for MR imaging have mostly focused on metal oxides such as superparamagnetic iron oxide NPs (SPIONs); however, in recent years, researchers have begun exploring M_xS_y-NPs as well [19,20]. Examples of such studies reporting the use of M_xS_y-NPs, wherein the MR contrast is brought about by the metal sulfide itself, are highlighted below.

Iron sulfide quantum dots (=FeS QDs) were synthesized via a biomimetic route using protein bovine serum albumin (=BSA) as a template. Nanoparticles based on FeS exhibit physicochemical properties similar to that of iron oxide nanoparticles as sulfur and oxygen are congeneric elements. However, iron sulfide (FeS, $Fe_{1-x}S$, FeS_2, Fe_3S_4) exist in more phases than iron oxide (Fe_3O_4, Fe_2O_3) showing more variability and also have a smaller band gap. The authors observed a strong NIR absorption which was exploited for photoacoustic imaging, whereas quantum confinement effects enabled fluorescence imaging. The longitudinal relaxation (=r_1) value of FeS QDs (5.35 mM^{-1} s^{-1}) was found to be higher than that of corresponding aggregates (0.2 mM^{-1} s^{-1}), which is attributed to the template-assisted synthesis [21]. The resulting QDs thus showed good dispersion, higher longitudinal relaxivity, extended rotational correlation time and lower magnetization in comparison to the clinically used gadolinium-based MRI contrast agent Gd-DTPA (r_1 = 3.1 mM^{-1} s^{-1}). As observed in Figure 1, the authors tested the MR, PA and fluorescence imaging ability of FeS QDs in vivo in 4T1 tumor-bearing mice post-intravenous (i.v.) administration [22]. As can be seen in Figure 1A,B, MR contrasts at 5 h post-administration is 1.8-fold higher as compared to pre-administration.

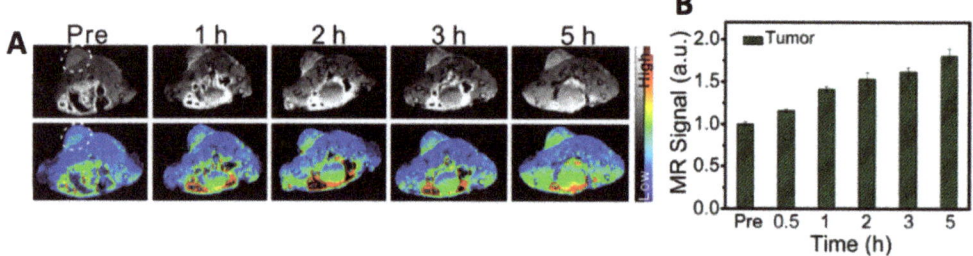

Figure 1. Representative images of (**A**) MR imaging and (**B**) its quantitative estimation. Reprinted with permission from Ref. [22]. Copyright © 2023, Elsevier.

A nanohybrid (=NH), based on the sulfides of bismuth and iron was prepared by Xiong et al. via biomineralization using BSA to yield Bi_2S_3/FeS_2@BSA NHs [23]. BSA acted as a source of sulfur, as a template for the synthesis and as a reducing agent, whereas Fe and Bi provided the contrast for MR and computed tomography (=CT) imaging, respectively. The X-ray absorption coefficient of the NHs is 8.02 HU mM^{-1} which increased in proportion to increasing concentrations of Bi. A similar trend was observed for MRI contrast and r_2, i.e., transverse relaxivity time was determined to 53.9 mM^{-1} s^{-1}. In vivo, Bi_2S_3/FeS_2@BSA NHs showed accumulation in the tumor with good CT and MR imaging contrast when injected intravenously in a 4T1 tumor-bearing mice [23]. Fu et al. exploited magnetocaloric and MR imaging properties of iron sulfide for imaging-guided thrombolysis in celiac vein thrombosis. The author's synthesized hydrophilic polyvinyl pyrrolidone-capped

Fe$_3$S$_4$-NPs with an r$_2$ value of 53.1 mM^{-1} s^{-1} [24]. Through simultaneous exposure to an alternating magnetic field (=AMF) and an 808 nm laser, the NP dispersion attained a temperature higher than when exposed to AMF or laser alone. In vitro, the synergistic thermal conversion resulted in near disappearance of the thrombus, whereas individual stimulation resulted in partial dissolution. When tested in a C57 mice model of deep vein thrombosis, it resulted in the reduction of thrombus, which was visualized by MR imaging. Unpaired 3D electrons in cobalt (Co) were utilized by Lv and colleagues for T$_2$-weighted MRI [25]. Therefore, the authors prepared hollow cobalt sulfide (Co$_3$S$_4$-) NPs which were coated with a shell of N-doped carbon and encapsulated the drug doxorubicin for therapeutic (chemotherapy, photothermal therapy and photodynamic therapy) and imaging (MRI and thermal imaging) applications [26]. The respective NPs showed a concentration-dependent increase in MR and thermal imaging contrast. In vivo, when tested in H22 tumor bearing mice, the nanoparticles showed a good contrast as compared to pre-treatment. Huang et al. synthesized Cu$_{2-x}$S@MnS core-shell NPs in which the Cu$_{2-x}$S-NPs are surrounded by a manganese sulfide (MnS) shell [27]. NIR absorption by CuS enabled photothermal treatment, whereas the presence of MnS facilitated light-triggered photodynamic therapy (PDT) and MRI. The NPs showed high photothermal conversion efficiency (47.9%) and ability to generate reactive oxygen species (=ROS) in the presence of hydrogen peroxide. With respect to MRI, T$_1$ contrast increased in proportion to the concentration of manganese and an r$_1$ value of 1.243 mM^{-1} s^{-1} was reported. Similarly, Chen et al. reported on the assembly of CuS-MnS$_2$ nanoflowers for MRI-guided photothermal-photodynamic therapy [28].

2.2. Computed Tomography

In CT imaging, differential tissue thicknesses and X-ray attenuations are exploited to generate three-dimensional and cross-sectional images [29]. High X-ray absorption as a consequence of high atomic numbers has resulted in the application of bismuth (Bi) and tungsten as CT contrast agents [30,31]. PEGylated-WS$_2$-NPs, i.e., polyethylene glycol (PEG)-coated tungsten disulfide NPs for CT-guided photothermal therapy (PTT) were prepared by Wang and colleagues [32]. The CT-imaging ability of the NPs was tested in 4T1 tumor-bearing mice using phosphate-buffered saline (=PBS)-treated mice as a control group. In conclusion, good photothermal stability and an effective use as CT contrast agents were reported. Similarly, Wang et al. introduced manganese dioxide (MnO$_2$-) coated mesoporous polydopamine nanosponges (=MPDA NSs) embedded with WS$_2$ nanodots (=ND), i.e., MPDA-WS$_2$@MnO$_2$ for multimodal imaging guided thermo-radiotherapy of cancer [33]. WS$_2$ NDs and MPDA NSs enabled radio-sensitization and PTT in addition to contrast for CT and multi-spectral optoacoustic tomography (=MSOT), respectively. The MnO$_2$ component provided MRI contrast and tumor hypoxia modulating properties. In all three imaging modalities, the contrast provided by MPDA-WS$_2$@MnO$_2$-NPs increased linearly with increasing concentration of the NPs. The authors reported a CT value of 35.3 HU L g^{-1} and a transverse relaxation value of 6.696 mM^{-1} S^{-1} at pH 6.5. Post intratumoral (=i.t.) and intravenous (=i.v.) administrations. In vivo, an 8- and 2.5-fold increase in signal intensity was observed for CT and MSOT imaging, respectively. Similar results were also observed for MRI.

Nosrati et al. used bismuth sulfide (Bi$_2$S$_3$-) NPs for combination therapy including chemotherapy and radiotherapy guided by CT imaging [34]. The Bi$_2$S$_3$-NPs were coated with BSA to improve their stability followed by curcumin encapsulation and functionalization with folic acid to yield Bi$_2$S$_3$@BSA-FA-CUR NPs. The NPs showed sustained release of curcumin, radio-sensitization effects and a linear increase in CT contrast with increasing Bi concentration. Similarly, Bi$_2$S$_3$@MSNs, i.e., bismuth sulfide NPs coated with mesoporous silica, were synthesized to enable drug delivery in addition to NIR-responsive PTT and CT imaging [35]. The presence of mesoporous pores in silica enabled high drug loadings up to 99%, whereas the presence of Bi resulted in a high photothermal conversion efficiency of 37%. Figure 2A shows the in vitro CT performance of Bi$_2$S$_3$@MSNs showing a linear increase with increasing Bi concentration [35]. As can be seen in the figure, the slope of iobitridol (25.63 HU L g^{-1}) is lower than that of Bi$_2$S$_3$@MSNs (32.83 HU L g^{-1}). In vivo, the authors evaluated the CT contrast to assess the active targeting potential of RGD (targeting ligand containing arginine(R)-glycine(G)-aspartate(D) triad) conjugated Bi$_2$S$_3$@MSNs. RGD–Bi$_2$S$_3$@MSNs show a good accumulation at the tumor site resulting in an increased CT signal from 2–24 h post-i.v. injection as compared to Bi$_2$S$_3$@MSNs (Figure 2B).

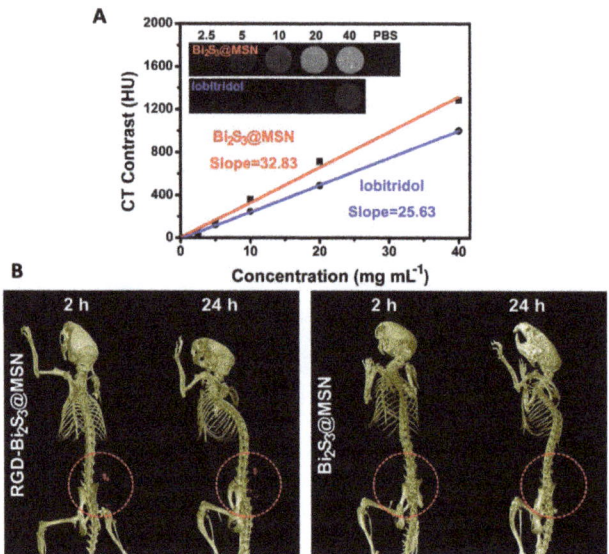

Figure 2. (**A**) In vitro CT performance of Bi$_2$S$_3$@MSNs in comparison with commercially available iobitridol. Inset: Suspensions of Bi$_2$S$_3$@MSNs and iobitridol at different concentrations showing CT contrast. (**B**) Representative CT images of UMR-106 tumor-bearing nude mice showing contrast provided by RGD–Bi$_2$S$_3$@MSN and Bi$_2$S$_3$@MSN captured 2 and 24 h post-treatment. The red circle highlights the tumor site. Reprinted with permission from Ref. [35]. Copyright © 2023 Wiley.

Wang et al. reported the synthesis of hydrophobic Cu$_3$BiS$_3$-NPs and their use for targeted photodynamic/photothermal therapy and CT/MR dual modal imaging [36]. Modifications to the NPs included coating with DSPE-PEG/DSPE-PEG-NH$_2$ (DSPE: 1, 2-Distearoyl-*sn*-glycero-3-phosphoethanolamine-Poly (ethylene glycol)) for hydrophilicity, conjugation of photosensitizer chlorin e6 (=Ce6) and functionalization with folic acid for targeting. The X-ray co-efficient value of Cu$_3$BiS$_3$-NPs was calculated as 17.7 HU mmol Bi/L, whereas r$_1$ relaxivity was found to be twice that of Gd-DTPA, which is a clinically used T$_1$-MRI contrast agent. In vivo, these translated into significant CT and MR contrast which peaked at 4–6 h post-i.v. injection via the tail vein. For MRI, a 281.6% increase in signal intensity was observed 6 h post-injection, whereas a quantitative CT value of 252.3 ± 25 HU was observed. Combined, the NPs were able to successfully accumulate at the tumor site and inhibit tumor growth in vivo [35]. In addition, Wang et al. discussed the use of rhenium disulfide (ReS$_2$-) NPs as gastrointestinal (=GI) tract and tumor imaging probes, due to their excellent X-ray and NIR absorption properties [37]. With respect to GI tract imaging, the ReS$_2$-NPs showed a higher signal-to-noise ratio with increasing X-ray energy 5 min post-oral administration in Kunming mice when compared to iohexol. Similar results were also observed in 4T1 tumor-bearing mice, when ReS$_2$-NPs were injected intratumorally, whereby the HU value increased from 30–50 to 110–150 in the tumor region [38].

2.3. Optical Imaging

When light is used to probe molecular and cellular interactions for visualization, it is called optical imaging [39]. Depending on the tissue composition, when light travels through it, photons may experience absorption, reflection or scattering. These interactions can be analyzed in different types of optical imaging techniques to yield unique spectral signatures [40]. For example, inelastic scattering of light is measured by Raman spectroscopy, whereas absorption followed by emission of light can be in fluorescence [19]. Optical imaging offers advantages such as the ability to image at the microscopic level and good spatial resolution but is limited by scattering of light in biological tissues. This is often overcome using imaging probes in the NIR region as there is lower absorption and scattering by soft tissue [40].

NPs exploited for optical imaging mostly include QDs, as their emission is often a function of their size and can be effectively tuned. Changes in the size of nanoparticles also leads to changes in their band gap which in turn influences their imaging properties. Optical bandgap, especially of semiconductor materials is inversely proportional to nanoparticle size distribution. Thus, size of QDs often plays an important role in imaging applications. The ability of M_xS_y-NPs to absorb in the second biological window, i.e., NIR-II (1000–1700 nm), thus enabling deep tissue penetration, better signal-to-noise ratio with reduced tissue auto-fluorescence has led to their widespread application in optical imaging [41]. M_xS_y-NPs studied for optical imaging include semiconductor metal-based QDs especially from group II–VI elements of the periodic table of the elements such as cadmium sulfide (=CdS) and zinc sulfide (=ZnS), respectively. Group I–VI semiconductor-based silver sulfide, i.e., Ag_2S-NPs are also being increasingly used in optical imaging due to properties like absorption in the second NIR window, high signal-to-background noise ratio and good resolution [42]. Examples of M_xS_y-NPs used for different types of optical imaging techniques are reported below.

Awasthi et al. prepared Ag_2S QDs for fluorescence imaging due to their favorable properties including high quantum yield, good photostability and biocompatibility [43]. To improve hydrophilicity and dispersion of the Ag_2S QDs, they were encapsulated in a PEGylated dendrimer to yield PEG-PATU-Ag_2S QDs [43]. When excited with a laser at 785 nm, the appropriate QDs exhibited fluorescence at 1110 nm and intensity of fluorescence improved when the QDs attained sizes greater than 25 nm. The authors also prepared A549 cancer cells labeled with Ag_2S QDs and intravenously injected them into BALB/c mice to test in vivo tracking ability of the QDs. As can be seen from Figure 3, 2 min post-administration, fluorescence signals were observed mainly from the liver which gradually decreased over time. About 30 min following administration, fluorescence signals spread throughout the body, thus showing the distribution of tumor cells in vivo. To probe the ability of Ag_2S QDs as a vascular imaging agent, PEG_{1000} was used for modification of the QDs followed by i.v. injection into BALB/c mice. After a few seconds post-administration, the main vascular system of the mouse was clearly visible using a real-time monitoring system (Figure 3D).

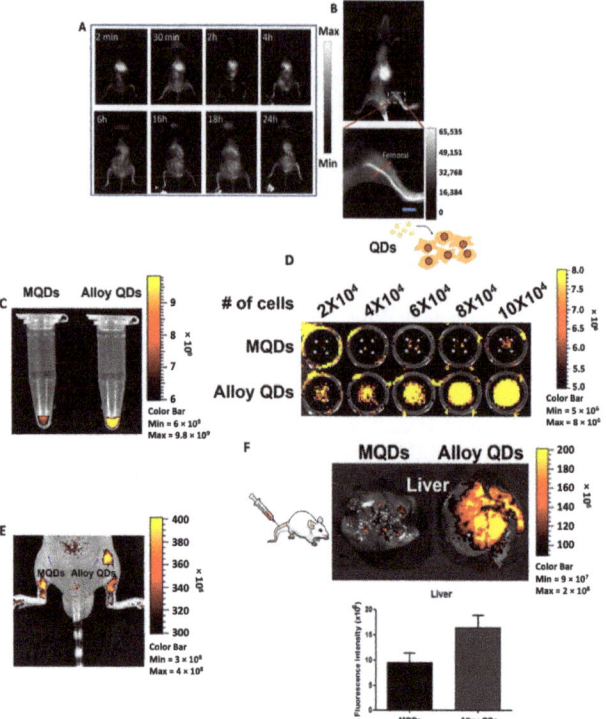

Figure 3. Representative images of NIR-II fluorescence imaging in BALB/c mice. (**A**) Full body distribution and (**B**) zoomed in image showing fluorescence from the femoral artery post-i.v. injection of Ag$_2$S QDs. [Reprinted with permission from Ref. [43]. Copyright © 2023, Royal Society of Chemistry]. (**C**) Higher fluorescence intensity observed from the Eppendorf tube containing the same concentration of alloy QDs as compared to MQDs. (**D**) Cell number dependent increase in fluorescence intensity observed in HeLa cells treated with alloy QDs as compared to MQDs. (**E**) Higher fluorescence intensity observed in vivo in mice treated with alloy QDs. (**F**) Images of liver captured 1 h post-treatment showed higher fluorescence in mice treated with alloy QDs. Reprinted with permission from Ref. [44]. Adopted from BioMed Central 2022.

Recently, silver/silver sulfide Janus NPs (=Ag/Ag$_2$S JNPs) for hydrogen peroxide (=H$_2$O$_2$) triggered NIR-II fluorescence imaging were reported by Zhang et al. [45]. In the presence of H$_2$O$_2$, the fluorescence of Ag/Ag$_2$S JNPs will be "turned on", whereas in its absence a nearly quenching effect was observed. This mechanism is attributed to an inhibited electron transfer between plasmonic Ag to semiconductor Ag$_2$S in the JNP when treated with H$_2$O$_2$ thus giving rise to electron deficient fluorescent Ag$_2$S. Because of the influence of H$_2$O$_2$ on plasmonic Ag, changes in morphology induced in the Ag/Ag$_2$S JNPs post-treatment by H$_2$O$_2$ was assessed. Ag/Ag$_2$S JNPs of size ~15 nm showed a decrease in size to ~10 nm which was in accordance with the mechanism wherein addition of H$_2$O$_2$ led to oxidation and eventual etching of plasmonic Ag in the JNP [46]. The authors also studied the increase in fluorescence intensity of Ag/Ag$_2$S JNPs treated with H$_2$O$_2$ and observed a 6-fold increase 24 h post-treatment. To confirm that fluorescence arises from the Ag$_2$S component, Ag and Ag$_2$S NPs were incubated separately with MCF-7 cells. An "always on" signal was observed in the cells in contrast to an "always off" signal solely with Ag NPs. To determine the in vivo H$_2$O$_2$-triggered fluorescing ability of Ag/Ag$_2$S JNP, they were injected intravenously in an AILI mice model of injured liver. PBS- and only Ag$_2$S NP-treated groups were chosen as control groups for the study. Whereas the Ag$_2$S-NP-treated group showed fluorescence that was "always on", Ag/Ag$_2$S JNP treated mice showed a gradual switch from off to on fluorescence signals with progressing liver injury. Harish et al. synthesized CdS QDs coated with the biopolymer chitosan to improve its stability and biocompatibility [47]. To test the effect of the chitosan coating, the viability of coated and bare CdS QDs were tested in human Jurkat and erythrocyte cell lines. A reduced cytotoxicity of chitosan-coated CdS QDs was found, as compared to the same concentration of solely CdS. Moreover, it was reported that coated QDs were readily taken up by cells as observed by fluorescence imaging analysis. Biocompatibility and uptake of chitosan-coated CdS QDs was attributed to reduced leaching of Cd^{2+} ions from the respective QDs leading otherwise to cytotoxic effects. In the presence of chitosan, released Cd^{2+} ions form coordination bonds with the amino groups of chitosan thus preventing contact with the cells. In another study, Xu et al. generated two cadmium telluride/cadmium sulfide (=CdTe/CdS) core-shell QDs emitting at 545 nm and 600 nm, respectively, to visualize distribution of two chemotherapeutic drugs in a tumor [48]. Coating of CdS over the core resulted in improved quantum efficiency, fluorescence lifetime, stability and biocompatibility of the QDs. The 5-Fluorouracil (=5-FU) and tamoxifen (=TAM) were encapsulated into CdTe/CdS QDs emitting at 545 nm and 600 nm, respectively. To test the effect of the drugs on the tumor resistant cell line MDA-MB-231, the authors conducted a set of experiments. In the first set, the cells were incubated only with QDs-5-FU and in the second set, the cells were incubated with QDs-TAM followed by QDs-5-FU. In the first experiment, green fluorescence of QDs-5-FU was observed only on the cell membrane, whereas in the second experiment green fluorescence was observed within the cell with orange-red fluorescence observed on the cell membrane.

An approach to improve the quantum yield for fluorescence imaging results from the accessibility of QDs in an alloyed core/shell structure containing ZnS in ref. [49]. In this study, Shim et al. modified CIS, i.e., CuInS$_2$ QDs, to form a ZnS-CIS alloyed core surrounded by a ZnS shell affording ZCIS/ZnS. The authors attributed this improvement to the suppression of defect states and electronic structure evolution which, in turn, increased radiative channels. In a similar study, alloy type core/shell CdSeZnS/ZnS QDs were synthesized by Kim and colleagues for bio-imaging applications [44]. The authors compared the quantum yield of the CdSeZnS/ZnS QDs (=alloy QDs) against conventional multilayer CdSe/CdS/ZnS QDs (=MQDs). For alloy QDs, a 1.5-fold higher quantum yield than that of MQDs was reported which significantly improved both in vitro and in vivo imaging (Figure 3C–F).

2.4. Photoacoustic Imaging

Photoacoustic imaging (=PAI) is a type of modified ultrasound imaging modality in which imaging signals are generated through acoustic (ultrasonic) waves caused by the photothermal effects of a PTT agent and can increase the spatial resolution and imaging depth in vivo [50]. The broad absorption by M$_x$S$_y$-NPs in NIR-I and NIR-II resulting from localized surface plasmon resonance has led to their applications as PTT agents and thus also as PAI contrast [51].

Liang et al. prepared glutathione (=GSH)-capped CuS NDs for PTT and PAI via a "one-pot" synthetic methodology [52]. Modification with GSH ensured good water dispersibility and size restriction of the NDs (<10 nm). Under irradiation by a 980 nm laser light, the NDs showed PA contrast three times greater than that of water with a minimal concentration of 1 mM Cu. In vitro studies were followed by in vivo testing in 4T1 tumor-bearing mice. Saline or GSH-CuS NDs were injected intratumorally as control or test, respectively, followed by irradiation at 900 nm. In a control experiment, a very weak PA signal indicating low intrinsic absorption by the tumor at 900 nm, was observed (Figure 4A). On the other hand, a good PA signal was observed in mice treated with GSH-CuS NDs with higher contrast observed in the intratumorally injected mice as evidenced by the enhanced permeation and retention (=EPR) effect and GSH coating on the surface of the NDs. Biomimetic CuS nanoprobes coated with a melanoma cell membrane (HCuSNP@B16F10) for PAI were made accessible by Wu et al. [53]. They loaded HCuSNP@B16F10 with indocyanine green (=ICG) and doxorubicin (=DOX) for PTT and chemotherapy studies. Cell membrane coating was confirmed by Western blotting, and cell viability remained 70% after incubation with 150 µg mL^{-1} for 24 h. In vivo HCuSNP@B16F10 showed a significant PA signal up to 4 h after i.v. injection. In another study, Ouyang and colleagues fabricated CuS nanoparticles trapped in a dendrimer functionalized with PEGylated-RGD (=RGD-CuS DENPs) peptide for PAI-guided PTT/gene therapy [54]. UV–Visible spectroscopy analysis showed good absorption by RGD-CuS DENPs in the 1000–1100 nm range with the CuS core having a diameter of 3.2 nm. The nanoparticles showed PAI contrast dependent on Cu concentration wherein PA signal peaked at 12 h post-intravenous injection in vivo. Figure 4C,D represent PAI obtained using FeS QDs fabricated by Yang et al. which shows a gradual increase in PAI contrast in vivo post-treatment with the QDs.

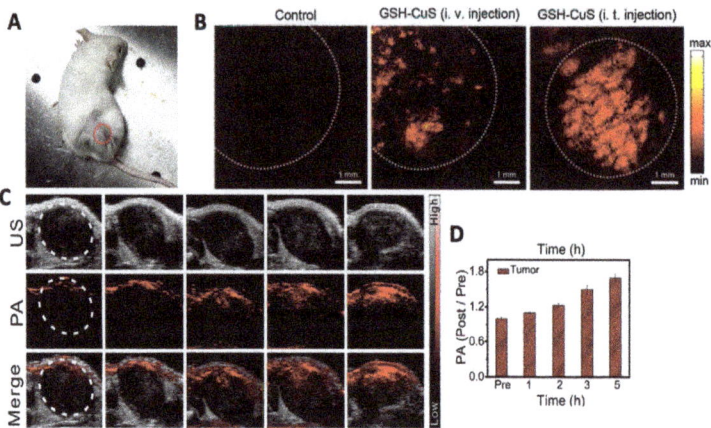

Figure 4. (**A**) 4T1 tumor bearing mice used for in vivo PAI. Laser scan section has been marked with a dotted circle. (**B**) Representative PA images taken before and after GSH-CuS NDs i.v. or i.t. injection Reprinted with permission from Ref. [52]. Copyright © 2023, Royal Society of Chemistry. (**C**) Representative images of PAI and (**D**) its quantitative estimation pre- and post- administration of FeS QDs at different time points. Tumor area has been demarcated with a white circle in (**B**,**C**). Reprinted with permission from Ref. [22]. Copyright © 2023, Elsevier.

In addition to X-ray absorption studies, strong NIR absorption has resulted in the application of Bi_2S_3 NPs for PAI as well. In this respect, Zhang et al. synthesized hollow Bi_2S_3 nanospheres with urchin-like rods (=U-BSHM) for spatio-temporal controlled drug release and PTT-PAI [55]. This was achieved by encapsulating the phase change material (=PCM) 1-tetradecanol and doxorubicin within the microspheres. Heat generated by U-BSHM-NPs under irradiation using an 808 nm laser melted the PCM, which in turn led to the release of DOX thus achieving controlled release. The authors reported a 65.37% release of DOX when U-BSHM-NPs attained a temperature of 43 °C or higher under laser irradiation. With respect to imaging, the NPs showed a concentration-dependent increase in the PA signal intensity by 808 nm laser irradiation. A significant PA signal was also observed when the NPs were irradiated with 700 and 900 nm lasers, respectively (Figure 4A). Zhao et al. synthesized ultra-small Bi_2S_3-NPs using self-assembled single-stranded DNA as a template and employed them imaging probe in myocardial infarction [56]. As a result, thereof, a good PA signal was found when tested in vivo. Similarly, Cheng et al. synthesized Bi_2S_3 nanorods (=NR) for PTT, radiotherapy, and dual modal PA/CT imaging [11]. In vivo, a significant PA signal post-i.v. injection of the NRs, which peaked 24 h post-treatment, was observed. With respect to CT imaging, the NRs showed an enhanced contrast as compared to the commercially available radiocontrast agent iopromide. The authors concluded that radiotherapy and PTT acted in synergism which inhibited tumor growth as well as metastasis. AgBiS$_2$-NDs coated with polyethyleneimine (=PEI) were developed by Lei and colleagues for theranostic applications such as PTT and dual modal PA/CT imaging [57]. PEI-AgBiS$_2$-NDs showed photothermal conversion efficiency of 35.2% which translated to a good PAI signal in vitro. With respect to CT imaging, the authors reported a slope higher than that of iobitridol which is a commercially available radiocontrast agent. The respective in vitro imaging results were correlated with in vivo observations and maximum signal intensity for CT/PA imaging was observed at 24 h post treatment.

MoS$_2$ which has an extinction co-efficient higher in comparison to gold nanorods (=AuNR) and a 7.8-fold higher NIR absorbance than that of graphene oxide is increasingly being used as an NIR absorbing probe with implications in biomedicine [58]. In order to improve the serum stability of MoS$_2$, Shin and colleagues synthesized hyaluronate (=HA) and MoS$_2$ conjugates (=HA-MoS$_2$) for PAI-guided PTT [59]. The size of MoS$_2$ nanoparticles increased from 61.9 nm to 85.9 nm after conjugation with HA. DLS studies revealed no significant changes in the mean hydrodynamic size of HA-MoS$_2$ after 7 days in comparison to MoS$_2$ alone, indicative of no aggregate formation and, thus, good stability. Liu et al. synthesized MoS$_2$ nanosheets conjugated with the dye ICG [60]. The conjugation led to a red shift in the absorption peak of MoS$_2$ from 675 nm to 800 nm for MoS$_2$-ICG. As a result, a 1.35- and 1.55-fold increase in signal intensity and signal-to-noise ratio were observed at 800 nm pulsed irradiation as compared to that of 675 nm, respectively. The improved PA signal intensity and penetration depth is explained to reduced tissue scattering and absorption at 800 nm. In another study, Au et al. developed nerve growth factor (NGF) targeted AuNR coated with MoS$_2$ nanosheets (=anti-NGF-MoS$_2$-AuNR) for PAI of osteoarthritis [61]. MoS$_2$ coated AuNR resulted in a 4-fold increase in PAI signal intensity and higher biocompaibility as compared to AuNR alone. Additionally, the authors also reported stable PA intensity and morphology of MoS$_2$ coated AuNR following irradiation for 30 min. In vivo when anti-NGF-MoS$_2$-AuNR were injected intravenously into Balb/c mice, PA signal peaked at 6 h post-treatment in the synovium of osteoarthritic knee. MoS$_2$ nanosheets modified with CuS nanoparticles were developed by Zhang and co-workers for PAI-guided chemo-PTT [62]. Colloidal stability and biocompatibility of the nanocomposites were improved by attachment of PEG-thiol (=PEG-SH). CuS-MoS$_2$-SH-PEG showed photothermal conversion efficiency higher than that of MoS$_2$ alone.

3. Applications of Metal Sulfide Nanoparticles in Photo- and Immuno-Therapy

3.1. Photothermal Therapy

Photothermal therapy (=PTT) is a non-invasive therapeutic strategy that uses photo-absorbents in the NIR region to induce hyperthermia (40–45 °C) in the tumor site. The NIR laser induces collateral thermal damage to the cancerous cells leading to cell death by apoptosis or by altering gene expression in cancerous cells [63].

CuS-NPs are an emerging class of photothermal agents that are biocompatible, have high extinction in the NIR range, are stable under laser irradiation and, are therefore considered to be better suited than the so far used gold (Au-) NPs [64,65]. The NIR absorption in CuS-NPs is due to d–d energy band transitions of Cu^{2+} ions and therefore their absorption wavelength remains unaffected by the surrounding biological environment. In one report, 980 nm NIR-light-driven CuS nanoplates were found to inhibit the growth of prostate cancer cells both in vivo and in vitro [66]. Respective CuS nanoplates were injected into the prostate tumor site under ultrasound guidance and PTT was performed. Lu et al. reported a platform for dual cancer therapy (photothermal and chemotherapy) based on PEGylated CuS@mSiO$_2$ nanocomposites [67]. The mesoporous silica allowed high payload capacity; however, this showed poor colloidal stability. Hence, polyethylene glycol grafting was carried out to improve the colloidal stability and enhance the EPR effect to deliver drugs to the target cells. Cheng et al. developed WS$_2$ nanosheets as PTT agent for bio-imaging and photothermal ablation of tumors [68]. The nanosheets were functionalized with PEG to enhance physiological stability and biocompatibility. The 4T1 cells were incubated with 0.1 mg ml^{-1} WS$_2$-PEG nanosheets for 6 h and irradiated by an 808 nm laser of varying power densities. The nanosheets effectively induced thermal ablation at a low dose (i.t., 2 mg kg^{-1}) and a higher dose (i.v. injection, 20 mg kg^{-1}) without causing any mortality (Figure 5). On similar lines, PVP-functionalized MoSe$_2$ nanosheets in a PNIPAM hydrogel with both a dual photo- and thermo-responsive behavior was effective towards HeLa cells [69]. Photo-thermal ablation of mammalian cells was also demonstrated by Chou et al. by using chemically exfoliated MoS$_2$-NPs at a very low concentration (<38 ppm) to effectively destruct the cancerous cells [70].

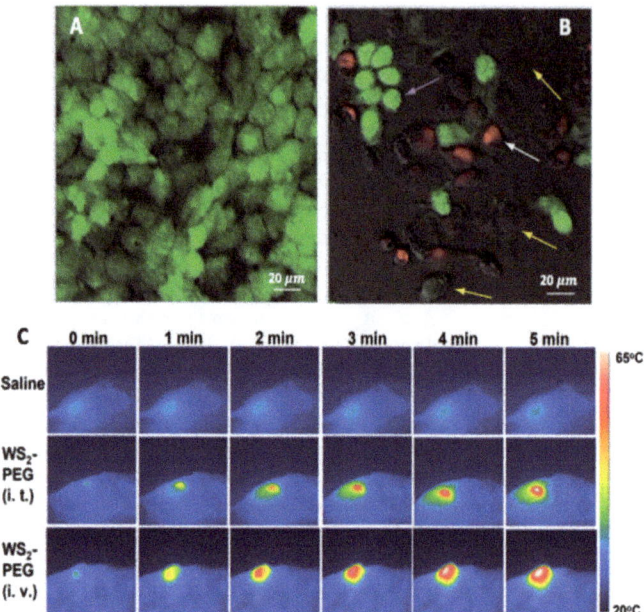

Figure 5. Microphotographs of HeLa cells with CuS-NPs. (**A**) without laser, cells were viable and had polygonal morphology. (**B**) with NIR laser irradiation at 24 W cm^{-2} for 5 min (purple arrows show shrinking of cells; yellow arrows show loss of cell viability by calcein-negative staining; white arrows show loss of cell membrane integrity by EthD-1 positive staining. (**C**) In vivo PTT in 4T1 tumor bearing mice with saline (top row), WS$_2$-PEG (middle row: i.t. low dose = 2 mg kg^{-1}), WS$_2$-PEG (bottom row: i.t. high dose = 20 mg kg^{-1}). The laser power density was 0.8 W cm^{-2}. Reprinted with permission from Refs. [64,68]. Copyright © 2023, Future Medicine and 2014, Wiley.

Qian et al. introduced PEGylated titanium disulfide (=TiS$_2$) as PTT agent for in vivo PAI-guided thermal ablation of cancer [71]. PEG was incorporated into the system to make the nanoparticles stable in polar solvents. The PTT agent exhibited strong NIR absorbance being able to destroy tumor cells. A multifunction theranostics platform, based on WS$_2$ QDs (3 nm), was synthesized to achieve simultaneous CT/PAI and synergistic PTT treatment of tumors, wherein the location of the tumor could be precisely observed and treated [72]. MRI-guided PTT was reported using iron sulfide nanoplates. Yang and coworkers prepared PEG-functionalized FeS nanoplates (=FeS-PEG) that exhibited high NIR absorption and superparamagnetism [73]. Highly effective in vivo PTT ablation in mice tumor was achieved using 20 mg kg^{-1} of FeS-PEG followed by 808 nm laser irradiation. MRI studies revealed accumulation of FeS-PEG NPs in the tumor cell and no toxicity was observed even at a higher dose.

Though metal sulfide NPs can be effectively used for photothermal ablation, however, poor photothermal conversion efficiency restricts their use for all practical applications.

To overcome this limitation, a combination of metal/metal sulfide NPs was designed [74]. Yang et al. reported surface plasmon-enhanced PTT using an Ag/CuS nanocomposite for effectively killing PC3 prostate cancer cells [75]. The nanocomposite was activated by a 980 nm laser at 0.6 W cm^{-2} for 5 min and, hence, an enhancement of CuS PTT efficacy was observed. This is attributed to the presence of surface plasmon resonance (=SPR) of the Ag-NPs that led to significant enhancement in the electric field near the surface, thereby increasing the rate of the transition process at the interfaces. Ding and coworkers studied the influence of dual plasmonic Au-Cu_9S_5-NPs on the photothermal transduction efficiency [76]. The nanocomposite exhibited localized SPR in both the visible and NIR region and the molar extinction coefficient of the composite was found to be 50% higher at 1064 nm than the individual counterparts. The composites were used for PTT on tumor-bearing mice at 100 ppm under 0.6 W cm^{-2} 1064 nm laser irradiation. Similar observations were reported by tuning localized SPR by applying Cu_5FeS_4-NPs to enhance the photothermal conversion efficiency up to 50.5% using an 808 nm laser [77].

3.2. Photodynamic Therapy

Photodynamic therapy (=PDT) is a clinically approved minimally invasive therapeutic modality in which a photosensitizer (=PS) is activated by a light of specific wavelength (laser) to generate singlet oxygen species ($^1O_2$) that destroys abnormal cells [78]. When the photosensitizer is excited, it transfers its energy to the molecular oxygen in tumor cells through a triplet state. During the process, cytotoxic singlet oxygen and other secondary molecules such as reactive oxygen species, super-oxides, etc., are formed via oxidation of cellular macromolecules. This event leads to necrosis or apoptosis of tumor cells [79]. Nanoparticles can be used as carriers of PS due to (i) easy functionalization with target molecules that increases biodistribution of PS, (ii) the higher surface area-to-volume ratio of NPs increasing the carrying capacity of PS, (iii) protect degradation of light-sensitive PS and enhance their circulation in bloodstream, and (iv) capability to incorporate other therapeutic or diagnostic modalities to PDT in the same system. M_xS_y-NPs have an edge over other NPs such as gold NPs for use in PDT, due to their strong absorption properties in the NIR region ranging from 700–1100 nm, high extinction coefficients and high fluorescence properties. Hence, the following sections will focus on M_xS_y-NPs that are widely applied in PDT.

Jia et al. used MoS_2 nanoplates for fluorescence imaging of ATP and PDT through ATP-mediated controllable to release $^1O_2$ under 660 nm laser irradiation [80]. Therefore, Ce6-aptamer was loaded on the MoS_2 nanoplates that especially responded to the ATPs in lysosomes and $^1O_2$ induced cell death through the lysosomal pathway. The studies exhibit the release of a single-stranded aptamer from the MoS_2 nanoplates and subsequent imaging of intracellular ATP and generation of singlet oxygen.

Plasmonic $Cu_{2-x}S$-NPs confirmed excellent surface plasmon absorption in the NIR region which mainly originates from the free holes of the unoccupied highest energy state of the valence bond [81]. This depends on the ratio of Cu:S and the crystal phase of the nanoparticles itself. Examples of plasmonic $Cu_{2-x}S$-NPs are $Cu_{31}S_{16}$ (monocyclic phase), Cu_9S_5 (cubic phase), Cu_7S_4 (orthorhombic phase), $Cu_{58}S_{32}$ (triclinic phase), and CuS (hexagonal phase or covellite). With decrease in the Cu:S ratio ($Cu_{2-x}S$ with $x > 0$), the concentration of free carriers increases inducing LSPR absorbance in the NIR area. $Cu_{2-x}S$-NPs enhanced the ROS generation in B16 cells under NIR radiation (808 nm, 0.6 W cm^{-2} for 5 min) [82]. Generation of hydroxyl radicals were detected by 5,5-dimethyl-1-pyrroline-N-oxide (=DMPO) spin-trapping adducts in electron spin resonance (=ESR) spectroscopy. The ROS generation was dependent on the concentration of the NPs and the laser power. From the ESR signal it can be concluded that the irradiation led to around 83% enhancement in •OH generation. In vivo, Cu(II) is reduced to Cu(I) by biomolecules such as ascorbic acid or glutathione, which reacts with hydrogen peroxide to form •OH species. Similar results were also obtained by other researchers [81,83–85].

Cheng and coworkers reported on the use of Bi_2S_3 nanorods for NIR-activated PDT [86]. The nanorods could be excited by a NIR laser to generate free holes in the valence band and electrons in the conduction band, which formed hydroxyl and superoxide radicals upon reaction with water and oxygen. Further, when the nanorods were associated with zinc protoporphyrin IX, a pronounced inhibitory effect of the tumor was observed under NIR irradiation. Lin et al. synthesized Co_9S_8 NDs and modified their surface with albumin to make them biocompatible [67]. Upon NIR irradiation, the NDs showed a marked time-dependent production of $^1O_2$ production with high photothermal conversion efficiency of 64%.

3.3. PTT-PDT Combinatorial Therapy

Photothermal and photodynamic therapies have an edge over conventional therapies including chemotherapy, surgery, and radiation due to high specificity, minimal invasion, and precise spatio-temporal selectivity [87]. Furthermore, in PTT and PDT, no extra targeting is required, however, tissue penetration of light is a concern. Heat conversion efficiency and formation of hypoxic environments in PTT and PDT are other concerns. PTT agents convert light energy into heat and eradicate tumors by hyperthermia, while PDT agents produce toxic reactive oxygen species to kill cancer cells. However, PTT generally requires high-power density lasers to produce enough heat and PDT requires the effective uptake of photosensitizers by cancer cells to induce tumor hypoxia. In other words, in PTT, self-protection of cancer cells induces heat shock response which weakens the PTT efficacy and on the other hand, in PDT, tissue hypoxia limits the PDT efficacy. Therefore, synergistic strategies by combining PTT and PDT in a single platform are now becoming important to overcome the concerns and gain improvised results of the therapies. Simultaneous hyperthermia and ROS are envisaged to cause cancer cell death and elimination of malignant tumors by PTT-PDT combinatorial therapy. In such cases, a single nanoplatform that can behave as both PTT and PDT agents are highly desirable. Following section deals with metal sulfide nanomaterials that are visualized to be PTT as well as PDT agents.

Song et al. designed bioconjugated MoS_2 nanosheets for combinatorial PTT-PDT in which bioconjugation was done with BSA to render biocompatibility to the nanosheets (Figure 6) [88]. The bioconjugated nanosheets produced both localized hyperthermia and $^1O_2$. A possible mechanism of the combinatorial effects can be explained by the following route: Firstly, when BSA-MoS_2 nanosheets are irradiated with an 808 nm laser at 0.8 W cm^{-2}, a rise in temperature (up to 48 °C in 4–5 min) takes place which then activates the dissolved oxygen to generate ROS (in the order $O_2 \rightarrow {}^1O_2 \rightarrow O_2^{\bullet-} \rightarrow {}^\bullet HO_2 \rightarrow H_2O_2 \rightarrow {}^\bullet OH$). Thus, BSA-$MoS_2$ nanosheets trigger ROS generation and enhance the phototherapy. In another study, following a similar mechanism, MoS_2 nanosheets in hydrogel were used as PTT and PDT agent along with chemotherapy [89]. Remarkable reduction in primary 4T1 breast tumors and distal lung metastatic nodules in vivo was observed. A mild photothermal heating was able to increase cell membrane permeability and cellular uptake of various agents such as photodynamic agents or chemotherapeutic drugs [90]. Similar results were obtained by Xu and his group wherein an IR-808 dye sensitized UCNP with Ce60-grafted MoS_2 nanosheets synergistically amplified the up-conversion efficiency and triggered the photosensitizer to produce large amounts of ROS [91].

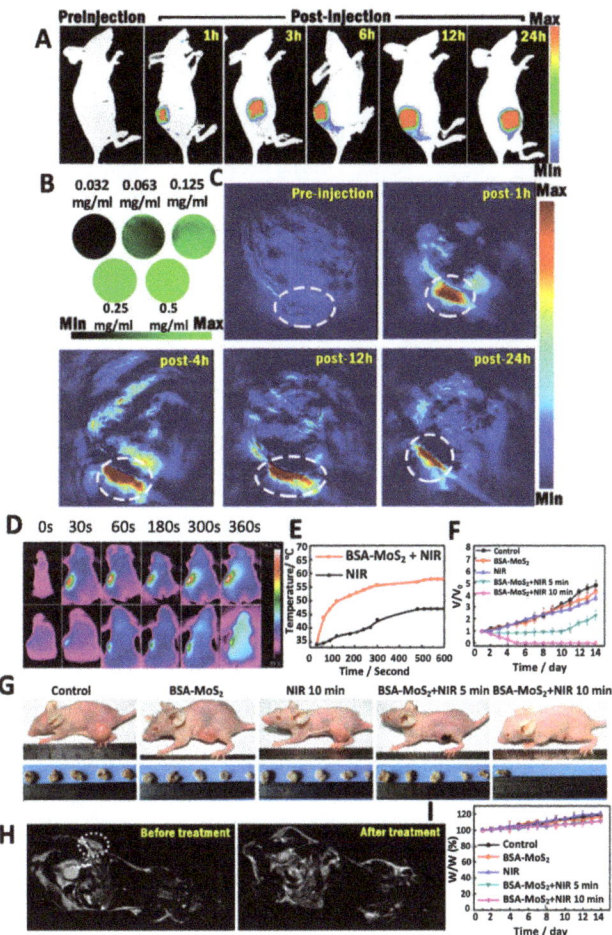

Figure 6. (**A**) Representative fluorescence images of BSA-MoS$_2$ treated tumor-bearing mice at different time points. (**B**) In vitro PAT imaging for different concentrations of BSA-MoS$_2$. (**C**) Representative PAT images of BSA-MoS$_2$ treated tumor-bearing mice at different time points with tumor area marked with a white dotted line. (**D**) Representative infrared images showing thermal profile of tumor-bearing mice treated with BSA-MoS$_2$ or PBS (control group) and their corresponding (**E**) temperature profile and (**F**) tumor volume. (**G**) Representative pictures of mice showing reduction in tumor size with respective treatments. (**H**) MR images of mice treated with BSA-MoS$_2$ before and post- treatment on the 14th day and corresponding (**I**) changes in body weight. Reprinted with permission from Ref. [88]. Copyright © 2023, Royal Society of Chemistry.

The combination of PDT and PTT was also demonstrated by Bharathiraja and coworkers where MBA-MD-231 cells were incubated with CuS-Ce6 NPs and exposed to an 808 nm laser light for 10 min at 2 W cm^{-2} [92]. MTT assay revealed synergistic cytotoxicity by the combination therapy rather than individual therapies. Similar observations were made by Wang's group [63]. Heat generation, due to photothermal efficacy of Cu$_{2-x}$S-NPs, was monitored in B16 cells by heat shock protein 70 (Hsp70) expression. The cells exposed to 100 s laser radiation (808 nm and 0.6 W cm^{-2}) showed significantly enhanced Hsp70 which is caused not only due to thermal stress but also due to elevated ROS levels [82]. Under NIR light and in tumor acidic regions, leaking of Cu(I) ions from the NP occurs, which react with the surrounding O_2 and H_2O_2 to form Cu(II) along with hydroxide and hydroxyl radicals that contribute to enhanced ROS [93]. Biocompatible PEGylated iron sulfide NPs (=FeS_2@C-PEG) were found to oxidize water to form O_2 under NIR exposure which improved the therapeutic efficacy of the NPs [94]. Formation of Fe(II) degraded the intracellular H_2O_2 to produce more ROS species that contributed to the combinatorial PTT-PDT. Zinc protoporphyrin IX (=ZP)-linked Bi_2S_3 nanorods provide active sites for binding heme oxygenase-1 (HO-1) that are overexpressed in solid tumors and suppressing the cellular antioxidant defense capability. The nanorods, upon NIR radiation, generated heat that facilitated an efficient electron–hole separation in ZP and Bi_2S_3 and produced ROS species. Once cells are attacked by ROS, the redox homeostasis is disturbed and HO-1 catalyzed the heme molecule to generate a series of antioxidants (biliverdin, carbon monoxide, and ferrous iron), which are the most potent endogenous scavengers of ROS. Here, ZP, as a potent HO-1 inhibitor, suppressed the HO-1 activity and strengthened the PDT effect. Under 808 nm laser irradiation (0.75 W cm^{-2}, 10 min), the nanorods exhibited photothermal conversion efficiency of 33.64%. The nanorods could accumulate in the 4T1 tumor and inhibit the HO-1 activity and enhance NIR-irradiated oxidative injury [86]. Cobalt chalcogenides also possess intrinsic peroxidase-like activity, high photothermal conversion efficiency and broad NIR absorption properties; however, it is challenging to synthesize biocompatible cobalt sulfide due to co-existence of both strongly reducible cobalt ions and oxidizable sulfide ions. Further, cobalt ions have strong affinity for oxygen and, therefore, it is difficult to exclude impurities such as cobalt oxide or cobalt hydroxide in the resultant NPs [95,96].

3.4. Combined Photo-Immunotherapy

Immunotherapy is a biological cancer treatment that makes use of substances from living organisms to treat cancer and help the immune system to fight cancer. Specifically, immunotherapy or immune activation involves production of cancer-fighting immune cells to identify and destroy cancerous cells. Immunotherapy includes checkpoint inhibitors, T-cell transfer, monoclonal antibodies, cancer vaccines and immune system modulators. In contrast to conventional therapies such as chemotherapy, radiotherapy, or surgery, that aim to destroy cancer cells along with healthy cells, immunotherapy aims to prevent the healthy cells and restore antitumor activity of the immune system. Research on delivery of immunotherapeutic agents by NPs showed minimization of adverse effects and maximization of the therapeutic index of immunotherapy [97]. Nanomaterial-based delivery of immunotherapeutics and biologicals (e.g., nucleic acids, antibodies, etc.) improves pharmacological properties of drugs such as solubility, and stability in physiological media. Assorted molecular-binding sites in nanomaterials help in shielding active drugs and biologics from degradation and macrophage clearance in blood after systemic administration. In other words, nanomaterials enhance bioavailability and control unwanted targeting which is significant in tumor management [98]. Further, the pharmacokinetic profile of the drug and their interaction with cells can also be modulated and controlled by the nanosystem [99]. Release of the drug or biologics can also be controlled and regulated by nanomaterials to enhance efficacy and reduce systemic toxicity. However, it is important to consider the structure and composition of NPs for active targeting of drugs or biologics and their release. Above all, nanotechnology offers possibilities of combining immunotherapy with chemo-, radio- or even photothermal and photodynamic therapies.

Several nanosystems ranging from carbon-, metal/metal oxide-, polymer- and lipid-based NPs are reported for specific delivery of immunotherapeutics to precisely target and control tumors [100,101]. However, very little literature is available on the use of metal sulfide NPs for immunotherapy. The following section will focus on metal sulfide-based NPs that are reported for immunotherapy along with other photothopies.

Guo and his group designed a light-induced transformative NP platform based on chitosan-coated hollow CuS-NPs that can assemble immunoadjuvants oligodeoxynucleotides containing the cytosineguanine (CpG) motifs [102]. The platform combined photothermal ablation and immunotherapy in which, upon laser excitation at 900 nm, the nanostructures broke and reassembled into polymer complexes which enhanced CpG tumor retention and uptake by plasmacytoid dendritic cells. It generates heat to ablate the tumor cells and releases the tumor antigens into the tumor sites, while the immunoadjuvants enhance antitumor immunity by promoting antigen uptake. The PTT synergistically acted with immunotherapy to enhance immune responses and made the tumor residues and metastases susceptible to immune-mediated killing. Similar observations were made by Chen et al. using core-shell CuS@PLGA-NPs in which the model antigen ovalbumin (OVA) was loaded [103]. On one hand, poly D, L-lactic-co-glycolic acid (=PLGA) made the system biocompatible and exhibit controlled biodegradation kinetics, and on the other hand, the CuS-NPs display favorable PTT by killing 4T1 tumor cells in vitro. Release of OVA and its further internalization into antigen-presenting cells (=APCs) induced the immune response. The heat conversion by CuS-NPs under NIR radiation not only triggered rapid release of OVA but also enhanced the cell membrane permeability that led to higher uptake of the antigen by the cells. Yan et al. reported synergistic PTT and immunotherapy driven by Cas9 ribonucleoprotein-loaded CuS-NPs to enhance the therapeutic effect on melanoma [104]. The NIR light triggered thermoresponsive CuS-NPs provide a platform to modify Cas9 ribonucleoprotein targeting PTPN2 for immunotherapy. Depletion of PTPN2 was observed after treatment with the targeted NPs which caused accumulation of infiltrating CD8 T lymphocytes in tumor mice. Also, the expression levels of interferons and cytokines (IFN-γ and TNF-α) was upregulated which sensitized the tumors to immunotherapy. Thus, tumor ablation along with immunogenic cell death induced by PTT amplified the anti-tumor efficacy. Similar integration of PTT and immunotherapy in a Cu_9S_5@$mSiO_2$ nanoagent was reported in a study by Zhou et al., in which the immune response of CpG effectively inhibited tumor metastasis [105]. Intracellular uptake of CpG promoted infiltration of cytotoxic T lymphocytes (=CTLs) in tumor tissue, which stimulated the production of IL-12, TNF-α and IFN-γ. Xu and coworkers verified adoptive macrophage therapy through CuS-NP regulation for antitumor effect in mice bearing B16F10 melanoma [106]. Within this study, bone-marrow-derived macrophages (=BMDMs) were incubated with PEGylated CuS-NP to promote cellular production of ROS through dynamin-related protein 1 (Drp1)-mediated mitochondrial fission. The high intracellular ROS level directs BMDMs polarization toward M1 phenotype by classical IKK-dependent NF-κB activation. Moreover, the CuS-NP-stimulated BMDMs downregulated PD-1 ligand expression and contributed to the promoted ability of phagocytosis and digestion. I.t. transfer of CuS-NP-redirected macrophages, triggered the local and systemic tumor-suppressive alterations, further enhancing the antitumor activity. On similar lines, MoS_2 nanosheets were functionalized with CpG and PEG to form nanoconjugates that upon NIR irradiation significantly enhanced intracellular accumulation of CpG [107]. The accumulation of CpG stimulated the production of proinflammatory cytokines and elevated immune response. The MoS_2 nanoconjugates also reduced proliferation of 4T1 cells when co-cultured with RAW264.7 (macrophage cells) upon NIR irradiation for 10 min at 2 W cm^{-2}. The increased uptake efficiency of CpG is attributed to the membrane permeability induced by laser irradiation.

MoS_2-NPs are able to induce low levels of the pro-inflammatory cytokines IL-1β, IL-6, IL-8, and TNF-α in human bronchial cells (NL-20) and activate antioxidant/detoxification defense mechanisms [108]. The low cytotoxicity of the MoS_2-NPs reflects the ability of the NPs to induce a favorable balance of cellular responses in vitro which can be extended to in vivo in future.

It can be inferred that the combination of photothermal therapy and immunotherapy can produce synergistic anti-tumor effects as well as reduce systemic toxicity [109]. Major applications of M_xS_y-NPs in photothermal therapy are due to their ability to convert NIR radiation into thermal energy which is subsequently used for ablation of cancer cells. However, it is important to achieve higher conversion efficiency so that the dose requirement is reduced. Moreover, integration of photothermal therapy with immunotherapy is essential to address cancer heterogeneity and adaptation.

4. Conclusions

Although M_xS_y-NPs have been researched as theranostic nanoplatforms over a decade, only a handful of reviews are highlighting recent developments and challenges in this field [9,10,20]. Metal sulfide NPs, specifically, transition metal dichalcogenides, have an array of desirable properties such as electronic band structure, tunable bandgap, luminescence, and Raman scattering, which can be tuned as per the end applications. However, because of the semiconductor behavior, they are intrinsically toxic which limits their use in biomedical applications. To address the concern, additional modifications of the appropriate nanomaterials are required to enhance biocompatibility and make them capable for their use as diagnostic tools or imbibe properties for applications such as drug delivery, sensing, etc. Further, metal sulfide NPs do not form very stable suspensions in polar solvents, for example, water, and therefore, their use in in vivo applications also remains a concern. Hence, proper NP functionalization is, therefore, required to provide colloidal stability to the respective NPs. Thus, selection of functional molecules (e.g., dyes, polymers, organic molecules including acids, small molecules such as hydroxyl, thiols, etc.) are crucial for facilitating interactions between the NPs and biological systems [110]. In many cases, functionalization may involve modification of atoms of the NPs present in the basal plane, kinks, edges or corners, which may change the electronic band structure of the NPs [111]. Voiry et al. reported that change in phase of sulfur- and selenium-based transition metal dichalcogenides from metal to semiconductor takes place when the NPs will be covalently functionalized with, for example, amides and methyl moieties, respectively [112]. Thus, designing synthesis and functionalization strategies of metal sulfide NPs are very important to meet the requirement of structural and chemical stability, dispersibility in physiological medium, uniformity in size distribution, and biocompatibility. In addition to functionalization, core–shell structures may also be developed to decrease leaching of toxic metals in cellular environments. This is especially true in heavy metal quantum dots such as lead sulfide (PbS), CdS, mercuric sulfide (HgS) offering excellent optical imaging properties but are limited due to their cytotoxicity. In such cases, formation of a shell over the core can impede direct contact of the heavy metals with cells and improve biocompatibility of the appropriate metal sulfide.

Though multimodal platforms (therapeutic and imaging) have proved beneficial for treatment of several diseases, overtreatment is emerging as a new concern. Minimizing the use of probe material and therapeutic dose, while maintaining the effectiveness of the platform, is crucial for patient's compliance. Integration of various functions in a nanosystem without changing individual properties can significantly synergize theranostic effects. It is also important to design a multimodal system of varying chemistries that would not only retain their individual functions, but also not interfere with the functions of other materials, which eventually can enhance the effectiveness of every component. Li et al. developed such a platform based on hydrophilic MnS@Bi_2S_3-PEG NPs which was successfully used as contrast agents for MRI, CT and PA-trimodal imaging moiety along with PTT and hyperthermia applying a single injection dose for tumor therapy. Hyperthermia significantly enhanced the efficacy of radiation and provided a unique platform to address the concern of overtreatment [113]. More such platforms would definitely prove beneficial; however, their short- and long-term efficacies and toxicities need to be evaluated.

Nanoparticle-based delivery of immunotherapeutics is significant in not only treating cancer but also developing immune defensive cells that can be used to identify and eliminate tumor cells. Due to limited toxicity and side-effects, immunotherapy can be used in conjunction with other interventions such as chemotherapy, radiation therapy, photothermia and hyperthermia. Several multifunctional nanomaterials have been explored as photoimmunotherapeutic agents to enhance phototherapy as well as carrier of immune adjuvants. Despite the progress, more research is required to understand the dynamic immune response and the molecular mechanism of NPs-immune interaction for promoting clinical translation of nano-immunotherapy. It is also important to consider the potential risk associated with overstimulation of the immune system that may lead to autoimmune toxicities. A balance between efficacy and safety rather than a strong anti-tumor immune response is required. Nevertheless, photoimmunotherapy has shown promising pre-clinical responses on various tumor models and therefore, has a potential for clinical translation.

Though there are proven reports of the versatility of M_xS_y-NP-based nanophototherapeutic platforms, clinical translation is a long way to go. More detailed understanding of degradations and metabolism of M_xS_y-NPs is required to validate their effectiveness with respect to degradation products of M_xS_y-NPs, metal metabolism, biodistribution, pharmacokinetic mechanism, fate, and elimination process. Nevertheless, advancements in research will have an impact on future phototherapeutic abilities of M_xS_y-NPs.

Author Contributions: Conceptualization, writing, data curation: A.S. and S.C.; writing—original draft preparation: A.S. and S.C.; writing —review and editing, supervision: H.L. and S.C. All authors have read and agreed to the published version of the manuscript.

Funding: This research received no external funding.

Institutional Review Board Statement: Not applicable.

Informed Consent Statement: Not applicable.

Data Availability Statement: Not applicable.

Conflicts of Interest: The authors declare no conflict of interest.

References

1. Khursheed, R.; Dua, K.; Vishwas, S.; Gulati, M.; Jha, N.; Aldhafeeri, G.M.; Alanazi, F.G.; Goh, B.H.; Gupta, G.; Paudel, K.R.; et al. Biomedical applications of metallic nanoparticles in cancer: Current status and future perspectives. *Biomed. Pharmacother.* **2022**, *150*, 112951. [CrossRef]
2. Rezic, I. Nanoparticles for biomedical Application and their synthesis. *Polymers* **2022**, *14*, 4961. [CrossRef]
3. Kim, D.; Kim, J.; Park, Y.I.; Lee, N.; Hyeon, T. Recent development of inorganic nanoparticles for biomedical imaging. *ACS Cent. Sci.* **2018**, *4*, 324–336. [CrossRef]
4. Mitchell, M.J.; Billingsley, M.M.; Haley, R.M.; Wechsler, M.E.; Peppas, N.A.; Langer, R. Engineering precision nanoparticles for drug delivery. *Nat. Rev. Drug Discov.* **2020**, *20*, 101–124. [CrossRef]
5. Li, N.; Sun, Q.; Yu, Z.; Gao, X.; Pan, W.; Wan, X.; Tang, B. Nuclear-targeted photothermal therapy prevents cancer recurrence with near-infrared triggered copper sulfide nanoparticles. *ACS Nano* **2018**, *12*, 5197–5206. [CrossRef]
6. Yi, X.; Chen, L.; Chen, J.; Maiti, D.; Chai, Z.; Liu, Z.; Yang, K. Biomimetic copper sulfide for chemo-radiotherapy: Enhanced uptake and reduced efflux of nanoparticles for tumor cells under ionizing radiation. *Adv. Funct. Mater.* **2018**, *28*, 11. [CrossRef]
7. Xie, C.; Cen, D.; Ren, Z.; Wang, Y.; Wu, Y.; Li, X.; Han, G.; Cai, X. FeS@BSA nanoclusters to enable H_2S-amplified ROS-based therapy with MRI guidance. *Adv. Sci.* **2020**, *7*, 1903512. [CrossRef]
8. Fei, W.; Zhang, M.; Fan, X.; Ye, Y.; Zhao, M.; Zheng, C.; Li, Y.; Zheng, X. Engineering of bioactive metal sulfide nanomaterials for cancer therapy. *J. Nanobiotechnol.* **2021**, *19*, 93. [CrossRef]
9. Argueta-Figueroa, L.; Martinez-Alvarez, O.; Santos-Cruz, J.; Garcia-Contreras, R.; Acosta-Torres, L.; de la Fuente-Hernandez, J.; Arenas-Arrocena, M. Nanomaterials made of non-toxic metallic sulfides: A systematic review of their potential biomedical applications. *Mater. Sci. Eng. C* **2017**, *76*, 1305–1315. [CrossRef]
10. Paca, A.M.; Ajibade, P.A. Metal sulfide semiconductor nanomaterials and polymer microgels for biomedical applications. *Int. J. Mol. Sci.* **2021**, *22*, 12294. [CrossRef]
11. Cheng, X.; Yong, Y.; Dai, Y.; Song, X.; Yang, G.; Pan, Y.; Ge, C. Enhanced radiotherapy using bismuth sulfide nanoagents combined with photo-thermal treatment. *Theranostics* **2017**, *7*, 4087–4098. [CrossRef]
12. Yi, H.; Zhou, X.; Zhou, C.; Yang, Q.; Jia, N. Liquid exfoliated biocompatible WS_2@BSA nanosheets with enhanced theranostic capacity. *Biomater. Sci.* **2021**, *9*, 148–156. [CrossRef]
13. Sun, X.; Fan, J.; Fu, C.; Yao, L.; Zhao, S.; Wang, J.; Xiao, J. WS_2 and MoS_2 biosensing platforms using peptides as probe biomolecules. *Sci. Rep.* **2017**, *7*, 10290. [CrossRef]
14. Anju, S.; Mohanan, P. Biomedical applications of transition metal dichalcogenides. *Synth. Met.* **2021**, *271*, 116610. [CrossRef]
15. Shetty, A.; Mishra, S.K.; De, A.; Chandra, S. Smart releasing CuS/ZnS nanocomposite dual drug carrier and photothermal agent for use as a theranostic tool for cancer therapy. *J. Drug Deliv. Sci. Technol.* **2022**, *70*, 103252. [CrossRef]
16. Reda, R.; Zanza, A.; Mazzoni, A.; Cicconetti, A.; Testarelli, L.; Di Nardo, D. An update of the possible applications of magnetic resonance imaging (MRI) in dentistry: A literature Review. *J. Imaging* **2021**, *7*, 75. [CrossRef]
17. Bouché, M.; Hsu, J.C.; Dong, Y.C.; Kim, J.; Taing, K.; Cormode, D.P. Recent advances in molecular imaging with gold nanoparticles. *Bioconjug. Chem.* **2020**, *31*, 303–314. [CrossRef]

18. Jeon, M.; Halbert, M.V.; Stephen, Z.R.; Zhang, M. Iron oxide nanoparticles as T_1 contrast agents for magnetic resonance imaging: Fundamentals, challenges, applications, and perspectives. *Adv. Mater.* **2020**, *33*, 1906539. [CrossRef]
19. Smith, L.; Byrne, H.L.; Waddington, D.; Kuncic, Z. Nanoparticles for MRI-guided radiation therapy: A review. *Cancer Nanotechnol.* **2022**, *13*, 38. [CrossRef]
20. Agarwal, V.; Chatterjee, K. Recent advances in the field of transition metal chalcogenides for biomedical applications. *Nanoscale* **2018**, *10*, 16365–16397. [CrossRef]
21. Yuan, Y.; Wang, L.; Gao, L. Nano-sized iron sulfide: Structure, synthesis, properties and biomedical applications. *Front. Chem.* **2020**, *8*, 818. [CrossRef]
22. Yang, W.; Xiang, C.; Xu, Y.; Chen, S.; Zeng, W.; Liu, K.; Jin, X.; Zhou, X.; Zhang, B. Albumin-constrained large-scale synthesis of renal clearable ferrous sulfide quantum dots for T1-weighted MR imaging and phototheranostics. *Biomaterials* **2020**, *255*, 120186. [CrossRef]
23. Xiong, Y.; Sun, F.; Liu, P.; Yang, Z.; Cao, J.; Liu, P.; Hu, J.; Xu, Z.; Yang, S. A biomimetic one-pot synthesis of versatile Bi_2S_3/FeS_2 theranostic nanohybrids for tumor-targeted photothermal therapy guided by CT/MR dual-modal imaging. *Chem. Eng. J.* **2019**, *378*, 122172. [CrossRef]
24. Fu, D.; Liu, J.; Ren, Q.; Ding, J.; Ding, H.; Chen, X.; Ge, X. Magnetic iron sulfide nanoparticles as thrombolytic agents for magnetocaloric therapy and photothermal therapy of thrombosis. *Front. Mater.* **2019**, *6*, 316. [CrossRef]
25. Li, Z.; Li, Z.; Chen, L.; Hu, Y.; Hu, S.; Miao, Z.; Sun, Y.; Besenbacher, F.; Yu, M. Polyethylene glycol-modified cobalt sulfide nanosheets for high-performance photothermal conversion and photoacoustic/magnetic resonance imaging. *Nano Res.* **2018**, *11*, 2436–2449. [CrossRef]
26. Lv, K.; Lin, H.; Qu, F. Biodegradable hollow Co_3S_4@N-doped carbon as enhanced PTT/PDT agent for multimodal MR/thermal imaging and synergistic antitumour therapy. *Chem. Eng. J.* **2020**, *392*, 124555. [CrossRef]
27. Huang, X.; Deng, G.; Han, Y.; Yang, G.; Zou, R.; Zhang, Z.; Sun, S.; Hu, J. Right Cu_{2-x}S@MnS core-shell nanoparticles as a photo/H_2O_2-responsive platform for effective cancer theranostics. *Adv. Sci.* **2019**, *6*, 1901461. [CrossRef]
28. Chen, W.; Wang, X.; Zhao, B.; Zhang, R.; Xie, Z.; He, Y.; Chen, A.; Xie, X.; Yao, K.; Zhong, M.; et al. CuS-MnS_2 nano-flowers for magnetic resonance imaging guided photothermal/photodynamic therapy of ovarian cancer through necroptosis. *Nanoscale* **2019**, *11*, 12983. [CrossRef]
29. Rawson, S.; Maksimcuka, J.; Withers, P.; Cartmell, S. X-ray computed tomography in life sciences. *BMC Biol.* **2020**, *18*, 21. [CrossRef]
30. Gomez, C.; Hallot, G.; Laurent, S.; Port, M. Medical applications of metallic bismuth nanoparticles. *Pharmaceutics* **2021**, *13*, 1793. [CrossRef]
31. Han, X.; Xu, K.; Taratula, O.; Farsad, K. Applications of nanoparticles in biomedical imaging. *Nanoscale* **2019**, *11*, 799–819. [CrossRef]
32. Wang, J.T.; Zhang, W.; Wang, W.B.; Wu, Y.J.; Zhou, L.; Cao, F. One-pot bottom-up fabrication of biocompatible PEGylated WS_2 nanoparticles for CT-guided photothermal therapy of tumors in vivo. *Biochem. Biophys. Res. Commun.* **2019**, *3*, 587–591. [CrossRef]
33. Wang, Y.; Song, S.; Lu, T.; Cheng, Y.; Song, Y.; Wang, S.; Tan, F.; Li, J.; Li, N. Oxygen-supplementing mesoporous polydopamine nanosponges with WS_2 QDs-embedded for CT/MSOT/MR imaging and thermoradiotherapy of hypoxic cancer. *Biomaterials* **2019**, *220*, 119405. [CrossRef]
34. Nosrati, H.; Charmi, J.; Salehiabar, M.; Abhari, F.; Danafar, H. Tumor targeted albumin coated bismuth sulfide nanoparticles (Bi_2S_3) as radiosensitizers and carriers of curcumin for enhanced chemoradiation therapy. *ACS Biomater. Sci. Eng.* **2019**, *5*, 4416–4424. [CrossRef]
35. Lu, Y.; Li, L.; Lin, Z.; Li, M.; Hu, X.; Zhang, Y.; Peng, M.; Xia, H.; Han, G. Enhancing osteosarcoma killing and CT imaging using ultrahigh drug loading and NIR-responsive bismuth sulfide@mesoporous silica nanoparticles. *Adv. Healthc. Mater.* **2018**, *7*, 1800602. [CrossRef]
36. Wang, Y.; Cai, D.; Wu, H.; Fu, Y.; Cao, Y.; Zhang, Y.; Wu, D.; Tian, Q.; Yang, S. Functionalized Cu_3BiS_3 nanoparticles for dual-modal imaging and targeted photothermal/photodynamic therapy. *Nanoscale* **2018**, *10*, 4452–4456. [CrossRef]
37. Miao, Z.H.; Lv, L.X.; Li, K.; Liu, P.Y.; Li, Z.; Yang, H.; Zhao, Q.; Chang, M.; Zhen, L.; Xu, C.Y. Liquid exfoliation of colloidal rhenium disulfide nanosheets as a multifunctional theranostic agent for in vivo photoacoustic/CT imaging and photothermal therapy. *Small* **2018**, *14*, 1703789. [CrossRef]
38. Wang, X.; Wang, J.; Pan, J.; Zhao, F.; Kan, D.; Cheng, R.; Zhang, X.; Sun, S.K. Rhenium sulfide nanoparticles as a biosafe spectral CT contrast agent for gastrointestinal tract imaging and tumor theranostics in vivo. *ACS Appl. Mater. Interfaces* **2019**, *11*, 33650–33658. [CrossRef]
39. Yoon, S.; Cheon, S.Y.; Park, S.; Lee, D.; Lee, Y.; Han, S.; Kim, M.; Koo, H. Recent advances in optical imaging through deep tissue: Imaging probes and techniques. *Biomater. Res.* **2022**, *26*, 57. [CrossRef]
40. Serrao, E.; Thakor, A.; Goh, V.; Gallagher, F. Functional and molecular imaging for personalized medicine in oncology. In *Grainger and Allison's Diagnostic Radiology*; Andreas, A., Ed.; Elsevier: Amsterdam, The Netherlands, 2020; pp. 1752–1765.
41. Zhang, N.N.; Lu, C.Y.; Chen, M.J.; Xu, X.L.; Shu, G.F.; Du, Y.Z.; Ji, J.S. Recent advances in near-infrared II imaging technology for biological detection. *J. Nanobiotechnol.* **2021**, *19*, 132. [CrossRef]
42. Hsu, J.C.; Cruz, E.D.; Lau, K.C.; Bouche, M.; Kim, J.; Maidment, A.D.; Cormode, D.P. Renally excretable and size-tunable silver sulfide nanoparticles for dual-energy mammography or computed tomography. *Chem. Mater.* **2019**, *31*, 7845–7854. [CrossRef]

43. Awasthi, P.; An, X.; Xiang, J.; Kalva, N.; Shen, Y.; Li, C. Facile synthesis of noncytotoxic PEGylated dendrimer encapsulated silver sulfide quantum dots for NIR-II biological imaging. *Nanoscale* **2020**, *12*, 5678–5684. [CrossRef]
44. Kim, J.; Hwang, D.W.; Jung, H.S.; Kim, K.W.; Pham, X.H.; Lee, S.H.; Byun, J.W.; Kim, W.; Kim, H.M.; Hahm, E.; et al. High-quantum yield alloy-typed core/shell CdSeZnS/ZnS quantum dots for bio-applications. *J. Nanobiotechnol.* **2022**, *20*, 22. [CrossRef]
45. Zhang, X.; Wang, W.; Su, L.; Ge, X.; Ye, J.; Zhao, C.; He, Y.; Yang, H.; Song, J.; Duan, H. Plasmonic-fluorescent Janus Ag/Ag$_2$S nanoparticles for in situ H$_2$O$_2$-activated NIR-II fluorescence imaging. *Nano Lett.* **2021**, *21*, 2625–2633. [CrossRef]
46. Wu, Z.; Tang, Y.; Chen, L.; Liu, L.; Huo, H.; Ye, J.; Ge, X.; Su, L.; Chen, Z.; Song, J. In-situ assembly of Janus nanoprobe for cancer activated NIR-II photoacoustic imaging and enhanced photodynamic therapy. *Anal. Chem.* **2022**, *94*, 10540–10548. [CrossRef]
47. Harish, R.; Nisha, K.D.; Prabhakaran, S.; Sridevi, B.; Harish, S.; Navaneethan, M.; Ponusamy, S.; Hayakawa, Y.; Vinniee, C.; Ganesh, M.R. Synthesis and cytotoxic assessment of chitosan coated CdS nanoparticles. *Appl. Surf. Sci.* **2020**, *499*, 143817. [CrossRef]
48. Xu, N.; Piao, M.; Arkin, K.; Ren, L.; Zhang, J.; Hao, J.; Zheng, Y.; Shang, Q. Imaging of water soluble CdTe/CdS core-shell quantum dots in inhibiting multidrug resistance of cancer cells. *Talanta* **2019**, *201*, 309–316. [CrossRef] [PubMed]
49. Shim, H.S.; Ko, M.; Jeong, S.; Shin, S.Y.; Park, S.M.; Do, Y.R.; Song, J.K. Enhancement mechanism of quantum yield in alloyed-core/shell structure of ZnS-CuInS2/ZnS quantum dots. *J. Phys. Chem. C* **2021**, *125*, 9965–9972. [CrossRef]
50. Liu, W.W.; Li, P.C. Photoacoustic imaging of cells in a three-dimensional microenvironment. *J. Biomed. Sci.* **2020**, *27*, 3. [CrossRef]
51. Huang, K.; Zhang, Y.; Lin, J.; Huang, P. Nanomaterials for photoacoustic imaging in the second near-infrared window. *Biomater. Sci.* **2019**, *7*, 472–479. [CrossRef] [PubMed]
52. Liang, G.; Jin, X.; Qin, H.; Xing, D. Glutathione-capped, renal-clearable CuS nanodots for photoacoustic imaging and photothermal therapy. *J. Mater. Chem. B* **2017**, *5*, 6366–6375. [CrossRef] [PubMed]
53. Wu, M.; Mei, T.; Lin, C.; Wang, Y.; Chen, J.; Le, W.; Sun, M.; Xu, J.; Dai, H.; Zhang, Y.; et al. Melanoma cell membrane biomimetic versatile CuS nanoprobes for homologous targeting photoacoustic imaging and photothermal chemotherapy. *Appl. Mater. Interfaces* **2020**, *12*, 16031–16039. [CrossRef]
54. Ouyang, Z.; Li, D.; Xiong, Z.; Song, C.; Gao, Y.; Liu, R.; Shen, M.; Shi, X. Antifouling dendrimer-entrapped copper sulfide nanoparticles enable photoacoustic imaging-guided targeted combination therapy of tumors and tumor metastasis. *ACS Appl. Mater. Interfaces* **2021**, *13*, 6069–6080. [CrossRef] [PubMed]
55. Zhang, C.; Li, D.; Pei, P.; Wang, W.; Chen, B.; Chu, Z.; Zha, Z.; Yang, X.; Wang, J.; Qian, H. Rod-based urchin-like hollow microspheres of Bi$_2$S$_3$: Facile synthesis, photo-controlled drug release for photoacoustic imaging and chemo-photothermal therapy of tumor ablation. *Biomaterials* **2020**, *237*, 119835. [CrossRef]
56. Zhao, P.; Li, B.; Li, Y.; Chen, L.; Wang, H.; Ye, L. DNA-templated ultrasmall bismuth sulfide nanoparticles for photoacoustic imaging of myocardial infarction. *J. Colloid Interface Sci.* **2022**, *615*, 475–488. [CrossRef]
57. Lei, P.; An, R.; Zheng, X.; Zhang, P.; Du, K.; Zhang, M.; Dong, L.; Gao, X.; Feng, J.; Zhang, H. Ultrafast synthesis of ultrasmall polyethyleneimine-protected AgBiS$_2$ nanodots by "rookie method" for in vivo dual-modal CT/PA imaging and simultaneous photothermal therapy. *Nanoscale* **2018**, *10*, 16765–16774. [CrossRef] [PubMed]
58. Santosh, K.C.; Longo, R.C.; Addou, R.; Wallace, R.M.; Cho, K. Impact of intrinsic atomic defects on the electronic structure of MoS2 monolayers. *Nanotechnology* **2014**, *25*, 375703. [CrossRef]
59. Shin, M.H.; Park, E.Y.; Han, S.; Jung, H.S.; Keum, D.H.; Lee, G.H.; Kim, T.; Kim, C.; Kim, K.S.; Yun, S.H.; et al. Multimodal cancer theranostics using hyaluronate-conjugated molybdenum disulfide. *Adv. Healthc. Mater.* **2018**, *8*, 8101036. [CrossRef]
60. Liu, C.; Chen, J.; Zhu, Y.; Gong, X.; Zheng, R.; Chen, N.; Chen, D.; Yan, H.; Zhang, P.; Zheng, H.; et al. Highly sensitive MoS$_2$-Indocyanine green hybrid for photoacoustic imaging of orthotopic brain glioma at deep site. *Nano-Micro Lett.* **2018**, *10*, 48. [CrossRef]
61. Au, M.T.; Shi, J.; Fan, Y.; Ni, J.; Wen, C.; Yang, M. Nerve growth factor-targeted molecular theranostics based on molybdenum disulfide nanosheet-coated gold nanorods (MoS2-AuNR) for osteoarthritis pain. *ACS Nano* **2021**, *15*, 11711–11723. [CrossRef]
62. Zhang, X.; Wu, J.; Williams, G.R.; Yang, Y.; Niu, S.; Qian, Q.; Zhu, L.M. Dual-responsive molybdenum disulfide/copper sulfide-based delivery systems for enhanced chemo-photothermal therapy. *J. Colloid Interface Sci.* **2019**, *539*, 433–441. [CrossRef]
63. Nomura, S.; Morimoto, Y.; Tsujimoto, H.; Arake, M.; Harada, M.; Saitoh, D.; Hara, I.; Ozeki, E.; Satoh, A.; Takayama, E.; et al. Highly reliable, targeted photothermal cancer therapy combined with thermal dosimetry using a near-infrared absorbent. *Sci. Rep.* **2020**, *10*, 9765. [CrossRef] [PubMed]
64. Li, Y.; Lu, W.; Huang, Q.; Huang, M.; Li, C.; Chen, W. Copper sulfide nanoparticles for photothermal ablation of tumor cells. *Nanomedicine* **2010**, *5*, 1161–1171. [CrossRef]
65. Jiapaer, Z.; Zhang, L.; Ma, W.; Liu, H.; Li, C.; Huang, W.; Shao, S. Disulfiram-loaded hollow copper sulfide nanoparticles show antitumor effects in preclinical models of colorectal cancer. *Biochem. Biophys. Res. Commun.* **2022**, *635*, 291–298. [CrossRef] [PubMed]
66. Chen, J.; Wang, Z.J.; Zhang, K.L.; Xu, Y.J.; Chen, Z.G.; Hu, X.Y. Selective castration-resistant prostate cancer photothermal ablation with copper sulfide nanoplates. *Urol. Technol. Eng.* **2019**, *125*, 248–255. [CrossRef] [PubMed]
67. Lu, F.; Wang, J.; Yang, L.; Zhu, J.J. A facile one-pot synthesis of colloidal stable, monodisperse, highly PEGylated CuS@mSiO$_2$ nanocomposites for the combination of photothermal therapy and chemotherapy. *Chem. Commun.* **2015**, *51*, 9447–9450. [CrossRef] [PubMed]

68. Cheng, L.; Liu, J.; Gu, X.; Gong, H.; Shi, X.; Liu, T.; Wang, C.; Wang, X.; Liu, G.; Xing, H.; et al. PEGylated WS_2 nanosheets as a multifunctional theranostic agent for in vivo dual-modal CT/Photoacoustic imaging guided photothermal therapy. *Adv. Mater.* **2014**, *26*, 1886–1893. [CrossRef]
69. Lei, Z.; Zhu, W.; Xu, S.; Ding, J.; Wan, J.; Wu, P. Hydrophilic $MoSe_2$ nanosheets as effective photothermal therapy agents and their application in smart devices. *ACS Appl. Mater. Interfaces* **2016**, *8*, 20900–20908. [CrossRef]
70. Chou, S.S.; Kaehr, B.; Kim, J.; Foley, B.M.; De, M.; Hopkins, P.E.; Huang, J.; Brinker, C.J.; Dravid, V.P. Chemically exfoliated MoS_2 as near-infrared photothermal agents. *Angew. Chem. Int. Ed.* **2013**, *52*, 4160–4164. [CrossRef]
71. Qian, X.; Shen, S.; Liu, T.; Cheng, L.; Liu, Z. Two-dimensional TiS_2 nanosheets for in vivo photoacoustic imaging and photothermal cancer therapy. *Nanoscale* **2015**, *7*, 6380–6387. [CrossRef]
72. Yong, Y.; Cheng, X.; Bao, T.; Zu, M.; Yan, L.; Yin, W.; Ge, C.; Wang, D.; Gu, Z.; Zhao, Y. Tungsten sulfide quantum dots as multifunctional nanotheranostics for in vivo dual-modal imaging guided photothermal/radiotherapy synergistic therapy. *ACS Nano* **2015**, *9*, 12451–12463. [CrossRef] [PubMed]
73. Yang, K.; Yang, G.; Chen, L.; Cheng, L.; Wang, L.; Ge, C.; Liu, Z. FeS nanoplates as a multifunctional nano-theranostic for magnetic resonance imaging guided photothermal therapy. *Biomaterials* **2015**, *38*, 1–9. [CrossRef] [PubMed]
74. Ma, L.; Liang, S.; Liu, X.L.; Yang, J.; Zhou, L.; Wang, Q.Q. Synthesis of dumbbell-like gold-metal sulfide core-shell nanorods with largely enhanced transverse plasmon resonance in visible region and efficiently improved photocatalytic activity. *Adv. Funct. Mater.* **2015**, *25*, 898–904. [CrossRef]
75. Yang, C.; Ma, L.; Zou, X.; Xiang, G.; Chen, W. Surface plasmon-enhanced Ag/CuS nanocomposites for cancer treatment. *Cancer Nanotechnol.* **2013**, *4*, 81–89. [CrossRef]
76. Ding, X.; Liow, C.H.; Zhang, M.; Huang, R.; Li, C.; Shen, H.; Liu, M.; Zou, Y.; Gao, N.; Zhang, Z.; et al. Surface plasmon resonance enhanced light absorption and photothermal therapy in the second near-infrared window. *J. Am. Chem. Soc.* **2014**, *136*, 15684–15693. [CrossRef]
77. Yuan, L.; Hu, W.; Zhang, H.; Chen, L.; Wang, J.; Wang, Q. Cu_5FeS_4 nanoparticles with tunable plasmon resonances for efficient photothermal therapy of cancers. *Front. Bioeng. Biotechnol.* **2020**, *8*, 21. [CrossRef]
78. Zhao, X.; Liu, J.; Fan, J.; Chao, H.; Peng, X. Recent progress in photosensitizers for overcoming challenges of photodynamic therapy: From molecular design to application. *Chem. Soc. Rev.* **2021**, *50*, 4185. [CrossRef] [PubMed]
79. Dolmans, D.; Fukumura, D.; Jain, R. Photodynamic therapy for cancer. *Nat. Rev. Cancer* **2003**, *3*, 380–387. [CrossRef]
80. Jia, L.; Ding, L.; Tian, J.; Bao, L.; Hu, Y.; Ju, H.; Yu, J.S. Aptamer loaded MoS_2 nanoplates as nanoprobe for detection of intracellular ATP and controllable photodynamic therapy. *Nanoscale* **2015**, *7*, 15953–15961. [CrossRef]
81. Huang, C.X.; Chen, H.J.; Li, F.; Wang, W.N.; Li, D.D.; Yang, X.Z.; Miao, Z.H.; Zha, Z.B.; Lu, Y.; Qian, H.S. Controlled synthesis of upconverting nanoparticles/CuS yolk–shell nanoparticles for in vitro synergistic photothermal and photodynamic therapy of cancer cells. *J. Mater. Chem. B* **2017**, *5*, 9487–9496. [CrossRef]
82. Wang, S.; Riedinger, A.; Li, H.; Fu, C.; Liu, H.; Li, L.; Liu, T.; Tan, L.; Barthel, M.J.; Pugliese, G.; et al. Plasmonic copper sulfide nanocrystals exhibiting near-infrared photothermal and photodynamic therapeutic effects. *ACS Nano* **2015**, *9*, 1788–1800. [CrossRef] [PubMed]
83. Wang, L.; Ma, X.; Cai, K.; Li, X. Morphological effect of copper sulfide nanoparticles on their near infrared laser activated photothermal and photodynamic performance. *Mater. Res. Express* **2019**, *6*, 105406. [CrossRef]
84. Gu, X.; Qiu, Y.; Lin, M.; Cui, K.; Chen, G.; Chen, Y.; Fan, C.; Zhang, Y.; Xu, L.; Chen, H.; et al. CuS nanoparticles as a photodynamic nanoswitch for abrogating bypass signaling to overcome gefitinib resistance. *Nano Lett.* **2019**, *19*, 3344–3352. [CrossRef]
85. Lin, S.; Wang, Y.; Chen, Z.; Li, L.; Zeng, J.; Dong, Q.; Wang, Y.; Chai, Z. Biomineralized enzyme-like cobalt sulfide nanodots for synergetic phototherapy with tumor multimodal imaging navigation. *ACS Sustain. Chem. Eng.* **2018**, *6*, 12061–12069. [CrossRef]
86. Cheng, Y.; Chang, Y.; Feng, Y.; Jian, H.; Wu, X.; Zheng, R.; Xu, K.; Zhang, H. Bismuth sulfide nanorods with retractable zinc protoporphyrin molecules for suppressing innate antioxidant defense system and strengthening phototherapeutic effects. *Adv. Mater.* **2019**, *31*, 8. [CrossRef] [PubMed]
87. Dias, L.D.; Buzzá, H.H.; Stringasci, M.D.; Bagnato, V.S. Recent advances in combined photothermal and photodynamic therapies against cancer using carbon nanomaterial platforms for in vivo studies. *Photochem* **2021**, *3*, 434–447. [CrossRef]
88. Song, C.; Yang, C.; Wang, F.; Ding, D.; Gao, Y.; Guo, W.; Yan, M.; Liu, S.; Guo, C. MoS_2-based multipurpose theranostic nanoplatform realizing dual-imaging-guided combination phototherapy to eliminate solid tumor via a liquefaction necrosis process. *J. Mater. Chem. B* **2017**, *5*, 9015–9024. [CrossRef]
89. Jin, R.; Yang, J.; Ding, P.; Li, C.; Zhang, B.; Chen, W.; Zhao, Y.D.; Cao, Y.; Liu, B. Antitumor Immunity triggered by photothermal therapy and photodynamic therapy of a 2D MoS_2 nanosheet-incorporated injectable polypeptide-engineered hydrogel combinated with chemotherapy for 4T1 breast tumor therapy. *Nanotechnology* **2020**, *31*, 205102. [CrossRef]
90. Liu, T.; Wang, C.; Cui, W.; Gong, H.; Liang, C.; Shi, X.; Li, Z.; Sun, B.; Liu, Z. Combined photothermal and photodynamic therapy delivered by PEGylated MoS_2 nanosheets. *Nanoscale* **2014**, *6*, 11219–221225. [CrossRef]
91. Xu, J.; Gulzar, A.; Liu, Y.; Bi, H.; Gai, S.; Liu, B.; Yang, D.; He, F.; Yang, P. Integration of IR-808 sensitized upconversion nanostructure and MoS_2 nanosheet for 808 nm NIR light triggered phototherapy and bioimaging. *Small* **2017**, *13*, 1701841. [CrossRef]
92. Bharathiraja, S.; Manivasagan, P.; Moorthy, M.S.; Bui, N.Q.; Lee, K.D.; Oh, J. Chlorin e6 conjugated copper sulfide nanoparticles for photodynamic combined photothermal therapy. *Photodiagn. Photodyn. Ther.* **2017**, *19*, 128–134. [CrossRef] [PubMed]

93. Li, M.; Wang, Y.; Lin, H.; Qu, F. Hollow CuS nanocube as nanocarrier for synergetic chemo/photothermal/photodynamic therapy. *Mater. Sci. Eng. C* **2019**, *96*, 591–598. [CrossRef] [PubMed]
94. Li, M.; Lin, H.; Qu, F. FeS$_2$@C-ICG-PEG nanostructure with intracellular O$_2$ generation for enhanced photo-dynamic/thermal therapy and imaging. *Chem. Eng. J.* **2020**, *384*, 123374. [CrossRef]
95. Hou, M.; Zhong, Y.; Zhang, L.; Xu, Z.; Kang, Y.; Xue, P. Polydopamine (PDA)-activated cobalt sulfide nanospheres responsive to tumor microenvironment (TME) for chemotherapeutic-enhanced photothermal therapy. *Chin. Chem. Lett.* **2021**, *32*, 1055–1060. [CrossRef]
96. Bao, S.J.; Li, Y.; Li, C.M.; Bao, Q.; Lu, Q.; Guo, J. Shape evolution and magnetic properties of cobalt sulfide. *Cryst. Growth Des.* **2008**, *8*, 3745–3749. [CrossRef]
97. Lim, S.; Park, J.; Shim, M.K.; Um, W.; Yoon, H.Y.; Ryu, J.H.; Lim, D.K.; Kim, K. Recent advances and challenges of repurposing nanoparticles-based drug delivery systems to enhance cancer immunotherapy. *Theranostics* **2019**, *9*, 7906–7923. [CrossRef]
98. Zhang, P.; Li, Y.; Tang, W.; Zhao, J.; Jing, L.; McHugh, K. Theranostic nanoparticles with disease-specific administration strategies. *NanoToday* **2022**, *42*, 101335. [CrossRef]
99. Chen, J.; Zhu, Y.; Wu, C.; Shi, J. Engineering lactate-modulating nanomedicines for cancer therapy. *Chem. Soc. Rev.* **2023**, *52*, 973–1000. [CrossRef]
100. Muluh, T.A.; Chen, Z.; Li, Y.; Xiong, K.; Jin, J.; Fu, S.; Wu, J. Enhancing cancer immunotherapy treatment goals by using nanoparticle delivery system. *Int. J. Nanomed.* **2021**, *16*, 2389–2404. [CrossRef]
101. Shao, K.; Singha, S.; Clemente-Casares, X.; Tsai, S.; Yang, Y.; Santamaria, P. Nanoparticle-Based Immunotherapy for Cancer. *ACS Nano* **2015**, *9*, 16–30. [CrossRef]
102. Guo, L.; Yan, D.D.; Yang, D.; Li, Y.; Wang, X.; Zalewski, O.; Yan, B.; Lu, W. Combinatorial photothermal and immuno cancer therapy using chitosan-coated hollow copper sulfide nanoparticles. *ACS Nano* **2014**, *8*, 5670–5681. [CrossRef]
103. Chen, Z.; Zhang, Q.; Zeng, L.; Zhang, J.; Liu, Z.; Zhang, M.; Zhang, X.; Xu, H.; Song, H.; Tao, C. Light-triggered OVA release based on CuS@poly(lactide-co-glycolide acid) nanoparticles for synergistic photothermal-immunotherapy of tumor. *Pharmacol. Res.* **2020**, *158*, 104902. [CrossRef] [PubMed]
104. Yan, T.; Yang, K.; Chen, C.; Zhou, Z.; Shen, P.; Jia, Y.; Xue, J.; Zhang, Z.; Shen, B.; Han, X. Synergistic photothermal cancer immunotherapy by Cas9 ribonucleoprotein-based copper sulfide nanotherapeutic platform targeting PTPN2. *Biomaterials* **2021**, *279*, 121233. [CrossRef] [PubMed]
105. Zhou, L.; Chen, L.; Hu, X.; Lu, Y.; Liu, Y.; Liu, W.; Sun, Y.; Yao, T.; Dong, C.; Shi, S. A Cu$_9$S$_5$ nanoparticle-based CpG delivery system for synergistic photothermal-, photodynamic- and immunotherapy. *Commun. Biol.* **2020**, *3*, 343. [CrossRef]
106. Xu, J.; Zheng, B.; Zhang, S.; Liao, X.; Tong, Q.; Wei, G.; Yu, S.; Chen, G.; Wu, A.; Gao, S.; et al. Copper sulfide nanoparticle-redirected macrophages for adoptive transfer therapy of melanoma. *Adv. Funct. Mater.* **2021**, *31*, 2008022. [CrossRef]
107. Han, Q.; Wang, X.; Jia, X.; Cai, S.; Liang, W.; Qin, Y.; Yang, R.; Wang, C. CpG loaded MoS$_2$ nanosheets as multifunctional agents for photothermal enhanced cancer immunotherapy. *Nanoscale* **2017**, *9*, 5927–5934. [CrossRef] [PubMed]
108. Pardo, M.; Shuster-Meiseles, T.; Levin-Zaidman, S.; Rudich, A.; Rudich, Y. Low cytotoxicity of inorganic nanotubes and fullerene-like nanostructures in human bronchial epithelial cells: Relation to inflammatory gene induction and antioxidant response. *Environ. Sci. Technol.* **2014**, *48*, 3457–3466. [CrossRef]
109. Zhang, W.; Zhang, C.C.; Wang, X.Y.; Li, L.; Chen, Q.Q.; Liu, W.W.; Cao, Y.; Ran, H.T. Light-responsive core−shell nanoplatform for bimodal imaging-guided photothermal therapy-primed cancer immunotherapy. *ACS Appl. Mater. Interfaces* **2020**, *12*, 48420–48431. [CrossRef]
110. Kalantar-Zadeh, K.; Ou, J.Z.; Daeneke, T.; Strano, M.S.; Pumera, M.; Gras, S.L. Two-dimensional transition metal dichalcogenides in biosystems. *Adv. Funct. Mater.* **2015**, *25*, 5086–5099. [CrossRef]
111. Ataca, C.; Ciraci, S. Functionalization of single-layer MoS$_2$ honeycomb structures. *J. Phys. Chem. C* **2011**, *115*, 13303–13311. [CrossRef]
112. Voiry, D.; Goswami, A.; Kappera, R.; Silva, C.; Kaplan, D.; Fujita, T.; Chen, M.; Asefa, T.; Chhowalla, M. Covalent functionalization of monolayered transition metal dichalcogenides by phase engineering. *Nat. Chem.* **2015**, *7*, 45–49. [CrossRef] [PubMed]
113. Li, Y.; Sun, Y.; Cao, T.; Su, Q.; Li, Z.; Huang, M.; Ouyang, R.; Chang, H.; Zhang, S.; Miao, Y. A cation-exchange controlled core-shell MnS@Bi$_2$S$_3$ theranostic platform for multimodal imaging guided radiation therapy with hyperthermia boost. *Nanoscale* **2017**, *9*, 14364–14375. [CrossRef] [PubMed]

Disclaimer/Publisher's Note: The statements, opinions and data contained in all publications are solely those of the individual author(s) and contributor(s) and not of MDPI and/or the editor(s). MDPI and/or the editor(s) disclaim responsibility for any injury to people or property resulting from any ideas, methods, instructions or products referred to in the content.

Review

A New Perspective for the Treatment of Alzheimer's Disease: Exosome-like Liposomes to Deliver Natural Compounds and RNA Therapies

Joana Ribeiro [1,2], Ivo Lopes [1] and Andreia Castro Gomes [1,2,*]

[1] Centre of Molecular and Environmental Biology (CBMA)/Aquatic Research Network (ARNET) Associate Laboratory, Universidade do Minho, Campus de Gualtar, 4710-057 Braga, Portugal; pg41579@alunos.uminho.pt (J.R.); ivo_lopes_@hotmail.com (I.L.)

[2] Institute of Science and Innovation for Sustainability (IB-S), Universidade do Minho, Campus de Gualtar, 4710-057 Braga, Portugal

* Correspondence: agomes@bio.uminho.pt; Tel.: +351-253-601511

Abstract: With the increment of the aging population in recent years, neurodegenerative diseases exert a major global disease burden, essentially as a result of the lack of treatments that stop the disease progression. Alzheimer's Disease (AD) is an example of a neurodegenerative disease that affects millions of people globally, with no effective treatment. Natural compounds have emerged as a viable therapy to fill a huge gap in AD management, and in recent years, mostly fueled by the COVID-19 pandemic, RNA-based therapeutics have become a hot topic in the treatment of several diseases. Treatments of AD face significant limitations due to the complex and interconnected pathways that lead to their hallmarks and also due to the necessity to cross the blood–brain barrier. Nanotechnology has contributed to surpassing this bottleneck in the treatment of AD by promoting safe and enhanced drug delivery to the brain. In particular, exosome-like nanoparticles, a hybrid delivery system combining exosomes and liposomes' advantageous features, are demonstrating great potential in the treatment of central nervous system diseases.

Keywords: neurodegenerative diseases; Alzheimer's disease; natural compounds; RNA therapy; blood-brain barrier; exosome-like liposomes

1. Introduction

Neurodegenerative diseases are a diversified group of conditions characterized by progressive degeneration of the function of the central nervous system (CNS) or peripheral nervous system, and the current therapeutic options do not provide a cure, only slowing down the disease progression. Together, neurodegenerative diseases exert a major burden in global healthcare systems, with dementia being a public health challenge in many developed countries, as aging is a strong risk factor [1].

Dementia is one of the highest global health crises of this century, with Alzheimer's disease (AD) being the most common form of dementia. In the United States, an estimated 6.7 million individuals aged 65 and older are living with AD in 2023, and the number is expected to reach 88 million by 2050 [2]. The estimated 2023 cost of caring for those with this disease is $345 billion. Between 2000 and 2019, the number of deaths from AD increased by 145%, while deaths from the number-one cause of death—heart disease—decreased by 7.3% [2].

AD is a progressive, irreversible neurodegenerative disease that leads to memory impairment, impacts cognition, and can ultimately affects behavior, speech, visuospatial orientation, and the motor system. This disease is characterized by two major pathological hallmarks: Progressive accumulation of amyloid beta (Aβ) plaques and neurofibrillary tangles (NFTs). Aβ damages neurons by interfering with neuron communication at synapses and NFTs block the transport of essential molecules for the normal function of neurons. Consequently, these lead to other complications such as oxidative stress, inflammation, and brain atrophy due to cell loss [2,3].

Natural compounds or extracts are a viable therapy to fill the huge gap in the treatment of this disease since they can target several hallmarks. However, for improved efficacy of these compounds, they can be administered by delivery systems. Moreover, mostly instigated by the recent COVID-19 pandemic, RNA-based therapies have become a topic of great interest to researchers and pharmaceutical companies. RNA therapies promise to change the current conventional drugs that are not capable to target and treat all types of diseases, and several clinical studies are ongoing for a variety of RNA-based therapeutics against various incurable diseases. RNA therapy presents several advantages such as cost effectiveness, manufacturing simplicity, and the ability to target previously inaccessible pathways [4,5]. RNA therapies such as microRNAs (miRNAs), small interfering RNAs (siRNAs), and messenger RNAs (mRNAs) seem to be some of the most promising molecules for the treatment of AD [6]. For the effective delivery of RNA, this molecule has to overcome several obstacles. Its hydrophilic, negatively charged properties make it difficult for the RNA molecule to passively diffuse across the cell's membrane, and so it has to undergo endocytosis and escape from the endosome to reach the cytoplasm. Furthermore, this molecule is highly susceptible to ribonucleases degradation and must have enhanced accumulation at targeted tissues [4].

In addition to the complex mechanisms that lead to AD, the blood–brain barrier (BBB) is known to be a particular reason for the lack of effective treatments for AD. The BBB is a physiological barrier constituted by blood vessels that vascularize the CNS and possess unique properties that allow precise control of the molecules allowed to enter the CNS [7].

Nanoparticles have been used to mitigate all of these hindrances in the delivery of RNA molecules into the brain. These drug delivery systems effectively protect RNA from degradation, enable the crossing of biological barriers, and allow a targeted accumulation and release [8]. In recent years, nanoparticles such as dendrimers, polymeric nanoparticles and gold nanoparticles, and carbon quantum dots have shown to be capable of crossing the BBB effectively [8]. Along with these nanoparticles, exosomes and liposomes are delivery systems with promising properties that allow them to cross the BBB. Exosomes are nano-sized extracellular vesicles (EVs) released into surrounding body fluids by their parental cells and carry cell-specific cargos of proteins, lipids, and genetic materials. These EVs can be selectively uptaken by neighboring or distant cells far from their release. On the other hand, liposomes are synthetic vesicles comprised of one or several concentric lipid bilayers surrounding an aqueous lumen that can be created from cholesterol and natural phospholipids or synthetic surfactants [9,10].

Exosome-like liposomes are a novel concept of nanoparticles that combine the advantages of both the exosomes and liposomes, creating a unique delivery system with several advantages such as the mimetic constituents of natural exosomes, high biocompatibility, small size, easy production, efficient transport and delivery of therapeutic compounds with low bioavailability (e.g., curcumin and RNA molecules), and the ability to load both hydrophilic and hydrophobic drugs [11].

2. Neurodegenerative Diseases

Neurodegenerative diseases are a heterogeneous group of neurological disorders characterized by cognitive, psychiatric, and motor deficits due to neuron loss [12]. In addition to these common features, there are also no current treatments to stop the advancement of the diseases. Some of the main reasons for the lack of effective treatment in neurodegenerative diseases are the limitations imposed by the BBB and the complex pathways that lead to the late diagnosis of the diseases [13].

Neurodegenerative diseases are characterized by (1) protein aggregation; (2) disruptive proteostasis; (3) neuroinflammation; (4) oxidative stress; (5) synaptic failure; and (6) neuronal death (Figure 1). The presence of protein aggregation is a key hallmark in a large variety of neurodegenerative disorders. These abnormally deposited proteins are found in brain regions that, when damaged, lead to physical vulnerability [1,14]. Several proteins are associated with neurodegenerative disorders:

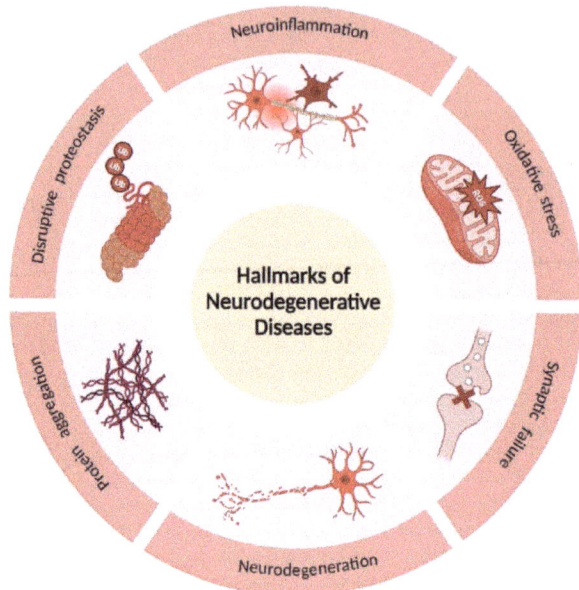

Figure 1. Hallmarks of neurodegenerative diseases. The scheme illustrates the six hallmarks of these disorders: Protein aggregation, disruptive proteostasis, neuroinflammation, oxidative stress, synaptic failure, and neuronal death. Created in BioRender.com (accessed on 25 May 2023).

- Tau protein—microtubule-associated protein—encoded by the microtubule-associated protein tau (MAPT) gene. Tau is substantially expressed in the cytoplasm of neurons and plays an important role primarily in the stabilization and assembly of axonal microtubules and also in a variety of physiological processes, which include axonal transport, signal transmission between neurons, neurogenesis, myelination, motor function, neuronal excitability, glucose metabolism, iron homeostasis, and DNA protection [15,16].
- Aβ—derives from the amyloid precursor protein (APP) and aggregates into amyloid plaques with Aβ polypeptides 40 and 42 amino acids long [17].
- Prion—another protein present in neurodegenerative diseases is the prion protein (PrP) encoded by the PRNP gene. In prion diseases, the prion protein misfolds, propagates, and aggregates rapidly, being responsible for spreading neurodegeneration between cells, and, consequently, brain regions [14].
- α-synuclein—this is a 140-amino-acid protein highly expressed in the brain, encoded by the α-synuclein (*SCNA*) gene [12].

Under healthy conditions, there are protein degradation systems to maintain protein homeostasis, with an important role in the clearance of toxic protein aggregates: The autophagy lysosomal pathway and the ubiquitin-proteosome system. However, these pathways lose activity in elderly individuals, contributing to the accumulation of toxic protein aggregates [14]. Glial cells (microglia and astrocyte) substitute peripheral immune cells' function in the brain. Microglia play a crucial role in defense functions in the brain, and these cells are activated in signs of pathogens or injury. In neurotoxic conditions with several aggregated proteins, microglia activation is induced, interacting with astrocytes and leading to inflammation [14,18]. Neurons are particularly susceptible to oxidative stress due to the high polyunsaturated fatty acid content in the membranes, high oxygen consumption, and low antioxidant defenses in the brain, inducing increased oxidation of proteins, nucleic acids, and lipids [1]. Synaptic failure has been described in various neurodegenerative diseases. Synapses are the functional part of the connection between neurons and the key physiological function of neurons. However, when pathogenic factors affect synapses, it suppresses the brain from learning and leads to memory impairment. Currently, these synaptic changes are the targets of many pharmacological interventions [19]. All these factors result in neuronal cell death and, in some neurodegenerative disorders, can result in brain volume loss [14].

Neurodegenerative diseases include Parkinson's disease (PD), amyotrophic lateral sclerosis (ALS), Huntington's disease (HD), and AD, which will be the focus of this review [1].

The brain of PD patients is affected by the presence of intra-neuronal inclusion bodies—Lewy bodies—and the accumulation of the protein α-synuclein, which spreads from one brain region to another. PD is characterized by movement disorder, motor function impairment, and other nonmotor symptoms, such as gastrointestinal issues and sleep disturbances [1,20].

ALS is a devastating, progressive disease, and the cause of this condition is unknown. This disease is characterized by a deficit in motor neurons in the spinal cord and the motor cortex of the cerebrum. Patients witness progressive muscle weakness and atrophy and respiratory failure due to the weakening of respiratory muscles, with a life expectancy of 15.8 months post-diagnosis [21,22].

HD is an inherited neurodegenerative disorder caused by a mutation in the Huntingtin gene, an abnormal trinucleotide expansion, which is translated into a mutant protein. The protein leads to the disruption of cellular molecular processes, which can involve both loss- and gain-of-function mechanisms. A hallmark of this disease is the degeneration of the striatum (caudate nucleus and putamen), with specific loss of efferent medium spiny neurons and brain shrinkage. This disorder is characterized by movement disturbance, cognitive decline, coordination loss, depression, obsessive–compulsive disorder, and other psychiatric symptoms [23–25].

Alzheimer's Disease

The pathological development of AD is complex and not yet fully understood [1]. At the microscopic level, the progression of this disease is characterized by the accumulation of two proteins: Aβ and Tau proteins, which aggregate into Aβ plaques and NFTs, respectively (Figure 2A) [26].

The presence of amyloid plaques is a consequence of an abnormal accumulation and deposition of Aβ, a product of the amyloid precursor protein (APP) [27]. APP is a transmembrane protein expressed in numerous human tissues, including the CNS, and can be cleaved by different proteases in the non-amyloidogenic or amyloidogenic processing of APP [28,29]. If the APP is cleaved by α-secretase, in the non-amyloidogenic pathway, it prevents the formation of the Aβ and, instead, soluble APPα fragments are formed, which are described to be non-cytotoxic [26]. The amyloidogenic pathway includes the combined action of β- and γ- secretases, which generate Aβ peptides with different C-terminal residues (Aβ_{40} and Aβ_{42}) [15,27]. The deposition and accumulation of these Aβ polypeptides lead to the formation of amyloid plaques in the brain, resulting in neuroinflammation and synaptic dysfunction [30–32]. NFTs are a consequence of the deposition of the abnormally phosphorylated Tau protein, leading to its aggregation [16]. The microtubule-binding protein Tau, a cytoskeletal protein produced by alternative splicing of the MAPT gene, has been identified as a key molecule in AD, but also in a series of neurodegenerative diseases referred to as tauopathies, in contrast to Aβ accumulation, which is a characteristic exclusive to AD [15,33,34]. Hyperphosphorylated Tau loses its functions in the synthesis and stabilization of microtubules. These proteins accumulate in neurites and neuronal cell bodies, where it develops into insoluble aggregates, NFTs. Tau can be secreted into the extracellular space either in its naked form or packaged in exosomes [35–37].

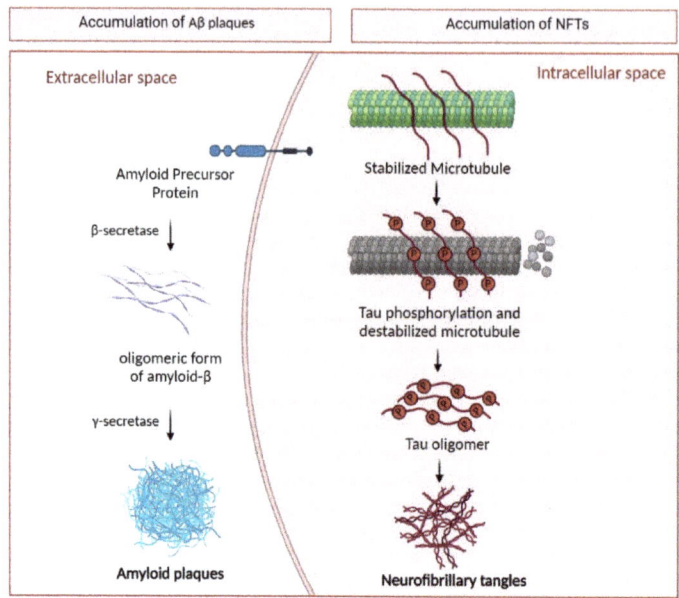
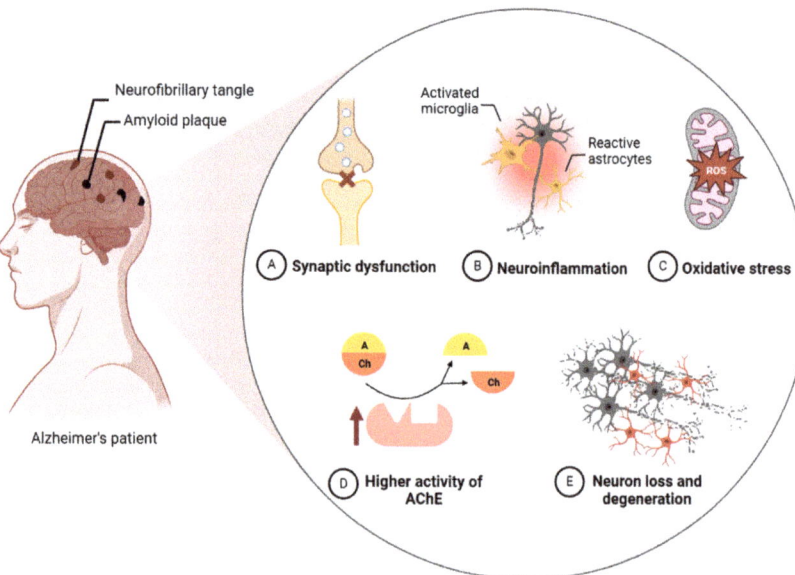

Figure 2. Alzheimer's disease hallmarks. (**A**) Pathway that leads to the formation of amyloid plaques and the mechanism responsible for the formation of neurofibrillary tangles. (**B**) Accumulation of amyloid plaques and neurofibrillary tangles provoke disorders such as synaptic dysfunction, neuroinflammation, oxidative stress, and higher activity of acetylcholinesterase. These result in neurodegeneration. Created in BioRender.com (accessed on 20 April 2023).

In addition the accumulation of these two proteins in the brain, there are other hallmarks in AD, such as synaptic impairment neuroinflammation, oxidative stress, and higher activity of acetylcholinesterase (AChE) (Figure 2B). Synaptic impairment is responsible for the reduction in presynaptic vesicle release and the decrease in glutamatergic receptors [33,38]. Additionally, microglia cells develop more susceptibility to stimulus and produce inflammatory cytokines and chemokines, leading to cytotoxic and pro-inflammatory events. Consequently, this contributes to the deterioration of the BBB and the inability to remove neurotoxic molecules such as Aβ plaques and NFTs from the CNS [39,40]. Damage to the mitochondrial structure, integrity, and biogenesis lead to excessive reactive oxygen species (ROS) production, causing damage to the cellular structure. The equilibrium of distinct neurotransmitter systems, such as acetylcholine (Ach), is crucial for healthy brain function [41,42]. In AD, alterations in the cholinergic system are often present, since there is a loss of cholinergic neurons that leads to an extensive decline of ACh, and consequently to a deficit of cholinergic transmission at the pre-synaptic level. This ACh decrease is due to the higher activity of AChE, the enzyme responsible for its degradation [43]. All of the mentioned microscopic hallmarks alter the brain at a macroscopic level, causing brain shrinkage with cortical thinning and atrophy, leading to decreased brain weight [3,27].

Currently, the therapeutic options for these patients are still very limited [44,45]. The main classes of drugs available to treat AD are acetylcholinesterase inhibitors, which include donepezil, rivastigmine, and galantamine. Although the equilibrium of different neurotransmitters, such as acetylcholine, plays a key role in normal brain function, the cholinergic system is not the only system affected by this pathology. Thus, cholinesterase inhibitors only treat symptoms and do not prevent disease progression. Treatments directed to other hallmarks, such as Aβ accumulation and Tau hyperphosphorylation, failed to provide effects [42,46]. Therefore, an innovative approach that can efficiently treat and control each hallmark of this multifactorial disease is required for effective AD treatment [47].

3. Natural Compounds for AD Treatment

Natural drugs are gaining increased interest from both scientific academia and pharmaceutical industries for the therapy of several diseases. In recent years, due to their multiple beneficial properties, more than 100 natural products have been proposed as a promising approach for AD therapy [48]. Many molecules, including lignans, flavonoids, tannins, polyphenols, triterpenes, sterols, and alkaloids, can act via different pathways since they have anti-inflammatory, anti-amyloidogenic, anticholinesterase, and anti-inflammatory properties and can reduce oxidative stress [49]. It is reasonable to speculate that the progression of AD could be slowed down or even prevented by natural products working on multiple pathological targets [50]. Table 1 presents an array of natural compounds or extracts that were researched for AD in the last fifteen years found on PubMed and Web of Science.

Table 1. Natural compounds or extracts that can be used for AD therapy.

Natural Compound	Role in AD	References
Eugenol	Rats were fed aluminum, a neurotoxic metal that leads to oxidative brain injury and enhanced lipid peroxidation, disruption of neurotrophic, cholinergic, and serotonergic functions, and induce apoptosis with ultimate neuronal and astrocyte damages. A neuroprotective role of eugenol against the aluminum effects was verified through its antioxidant, antiapoptotic potential and its neurotrophic properties.	[51]
Menthol	Menthol inhalation by mice (1 week per month, for 6 months) prevented cognitive impairment in the APP/PS1 mouse model of Alzheimer's.	[52]
Chrysin	Chrysin showed the ability to act as a membrane shield against early oxidative events mediated by O_2^- and other ROS that contribute to neuronal death triggered by $AlCl_3$ exposure, showing chrysin's neuroprotective action.	[53]

Table 1. *Cont.*

Natural Compound	Role in AD	References
Rosmarinic acid	Suppresses Aβ accumulation in mice.	[54]
Ginkgo biloba	Ginkgo biloba improves microcirculation, inhibits the expression of inflammatory factors, and reduces inflammatory damage to neurons, thereby improving the spatial exploration memory of dementia model rats.	[55]
Resveratrol	Multiple studies demonstrated that resveratrol has neuroprotective, anti-inflammatory, and antioxidant characteristics and the ability to minimize Aβ peptide aggregation and toxicity in the hippocampus of Alzheimer's patients, stimulating neurogenesis and inhibiting hippocampal degeneration. Furthermore, resveratrol's antioxidant effect promotes neuronal development by activating the silent information regulator-1, which can protect against the detrimental effects of oxidative stress.	[56]
Huperzine A	Huperzine A is natural, potent, highly specific reversible inhibitor of acetylcholinesterase, with the ability to cross the BBB.	[57]
Brahmi	The neuroprotective properties of Brahmi include the reduction of ROS and neuroinflammation, the inhibition of the aggregation of Aβ and the improvement of cognitive and learning behavior.	[58]
Uncaria tomentosa	Inhibits plaques and tangles formation.	[59]
Berberine	Berberine has antioxidant activity and promotes AChE and monoamine oxidase inhibition. Berberine has been shown to improve memory, lower Aβ and APP concentration, and diminish Aβ plaque accumulation.	[60]
Quercetin	Behavioral and biochemical tests confirm that quercetin promotes the reduction in oxidative stress and increased cognition in zebrafish AD models induced with aluminum chloride.	[61]
Betaine	Betaine has been shown to decrease homocysteine levels and Aβ toxicity in *Caenorhabditis elegans* AD model.	[62]
Curcumin	Curcumin is known to be a potent antioxidant, anti-inflammatory and anti-amyloidogenic compound, that plays a beneficial role in treating AD through several mechanisms. Curcumin can promote a significant reduction of Aβ oligomers and fibril formation.	[46]
Crocin	Crocin, the main constituent of *Crocus sativus* L., has a multifunctional role in protecting brain cells, modulating aggregation of Aβ and Tau proteins, attenuating cognitive and memory impairments, and improving oxidative stress.	[63]
Withania somnifera	Withania somnifera extract can protect against Aβ peptide- and acrolein-induced toxicity. Treatment with this extract significantly protected against Aβ and acrolein, in various cell survival assays with the human neuroblastoma cell line SK-N-SH, significantly reduced the generation of ROS and was demonstrated to be a potent inhibitor of AChE activity.	[64]
Poncirus trifoliate	The extract of *Poncirus trifoliate* is a naturally occurring AChE inhibitor. It showed a 47.31% inhibitory effect on the activity of acetylcholine.	[65]
Convolvulus pluricaulis	*Convolvulus pluricaulis* prevented aluminum-induced neurotoxicity in rat cerebral cortex.	[66]
α-Cyperone	α-Cyperone binds and interacts with tubulin, being capable of destabilizing microtubule polymerization. The effect of this interaction could result in reduction of inflammation.	[67]

Table 1. Cont.

Natural Compound	Role in AD	References
Andrographolide	Andrographolide has beneficial effects in the recovery of spatial memory and learning performance, recovery of synaptic basal transmission, partial or complete protection of certain synaptic proteins and shows a specific neuroprotective effect, that includes the reduction of phosphorylated Tau and Aβ aggregate maturation, in aged degus.	[68]
Apigenin	Apigenin has been shown to have anti-inflammatory and neuroprotective properties in a number of cell and animal models. This compound is also able to protect human induced pluripotent stem cell-derived AD neurons via multiple pathways, by reducing the frequency of spontaneous Ca^{2+} signals and significantly reducing caspase-3/7 mediated apoptosis.	[69]
Baicalein	Baicalein has antioxidant and anti-inflammatory effects.	[70]
Carvacrol	Carvacrol possesses anti-AChE, antioxidant, and neuroprotective properties. This compound alleviated Aβ-induced deficits by reducing cellular neurotoxicity and oxidative stress in the SH-SY5Y cell line, and by reducing oxidative stress and memory impairment in a rat model of AD.	[71]
Decursin/Decursinol angelate	Decursin and decursinol angelate increase cellular resistance to Aβ-induced oxidative injury in PC12 cells.	[72]
Genistein	In vivo studies have shown that genistein improves brain function, antagonizes the toxicity of Aβ and has neuroprotective effects.	[73]
Wogonin	Wogonin has various neuroprotective and neurotrophic activities, such as inducing neurite outgrowth.	[74]
Rutin	Rutin is antioxidant, anti-inflammatory, and has the capacity of reducing Aβ oligomer activities.	[75]
Luteolin	Luteolin has the capacity to cross the BBB and can inhibit β- and γ-secretase to decrease Aβ. It can also reduce neuroinflammation and attenuate the phosphorylation of Tau.	[76]
Linalool	A linalool-treated mice model of AD showed improved learning and spatial memory. This compound reverses the histopathological hallmarks of AD and restores cognitive and emotional functions via an anti-inflammatory effect.	[77]
Asiatic acid	Pre-treatment with Asiatic Acid enhanced cell viability, attenuated rotenone-induced ROS, mitochondrial membrane dysfunction and apoptosis regulating AKT/GSK-3β signaling pathway, after aluminum maltolate neurotoxicity induction in SH-SY5Y neuroblastoma cells.	[78]

Overcoming Limitations of Natural Compounds with Delivery Systems

In addition to the multiple beneficial properties of natural products, these display several limitations such as low hydrophilicity, rapid metabolism and degradation, low bioavailability, reduced targeting, susceptibility to physiological media, and poor permeability through lipid bilayers. Consequently, in vivo, natural drugs require a high-dose administration beyond a safe range, to result in an effective and safe bioavailability [48,79]. Nanotechnology represents a new method to overcome these challenges. Delivery systems for natural compounds lead to the enhancement of pharmacological activity by improving the stability of drugs in vivo, bioavailability, and controlled release, increasing the accumulation of active ingredients in target sites, promoting the solubility of insoluble drugs, and reducing the required doses to produce therapeutic effects [79]. Nanotechnology offers multiple advantages in the delivery of natural products, since by this method, these drugs can exert their therapeutic effect in the treatment of AD (Figure 3) [80].

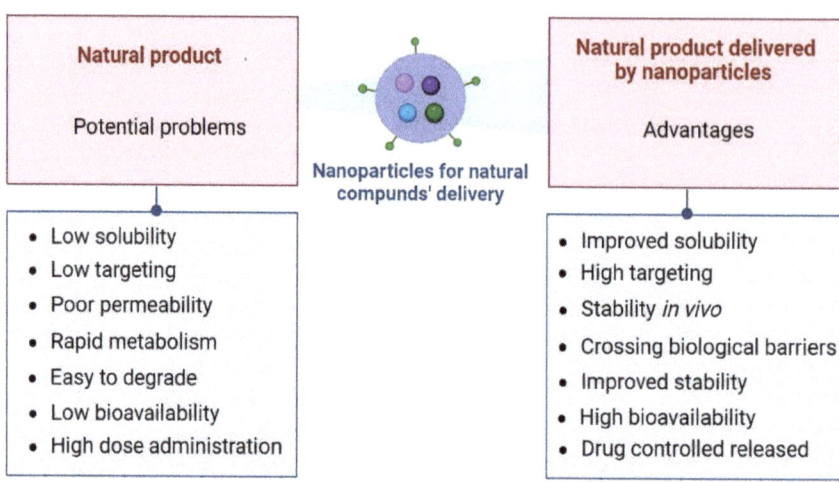

Figure 3. Potential problems of natural products in vivo and advantages of natural products delivered by nanoparticles. Created in BioRender.com (accessed on 5 May 2023).

4. RNAs as a Promising Tool in the Treatment of AD

The small-molecule- and protein-based therapies that interact with a particular biologic molecule to obtain a pharmacological response to control a disease have been successfully dealing with many diseases in past years [81]. However, this conventional pharmacotherapy has several limitations. Protein-based medications primarily target proteins to inhibit their activity, and only ~1.5% of the human genome encodes proteins. Consequently, the range of disease targets of this drug is limited and unable to meet the required demands [5,82]. Most protein-based drugs are too large to enter their target cells and therefore are only effective when their target molecule is extracellular, demonstrating difficult tissue penetration [82,83].

The study of RNA therapeutics started decades ago, leading to a long scientific journey. However, just recently, this field of research has developed dramatically as a result of the response to the COVID-19 pandemic, which revealed how RNA-based therapeutics could lead to a new era of different and accessible new technology in combatting a wide range of diseases [40,84]. In addition to the inherent instability of RNA, these molecules possess particular features and versatility, with multiple advantages over protein and DNA-based drugs. These characteristics include the inducement of protein coding, binding specificity to target molecules, and inhibition of protein translation. Additionally, RNA has the capacity to recognize a wide range of ligands, targeting almost any genetic component within the cell that is out of reach for the most established drug models [81,84].

RNA can be modified in the base, backbone, and sugar, increasing target affinity and preventing nuclease digestion [85]. These modifications make them completely different from cellular RNAs transcribed from the genome [81]. Thus, unlike DNA-based therapeutics, which must cross the cytoplasmic and nuclear membrane and can integrate the host genome and cause a mutation, RNA therapeutics have no risk of chromosomal integration, exhibiting a safer profile [5,86]. Another important advantage of RNA therapy is its long-lasting effects when using, for example, siRNAs as a drug, benefiting patients who cannot receive frequent treatments [87]. The fast production of RNA-based therapies is also a distinct advantage, considering the increased knowledge that will provide faster and easier design of RNA molecules when compared to the process to produce novel protein-based drugs, which takes years. The vaccine's rapid production and successful reduction of the severity of the disease in infected people during the COVID-19 pandemic is evidence of how quickly this type of therapy can be developed and implemented [86].

A long scientific journey has led to prominent technological advances in the RNA field, and several new types of RNA molecules have been discovered, leading to the development of several studies designed to implement more effective treatments for diseases that still remain without any cure [40]. RNA-based therapeutics can be classified into five different categories: (1) mrna, which encodes for proteins; (2) siRNA, which are double-stranded and primarily cause translational repression of their target protein; (3) mRNAs, which are small RNAs that can either inhibit protein synthesis when they bind to an mRNA target (miRNA mimics) or free up mRNA by binding to the miRNA that represses the translation of that particular mRNA (miRNA inhibitors); (4) antisense oligonucleotides, which are small (~15–25 nucleotides) single-stranded RNAs that can either promote or repress target expression; and (5) aptamers, which are short single-stranded nucleic acids that form secondary and tertiary structures that inhibit several types of target molecules, including proteins [83,84,88].

The use of nucleic acid therapy has limitless potential to treat not only neurological diseases but also a considerable range of disorders since preclinical studies in cellular and animal models proved that mRNAs and short RNAs can be a new class of medicine [89]. Considering the unmet need for effective treatments for AD patients, it is crucial to evaluate diverse therapeutic targets and strategies to cure this disease [44]. In recent years, Aβ and Tau proteins had been the principal targets for researchers. Despite being well-documented hallmarks of AD, treatments involving the regulation of these proteins, unfortunately, are still an unsuccessful strategy [45,90]. In this review, we are going to focus on RNA therapies such as siRNA, miRNA, and mRNA for the treatment of AD (Figure 4) and describe some examples in Table 2.

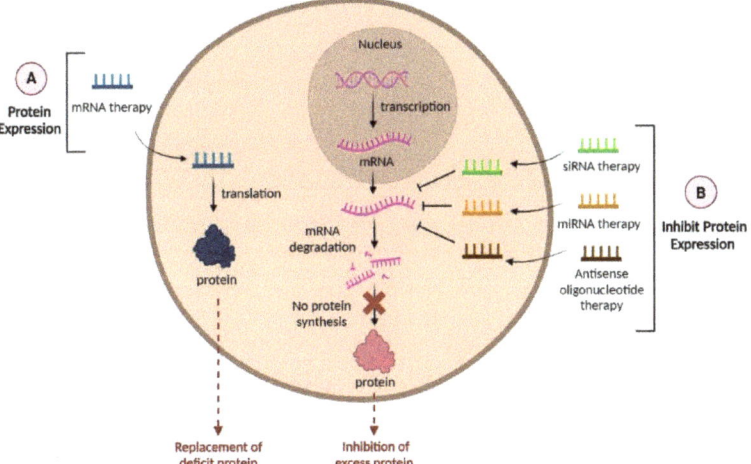

Figure 4. Possible RNA-based therapeutic strategies for AD. (**A**) mRNA leads to protein expression and (**B**) siRNA, miRNA, and antisense oligonucleotide inhibit protein expression. Created in BioRender.com (accessed on 20 April 2023).

- miRNA

miRNAs are small non-coding single-stranded RNAs approximately ~22 nucleotides in length [91]. Endogenous miRNAs are essential in cell development and play a key role in post-transcriptional gene regulation since they regulate the expression of multiple mRNAs both by promoting mRNA degradation and blocking the translation of multiple target mRNAs, inhibiting protein synthesis [84]. They can regulate mRNA translation by binding to the 3′ untranslated region, allowing the reduction in the amount of target protein, instead of only inhibiting its activity [40,92]. In addition to gene expression regulation, miRNA can also act as signaling molecules for intercellular communication, revealing that it can be packaged into exosomes to exert this function [93]. The miRNA therapeutic strategy could be categorized into two types: miRNA mimics and miRNA inhibitors [5,82]. miRNA mimics are synthetic RNA molecules that are designed to act as endogenous miRNA to silence genes and can be applied when increased levels of mRNA are prevalent. In contrast, miRNA inhibitors are synthetic ssRNA molecules and can interrupt the miRNA function via sequence-specific binding to mature miRNA, without causing gene silencing. This option is interesting when protein synthesis restoration is needed [82]. These particular features allow miRNA to target multiple sites of various molecularly deregulated cascades in disease conditions, similar to what occurs in AD [40,91]. miRNAs molecules have crucial functions in the nervous system, such as neuronal differentiation, neurite outgrowth, and synaptic plasticity, and are responsive to neuropathological processes, including oxidative stress, neuroinflammation, and protein aggregation. This proves that miRNAs are key molecules in AD, and the dysfunction of miRNAs in this neurological disease is being recognized [45,91].

- siRNA

siRNAs are short, synthetic double-stranded RNA oligonucleotides (20–25 nt) that take advantage of the RNA interference pathway to silence gene expression by targeting their complementary mRNA [83,84]. siRNAs offer promising therapeutics for brain disease treatment by directly blocking causative gene expression with high targeting specificity, requiring low effective dosages, and benefiting from a relatively simple drug development process [94]. Several siRNA-based therapeutics were already approved by the FDA for other diseases, supporting their potential use for AD therapeutics [40].

- mRNA

mRNA is a type of single-stranded RNA involved in protein synthesis. The role of mRNA is to carry protein information from the DNA in a cell's nucleus to the cell's cytoplasm. Compared to other RNA therapies, mRNA can provide advantages such as (1) safety, (2) effectiveness, particularly in slowly dividing or non-dividing cells such as neural cells, and (3) better control of protein expression [40].

Table 2. miRNA, siRNA, and mRNA therapeutic applications in Alzheimer's disease.

Types of miRNA	Role in AD	References
miR-101	Significantly reduced the expression of a reporter under control of APP 3′-UTR in HeLa cells.	[95]
miR-106b	Overexpression of miR-106b inhibited $A\beta_{1-42}$-induced tau phosphorylation at Tyr18 in SH-SY5Y cells stably expressing Tau.	[96]
miR-137	miR-137 inhibited increased expression levels of p-tau induced by $A\beta_{1-42}$ in SH-SY5Y and inhibited the hyperphosphorylation of Tau protein in a transgenic mouse model of AD.	[97]
miR-219	In a *Drosophila* model that produces human Tau, reduction of miR-219 exacerbated Tau toxicity, while overexpression of miR-219 partially annulled toxic effects.	[98]
miR-17	miR-17 inhibits elevated miR-17 in adult AD (5xFAD) mice microglia improves Aβ degradation.	[99]
miR-20b-5p	Treatment with miR-20b-5p reduced APP mRNA and protein levels in cultured human neuronal cells.	[100]
miR-29c	Over-expression of miR-29c in SH-SY5Y, HEK-293T cell lines and miR-29c in transgenic mice downregulated BACE1 protein levels.	[101]
miR-298	miR-298 is a repressor of APP, BACE1, and the two primary forms of Aβ (Aβ40 and Aβ42) in a primary human cell culture model. Thus, miR-298 significantly reduced levels of ~55 and 50 kDa forms of the Tau protein without significant alterations of total Tau or other forms.	[102]

Table 2. *Cont.*

	Role in AD	References
miR-485-5p	miR-485-5p overexpression facilitated the learning and memory capabilities of APP/PS1 mice and promoted pericyte viability and prohibited pericyte apoptosis in this model.	[103]
miR-9-5p	miR-9-5p overexpression inhibited $A\beta_{25-35}$-induced mitochondrial dysfunction, cell apoptosis, and oxidative stress by regulating GSK-3β expression in HT22 cells.	[104]
miR-132	miR-132 inhibited hippocampal iNOS expression and oxidative stress by inhibiting MAPK1 expression to improve the cognitive function of rats with AD.	[105]
miR-153	Using miR-153 transgenic mouse model, was verified that miR-153 downregulated the expression of APP and APLP2 protein in vivo.	[106]
Targeted gene silencing by siRNA		
Tau	siRNA against MAPT can effectively suppress tau expression in vitro and in vivo without a specific delivery agent.	[107]
BACE1	Polymeric siRNA nanomedicine targeting BACE1 in APP/PS1 transgenic AD mouse model can efficiently penetrate the BBB via glycemia-controlled glucose transporter-1–mediated transport, ensuring that siRNAs decrease BACE1 expression.	[94]
Presenilin1 (PS1)	Downregulation of PS1 and $A\beta_{42}$ in IMR32 cells transfected with siRNA against PS1 was verified.	[108]
APP	Infusion of siRNAs that down-regulated mouse APP protein levels into the ventricular system for 2 weeks down-regulated APP mRNA in mouse brain.	[109]
Proteins encoded by mRNA		
mRNA encoding neprilysin	Neprilysin plays a major role in the clearance of Aβ in the brain. New mRNA therapeutic strategy utilizing mRNA encoding the mouse neprilysin protein has been shown to decrease Aβ deposition and prevent pathogenic changes in the brain.	[110]

Overcoming Limitations of RNA Therapies with Delivery Systems

In spite of being a hot research topic in the present day, the development of novel RNA therapeutics has proven to be highly challenging in the past two decades. Some of the major disadvantages that stop RNA from being a clinical success are its instability for in vivo application since the human body has several intrinsic defense systems to protect the cells against exogenous molecules, such as ribonucleases (RNases). RNA's low targeted tissue accumulation decreases its therapeutic efficacy, requiring high therapeutic RNA doses that can induce toxicity [111]. Early degradation of naked RNAs is another hurdle for their therapeutic efficiency since they tend to be promptly eliminated from the body via renal or hepatic clearance, shortly after the systemic administration [82,111]. For reference, the half-life of naked siRNA is approximately 15 min, and the half-life of naked mRNA can vary between 2 and 25 min, depending on the presence of 5′ capping, the length of the 3′ poly-A tail, and the RNA secondary structure [112]. Finally, being a large and negatively charged molecule, it is difficult to deliver into the cellular cytoplasm, where it exerts its action. Even if cellular internalization of RNA occurs, there is also the risk that RNA cannot escape the endosomal pathway, with only, for example, ~1–2% of siRNAs uptaken by the cells escaping the endosome [81,82].

These obstacles have been considerably surpassed, thanks to the recent advancements in research areas such as RNA biology and nanotechnology that allowed the development of new materials and technologies for the delivery of RNA molecules [81,84]. These new advances transformed RNA technology into a novel therapeutic too since RNA can now be safely transported and delivered to the target thanks to delivery systems [82] (Figure 5).

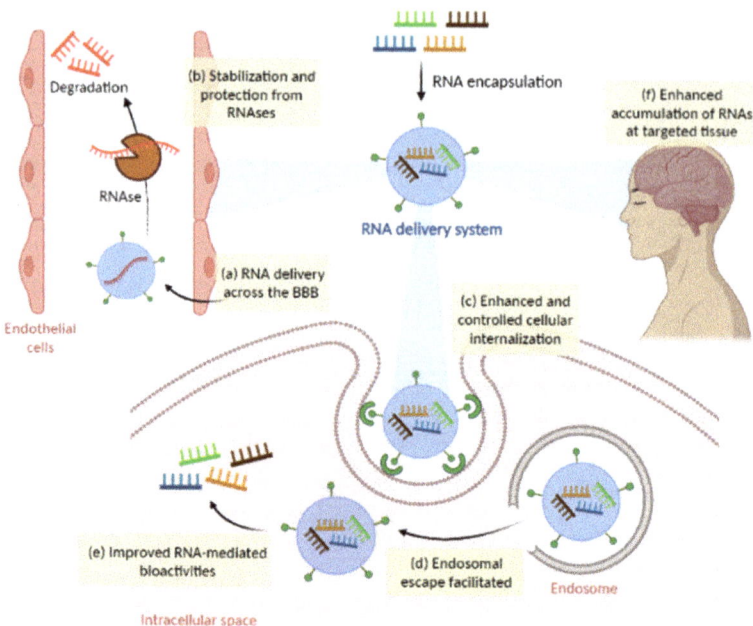

Figure 5. How RNA delivery systems can overcome the physiological obstacles faced by naked RNAs. (**a**) RNA can be delivered across the BBB; (**b**) protection of RNAs from degradation by ribonucleases (RNAses); (**c**) enhanced cellular internalization by controlled or specific pathways; (**d**) intracellular endosomal degradation can be avoided by endosomal escape; (**e**) improved RNA-mediated bioactivities; (**f**) enhanced accumulation of RNA in target tissue. Created in BioRender.com (accessed on 20 April 2023).

RNA delivery is key for the treatment of diseases such as neurological disorders, as it gives RNA the ability to target diseases that cannot be treated with other conventional drug groups by encapsulating these molecules in delivery vectors. These RNA vectors have the capacity to effectively protect this molecule from biodegradation, increase bioavailability, solubility and permeation, surpass biological barriers, and promote targeted delivery and release. Hence, RNA nanoencapsulation potentiates the treatment of diseases by silencing genes or expressing therapeutic proteins [113,114].

5. Nanoparticles and the BBB

Delivery systems can overcome RNA's limitations, becoming a successful therapy, particularly for facilitating the crossing of the BBB, which is the main culprit for the shortage of new and effective treatments for AD [115].

In this review, we will explain the BBB anatomic composition and characteristics, as well as the BBB pathways into the CNS. Subsequently, we will discuss exosomes, liposomes, and exosome-like liposomes as possible tools for RNA transport across the BBB.

From an anatomical point of view, the BBB is composed of different cell types such as endothelial cells, pericytes, and astrocytes (Figure 6A) [116]. In between the endothelial cells, there are tight junctions, which are surrounded by a thin basal membrane and astrocytes vascular feet. These highly restrictive tight junctions are a key BBB feature since they are responsible for the barrier properties and limit the transfer of almost all drugs [116,117]. Pericytes cover 20% of the outer surface of endothelial cells and are responsible for the regulation of the blood flow in the brain capillary through contraction and relaxation. The astrocytes are glial cells that connect the brain capillary and neurons and also maintain BBB functions by providing nutrients to neurons and protecting the brain from oxidative stress and metal toxicity [117]. Additionally, the basal membrane provides structural support around the pericytes and endothelial cells [118].

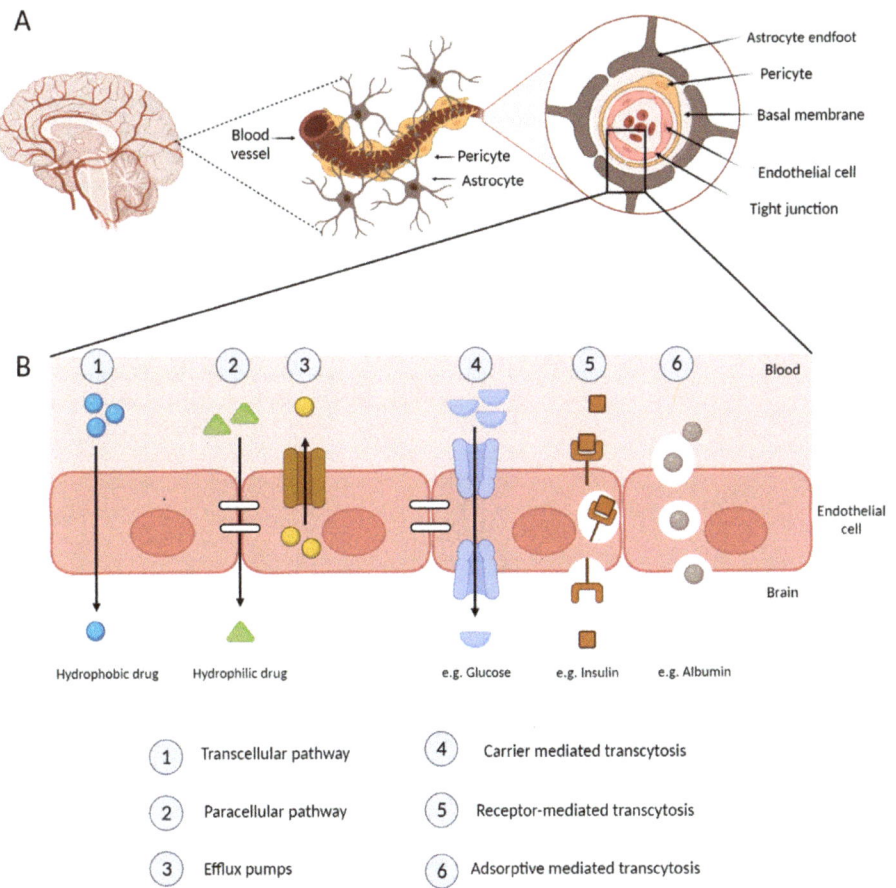

Figure 6. Structural representation of the BBB anatomy (**A**) and possible pathways of entrance into the central nervous system (**B**). Created in BioRender.com (accessed on 20 April 2023).

This specialized barrier acts as an interface with the capacity to regulate the entry of plasma components, red blood cells, and leukocytes into the CNS and ensures the export of potentially neurotoxic molecules from the brain to the blood [119,120]. This barrier strictly controls the molecule movements between the blood and the brain, regulating the homeostasis of the nervous system [119]. Moreover, more than 98% of all small-molecule drugs and approximately 100% of biological drugs are incapable of crossing the BBB. Additionally, water-soluble molecules in the blood are prevented from entering the CNS, while lipid-soluble molecules are reduced by enzymes or efflux pumps [115,121].

The BBB possesses several permanently active transport mechanisms to ensure the transport of nutrients into the CNS while excluding blood-borne molecules that could be detrimental [122]. On one hand, these BBB properties are proof of its vital role in maintaining the specialized microenvironment of the brain tissue. On the other hand, these features make CNS access one of the most difficult of the body, limiting the development of novel effective drugs to treat AD [119]. Even though the BBB is a strict barrier for the circulation of molecules between the blood and the CNS, there are a few pathways that allow the delivery of essential molecules that maintain brain homeostasis (Figure 5B). These include the transcellular pathway, the paracellular pathway, efflux pumps, carrier-mediated transcytosis, receptor-mediated transcytosis, and adsorptive-mediated transcytosis [114].

In a healthy BBB, transcellular diffusion (Figure 6B1) consists of the diffusion of solute particles through the endothelial cells. Particles transported through this route are small lipophilic molecules that penetrate through the cells. On the other hand, paracellular transport (Figure 5B is restricted by the tight junctions between the endothelial cells, allowing only hydrophilic molecules to pass, with a molecular weight < 500 Da.

Efflux pumps (Figure 6B3) are a set of proteins responsible for limiting the accumulation of various potentially toxic molecules, and eventually, for expelling these molecules from the brain. These proteins are a limiting factor for the delivery of bioactive compounds to the brain.

Carrier-mediated transcytosis (Figure 6B4) consists of active transport with the support of carrier proteins such as the glucose transporter isoform (GLUT-1) and the large amino acid transporter (LAT), allowing entrance to glucose or amino acids. The transport of these molecules occurs when they bind to the protein on the blood side of the BBB, and a subsequent conformational change allows their transport into the brain [114,123].

Receptor-mediated transcytosis (Figure 6B5) is a specialized transport system by which endogenous molecules can cross the BBB through receptors present on the cell surface. This type of transport relies on the following mechanisms: Endocytosis, intracellular vesicular trafficking, and exocytosis. Active components bind to their specific receptors on the luminal side of the endothelial cells, and an intracellular vesicle is formed through membrane invagination. The formed vesicles cross the cell to release the ligand at the basolateral side via exocytosis. The most common receptors involved in this process are the transferrin receptor (TfR), the insulin and insulin-like growth factor receptors, the low-density lipoprotein receptor (LDLR), the low-density lipoprotein-receptor-related protein 1 and 2 (LRP1 and LRP2), the scavenger receptor class B type I (SR-B1), the leptin receptor, the albumin receptor, and the lactoferrin receptor. Receptor-mediated transcytosis is one of the most promising pathways for nanoparticle drug delivery through the BBB [114,122].

Adsorptive-mediated transcytosis (Figure 6B6) is another important BBB-crossing pathway, without the involvement of specific plasma-membrane receptors. The basic mechanism is responsible for the transport of charged particles by taking advantage of the electrostatic interactions between the positively charged drug carriers and the negatively charged luminal membrane of the brain endothelial cells. This transport has lower affinity but higher capacity compared to receptor-mediated transcytosis [114,122,123].

There is a crucial need for an ideal and safe approach to effectively carry pharmaceutical agents in a target-specific and sustained-release manner into the CNS, without disrupting the BBB [122,123]. A promising approach is taking advantage of receptor-mediated transcytosis for drug delivery to the brain with the help of ligand-functionalized nanoparticles. Nanoparticles are gaining popularity as drug carriers for the treatment of neurological disorders, due to their small size and unique physical properties [124]. By virtue of their biochemical composition, lipidic nanoparticles provide biomimicking and bio-degradable platforms. As a result of their lipophilicity and size, nanosized particles such as exosomes and liposomes are promising drug delivery carriers to increase the penetration of the BBB [113,125].

5.1. Exosomes

Exosomes are a subset of EVs with a diameter ranging from 30 to 100 nm, and their composition includes lipids, proteins, and nucleic acids [126,127]. Their lipid content includes sphingomyelin, phosphatidylserine, cholesterol, and ceramide or derivatives. Exosomes also carry non-specific proteins (e.g., cytoplasmic enzymes, cytosolic proteins, heat shock proteins, and transferring proteins) and specific proteins that differ from one exosome to another, depending on their origin. The genetic material of these vesicles includes microRNAs, mRNAs, long non-coding RNAs, and DNA fragments [126]. Exosomes are also enriched in late endosome components such as CD63, CD9, and CD81 since they originate from the endocytic compartment of the producer cell in a process that generates multi-vesicular endosomes (MVEs), which subsequently fuse with the plasma membrane to release exosomes into the extracellular space.

Exosomes yield information that can reflect the phenotype of the parental cell since these vesicles carry distinct RNA and protein cargoes that allow the identification of their parental cells, as well as cell-specific or tissue-specific factors that can be used to determine their site of origin [128,129]. Moreover, when exosomes are secreted, undesirable proteins and other molecules are discarded, making these vesicles a compartment of cellular debris for subsequent disposal [130].

Originally, exosomes were primarily described as being for the elimination of excessive and unnecessary molecules from the cells. However, in the last decade, it has been shown that they have other key functions in both physiological and pathological processes. Regarding the physiological roles, exosomes play an important role in intercellular communication since they are able to deliver a number of bioactive cargos to near or distant target cells [126]. Moreover, exosomes display a role in tissue homeostasis and have anti-inflammatory functions. An example of their function is neuronal communication via the secretion of exosomes, which can contribute to a range of neurobiological functions, including synaptic plasticity. In relation to pathological functions, they control the expansion and progression of diseases, such as cancer and neurodegenerative diseases [130]. Exosomes are involved in the complex mechanisms of secretion, spread, and degradation of the Aβ and Tau proteins and are, especially, involved in Tau propagation between neuronal cells [131].

The unique properties of exosomes include their small size, durability, stability, potential cell selectivity, low immunogenicity, and ideal biocompatibility. These properties make them good candidates to be a therapeutic delivery system. Compared to traditional therapeutic drugs, exosomes have a higher potential to pass through the BBB, which helps the drugs they might carry to reach the CNS. Moreover, since exosomes can be isolated from all body fluids, they can be candidates for analysis as part of a non-invasive liquid biopsy [129–131].

5.2. Liposomes

In the last two decades, lipid-based nanoparticles (LNPs), especially liposomes, have been attractive nanometric delivery systems for being the most well-studied nonviral platforms for the delivery of RNA molecules and achieving significant clinical success [113,132] highlighted by the highly effective mRNA COVID-19 vaccines [132,133].

There are several types of LNPs such as liposomes, niosomes, transfersomes, nanoemulsions, solid lipid nanoparticles (SLNs), lipid nanocapsules (LNCs), nanostructured lipid carriers (NLCs), lipid-based micelles, core–shell lipid nanoparticles (CLNs), and hybrid lipid-polymeric nanoparticles [132,134]. LNPs range in size from 40 nm to 1000 nm and are colloidal lipophilic systems constituted by four main components: A pH-sensitive cationic lipid, a helper lipid, cholesterol, and a PEG-lipid. The cationic lipid is a synthetic lipid, constituted by a hydrophilic head with a protonable tertiary amino group (pKa 6–6.7) and a long hydrophobic tail. Cholesterol is incorporated into the LNPs formulation with the goal of increasing their flexibility, whilst the helper phospholipid assists in the process of endosomal escape and contributes to the stability of LNPs [135,136]. Finally, the insertion of a short-chain PEG-lipid derivative (normally of 14 carbon atoms) is essential to maximize the ex vivo stability and control the particle size before administration [137].

Due to their resemblance to biological and natural components, these nano systems show tremendous promise as carriers for therapeutic applications. The main advantages of LNPs over other nanoparticles are their low toxicity and biocompatibility, biodegradability, safety, high mechanical and chemical versatility, and the capacity to protect the active ingredient from degradation processes induced by external factors. Along with these features, these lipid-based nanoparticles can incorporate the delivery of both hydrophobic and hydrophilic molecules and most of their preparation methods can be easily scaled up. Additionally, because of their lipophilicity, LNPs possess the ability to overcome difficult physiological barriers, such as the BBB, even without surface modification [124,134].

Among all the nano-based drug delivery systems, liposomes are the most biocompatible and least toxic since they are composed of phospholipids and cholesterol, the main components of cell membranes. Liposomes are an extremely versatile nanocarrier platform that has the capacity to load multiple drugs, provide protection from degradation, have controlled and targeted drug release, and enhance drug endocytosis into cells [11]. Additionally, these nanoparticles are able to incorporate both hydrophilic and hydrophobic therapeutic agents. The hydrophilic compounds may either be entrapped into the aqueous core of the liposomes or be located at the interface between the lipid bilayer and the external water phase, while the hydrophobic compounds are generally entrapped in the hydrophobic core of their lipid bilayers. The positively charged lipids in the liposome's constitution allow the electrostatic interaction with negatively charged nucleic acids, such as DNA and RNA, in gene delivery applications [138].

There is evidence that lipid-based nanoplatforms will play a key role in the development of RNA neuro-therapies, with liposomes being one of the main lipidic platforms for RNA delivery to the CNS [113]. In order to enhance drug delivery into the CNS, the liposome surface can be modified by the inclusion of biologically active ligands, such as peptides, polysaccharides, antibodies, or aptamers, which specifically bind to receptors expressed on the surface of the brain endothelial cells, facilitating their binding and transport across the BBB. The addition of polyethylene glycol (PEG) offers superficial protection for the liposomes by avoiding binding with plasma proteins, and hence, preventing their opsonization and subsequent clearance. PEGylation of liposomes prolongs their circulation time in the body and also plays a crucial role in brain drug delivery, allowing liposomes to cross the BBB [138].

5.3. Exosome-like Liposomes as a Novel Strategy

Scientists have explored various nanomaterials for targeted delivery through the BBB with significant efficacy, such as dendrimers, polymeric nanoparticles, gold nanoparticles, carbon quantum dots, and exosome-like liposomes [8,139]. First, in this review, we will summarize the exosome-like liposomes' competition and then focus only on this delivery system.

- Dendrimers: Dendrimers are highly branched, characterized by defined molecular weights and specific encapsulation properties. This type of delivery system is composed of symmetrical polymeric macromolecules with a large number of reactive surface groups, with three distinctive architectural components: An interior core, an interior layer consisting of repeating units radially attached to the inner core, and functional end groups on the outside layer. Because of these unique features, dendrimers can cross-impair the BBB and target astrocytes and microglia after systemic administration in animal models [139].
- Polymeric nanoparticles: Polymeric nanoparticles can be produced from synthetic or natural polymers. However, to be applied in brain drug delivery, these nanoparticles need to be biodegradable and biocompatible. PBCA, PLA, and PLGA nanoparticles are nanoparticles able to cross the BBB. These nanocarriers possess controlled drug release, targeting efficiency, and can avoid phagocytosis by the reticuloendothelial system, thus improving the concentration of drugs in the brain [140].
- Gold nanoparticles: Nanoparticles (mostly < 10 nm in size) composed of a gold core and with covalently or non-covalently attached surface ligands. Multiple in vivo studies on rodents have shown that low amounts of this delivery system were able to cross the BBB. However, greater amounts of the administered dose were found in the liver and in the blood [8]. Additionally, Sela et al. proved that gold nanoparticles could penetrate the BBB of rats without the use of an external field or surface modification and were found to be distributed uniformly in both the hypothalamus and hippocampus indicating there is no selective binding in these regions of the brain [141].
- Carbon quantum dots: This delivery system retains a polymeric core structure and various functional groups on the surface, facilitating their conjugation with drug molecules for specific delivery. This is a carrier with several efficient features for BBB crossing such as excellent biocompatibility and low toxicity due to the lack of

metal elements, small size, and photoluminescence, which can be utilized to track the penetration of CDs through the BBB [117].

In addition to the fact that these nanoparticles are promising in overcoming the BBB, exosome-like liposomes are also delivery systems with unique properties to achieve this goal. Exosomes and liposomes are promising nanocarriers with unique properties. However, these delivery systems have some limitations. For instance, liposomal targeting efficiency is limited, and they can induce immunogenicity [11], whilst efficient and reliable isolation and purification are needed for the clinical application of exosomes. It is also crucial to identify appropriate strategies to increase loading capacity and specificity to use exosomes as a nanocarrier [10,142]. When comparing the advantages and disadvantages of exosomes and liposomes, it becomes evident that the two systems are complementary, since the advantages of the one system mitigate the disadvantages of the other, and vice-versa (Figure 7). Therefore, the development of a system that incorporates the desirable features of the two carriers into one hybrid delivery system, led to an innovative carrier for drug delivery applications, the exosome-like liposome (Figure 7) [11,143]. These novel exosome-like liposomes are formulated with a lipid composition that mimics that of exosomes, which impart them with some of the desirable characteristics of exosomes, such as enhanced passive targeting, biocompatibility and RES evasion, and the ability to cross biological barriers, whilst allowing much higher encapsulation efficiencies and larger-scale production in good manufacturing practice that the use of exosomes themselves does not allow.

Thus, particle size plays a major role in drug delivery efficacy, including in the therapeutic effect achieved. The endothelial cells' slit width is 200 to 500 nm. Consequently, for the long-term circulation of these nanoparticles, they need to have a size that does not exceed 200 nm, leading to better brain drug delivery across the BBB. Moreover, the fact that exosome-like liposomes are coated with this suitable hydrophilic polymer leads to advantages such as escaping phagocytosis, avoiding opsonization, and further increasing the blood circulation time, since the hydrophobic nature of the nanoparticle containing the drug is shielded. These unique characteristics drive researchers to believe that this type of carrier can have a key impact in crossing the BBB, helping in the therapy of neurodegenerative diseases such as AD [123].

Additionally, exosome-like liposomes were highly efficient at encapsulating curcumin, with encapsulation efficiencies ranging between 85 and 94%, and were effective at delivering curcumin into neuronal cells to promote a superior neuroprotective effect after oxidative insult compared to free curcumin at the same concentrations. Thus, these delivery systems have shown to be highly biocompatible, without significantly affecting cell viability or causing hemolysis. Finally, the exosome-like liposomes have been shown to be internalized by zebrafish embryos and accumulate in lipid-rich zones, such as the brain and yolk sac, also in stages where the BBB is already formed [144]. In Table 3, we demonstrate other crucial examples of applications of exosome-like liposomes in vivo and in vitro.

Table 3. Examples of applications of exosome-like liposomes.

Exosome-like Liposomes Applications	References
Exosome mimetics-mediated gene-activated matrix encapsulating the plasmid of vascular endothelial growth factor (VEGF) was able to sustainably deliver *VEGF* gene and significantly enhance the vascularized osteogenesis in vivo.	[145]
PSMA-exosome mimetics showed increased cellular internalization in PSMA-positive PC cell lines (LNCaP and C4-2B) and higher tumor targeting was observed in solid C4-2B tumors, following intravenous administration, confirming their targeting ability in vivo.	[146]
Exosome mimetics are reported for bone targeting involving the introduction of hydroxyapatite-binding moieties through bioorthogonal functionalization. Bone-binding ability of the engineered exosome mimetics is verified with hydroxyapatite-coated scaffolds and an ex vivo bone-binding assay.	[147]
Administration of mesenchymal stem cells-exosome mimetics in conjunction with an injectable chitosan hydrogel into mouse nonhealing calvarial defects demonstrated robust bone regeneration.	[148]

Table 3. *Cont.*

Exosome-like Liposomes Applications	References
Bone marrow mesenchymal stem cells were sequentially extruded to generate exosome-mimetic to encapsulate doxorubicin to treat osteosarcoma. The results showed that demonstrated significantly more potent tumor inhibition activity and fewer side effects than free doxorubicin.	[149]
In vitro, chemotherapeutic drug-loaded exosome-mimetics induced TNF-R-stimulated endothelial cell death in a dose-dependent manner. In vivo, experiments in mice showed that the chemotherapeutic drug-loaded exosome-mimetics traffic to tumor tissue and reduce tumor growth without the adverse effects observed with equipotent free drug.	[150]
Multifunctional exosomes-mimetics decorated with angiopep-2 (Ang-EM) incorporating Docetaxel, for enhancing glioblastoma drug delivery by manipulating protein corona, Ang-EM showed enhanced BBB penetration ability and targeting ability to the gioblastoma. Ang-EM-mediated delivery increased the concentration of docetaxel in the tumor area.	[151]
A designed lung-targeting liposomal nanovesicle carrying miR-29a-3p that mimics the exosomes, significantly down-regulated collagen I secretion by lung fibroblasts in vivo, thus alleviating the establishment of a pro-metastatic environment for circulating lung tumor cells.	[152]

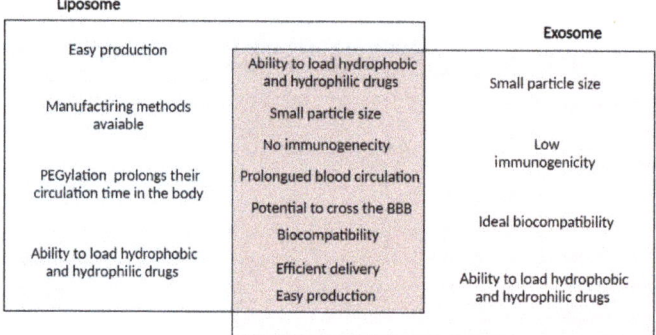

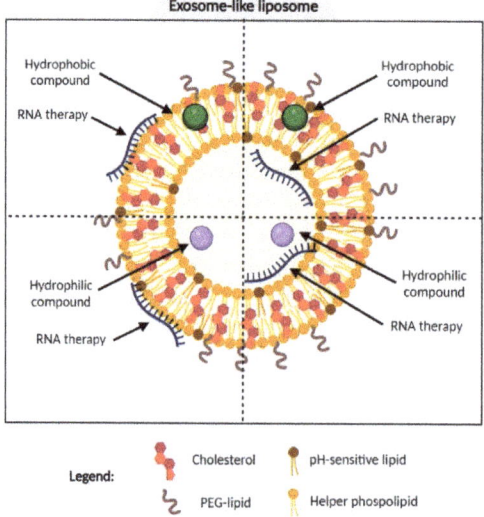

Figure 7. Exosome-like liposomes combine advantages of liposomes and exosomes, with the capacity to be loaded with different types of molecules due to their constitution. Created in BioRender.com (accessed on 20 April 2023).

6. Conclusions

AD is a neurodegenerative disease that has a tremendous impact on people's quality of life, all around an aging world. This burden is further impacted by the lack of therapies, due to the complex mechanisms that lead to the disease onset and also due to limitations that the BBB imposes on the entrance and maintenance of therapeutic molecules in the CNS.

However, the emergence of RNA as an especially versatile tool for the treatment of several diseases has opened multiple new possibilities for the development of effective treatments for AD.

The accelerated growth of RNA therapies requires the development of efficient delivery systems to transport a selectively deliver the RNA molecules into their target cells or tissues, due to the low bioavailability of these molecules.

Nanoparticles have provided new and safer avenues for the delivery of natural compounds and RNA. In this field, novel exosome-like liposomes are positioning themselves as a prime solution for the vehiculation of therapeutical compounds to the CNS, due to the combination of the favorable characteristics of exosomes with those of liposomes, which entitle scalable production in good manufacturing practices, which is currently unfeasible for the exosomes themselves.

7. Patents

The authors are co-inventors in a patent application (EXOSOME-MIMETIC LIPOSOME COMPOSITION AND USE, Instituto Nacional da Propriedade Industrial PPP No. 116560 P, 3 July 2020).

Author Contributions: Conceptualization, J.R., I.L. and A.C.G.; investigation, formal analysis, J.R.; writing—original draft preparation, J.R.; writing—review and editing, J.R., I.L. and A.C.G.; supervision, I.L. and A.C.G.; funding acquisition, A.C.G. All authors have read and agreed to the published version of the manuscript.

Funding: This work was supported by the strategic program UID/BIA/04050/2020 funded by National funds through the Fundação para a Ciência e Tecnologia I.P. Ivo Lopes is a recipient of a scholarship from the Direção Regional da Ciência e Tecnologia, Governo Regional dos Açores (M3.1.a/F/128/2015).

Institutional Review Board Statement: Not applicable.

Informed Consent Statement: Not applicable.

Data Availability Statement: Data will be made available upon request.

Acknowledgments: We acknowledge Mario Fernandes for support in revisions.

Conflicts of Interest: The authors declare no conflict of interest.

References

1. Bloomingdale, P.; Karelina, T.; Ramakrishnan, V.; Bakshi, S.; Véronneau-Veilleux, F.; Moye, M.; Sekiguchi, K.; Meno-Tetang, G.; Mohan, A.; Maithreye, R.; et al. Hallmarks of neurodegenerative disease: A systems pharmacology perspective. *CPT Pharmacomet. Syst. Pharmacol.* **2022**, *11*, 1399–1429. [CrossRef]
2. Alzheimer's Association. 2023 Alzheimer's disease facts and figures. *Alzheimer's Dement.* **2023**, *19*, 1598–1695. [CrossRef]
3. De Ture, M.A.; Dickson, D.W. The neuropathological diagnosis of Alzheimer's disease. *Mol. Neurodegener.* **2019**, *14*, 32. [CrossRef] [PubMed]
4. Zhu, Y.; Zhu, L.; Wang, X.; Jin, H. RNA-based therapeutics: An overview and prospectus. *Cell Death Dis.* **2022**, *13*, 644. [CrossRef]
5. Damase, T.R.; Sukhovershin, R.; Boada, C.; Taraballi, F.; Pettigrew, R.I.; Cooke, J.P. The Limitless Future of RNA Therapeutics. *Front. Bioeng. Biotechnol.* **2021**, *9*, 628137. [CrossRef] [PubMed]
6. Luo, M.; Lee, L.K.C.; Peng, B.; Choi, C.H.J.; Tong, W.Y.; Voelcker, N.H. Delivering the Promise of Gene Therapy with Nanomedicines in Treating Central Nervous System Diseases. *Adv. Sci.* **2022**, *9*, 2201740. [CrossRef]
7. Daneman, R.; Prat, A. The Blood–Brain Barrier. *Cold Spring Harb. Perspect. Biol.* **2015**, *7*, a020412. [CrossRef]
8. Lombardo, S.M.; Schneider, M.; Türeli, A.E.; Günday Türeli, N. Key for crossing the BBB with nanoparticles: The rational design. *Beilstein J. Nanotechnol.* **2020**, *11*, 866–883. [CrossRef]

9. Akbarzadeh, A.; Rezaei-Sadabady, R.; Davaran, S.; Joo, S.W.; Zarghami, N.; Hanifehpour, Y.; Samiei, M.; Kouhi, M.; Nejati-Koshki, K. Liposome: Classification, preparation, and applications. *Nanoscale Res. Lett.* **2013**, *8*, 102. [CrossRef]
10. Zhang, Y.; Liu, Y.; Liu, H.; Tang, W.H. Exosomes: Biogenesis, biologic function and clinical potential. *Cell Biosci.* **2019**, *9*, 19. [CrossRef]
11. Antimisiaris, S.; Mourtas, S.; Marazioti, A. Exosomes and Exosome-Inspired Vesicles for Targeted Drug Delivery. *Pharmaceutics* **2018**, *10*, 218. [CrossRef] [PubMed]
12. Forrest, S.L.; Kovacs, G.G. Current Concepts of Mixed Pathologies in Neurodegenerative Diseases. *Can. J. Neurol. Sci. J. Can. Sci. Neurol.* **2022**, *50*, 329–345. [CrossRef]
13. Masoudi Asil, S.; Ahlawat, J.; Guillama Barroso, G.; Narayan, M. Nanomaterial based drug delivery systems for the treatment of neurodegenerative diseases. *Biomater. Sci.* **2020**, *8*, 4109–4128. [CrossRef] [PubMed]
14. Wilson, D.M.; Cookson, M.R.; Van Den Bosch, L.; Zetterberg, H.; Holtzman, D.M.; Dewachter, I. Hallmarks of neurodegenerative diseases. *Cell* **2023**, *186*, 693–714. [CrossRef]
15. Roda, A.; Serra-Mir, G.; Montoliu-Gaya, L.; Tiessler, L.; Villegas, S. Amyloid-beta peptide and tau protein crosstalk in Alzheimer's disease. *Neural Regen. Res.* **2022**, *17*, 1666. [CrossRef]
16. Liang, S.Y.; Wang, Z.T.; Tan, L.; Yu, J.T. Tau Toxicity in Neurodegeneration. *Mol. Neurobiol.* **2022**, *59*, 3617–3634. [CrossRef]
17. Yu, H.; Wu, J. Amyloid-β: A double agent in Alzheimer's disease? *Biomed. Pharmacother.* **2021**, *139*, 111575. [CrossRef] [PubMed]
18. Xiao, Y.; Wang, S.-K.; Zhang, Y.; Rostami, A.; Kenkare, A.; Casella, G.; Yuan, Z.-Q.; Li, X. Role of extracellular vesicles in neurodegenerative diseases. *Prog. Neurobiol.* **2021**, *201*, 102022. [CrossRef] [PubMed]
19. Wei, Z.; Wei, M.; Yang, X.; Xu, Y.; Gao, S.; Ren, K. Synaptic Secretion and Beyond: Targeting Synapse and Neurotransmitters to Treat Neurodegenerative Diseases. *Oxidative Med. Cell. Longev.* **2022**, *2022*, 9176923. [CrossRef]
20. Van Den Berge, N.; Ulusoy, A. Animal models of brain-first and body-first Parkinson's disease. *Neurobiol. Dis.* **2022**, *163*, 105599. [CrossRef]
21. Sun, X.; Song, J.; Huang, H.; Chen, H.; Qian, K. Modeling hallmark pathology using motor neurons derived from the family and sporadic amyotrophic lateral sclerosis patient-specific iPS cells. *Stem Cell Res. Ther.* **2018**, *9*, 315. [CrossRef]
22. Xiong, L.; McCoy, M.; Komuro, H.; West, X.Z.; Yakubenko, V.; Gao, D.; Dudiki, T.; Milo, A.; Chen, J.; Podrez, E.A.; et al. Inflammation-dependent oxidative stress metabolites as a hallmark of amyotrophic lateral sclerosis. *Free. Radic. Biol. Med.* **2022**, *178*, 125–133. [CrossRef]
23. Mehler, M.F.; Petronglo, J.R.; Arteaga-Bracho, E.E.; Gulinello, M.E.; Winchester, M.L.; Pichamoorthy, N.; Young, S.K.; DeJesus, C.D.; Ishtiaq, H.; Gokhan, S.; et al. Loss-of-Huntingtin in Medial and Lateral Ganglionic Lineages Differentially Disrupts Regional Interneuron and Projection Neuron Subtypes and Promotes Huntington's Disease-Associated Behavioral, Cellular, and Pathological Hallmarks. *J. Neurosci.* **2019**, *39*, 1892–1909. [CrossRef] [PubMed]
24. Jimenez-Sanchez, M.; Licitra, F.; Underwood, B.R.; Rubinsztein, D.C. Huntington's Disease: Mechanisms of Pathogenesis and Therapeutic Strategies. *Cold Spring Harb. Perspect. Med.* **2017**, *7*, a024240. [CrossRef] [PubMed]
25. Machiela, E.; Southwell, A.L. Biological Aging and the Cellular Pathogenesis of Huntington's Disease. *J. Huntington's Dis.* **2020**, *9*, 115–128. [CrossRef]
26. Gallego Villarejo, L.; Bachmann, L.; Marks, D.; Brachthäuser, M.; Geidies, A.; Müller, T. Role of Intracellular Amyloid β as Pathway Modulator, Biomarker, and Therapy Target. *Int. J. Mol. Sci.* **2022**, *23*, 4656. [CrossRef] [PubMed]
27. Chen, X.Q.; Mobley, W.C. Alzheimer Disease Pathogenesis: Insights from Molecular and Cellular Biology Studies of Oligomeric Aβ and Tau Species. *Front. Neurosci.* **2019**, *13*, 659. [CrossRef] [PubMed]
28. Guo, Y.; Wang, Q.; Chen, S.; Xu, C. Functions of amyloid precursor protein in metabolic diseases. *Metabolism* **2021**, *115*, 154454. [CrossRef] [PubMed]
29. Liu, X.; Liu, Y.; Ji, S. Secretases Related to Amyloid Precursor Protein Processing. *Membranes* **2021**, *11*, 983. [CrossRef] [PubMed]
30. Zhang, T.; Chen, D.; Lee, T.H. Phosphorylation Signaling in APP Processing in Alzheimer's Disease. *Int. J. Mol. Sci.* **2019**, *21*, 209. [CrossRef] [PubMed]
31. Lane, C.A.; Hardy, J.; Schott, J.M. Alzheimer's disease. *Eur. J. Neurol.* **2018**, *25*, 59–70. [CrossRef] [PubMed]
32. Beera, A.M.; Seethamraju, S.M.; Nori, L.P. Alzheimer's Disease: Perspective on Therapeutic Options and Recent Hallmarks in Clinical Research. *Int. J. Pharm. Res. Allied Sci.* **2021**, *10*, 110–120. [CrossRef]
33. Wu, M.; Zhang, M.; Yin, X.; Chen, K.; Hu, Z.; Zhou, Q.; Cao, X.; Chen, Z.; Liu, D. The role of pathological tau in synaptic dysfunction in Alzheimer's diseases. *Transl. Neurodegener.* **2021**, *10*, 45. [CrossRef] [PubMed]
34. Chu, D.; Liu, F. Pathological Changes of Tau Related to Alzheimer's Disease. *ACS Chem. Neurosci.* **2019**, *10*, 931–944. [CrossRef]
35. Zhang, H.; Cao, Y.; Ma, L.; Wei, Y.; Li, H. Possible Mechanisms of Tau Spread and Toxicity in Alzheimer's Disease. *Front. Cell Dev. Biol.* **2021**, *9*, 707268. [CrossRef]
36. Fleeman, R.M.; Proctor, E.A. Astrocytic Propagation of Tau in the Context of Alzheimer's Disease. *Front. Cell Neurosci.* **2021**, *15*, 645233. [CrossRef]
37. Silva, M.V.F.; Loures, C.D.M.G.; Alves, L.C.V.; de Souza, L.C.; Borges, K.B.G.; Carvalho, M.D.G. Alzheimer's disease: Risk factors and potentially protective measures. *J. Biomed. Sci.* **2019**, *26*, 33. [CrossRef] [PubMed]
38. Rawat, P.; Sehar, U.; Bisht, J.; Selman, A.; Culberson, J.; Reddy, P.H. Phosphorylated Tau in Alzheimer's Disease and Other Tauopathies. *Int. J. Mol. Sci.* **2022**, *23*, 12841. [CrossRef]

39. DeVos, S.L.; Corjuc, B.T.; Oakley, D.H.; Nobuhara, C.K.; Bannon, R.N.; Chase, A.; Commins, C.; Gonzalez, J.A.; Dooley, P.M.; Frosch, M.P.; et al. Synaptic Tau Seeding Precedes Tau Pathology in Human Alzheimer's Disease Brain. *Front. Neurosci.* **2018**, *12*, 267. [CrossRef] [PubMed]
40. Riscado, M.; Baptista, B.; Sousa, F. New RNA-Based Breakthroughs in Alzheimer's Disease Diagnosis and Therapeutics. *Pharmaceutics* **2021**, *13*, 1397.
41. Plascencia-Villa, G.; Perry, G. Neuropathologic Changes Provide Insights into Key Mechanisms of Alzheimer Disease and Related Dementia. *Am. J. Pathol.* **2022**, *192*, 1340–1346. [CrossRef] [PubMed]
42. Marucci, G.; Buccioni, M.; Ben, D.D.; Lambertucci, C.; Volpini, R.; Amenta, F. Efficacy of acetylcholinesterase inhibitors in Alzheimer's disease. *Neuropharmacology* **2021**, *190*, 108352. [CrossRef] [PubMed]
43. Vecchio, I.; Sorrentino, L.; Paoletti, A.; Marra, R.; Arbitrio, M. The State of The Art on Acetylcholinesterase Inhibitors in the Treatment of Alzheimer's Disease. *J. Cent. Nerv. Syst. Dis.* **2021**, *13*, 117957352110291. [CrossRef]
44. Bennett, C.F.; Kordasiewicz, H.B.; Cleveland, D.W. Antisense Drugs Make Sense for Neurological Diseases. *Annu. Rev. Pharmacol. Toxicol.* **2021**, *61*, 831–852. [CrossRef]
45. Angelucci, F.; Cechova, K.; Valis, M.; Kuca, K.; Zhang, B.; Hort, J. MicroRNAs in Alzheimer's Disease: Diagnostic Markers or Therapeutic Agents? *Front. Pharmacol.* **2019**, *10*, 665. [CrossRef]
46. Noori, T.; Dehpour, A.R.; Sureda, A.; Sobarzo-Sanchez, E.; Shirooie, S. Role of natural products for the treatment of Alzheimer's disease. *Eur. J. Pharmacol.* **2021**, *898*, 173974. [CrossRef] [PubMed]
47. Lee, C.Y.; Ryu, I.S.; Ryu, J.H.; Cho, H.J. miRNAs as Therapeutic Tools in Alzheimer's Disease. *Int. J. Mol. Sci.* **2021**, *22*, 13012. [CrossRef] [PubMed]
48. Ramalho, M.J.; Andrade, S.; Loureiro, J.A.; do Carmo Pereira, M. Nanotechnology to improve the Alzheimer's disease therapy with natural compounds. *Drug Deliv. Transl. Res.* **2020**, *10*, 380–402. [CrossRef] [PubMed]
49. Alhazmi, H.A.; Albratty, M. An update on the novel and approved drugs for Alzheimer disease. *Saudi Pharm. J.* **2022**, *30*, 1755–1764. [CrossRef] [PubMed]
50. Chen, X.; Drew, J.; Berney, W.; Lei, W. Neuroprotective Natural Products for Alzheimer's Disease. *Cells* **2021**, *10*, 1309. [CrossRef] [PubMed]
51. Said, M.M.; Rabo, M.M.A. Neuroprotective effects of eugenol against aluminiuminduced toxicity in the rat brain. *Arch. Ind. Hyg. Toxicol.* **2017**, *68*, 27–37. [CrossRef]
52. Casares, N.; Alfaro, M.; Cuadrado-Tejedor, M.; Lasarte-Cia, A.; Navarro, F.; Vivas, I.; Espelosin, M.; Cartas-Cejudo, P.; Fernández-Irigoyen, J.; Santamaría, E.; et al. Improvement of cognitive function in wild-type and Alzheimer's disease mouse models by the immunomodulatory properties of menthol inhalation or by depletion of T regulatory cells. *Front. Immunol.* **2023**, *14*, 1130044. [CrossRef]
53. Campos, H.M.; da Costa, M.; Moreira, L.K.d.S.; Neri, H.F.d.S.; da Silva, C.R.B.; Pruccoli, L.; dos Santos, F.C.A.; Costa, E.A.; Tarozzi, A.; Ghedini, P.C. Protective effects of chrysin against the neurotoxicity induced by aluminium: In vitro and in vivo studies. *Toxicology* **2022**, *465*, 153033. [CrossRef] [PubMed]
54. Hase, T.; Shishido, S.; Yamamoto, S.; Yamashita, R.; Nukima, H.; Taira, S.; Toyoda, T.; Abe, K.; Hamaguchi, T.; Ono, K.; et al. Rosmarinic acid suppresses Alzheimer's disease development by reducing amyloid β aggregation by increasing monoamine secretion. *Sci. Rep.* **2019**, *9*, 8711. [CrossRef] [PubMed]
55. Guan, X.; Xu, J.; Liu, J.; Wu, J.; Chen, L. *Ginkgo biloba* preparation prevents and treats senile dementia by inhibiting neuroinflammatory responses. *Trop. J. Pharm. Res.* **2019**, *17*, 1961. [CrossRef]
56. Islam, F.; Nafady, M.H.; Islam, R.; Saha, S.; Rashid, S.; Akter, A.; Or-Rashid, H.; Akhtar, M.F.; Perveen, A.; Ashraf, G.M.; et al. Resveratrol and neuroprotection: An insight into prospective therapeutic approaches against Alzheimer's disease from bench to bedside. *Mol. Neurobiol.* **2022**, *59*, 4384–4404. [CrossRef] [PubMed]
57. Villegas, C.; Perez, R.; Petiz, L.L.; Glaser, T.; Ulrich, H.; Paz, C. Ginkgolides and Huperzine A for complementary treatment of Alzheimer's disease. *IUBMB Life* **2022**, *74*, 763–779. [CrossRef]
58. Dubey, T.; Chinnathambi, S. Brahmi (*Bacopa monnieri*): An ayurvedic herb against the Alzheimer's disease. *Arch. Biochem. Biophys.* **2019**, *676*, 108153. [CrossRef]
59. Snow, A.D.; Castillo, G.M.; Nguyen, B.P.; Choi, P.Y.; Cummings, J.A.; Cam, J.; Hu, Q.; Lake, T.; Pan, W.; Kastin, A.J.; et al. The Amazon rain forest plant *Uncaria tomentosa* (cat's claw) and its specific proanthocyanidin constituents are potent inhibitors and reducers of both brain plaques and tangles. *Sci. Rep.* **2019**, *9*, 561. [CrossRef]
60. Huang, M.; Jiang, X.; Liang, Y.; Liu, Q.; Chen, S.; Guo, Y. Berberine improves cognitive impairment by promoting autophagic clearance and inhibiting production of β-amyloid in APP/tau/PS1 mouse model of Alzheimer's disease. *Exp. Gerontol.* **2017**, *91*, 25–33. [CrossRef] [PubMed]
61. Mani, R.J.; Mittal, K.; Katare, D.P. Protective Effects of Quercetin in Zebrafish Model of Alzheimer's Disease. *Asian J. Pharm.* **2018**, *12*, S660.
62. Leiteritz, A.; Dilberger, B.; Wenzel, U.; Fitzenberger, E. Betaine reduces β-amyloid-induced paralysis through activation of cystathionine-β-synthase in an Alzheimer model of *Caenorhabditis elegans*. *Genes Nutr.* **2018**, *13*, 21. [CrossRef]
63. Finley, J.W.; Gao, S. A Perspective on *Crocus sativus* L. (Saffron) Constituent Crocin: A Potent Water-Soluble Antioxidant and Potential Therapy for Alzheimer's Disease. *J. Agric. Food Chem.* **2017**, *65*, 1005–1020. [CrossRef] [PubMed]

64. Singh, M.; Ramassamy, C. In vitro screening of neuroprotective activity of Indian medicinal plant *Withania somnifera*. *J. Nutr. Sci.* **2017**, *6*, e54. [CrossRef] [PubMed]
65. Kim, J.K.; Bae, H.; Kim, M.-J.; Choi, S.J.; Cho, H.Y.; Hwang, H.-J.; Kim, Y.J.; Lim, S.T.; Kim, E.K.; Kim, H.K.; et al. Inhibitory Effect of *Poncirus trifoliate* on Acetylcholinesterase and Attenuating Activity against Trimethyltin-Induced Learning and Memory Impairment. *Biosci. Biotechnol. Biochem.* **2009**, *73*, 1105–1112. [CrossRef] [PubMed]
66. Bihaqi, S.W.; Sharma, M.; Singh, A.P.; Tiwari, M. Neuroprotective role of *Convolvulus pluricaulis* on aluminium induced neurotoxicity in rat brain. *J. Ethnopharmacol.* **2009**, *124*, 409–415. [CrossRef]
67. Azimi, A.; Ghaffari, S.M.; Riazi, G.H.; Arab, S.S.; Tavakol, M.M.; Pooyan, S. α-Cyperone of *Cyperus rotundus* is an effective candidate for reduction of inflammation by destabilization of microtubule fibers in brain. *J. Ethnopharmacol.* **2016**, *194*, 219–227. [CrossRef]
68. Rivera, D.S.; Lindsay, C.; Codocedo, J.F.; Morel, I.; Pinto, C.; Cisternas, P.; Bozinovic, F.; Inestrosa, N.C. Andrographolide recovers cognitive impairment in a natural model of Alzheimer's disease (*Octodon degus*). *Neurobiol. Aging* **2016**, *46*, 204–220. [CrossRef] [PubMed]
69. Balez, R.; Steiner, N.; Engel, M.; Muñoz, S.S.; Lum, J.S.; Wu, Y.; Wang, D.; Vallotton, P.; Sachdev, P.; O'connor, M.; et al. Neuroprotective effects of apigenin against inflammation, neuronal excitability and apoptosis in an induced pluripotent stem cell model of Alzheimer's disease. *Sci. Rep.* **2016**, *6*, 31450. [CrossRef] [PubMed]
70. Shi, J.; Li, Y.; Zhang, Y.; Chen, J.; Gao, J.; Zhang, T.; Shang, X.; Zhang, X. Baicalein Ameliorates Aβ-Induced Memory Deficits and Neuronal Atrophy via Inhibition of PDE2 and PDE4. *Front. Pharmacol.* **2021**, *12*, 794458. [CrossRef]
71. Celik Topkara, K.; Kilinc, E.; Cetinkaya, A.; Saylan, A.; Demir, S. Therapeutic effects of carvacrol on beta-amyloid-induced impairments in in vitro and in vivo models of Alzheimer's disease. *Eur. J. Neurosci.* **2022**, *56*, 5714–5726. [CrossRef] [PubMed]
72. Li, L.; Li, W.; Jung, S.W.; Lee, Y.W.; Kim, Y.H. Protective Effects of Decursin and Decursinol Angelate against Amyloid β-Protein-Induced Oxidative Stress in the PC12 Cell Line: The Role of Nrf2 and Antioxidant Enzymes. *Biosci. Biotechnol. Biochem.* **2011**, *75*, 434–442. [CrossRef]
73. Duan, X.; Li, Y.; Xu, F.; Ding, H. Study on the neuroprotective effects of Genistein on Alzheimer's disease. *Brain Behav.* **2021**, *11*, e02100. [CrossRef] [PubMed]
74. Huang, D.S.; Yu, Y.C.; Wu, C.H.; Lin, J.Y. Protective Effects of Wogonin against Alzheimer's Disease by Inhibition of Amyloidogenic Pathway. *Evid.-Based Complement. Altern. Med.* **2017**, *2017*, 3545169. [CrossRef]
75. Xu, P.-X.; Wang, S.-W.; Yu, X.-L.; Su, Y.-J.; Wang, T.; Zhou, W.-W.; Zhang, H.; Wang, Y.-J.; Liu, R.-T. Rutin improves spatial memory in Alzheimer's disease transgenic mice by reducing Aβ oligomer level and attenuating oxidative stress and neuroinflammation. *Behav. Brain Res.* **2014**, *264*, 173–180. [CrossRef] [PubMed]
76. Daily, J.W.; Kang, S.; Park, S. Protection against Alzheimer's disease by luteolin: Role of brain glucose regulation, anti-inflammatory activity, and the gut microbiota-liver-brain axis. *BioFactors* **2021**, *47*, 218–231. [CrossRef] [PubMed]
77. Sabogal-Guáqueta, A.M.; Osorio, E.; Cardona-Gómez, G.P. Linalool reverses neuropathological and behavioral impairments in old triple transgenic Alzheimer's mice. *Neuropharmacology* **2016**, *102*, 111–120. [CrossRef]
78. Asiatic Acid Nullified Aluminium Toxicity in In Vitro Model of Alzheimer's Disease. Available online: https://www.imrpress.com/journal/FBE/10/2/10.2741/E823 (accessed on 30 May 2023).
79. Li, Z.; Zhao, T.; Li, J.; Yu, Q.; Feng, Y.; Xie, Y.; Sun, P. Nanomedicine Based on Natural Products: Improving Clinical Application Potential. *J. Nanomater.* **2022**, *2022*, 3066613. [CrossRef]
80. Woon, C.K.; Hui, W.K.; Abas, R.; Haron, M.H.; Das, S.; Lin, T.S. Natural Product-based Nanomedicine: Recent Advances and Issues for the Treatment of Alzheimer's Disease. *Curr. Neuropharmacol.* **2022**, *20*, 1498–1518. [CrossRef]
81. Yu, A.M.; Jian, C.; Yu, A.H.; Tu, M.J. RNA therapy: Are we using the right molecules? *Pharmacol. Ther.* **2019**, *196*, 91–104. [CrossRef]
82. Shin, H.; Park, S.-J.; Yim, Y.; Kim, J.; Choi, C.; Won, C.; Min, D.-H. Recent Advances in RNA Therapeutics and RNA Delivery Systems Based on Nanoparticles. *Adv. Ther.* **2018**, *1*, 1800065. [CrossRef]
83. Zogg, H.; Singh, R.; Ro, S. Current Advances in RNA Therapeutics for Human Diseases. *Int. J. Mol. Sci.* **2022**, *23*, 2736. [CrossRef] [PubMed]
84. Mollocana-Lara, E.C.; Ni, M.; Agathos, S.N.; Gonzales-Zubiate, F.A. The infinite possibilities of RNA therapeutics. *J. Ind. Microbiol. Biotechnol.* **2021**, *48*, kuab063. [CrossRef]
85. DeLong, R. Ushering in a new era of RNA-based therapies. *Commun. Biol.* **2021**, *4*, 577. [CrossRef] [PubMed]
86. Kim, Y.K. RNA therapy: Rich history, various applications and unlimited future prospects. *Exp. Mol. Med.* **2022**, *54*, 455–465. [CrossRef] [PubMed]
87. Kim, Y.K. RNA Therapy: Current Status and Future Potential. *Chonnam Med. J.* **2020**, *56*, 87. [CrossRef] [PubMed]
88. Anthony, K. RNA-based therapeutics for neurological diseases. *RNA Biol.* **2022**, *19*, 176–190. [CrossRef]
89. Lee, M.J.; Lee, I.; Wang, K. Recent Advances in RNA Therapy and Its Carriers to Treat the Single-Gene Neurological Disorders. *Biomedicines* **2022**, *10*, 158. [CrossRef] [PubMed]
90. Jurcău, M.C.; Andronie-Cioara, F.L.; Jurcău, A.; Marcu, F.; Țiț, D.M.; Pașcalău, N.; Nistor-Cseppentö, D.C. The Link between Oxidative Stress, Mitochondrial Dysfunction and Neuroinflammation in the Pathophysiology of Alzheimer's Disease: Therapeutic Implications and Future Perspectives. *Antioxidants* **2022**, *11*, 2167. [CrossRef] [PubMed]

1. Walgrave, H.; Zhou, L.; De Strooper, B.; Salta, E. The promise of microRNA-based therapies in Alzheimer's disease: Challenges and perspectives. *Mol. Neurodegener.* **2021**, *16*, 76. [CrossRef] [PubMed]
2. Kreth, S.; Hübner, M.; Hinske, L.C. MicroRNAs as Clinical Biomarkers and Therapeutic Tools in Perioperative Medicine. *Obstet. Anesth. Dig.* **2018**, *126*, 670–681. [CrossRef]
3. Martier, R.; Konstantinova, P. Gene Therapy for Neurodegenerative Diseases: Slowing down the Ticking Clock. *Front. Neurosci.* **2020**, *14*, 580179. [CrossRef] [PubMed]
4. Zhou, Y.; Zhu, F.; Liu, Y.; Zheng, M.; Wang, Y.; Zhang, D.; Anraku, Y.; Zou, Y.; Li, J.; Wu, H.; et al. Blood-brain barrier–penetrating siRNA nanomedicine for Alzheimer's disease therapy. *Sci. Adv.* **2020**, *6*, eabc7031. [CrossRef] [PubMed]
5. Long, J.M.; Lahiri, D.K. MicroRNA-101 downregulates Alzheimer's amyloid-β precursor protein levels in human cell cultures and is differentially expressed. *Biochem. Biophys. Res. Commun.* **2011**, *404*, 889–895. [CrossRef] [PubMed]
6. Liu, W.; Zhao, J.; Lu, G. miR-106b inhibits tau phosphorylation at Tyr18 by targeting Fyn in a model of Alzheimer's disease. *Biochem. Biophys. Res. Commun.* **2016**, *478*, 852–857. [CrossRef] [PubMed]
7. Jiang, Y.; Xu, B.; Chen, J.; Sui, Y.; Ren, L.; Li, J.; Zhang, H.; Guo, L.; Sun, X. Micro-RNA-137 Inhibits Tau Hyperphosphorylation in Alzheimer's Disease and Targets the CACNA1C Gene in Transgenic Mice and Human Neuroblastoma SH-SY5Y Cells. *Med. Sci. Monit.* **2018**, *24*, 5635–5644. [CrossRef]
8. Santa-Maria, I.; Alaniz, M.E.; Renwick, N.; Cela, C.; Fulga, T.A.; Van Vactor, D.; Tuschl, T.; Clark, L.N.; Shelanski, M.L.; McCabe, B.D.; et al. Dysregulation of microRNA-219 promotes neurodegeneration through post-transcriptional regulation of tau. *J. Clin. Investig.* **2015**, *125*, 681–686. [CrossRef]
9. Estfanous, S.; Daily, K.P.; Eltobgy, M.; Deems, N.P.; Anne, M.N.K.; Krause, K.; Badr, A.; Hamilton, K.; Carafice, C.; Hegazi, A.; et al. Elevated Expression of MiR-17 in Microglia of Alzheimer's Disease Patients Abrogates Autophagy-Mediated Amyloid-β Degradation. *Front. Immunol.* **2021**, *12*, 705581. [CrossRef] [PubMed]
10. Wang, R.; Chopra, N.; Nho, K.; Maloney, B.; Obukhov, A.G.; Nelson, P.T.; Counts, S.E.; Lahiri, D.K. Human microRNA (miR-20b-5p) modulates Alzheimer's disease pathways and neuronal function, and a specific polymorphism close to the MIR20B gene influences Alzheimer's biomarkers. *Mol. Psychiatry* **2022**, *27*, 1256–1273. [CrossRef]
11. Zong, Y.; Wang, H.; Dong, W.; Quan, X.; Zhu, H.; Xu, Y.; Huang, L.; Ma, C.; Qin, C. miR-29c regulates BACE1 protein expression. *Brain Res.* **2011**, *1395*, 108–115. [CrossRef]
12. Chopra, N.; Wang, R.; Maloney, B.; Nho, K.; Beck, J.S.; Pourshafie, N.; Niculescu, A.; Saykin, A.J.; Rinaldi, C.; Counts, S.E.; et al. MicroRNA-298 reduces levels of human amyloid-β precursor protein (APP), β-site APP-converting enzyme 1 (BACE1) and specific tau protein moieties. *Mol. Psychiatry* **2021**, *26*, 5636–5657. [CrossRef] [PubMed]
13. He, C.; Su, C.; Zhang, W.; Wan, Q. miR-485-5p alleviates Alzheimer's disease progression by targeting PACS1. *Transl. Neurosci.* **2021**, *12*, 335–345. [CrossRef] [PubMed]
14. Liu, J.; Zuo, X.; Han, J.; Dai, Q.; Xu, H.; Liu, Y.; Cui, S. MiR-9-5p inhibits mitochondrial damage and oxidative stress in AD cell models by targeting GSK-3β. *Biosci. Biotechnol. Biochem.* **2020**, *84*, 2273–2280. [CrossRef] [PubMed]
15. Deng, Y.; Zhang, J.; Sun, X.; Ma, G.; Luo, G.; Miao, Z.; Song, L. miR-132 improves the cognitive function of rats with Alzheimer's disease by inhibiting the MAPK1 signal pathway. *Exp. Ther. Med.* **2020**, *20*, 159. [CrossRef]
16. Liang, C.; Zhu, H.; Xu, Y.; Huang, L.; Ma, C.; Deng, W.; Liu, Y.; Qin, C. MicroRNA-153 negatively regulates the expression of amyloid precursor protein and amyloid precursor-like protein 2. *Brain Res.* **2012**, *1455*, 103–113. [CrossRef] [PubMed]
17. Xu, H.; Rosler, T.W.; Carlsson, T.; Andrade, A.; Fiala, O.; Hollerhage, M.; Oertel, W.H.; Goedert, M.; Aigner, A.; Hoglinger, G.U. Tau Silencing by siRNA in the P301S Mouse Model of Tauopathy. *Curr. Gene Ther.* **2014**, *14*, 343–351. [CrossRef]
18. Kandimalla, R.J.; Wani, W.Y.; Bk, B.; Gill, K.D. siRNA against presenilin 1 (PS1) down regulates amyloid b42 production in IMR-32 cells. *J. Biomed. Sci.* **2012**, *19*, 2. [CrossRef]
19. Senechal, Y.; Kelly, P.H.; Cryan, J.F.; Natt, F.; Dev, K.K. Amyloid precursor protein knockdown by siRNA impairs spontaneous alternation in adult mice: In vivo knockdown of APP by RNAi. *J. Neurochem.* **2007**, *102*, 1928–1940. [CrossRef] [PubMed]
20. Lin, C.Y.; Perche, F.; Ikegami, M.; Uchida, S.; Kataoka, K.; Itaka, K. Messenger RNA-based therapeutics for brain diseases: An animal study for augmenting clearance of beta-amyloid by intracerebral administration of neprilysin mRNA loaded in polyplex nanomicelles. *J. Control. Release* **2016**, *235*, 268–275. [CrossRef]
21. Lim, S.A.; Cox, A.; Tung, M.; Chung, E.J. Clinical progress of nanomedicine-based RNA therapies. *Bioact. Mater.* **2022**, *12*, 203–213. [CrossRef] [PubMed]
22. Gorshkov, A.; Purvinsh, L.; Brodskaia, A.; Vasin, A. Exosomes as Natural Nanocarriers for RNA-Based Therapy and Prophylaxis. *Nanomaterials* **2022**, *12*, 524. [CrossRef] [PubMed]
23. Tsakiri, M.; Zivko, C.; Demetzos, C.; Mahairaki, V. Lipid-based nanoparticles and RNA as innovative neuro-therapeutics. *Front. Pharmacol.* **2022**, *13*, 900610. [CrossRef] [PubMed]
24. Fernandes, F.; Dias-Teixeira, M.; Delerue-Matos, C.; Grosso, C. Critical Review of Lipid-Based Nanoparticles as Carriers of Neuroprotective Drugs and Extracts. *Nanomaterials* **2021**, *11*, 563. [CrossRef] [PubMed]
25. Pardridge, W.M. Treatment of Alzheimer's Disease and Blood–Brain Barrier Drug Delivery. *Pharmaceuticals* **2020**, *13*, 394. [CrossRef]
26. Wohlfart, S.; Gelperina, S.; Kreuter, J. Transport of drugs across the blood–brain barrier by nanoparticles. *J. Control. Release* **2012**, *161*, 264–273. [CrossRef] [PubMed]

117. Zhou, Y.; Peng, Z.; Seven, E.S.; Leblanc, R.M. Crossing the blood-brain barrier with nanoparticles. *J. Control. Release* **2018**, *270*, 290–303. [CrossRef] [PubMed]
118. Bors, L.; Erdő, F. Overcoming the Blood–Brain Barrier. Challenges and Tricks for CNS Drug Delivery. *Sci. Pharm.* **2019**, *87*, 6. [CrossRef]
119. Wong, K.H.; Riaz, M.K.; Xie, Y.; Zhang, X.; Liu, Q.; Chen, H.; Bian, Z.; Chen, X.; Lu, A.; Yang, Z. Review of Current Strategies for Delivering Alzheimer's Disease Drugs across the Blood-Brain Barrier. *Int. J. Mol. Sci.* **2019**, *20*, 381. [CrossRef] [PubMed]
120. Zenaro, E.; Piacentino, G.; Constantin, G. The blood-brain barrier in Alzheimer's disease. *Neurobiol. Dis.* **2017**, *107*, 41–56. [CrossRef] [PubMed]
121. Sharma, C.; Woo, H.; Kim, S.R. Addressing Blood–Brain Barrier Impairment in Alzheimer's Disease. *Biomedicines* **2022**, *10*, 742. [CrossRef] [PubMed]
122. Juhairiyah, F.; de Lange, E.C.M. Understanding Drug Delivery to the Brain Using Liposome-Based Strategies: Studies that Provide Mechanistic Insights Are Essential. *AAPS J.* **2021**, *23*, 114. [CrossRef] [PubMed]
123. Satapathy, M.K.; Yen, T.-L.; Jan, J.-S.; Tang, R.-D.; Wang, J.-Y.; Taliyan, R.; Yang, C.-H. Solid Lipid Nanoparticles (SLNs): An Advanced Drug Delivery System Targeting Brain through BBB. *Pharmaceutics* **2021**, *13*, 1183. [CrossRef] [PubMed]
124. Musielak, E.; Feliczak-Guzik, A.; Nowak, I. Synthesis and Potential Applications of Lipid Nanoparticles in Medicine. *Materials* **2022**, *15*, 682. [CrossRef]
125. Heidarzadeh, M.; Gürsoy-Özdemir, Y.; Kaya, M.; Abriz, A.E.; Zarebkohan, A.; Rahbarghazi, R.; Sokullu, E. Exosomal delivery of therapeutic modulators through the blood–brain barrier; promise and pitfalls. *Cell Biosci.* **2021**, *11*, 142. [CrossRef]
126. Negahdaripour, M.; Vakili, B.; Nezafat, N. Exosome-based vaccines and their position in next generation vaccines. *Int. Immunopharmacol.* **2022**, *113*, 109265. [CrossRef] [PubMed]
127. Jafari, D.; Malih, S.; Eini, M.; Jafari, R.; Gholipourmalekabadi, M.; Sadeghizadeh, M.; Samadikuchaksaraei, A. Improvement, scaling-up, and downstream analysis of exosome production. *Crit. Rev. Biotechnol.* **2020**, *40*, 1098–1112. [CrossRef] [PubMed]
128. Hu, Q.; Su, H.; Li, J.; Lyon, C.; Tang, W.; Wan, M.; Hu, T.Y. Clinical applications of exosome membrane proteins. *Precis. Clin. Med.* **2020**, *3*, 54–66. [CrossRef] [PubMed]
129. Ludwig, N.; Whiteside, T.L.; Reichert, T.E. Challenges in Exosome Isolation and Analysis in Health and Disease. *Int. J. Mol. Sci.* **2019**, *20*, 4684. [CrossRef] [PubMed]
130. Rashed, M.H.; Bayraktar, E.; Helal, G.K.; Abd-Ellah, M.F.; Amero, P.; Chavez-Reyes, A.; Rodriguez-Aguayo, C. Exosomes: From Garbage Bins to Promising Therapeutic Targets. *Int. J. Mol. Sci.* **2017**, *18*, 538. [CrossRef]
131. Zhang, N.; He, F.; Li, T.; Chen, J.; Jiang, L.; Ouyang, X.-P.; Zuo, L. Role of Exosomes in Brain Diseases. *Front. Cell Neurosci.* **2021**, *15*, 743353. [CrossRef] [PubMed]
132. Xu, L.; Wang, X.; Liu, Y.; Yang, G.; Falconer, R.J.; Zhao, C.X. Lipid Nanoparticles for Drug Delivery. *Adv. NanoBiomed Res.* **2022**, *2*, 2100109. [CrossRef]
133. Feng, R.; Patil, S.; Zhao, X.; Miao, Z.; Qian, A. RNA Therapeutics—Research and Clinical Advancements. *Front. Mol. Biosci.* **2021**, *8*, 710738. [CrossRef]
134. Burduşel, A.C.; Andronescu, E. Lipid Nanoparticles and Liposomes for Bone Diseases Treatment. *Biomedicines* **2022**, *10*, 3158. [CrossRef]
135. Duan, Y.; Dhar, A.; Patel, C.; Khimani, M.; Neogi, S.; Sharma, P.; Kumar, N.S.; Vekariya, R.L. A brief review on solid lipid nanoparticles: Part and parcel of contemporary drug delivery systems. *RSC Adv.* **2020**, *10*, 26777–26791. [CrossRef] [PubMed]
136. Tenchov, R.; Bird, R.; Curtze, A.E.; Zhou, Q. Lipid Nanoparticles─From Liposomes to mRNA Vaccine Delivery, a Landscape of Research Diversity and Advancement. *ACS Nano* **2021**, *15*, 16982–17015. [CrossRef]
137. Abdellatif, A.A.; Younis, M.A.; Alsowinea, A.F.; Abdallah, E.M.; Abdel-Bakky, M.S.; Al-Subaiyel, A.; Hassan, Y.A.; Tawfeek, H.M. Lipid nanoparticles technology in vaccines: Shaping the future of prophylactic medicine. *Colloids Surf. B Biointerfaces* **2023**, *222*, 113111. [CrossRef] [PubMed]
138. Vieira, D.; Gamarra, L. Getting into the brain: Liposome-based strategies for effective drug delivery across the blood–brain barrier. *Int. J. Nanomed.* **2016**, *11*, 5381–5414. [CrossRef]
139. Gauro, R.; Nandave, M.; Jain, V.K.; Jain, K. Advances in dendrimer-mediated targeted drug delivery to the brain. *J. Nanoparticle Res.* **2021**, *23*, 76. [CrossRef]
140. Teleanu, D.; Chircov, C.; Grumezescu, A.; Volceanov, A.; Teleanu, R. Blood-Brain Delivery Methods Using Nanotechnology. *Pharmaceutics* **2018**, *10*, 269. [CrossRef] [PubMed]
141. Sela, H.; Cohen, H.; Elia, P.; Zach, R.; Karpas, Z.; Zeiri, Y. Spontaneous penetration of gold nanoparticles through the blood brain barrier (BBB). *J. Nanobiotechnol.* **2015**, *13*, 71. [CrossRef]
142. Li, X.; Corbett, A.L.; Taatizadeh, E.; Tasnim, N.; Little, J.P.; Garnis, C.; Daugaard, M.; Guns, E.; Hoorfar, M.; Li, I.T.S. Challenges and opportunities in exosome research—Perspectives from biology, engineering, and cancer therapy. *APL Bioeng.* **2019**, *3*, 011503. [CrossRef] [PubMed]
143. Schiffelers, R.; Kooijmans, S.; Vader, P.; Dommelen, V.; Solinge, V. Exosome mimetics: A novel class of drug delivery systems. *Int. J. Nanomed.* **2012**, *7*, 1525–1541. [CrossRef]
144. Fernandes, M.; Lopes, I.; Magalhães, L.; Sárria, M.P.; Machado, R.; Sousa, J.C.; Botelho, C.; Teixeira, J.; Gomes, A.C. Novel concept of exosome-like liposomes for the treatment of Alzheimer's disease. *J. Control. Release* **2021**, *336*, 130–143. [CrossRef]

45. Zha, Y.; Lin, T.; Li, Y.; Zhang, X.; Wang, Z.; Li, Z.; Ye, Y.; Wang, B.; Zhang, S.; Wang, J. Exosome-mimetics as an engineered gene-activated matrix induces in-situ vascularized osteogenesis. *Biomaterials* **2020**, *247*, 119985. [CrossRef] [PubMed]
46. Severic, M.; Ma, G.; Pereira, S.G.T.; Ruiz, A.; Cheung, C.C.; Al-Jamal, W.T. Genetically-engineered anti-PSMA exosome mimetics targeting advanced prostate cancer in vitro and in vivo. *J. Control. Release* **2021**, *330*, 101–110. [CrossRef]
47. Lee, C.-S.; Fan, J.; Hwang, H.S.; Kim, S.; Chen, C.; Kang, M.; Aghaloo, T.; James, A.W.; Lee, M. Bone-Targeting Exosome Mimetics Engineered by Bioorthogonal Surface Functionalization for Bone Tissue Engineering. *Nano Lett.* **2023**, *23*, 1202–1210. [CrossRef] [PubMed]
48. Fan, J.; Lee, C.-S.; Kim, S.; Chen, C.; Aghaloo, T.; Lee, M. Generation of Small RNA-Modulated Exosome Mimetics for Bone Regeneration. *ACS Nano* **2020**, *14*, 11973–11984. [CrossRef] [PubMed]
49. Wang, J.; Li, M.; Jin, L.; Guo, P.; Zhang, Z.; Zhanghuang, C.; Tan, X.; Mi, T.; Liu, J.; Wu, X.; et al. Exosome mimetics derived from bone marrow mesenchymal stem cells deliver doxorubicin to osteosarcoma in vitro and in vivo. *Drug Deliv.* **2022**, *29*, 3291–3303. [CrossRef]
50. Jang, S.C.; Kim, O.Y.; Yoon, C.M.; Choi, D.-S.; Roh, T.-Y.; Park, J.; Nilsson, J.; Lötvall, J.; Kim, Y.-K.; Gho, Y.S. Bioinspired Exosome-Mimetic Nanovesicles for Targeted Delivery of Chemotherapeutics to Malignant Tumors. *ACS Nano* **2013**, *7*, 7698–7710. [CrossRef]
51. Wu, J.-Y.; Li, Y.-J.; Wang, J.; Hu, X.-B.; Huang, S.; Luo, S.; Xiang, D.-X. Multifunctional exosome-mimetics for targeted anti-glioblastoma therapy by manipulating protein corona. *J. Nanobiotechnol.* **2021**, *19*, 405. [CrossRef]
52. Yan, Y.; Du, C.; Duan, X.; Yao, X.; Wan, J.; Jiang, Z.; Qin, Z.; Li, W.; Pan, L.; Gu, Z.; et al. Inhibiting collagen I production and tumor cell colonization in the lung via miR-29a-3p loading of exosome-/liposome-based nanovesicles. *Acta Pharm. Sin. B* **2022**, *12*, 939–951. [CrossRef]

Disclaimer/Publisher's Note: The statements, opinions and data contained in all publications are solely those of the individual author(s) and contributor(s) and not of MDPI and/or the editor(s). MDPI and/or the editor(s) disclaim responsibility for any injury to people or property resulting from any ideas, methods, instructions or products referred to in the content.

Review

Nucleic Acid Probes in Bio-Imaging and Diagnostics: Recent Advances in ODN-Based Fluorescent and Surface-Enhanced Raman Scattering Nanoparticle and Nanostructured Systems

Monica-Cornelia Sardaru [1,2], Narcisa-Laura Marangoci [1], Rosanna Palumbo [3], Giovanni N. Roviello [3,*] and Alexandru Rotaru [1,4,*]

1 "Petru Poni" Institute of Macromolecular Chemistry, Romanian Academy, Centre of Advanced Research in Bionanoconjugates and Biopolymers, Grigore Ghica Voda Alley 41 A, 700487 Iasi, Romania
2 The Research Institute of the University of Bucharest (ICUB), 90 Sos. Panduri, 050663 Bucharest, Romania
3 Institute of Biostructures and Bioimaging, Italian National Council for Research (IBB-CNR), Area di Ricerca Site and Headquarters, Via Pietro Castellino 111, 80131 Naples, Italy
4 Institute for Research, Innovation and Technology Transfer, UPS "Ion Creanga", Ion Creanga Str. 1, MD2069 Chisinau, Moldova
* Correspondence: giroviel@unina.it (G.N.R.); rotaru.alexandru@icmpp.ro (A.R.)

Abstract: Raman nanoparticle probes are a potent class of optical labels for the interrogation of pathological and physiological processes in cells, bioassays, and tissues. Herein, we review the recent advancements in fluorescent and Raman imaging using oligodeoxyribonucleotide (ODN)-based nanoparticles and nanostructures, which show promise as effective tools for live-cell analysis. These nanodevices can be used to investigate a vast number of biological processes occurring at various levels, starting from those involving organelles, cells, tissues, and whole living organisms. ODN-based fluorescent and Raman probes have contributed to the achievement of significant advancements in the comprehension of the role played by specific analytes in pathological processes and have inaugurated new possibilities for diagnosing health conditions. The technological implications that have emerged from the studies herein described could open new avenues for innovative diagnostics aimed at identifying socially relevant diseases like cancer through the utilization of intracellular markers and/or guide surgical procedures based on fluorescent or Raman imaging. Particularly complex probe structures have been developed within the past five years, creating a versatile toolbox for live-cell analysis, with each tool possessing its own strengths and limitations for specific studies. Analyzing the literature reports in the field, we predict that the development of ODN-based fluorescent and Raman probes will continue in the near future, disclosing novel ideas on their application in therapeutic and diagnostic strategies.

Keywords: oligodeoxyribonucleotides; fluorescence probes; Raman probes; bio-imaging

Citation: Sardaru, M.-C.; Marangoci, N.-L.; Palumbo, R.; Roviello, G.N.; Rotaru, A. Nucleic Acid Probes in Bio-Imaging and Diagnostics: Recent Advances in ODN-Based Fluorescent and Surface-Enhanced Raman Scattering Nanoparticle and Nanostructured Systems. *Molecules* 2023, 28, 3561. https://doi.org/10.3390/molecules28083561

Academic Editors: Sudeshna Chandra and Heinrich Lang

Received: 6 March 2023
Revised: 12 April 2023
Accepted: 13 April 2023
Published: 18 April 2023

Copyright: © 2023 by the authors. Licensee MDPI, Basel, Switzerland. This article is an open access article distributed under the terms and conditions of the Creative Commons Attribution (CC BY) license (https://creativecommons.org/licenses/by/4.0/).

1. Introduction

One of the most effective strategies aimed at understanding the continuous functioning of cells and related physiological processes consists of the comprehensive visualization of basic events or mechanisms occurring within cells [1–3] and living organisms [4]. To monitor biomolecules and their biological roles in a cell, fluorescent and—more recently—Raman spectroscopy signals are commonly employed [5–9]. Nanoparticle probes utilizing Raman properties are associated with unique spectral signatures endowed with narrow peaks that are ideal for the simultaneous detection of multiple targets. Raman emission is much weaker than that associated with fluorescence, but significant signal enhancements can be obtained by adsorbing Raman-active probes onto metal surfaces or nanoparticles [10,11]. In view to gain a deeper understanding and expand the range of monitored events, improved fluorescent and Raman probes with specific designs are required in order to study precise interactions or decrease the limit of detection (LOD). To achieve high specificity in both fluorescent and Raman probes, oligodeoxyribonucleotides (ODNs)—short strands of nucleic acid molecules that have been implemented in numerous medical and scientific applications—can be utilized [12–19]. Specially designed ODNs have the ability to identify a diverse range of biomedically-relevant targets, such as sequence-specific recognition of nucleic acids within living cells, proteins, small organic molecules, or even metals [20–23]. Precise identification of specific molecular targets for reliable fluorescent or Raman on-and-off readings and observations is heavily reliant on the design of ODN probes. Excellent reviews have previously covered various ODN probes whose intensity of readouts is regulated in a sequence-specific manner [24–26]. However, several unsolved issues remain therein about the need for: (i) different types of fluorescent emitting entities (fluorophores or quantum dots) in an efficacious probe; (ii) a more complex design of noble metal surface for fluorescent quenching or efficient hot spots for sequence-specific Raman (SERS) effects; and (iii) probes requiring the formation of more complex designs for imaging or detecting targets at even lower limits. Limitations of the SERS method include: (i) the need for an intimate contact between the analyte and the enhancing surface; (ii) substrates degrade over time, reducing the SERS signal; (iii) limited selectivity of the substrate for a given analyte; (iv) Substrate reusability is limited; (v) problems with uniformity and reproducibility of the SERS signal within the substrate.

In order to develop next-generation fluorescent and Raman probes for bio-imaging and sensing, a new concept for ODN-based probes must be established, which could also address the problems mentioned above. Hence, this review focuses on the recent advancements of SERS and fluorescence-imaging probes covering the period between 1 January 2018 and 31 December 2022. We particularly focused our attention on the reported original designs of nanoparticle-based ODN probes (including both gold and gold-free nanoprobes), as well as aspects of DNA nanotechnology which are winning approaches for efficient fluorescent and Raman bio-imaging.

2. Nanoparticle-Based ODN Conjugates for Fluorescent Bio-Imaging

The utilization of nanoparticle-based ODN conjugates has become a powerful technique in fluorescent bio-imaging. These conjugates are frequently used to deliver oligonucleotides to particular cells or tissues in various bio-imaging applications [1]. In view to create nanoparticle-based ODN conjugates, various strategies can be employed, including hybridization and covalent conjugation. Once functionalized, the conjugates are used to deliver fluorescent dyes or other imaging agents to specific cells or tissues, allowing for highly localized imaging thanks to the high specificity of the oligonucleotides for their target molecules. This specificity is beneficial in studying biomolecule dynamics in living cells and for imaging specific cells or tissues in vivo. Using nanoparticle-based ODN conjugates is an effective method for fluorescent bio-imaging, which provides localized and highly specific imaging capabilities. Researchers are actively exploring the use of gold and gold-free nanoparticles as carriers for these conjugates, opening up new possibilities for the development of cancer diagnostics and therapies [27–32]. The following are some selected reports on gold-based and gold-free nanoparticle-based ODN conjugates that successfully bound to specific receptors on cells, allowing the nanoparticles to be taken up by the cells and deliver fluorescent dyes or other imaging agents, resulting in highly specific imaging of the cancer cells.

2.1. Gold ODN-Conjugated Nanoparticles for Fluorescent Bio-Imaging

Gold nanoparticles (AuNPs) have been widely used in bio-imaging applications due to their high biocompatibility, strong absorption and scattering properties, and easy functionalization with oligonucleotides. AuNPs can be functionalized with oligonucleotides through various methods, such as hybridization or covalent conjugation. The densely arranged ODN shell, also known as spherical nucleic acid probes (SNAPs) [33,34], boasts several advantages over traditional nucleic acid chains. Once functionalized, the AuNPs can be targeted to specific receptors on cells, allowing for the selective uptake of the particles and subsequent delivery of imaging agents to specific cells or tissues. Overall, gold nanoparticles conjugated with oligonucleotides offer a powerful approach for fluorescent bio-imaging, providing high specificity and localized imaging of specific cells or tissues, and have potential applications in cancer diagnostics and therapies. Recently, Gao et al. [35] reported a concept of tumor-cell detection that bypasses genotypic and phenotypic features of different tumor types. Combining spherical nucleic acids with molecular beacons (SNAB technology) enabled the detection of tumor cells, distinguishing them from normal cells by examining the telomerase activity. The molecular beacon consisting of a long telomerase primer for the specific recognition of the telomerase catalytic core [36,37] and a short fluorophore-modified DNA strand were pre-hybridized before immobilizing onto AuNPs surface via the Au-thiol linkage. In the presence of the telomerase, the primer was effectively recognized by its catalytic core, which led to the formation of a more thermodynamically stable DNA hairpin structure. The fluorophore-labeled strand, subsequently displaced into the solution, restored its strong fluorescence for detection. The proposed SNAB probe can be used in multiple platforms, including single-cell imaging and solution-based assays, as well as in vivo solid tumor imaging, making it a versatile tool. The authors believe the SNAB technology will have a great impact on cancer diagnosis, therapeutic response assessment, and image-guided surgery. Au-based SNAPs, which use AuNPs as cores and densely-modified nucleic acid chains secured via an Au-S bond, have been used for diagnostic and therapeutic applications in the case of various diseases [38–40]. Unfortunately, the presence of biothiols in living cells can dislocate the nucleic acid chains from the AuNPs surface and strongly reduce their theranostic performance. Li and Tang et al. [41] reported a strategy to overcome this impediment by designing a selenol terminal functionalized molecular beacon building up an Au-Se bond-based SNAP (SNAP-Se) for bio-imaging. The performed experiments showed the successful creation and usage of SNAP-Se, and its ability to prevent false-positive signals while imaging biomarkers in living cells.

In another approach, Liu and co-workers [42] reported the application of SNAPs to overcome the production of non-specific responses in healthy tissues. The authors designed a DNA nanosphere system to specifically sense cancer biomarker flap endonuclease 1 (FEN1) and spatiotemporally modulate drug release. The gold-based nanostar-conjugated FEN1 substrate acted as a spherical nucleic acid and produced a fluorescent signal when stimulated by FEN1 for diagnosis. Subsequently, the nanoflare prompted a controlled release of drugs at desired sites by using external NIR light. This DNA nanosphere exhibited good sensitivity, stability, and specificity toward FEN1 assay and can be used as a precision theranostic agent for targeted and controlled drug delivery. The proposed approach provides a reliable way to image FEN1 both in vitro and in vivo and serves as an efficacious tool for precision medicine. Successful use of fluorescent SNAPs, which use poly-adenine (polyA) tails on AuNPs to detect mRNA within cells, was reported by Wang et al. [43]. By adjusting the loading density of DNA on the gold nanoparticle interface, the sensitivity of the probes could be easily adjusted. AuNPs with polyA-tailed recognition sequences were hybridized into fluorescent "reporter" strands, creating fluorescence-quenched SNAP probes. When exposed to the target gene, the "reporter" strands were released from the SNAP through strand displacement, and fluorescence was recovered. With a 20 bases-long polyA tail, the detection limit of the probes was 0.31 nM, which is approximately 55 times lower than that observed for thiolated probes without any surface density regulations. These probes can be used for quantitative intracellular mRNA detection and imaging within 2 h, indicating their potential application in rapid and sensitive intracellular target imaging.

Li and Xu et al. [44] proposed a smart DNAzyme nanodevice allowing for control of its activity in living cells and in situ simultaneous visualization of metal ions (Zn^{2+} and Pb^{2+}). The proposed nanodevice was composed of 18 nm AuNPs decorated with acid-switchable DNA (SW-DNA) and DNAzyme precursors (DPs), which can precisely respond to pH changes in the 4.5–7.0 range. Before being transported into cells, the three-strand hybridization of DPs kept the DNAzymes inactive. Once the nanodevice entered living cells, the SW-DNA changed its topology from linear to triplex in the acidic intracellular compartments (lysosomes, pH~4.5–5.0). Consequently, the strands hybridized with SW-DNA were liberated and reacted with DPs to form the active DNAzyme, which allowed for the multi-imaging of intracellular metal ions. This platform has the potential to be used as a promising method for realizing different acid-switchable nanotools for visual analysis of multiple biological targets in living cells.

2.2. Hybrid Gold-Inorganic ODN-Conjugated Nanoparticles for Fluorescent Bio-Imaging

Along with AuNPs, hybrid materials like gold-coated silica, gold-coated iron oxide, and gold-coated mesoporous silica nanoparticles were also successfully used as platforms for fluorescent bio-imaging. Kuang et al. [45] reported a conceptually new method for sensitive and reliable detecting and measuring of microRNA (miRNA) using a complex made of a zirconium metal-organic framework (ZrMOF) and gold clusters functionalized with two fluorescent dyes: Quasar and Cyanine5.5. This complex was able to detect the intracellular miRNA target by measuring changes in fluorescence properties. When the miRNA-21, which is overexpressed in cancer cells, was present, the fluorescence of Cyanine5.5 decreased while the fluorescence of Quasar increased. This change in the fluorescence ratio allowed for the detection of miRNA-21 in the range of 0.006 to 67.988 amol/ngRNA with a LOD of 4.51 zmol/ngRNA in living cells. The proposed approach offers new opportunities for the quantification of miRNAs for concomitant diagnoses and treatments of early-stage cancers [46,47]. Ma et al. [48] have recently reported a new nucleic acid multicolor fluorescent probe using silica-coated symmetric gold nanostars (S-AuNSs@SiO$_2$) for highly sensitive, in situ real-time imaging of P53 mRNA, Bax mRNA, and cytochrome c (Cyt c) during T-2 toxin-induced apoptosis. The probe design included first the attachment of carboxyl group-modified nucleic acid chains to the surface of S-AuNSs@SiO$_2$ using an amide-forming reaction. The complementary chains of the targeted mRNA and the aptamer of targeted Cyt c were then modified with different fluorophores and successfully hybridized on the S-AuNSs@SiO$_2$ surface. In the presence of the targets, the fluorescent chains selectively bound to the targets, resulting in revived fluorescence. The probes based on S-AuNSs exhibited excellent performance due to the presence of 20 symmetric "hot spots". Additionally, the amide-bonded probe showed exceptional anti-interference capability against biological agents such as nucleases and biothiols. During real-time fluorescence imaging of T-2 toxin-induced apoptosis, sequential fluorescence signals of P53 mRNA, Bax mRNA, and Cyt c were observed. The reported S-AuNSs@SiO$_2$ probe not only provides a new tool for monitoring apoptosis pathway cascades in real time but is also of potential utility for disease diagnosis and pharmaceutical medicine.

2.3. Gold-Free ODN-Conjugated Nanoparticles

Aiming at an early diagnosis and consequent treatment of cancer, telomerase activity detection, and in situ imaging are essential steps to be accomplished. Ma and Huang et al. [49] proposed a new quantum dot-based nano-beacon (QD-BHQ2) for sensitive and visible detection of telomerase activity both in vitro and in living cells. The CdTe:Zn^{2+} QDs were functionalized with a hairpin BHQ2-DNA, and the fluorescence signals were subsequently restored via the hybridization of the hairpin DNA with products of the telomerase reaction. This was achieved by incubating the telomerase primer (TS primer), the deoxynucleotide triphosphates, and the telomerase extract. This simple method for detecting the telomerase activity possessed a linear range from 10 to 600 HeLa cells and a detection limit of 1 cell. The tested QD-BHQ2 presented several advantages, such as good biocompatibility, photo-stability, and specificity properties, and is envisaged to function as a powerful platform for nucleic acid detection and cell imaging. The QD strategy was also investigated by Zheng and Wang et al. [50], who have developed multifunctional carboxymethyl cellulose (CMC)-based nanohydrogels using near-infrared DNA-templated CdTeSe quantum dots (DNA-CdTeSe QDs) as building blocks. These nanocarriers can address the challenges of precise treatment and serious side effects in cancer theranostics by actively targeting tumors, tracking fluorescence, releasing drugs in a controlled manner, and regulating genes. The nanohydrogels were formed by crosslinking single-stranded DNA containing miRNA of complementary sequences and cysteine with antinucleolin aptamer DNA (AS1411)-modified CMC and DNA-CdTeSe QD-modified CMC chains. The hydrogels, which successfully incorporated doxorubicin (DOX) as a model anticancer drug, showed an average diameter of 150 nm, allowing them to be used for tumor targeting and for DOX releasing by activating both glutathione (GSH) and miRNA in the tumor microenvironment. The CdTeSe QDs trapped in the nanohydrogels acted as fluorophores for bio-imaging during the diagnosis and treatment process. This multifunctional delivery system provided a promising nanosystem for tumor imaging and precise therapy, effectively reducing side effects and improving treatment in the clinical therapy of tumors. Zhao et al. [51] investigated a convenient method for detecting cellular miRNA in living cells using a fluorescent amplification strategy. The approach was based on catalytic hairpin assembly (CHA), which indirectly attaches through covalent bonding to Fe_3O_4@C nanoparticles via short single-stranded DNA. This strategy integrates the highly quenching efficiency of Fe_3O_4@C nanoparticles with low background, non-enzyme target-active releasing for signal amplification, and a ssDNA-assisted fluorescent group-fueled chain releasing from Fe_3O_4@C nanoparticles with increased fluorescence response. The platform is not only highly sensitive, with a 0.450–190 pM concentration range, but also significantly specific, detecting miRNA-20a with the ability to distinguish a single mismatched base. Remarkably, the CHA-Fe_3O_4@C strategy was also successfully used for imaging visualization of miRNA-20a in living cells. Huo and Ding et al. [52] introduced a new probe design strategy called nano-amplicon comparator (NAC) and demonstrated its applicability for intracellular miRNA imaging. The NAC design combined the robustness of spherical nucleic acids, the sensitivity given by the CHA, and the consistency due to upconversion nanoparticles (UNP). NaYF4:Er/Gd/Yb@NaGdF4 luminescent core-shell nanoparticles with a size of 26 nm were selected as UNP and functionalized with specifically designed CHA sequences. The NAC probe responded to the target miRNA and generated a complex of UNP-hairpin DNA/fluorophore as a quantitative image for UNP-to-organic-fluorophore luminescent resonance energy transfer (LRET) imaging against a native UNP emission reference channel. By applying this strategy to miR-21, it was possible to monitor miRNA expression levels across different cell lines and under diverse external stimuli. Wang and Wei et al. [53] introduced a moderate biomineralization strategy to synthesize Y-shaped DNA@$Cu_3(PO_4)_2$ (Y-DNA@CuP) hybrid nanoflowers as DNA-inorganic hybrid nanomaterials for cell uptake. Y-DNA with a loop structure was utilized as a biomineralization template and also served as the recognition unit for the thymidine kinase 1 (TK1) mRNA. The Y-DNA probe was capable of detecting TK1 mRNA target sequences linearly in a range of concentrations from 2 to 150 nM with a low limit of detection, i.e., 0.56 nM. Interestingly, the presence of Y-DNA reduced the size of $Cu_3(PO_4)_2$ particles, making them suitable for use as gene nanocarriers within cells. It was shown that once inside the cells, the Y-DNA@CuP nanoflowers dissolved, releasing the cargo of Y-shaped DNA. Therefore, the intracellular TK1 mRNA hybridized with the loop region of the Y-DNA probe, which caused the dissociation of the Cy3-labeled loop strand and turned on the red fluorescence signal. Thus, the real-time imaging of intracellular TK1 mRNA enabled the assessment of tumor cells before and after therapeutic treatments, including the administration of β-estradiol and tamoxifen. In the line of gold-free approaches in bio-imaging, Xu and Tian et al. [54] reported a conceptually new strategy that used an AND logic gated-DNA nanodevice based on a nucleic acid probe and polymer-modified MnO_2 nanosheets to detect glutathione (GSH) and miRNA-21 signals in a tumor-responsive manner. This nanodevice can release significantly amplified fluorescence and magnetic resonance (MR) signals when it detects high levels of miRNA and GSH in tumor cells. While the fluorescence signal results were quenched, the MR signal remained at the background level in the case of the low levels of miRNA and what the GSH had in normal cells, which reduced false-positive signals by more than 50%. The nanodevice proved capable of effectively killing tumor cells under the guidance of miRNA profiling and MR imaging, and it is also endowed with glucose oxidase-like and catalase-like activities. The presented system was an innovative tumor-responsive theranostic DNA nanodevice that provided new insights into the design of smart theranostic strategies finalized to potentially relieve conditions like hypoxia and starvation in tumors.

3. Self-Assembled ODN-Nanostructure Conjugates for Fluorescent Bio-Imaging

The successful intracellular transportation of nucleic acids is crucial for innumerable biological processes, and from a theranostic perspective, it enables sensitive detection and gene control. Hence, DNA nanotechnology provides an excellent solution for accurately assembling nucleic acids, which makes it a valuable tool in the development of effective nucleic acid carriers for intracellular delivery. Wu and Jiang et al. [55] developed a new technique to efficiently deliver nucleic acids using a four-arm DNA nanostructure called a protein-scaffolded DNA tetrad. This chimeric structure was realized using streptavidin and four biotinylated hairpin DNA probes. The DNA tetrads were easy to prepare and allowed for precise control of the probe structure. These DNA nanosystems were found to be highly efficient in delivering DNA probes into cells and allowed their DNA cargoes to escape lysosome entrapment once internalized. In order to detect miRNA with high sensitivity and spatial resolution, the crosslinking hybridization chain reaction (cHCR) technique was employed, generating crosslinked hydrogel networks that specifically targeted miRNA. This was the first report of HCR amplification realized on nanostructures. The cHCR was designed using fluorescence resonance energy transfer (FRET) technology, which provided improved precision and allowed for dual-emission ratiometric imaging to avoid false signals. The DNA tetrad-based cHCR was found to be highly effective in ultrasensitive and accurate miRNA imaging in living cells, providing a potential tool for quantitative measurement of intracellular miRNA. This new technique was revealed to be a powerful strategy for nucleic acid delivery and low-level biomarker discovery. A similar strategy was utilized by the same authors [56] to develop a novel protein scaffold called recombinant fusion streptavidin, which combines DNA nanotetrads for highly efficient delivery of nucleic acids and imaging of telomerase activity in living cells using cHCR. The recombinant streptavidin protein, fused with multiple nuclear localization and nuclear export signals (NLS and NES, respectively), was obtained through *Escherichia coli* expression. The resulting NLS-carrying protein was easily connected with four biotinylated DNA probes, forming a well-defined DNA tetrad nanostructure through high-affinity non-covalent interactions. Similar to the previously mentioned nanosystems realized by the same group, these DNA nanotetrads were also effective in delivering nucleic acids within cells, which occurred through a caveolar-mediated endocytosis pathway, enabling them to avoid degradation in lysosomes. Additionally, the nanotetrads allowed for efficient cHCR assembly in response to telomerase, leading to highly sensitive detection and imaging of telomerase with a detection limit as low as 90 HeLa cells/mL both in vitro and in cellulo. The brightness of the fluorescence obtained during live-cell imaging was found to be dynamically correlated to both the telomerase activity and the inhibitor concentrations. Overall, this strategy provided a highly efficient method for delivering nucleic acids and imaging biomarkers in living cells. Kjems et al. [57] also took advantage of DNA nanotechnology and presented a nanoscaffold (~16 kDa) made entirely of nucleic acid building blocks, which was successfully tested by in vitro and in vivo experiments. The nanoscaffold was flexible and could be customized by attaching various biomolecules to its four modules and proved to be able to target liver cells with high efficiency. This nanosystem was self-assembled with chemically conjugated functionalities and offered complete control of stoichiometry and site specificity. The ODN-based nanoscaffold represented a flexible technology for theranostics because of its easy customization, small size, and high in sero and thermal stability. Its unique properties allowed for the possibility of rapid on-site combination of imaging agents or drugs based on larger libraries of functionalized oligonucleotide modules. Guo et al. [58] presented a new approach for examining transcriptomics and genomics in a spatial context. The proposed method makes use of fluorescence in situ hybridization (FISH) to detect each nucleic acid molecule as a fluorescent dot within the cell and employs cycles of hybridization, imaging, and photobleaching to identify the nucleic acids based on their unique color sequences (Figure 1). The authors demonstrated the effectiveness of this technique by accurately quantifying transcripts or genomic loci in single cells with either two fluorophores and 16 C-FISH cycles or three fluorophores and nine C-FISH cycles without any error correction. These findings suggest that the developed C-FISH method has the potential to accurately profile tens of thousands of different transcripts or genomic loci in individual cells within their natural environment.

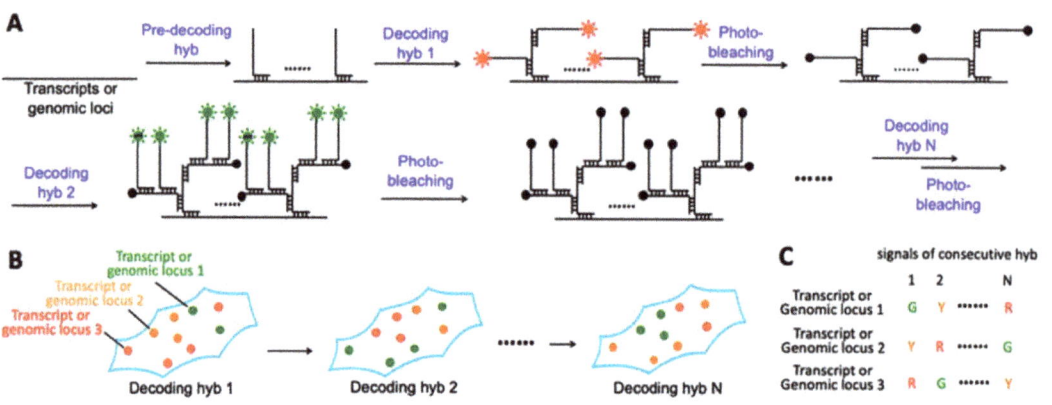

Figure 1. Figure taken from Xiao et al. 2020 [58]: C-FISH is a useful strategy for genomics analysis and spatial transcriptomics a. (**A**) Each C-FISH cycle includes 3 most important steps, consisting of probe hybridization, fluorescence imaging, and photobleaching. Excited fluorophores are symbolized by green and red sun-like symbols, while photobleached fluorophores are represented with black solid dots. (**B**) Each nucleic acid molecule can be visualized by a microscope as a fluorescent spot in each cycle of the analysis. (**C**) The various nucleic acid identities can be determined by the different corresponding color sequences, with their locations remaining throughout all the analysis cycles.

An interesting approach was presented by Yang and Xu et al. [59], who described a self-assembled DNA/RNA-based nanocarrier called nucleic acid nanosphere (NS), which used four specific oligonucleotides (denominated THp, H1, H2, H3) to detect miRNAs inside living cells. The NS system was stable, compatible with living cells, and responsive to RNase H, which allowed the sensing components to be delivered on demand. When the target miRNA was present, the structure used a process called a catalytic hairpin assembly and hybridization chain reaction to create a strong fluorescent signal. The researchers used miRNA 155 as an example and successfully detected it both in vitro and in tumor cells, demonstrating the potential of their NS nanosystems for miRNA detection in tumor screening strategies.

Taking advantage of the precision of DNA self-assembled nanostructures, Tian et al. [60] proposed the use of DNA nanotechnology in aid of the non-invasive imaging of brain tumors by developing a near-infrared II (NIR-II) emitting nanofluorophore able to cross the brain-blood barrier (BBB). The researchers synthesized a DNA block copolymer, PS-b-DNA, using a solid-phase "click" reaction and found that its self-assembled structure has exceptional cluster effects, particularly in BBB crossing. For the first time, PS-b-DNA was used as an amphiphilic matrix to create the NIR-II nanofluorophore, which was then applied in in vivo bio-imaging. The DNA-based nanofluorophore resulted in a 3.8-fold increase in the NIR-II fluorescence signal for glioblastoma imaging compared to its PEG-based counterpart. This increase in imaging resolution can greatly benefit the diagnosis and treatment of brain tumors.

The complexity of biological samples typically under analysis requires fluorescence probes for nucleic acid detection, both highly specific and sensitive, due to the very low concentrations of nucleic acids present in the samples. In this context, Tian et al. [61] investigated an innovative type of gold-free SNAPs, which used fluorescent π-conjugated polymers (FCPs) as a light-harvesting antenna to improve the performance of nucleic acid detection. Specifically, DNA-grafted FCPs were created and self-assembled into FCP-SNAP structures, with the size and light-harvesting capabilities of the SNAPs adjustable by suitably changing the hydrophobicity character of the copolymer. Larger FCP-SNAPs were revealed to be more efficient at signal amplification, resulting in up to 37-fold signal amplification and a detection limit of 1.7 pM in miRNA detection when the FCP-SNAP was optimized. The FCP-SNAP was then used for amplified in situ nucleic acid detection and imaging at the single-cell level.

4. ODN-Nanostructure Conjugates for Raman Bio-Imaging

The field of Raman imaging has recently gained a significant amount of attention, primarily due to two reasons. Firstly, the label-free Raman techniques have proven to be effective, and secondly, the distinctive characteristics of nanoparticles have made them ideal contrast agents and Raman reporters for detecting proteins and nucleic acid targets [62,63]. In addition, both the resonance Raman effect and SERS can greatly enhance the signal of the probe, making it detectable. A reliable SERS-based imaging or sensing method requires a sturdy SERS substrate that offers significant extinction cross-sections to improve SERS performance while also possessing strong chemical stability and biocompatibility for biomedical use. To detect SERS signals from Raman active molecules, different types of nanomaterials have been developed, such as gold nanoparticles, gold nanowires, gold nanostars, hollow gold nanoshells, and other types of nanoparticles [64,65].

Li et al. [66] proposed a new type of SERS nanosensor to measure the activity of endonuclease in both living cells and in vitro. The sensor was based on alloyed Au/Ag nanoparticles (Au/AgNPs) that were optimized for their plasmonic properties, chemical stability, and low toxicity. Aiming at detecting the endonuclease activity, the authors attached single-stranded DNA (ssDNA) molecules to the Au/AgNPs, with one type of ssDNA carrying a 3-(4-(phenylethynyl)benzylthio)propanoic acid (PEB, Figure 2a) moiety able to respond to endonuclease activity releasing from the particle surface, and another type of ssDNA carrying 4-thiol phenylacetylene (TPA, Figure 2b) as an internal standard.

When the endonuclease was present, it provoked the cleavage of the ssDNA, which led to a decrease in the SERS signal at 2215 cm^{-1} from PEB, while the SERS signal at 1983 cm^{-1} from TPA remained unchanged. By measuring the ratio of the two signals, with a detection limit of 0.056 U/mL, the activity of the endonuclease could be quantitatively determined. The nanosensor was found to be biocompatible and usable in living cells, which was successfully accomplished by testing the endonuclease activity both inside and outside of living cells.

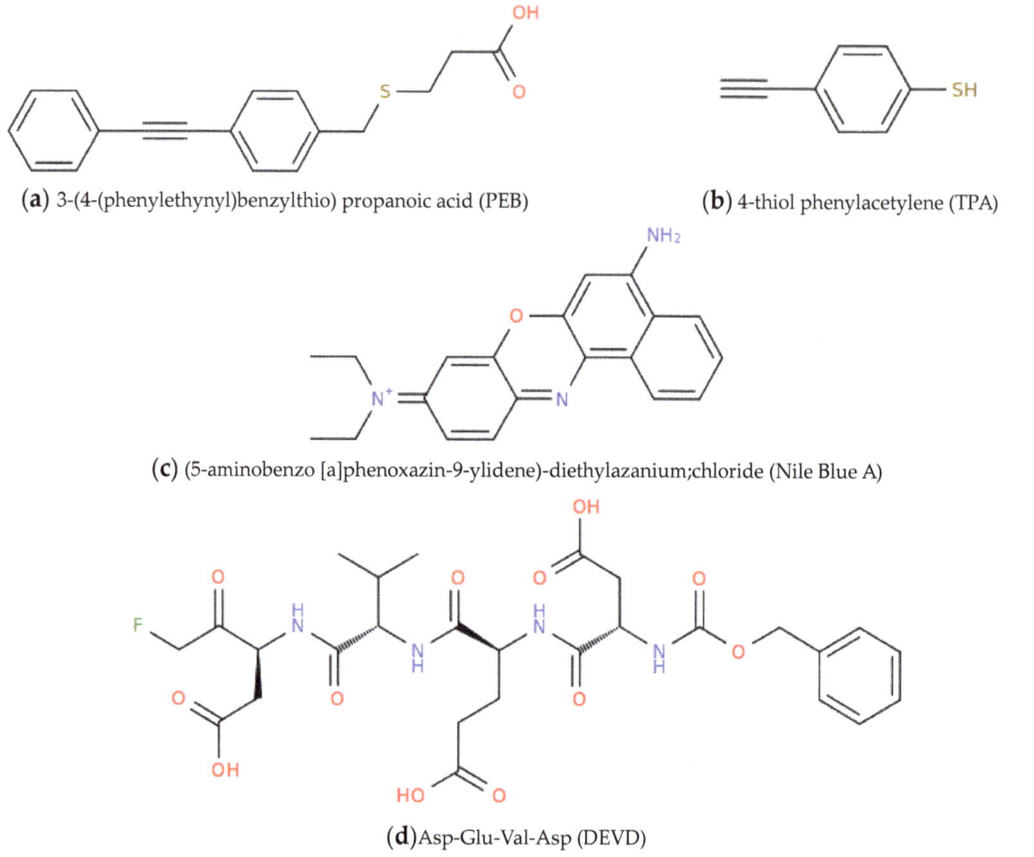

(a) 3-(4-(phenylethynyl)benzylthio) propanoic acid (PEB)

(b) 4-thiol phenylacetylene (TPA)

(c) (5-aminobenzo [a]phenoxazin-9-ylidene)-diethylazanium;chloride (Nile Blue A)

(d) Asp-Glu-Val-Asp (DEVD)

Figure 2. The endonuclease activity-responsive molecule PEB (**a**) and the internal standard TPA (**b**) anchored to the ssDNA molecules attached to Au/AgNPs by Li et al. [66] The Nile Blue A dye (**c**) used by Bi, Cao, et al. in SERS probes for caspase-3 detection [67]. The DEVD caspase-3 specific substrate (**d**).

Bearing in mind that the conventional methods for analyzing biomarkers related to cell differentiation require a large number of cells or cell lysates, resulting in the loss of cell sources and making real-time monitoring very difficult, Cao et al. [68] devoted their efforts to developing an ultrasensitive SERS method for detecting and imaging miR-144-3p in the osteogenic differentiation of bone marrow mesenchymal stem cells (BMSCs) using a gold nanocage (GNC)-hairpin DNA1 (hpDNA1)-hpDNA2-GNC assembly. The SERS strategy was designed using the finite-difference time-domain method to enhance electromagnetic intensities, and the hpDNA-conjugated GNC probes were prepared in order to recognize the target miRNA and distinguish it from the other nucleic acids. The method showed excellent sensitivity and selectivity characteristics toward miR-144-3p with a limit of detection of 13.6 aM and a broad range from 100 aM to 100 pM in cell lysates. The designed nanoprobes were not cytotoxic and were observed to only respond in BMSCs that underwent osteogenic differentiation, as well as in living undifferentiated bone marrow-derived mesenchymal stem/stromal cells, but not in osteoblasts. The accuracy of SERS was confirmed by a quantitative real-time polymerase chain reaction experiment. The proposed nanoprobes were capable of long-term tracking of the dynamic expression of miR-144-3p (up to 3 weeks) in differentiating BMSCs, demonstrating the potential of SERS in stem cell identification, differentiation, and isolation of specific cell types, as well as in biomedical diagnosis.

Accurate detection, imaging, and monitoring of intracellular caspase-3 levels are crucial for comprehending cell apoptosis and studying the progression of caspase-3-related cervical cancer. Bi and Cao et al. [67] developed a convenient SERS probe to detect caspase-3 during cervical cancer cell apoptosis. The probe consisted of gold nanoboxes modified with Nile Blue A (Figure 2c) as a Raman reporter and a caspase-3-specified peptide as a molecular cross-linker. When caspase-3 was present, the substrate peptide was cleaved, causing the Au nanoboxes-NBA-peptide to assemble into aggregates, resulting in a significant SERS signal enhancement. Finite-difference time-domain simulation showed that hot spots were mainly located in the nanogaps of the aggregated Au nanoboxes, proving the rationality of this signal amplification method. The SERS probes were highly reproducible and selective toward caspase-3, with a detection limit of 0.127 fM and a dynamic range from 1 fM to 1 nM. The probes showed no cytotoxicity within the explored concentration range. HeLa cells treated with doxorubicin to induce long-term apoptosis were monitored using SERS. The activity of caspase-3 increased with the prolongation of apoptosis time, and the SERS results were consistent with those of the western blotting assay. This kind of probe was envisaged to have significant potential for detecting enzymatic activities in cell physiological processes. Guo et al. [69] proposed a self-assembled tetrahedron probe for the simultaneous detection of telomerase and epithelial cell-adhesion molecule (EpCAM) in living cells using SERS. The probe was composed of a nucleic acid aptamer and encoded internal reference nanoparticles. When the target was present, the AuNPs, which were modified with corresponding tags, dissociated and resulted in decreased SERS signals. The ratio of Raman intensity at specific frequencies compared to the internal reference was used to quantify the levels of telomerase and EpCAM, which contributed to eliminating the background noise. The linear relationship between the ratio and the levels of TE and EpCAM was good and consistent with Raman confocal imaging. The LOD for TE and EpCAM was 7.6×10^{-16} IU and 0.53 pg/mL, respectively. The probe was also confirmed to be versatile and specific in the different cell lines. This study provided a highly sensitive and reliable approach for the in situ detection of biomarkers and a useful approach for SERS-based tetrahedron-based early diagnosis of cancer. The non-enzymatic isothermal amplification technique is a useful method for detecting miRNAs, but its limited sensitivity, slow speed, and low efficiency have hindered its practical use. To overcome these limitations, Bi and co-workers [70] reported a new method called DNA tetrahedron-mediated branched catalytic hairpin assembly (DTM-bCHA). This method involved the dynamic creation of hyperbranched DNA structures using DNA tetrahedrons and HDNA probes. Compared to traditional methods, the reaction time was 11.1 times faster due to an increased collision probability among the involved reactants. Additionally, the method allowed for ultrasensitive detection and in situ imaging of microRNA-21 in different living cell models by assembling Raman reporter DTNB-functionalized gold nanoparticles with hyperbranched DNAs. Furthermore, this method allowed for the construction of a series of two-input molecular logic gates, including INHIBIT, AND, OR, and NOR gates, in response to miRNA-21 and miRNA-155, which enabled the simultaneous analysis of multiple biomarkers. Overall, the DTM-bCHA-based isothermal amplification strategy disclosed new scenarios in complex DNA nanostructure development that can be applied to clinical diagnosis and bioanalysis.

Liu et al. [71] suggested a core-satellite (CS) nanostructure designed using miRNA-triggered catalytic hairpin assembly (CHA). The CS nanostructure was made up of plasmonic Au nanodumbbells as the core and AuNPs as the satellites. This plasmonic CS nanostructure was associated with an enhanced electromagnetic field compared to that of Au NPs-Au nanorods CS and that possessed by AuNPs alone. By using an "off-to-on" SERS strategy, the CS nanostructure allowed the detection of miRNA targets in a wide linear range (10^{-19}–10^{-9} M) with a LOD as low as 0.85 aM in vitro. Furthermore, the same CS nanostructure could accurately and sensitively detect miRNAs in different cell lines with varying levels of miRNA expression. This proposed SERS platform was effective in improving the sensitivity of SERS by engineering metallic nanoparticle aggregates with strong electromagnetic fields, and potential applications in the precise and quantitative detection of significant intracellular molecules were proposed for this nanosystem.

Urinary exosomal miRNAs were proposed as biomarkers for early-stage diagnosis and prognosis prediction of prostate cancer (PC) due to their characteristics of inherent stability, non-invasiveness, and representation of cell status. However, accurately detecting these miRNAs in urine is challenging because of their low abundance and high sequence homology with other RNAs. To address this issue, Sim and co-workers [72] developed a new sensing platform using SERS to detect urinary exosomal miRNAs. This platform was based on a three-dimensional (3D) hierarchical plasmonic nano-architecture, which created multiple plasmonic hot spots by self-assembling head-flocked gold nanopillars and target-complementary DNA probes-conjugated gold nanoparticles in the presence of the miRNA target. This 3D SERS biosensor achieved a detection limit of ~10 aM for miR-10a and miR-21, which corresponds to an over 1000-fold higher sensitivity with respect to previously reported miRNA sensors, without requiring any labeling or pre-treatment steps. Clinical validation was also performed using urinary samples, which showed that the 3D SERS sensor could accurately discriminate PC-affected patients from healthy individuals with a diagnostic accuracy of 0.93 based on the differential expression level of the urinary exosomal miRNAs. Overall, this SERS sensor based on 3D hierarchical nano-architectures proved to be a fast, accurate, and easy strategy applicable for measuring miRNA expression and could aid in the diagnosis of various diseases.

Zeng et al. [73] developed a novel readout technique for biomedical analysis called 'Click' SERS. This technique, based on Raman scattered light splices from nanoparticle assemblies, made use of triple bond-containing reporters to create a single, narrow emission (1–2 nm). The resulting output was dynamic and could be controlled by splicing together SERS-active nanoparticles in a way similar to click chemistry. Unlike conventional readout protocols, which rely on a single code related to a single target, the 'Click' SERS method was based on the number of combinatorial emissions and, thus, results were very intuitive, predictable, and uniquely identifiable. With this technique, 10 different biomarkers were detected simultaneously in a single scan, and accurate cellular imaging was achieved with double exposure. By using the 'Click' SERS method, it was demonstrated that multiple single-band Raman scattering could be used as an authentic optical analysis method in biomedicine. Although detecting biomolecules homogeneously has been important in clinical assays, it is challenging to achieve their precise in situ imaging. Additionally, problems such as nonspecific adsorption between probes and biomolecules and low sensitivity remain still widely unsolved. Another 'Click' SERS strategy was developed by Shen et al. [74] to overcome these challenges and consequently enable highly selective homogeneous detection of biomolecules through simultaneously enhancing dual SERS emissions. The above-cited detection of caspase-3 is an example of how this strategy can be used for the highly selective detection of biomolecular targets and their precise intracellular imaging during cell apoptosis. The strategy of Shen et al. involved modifying polyA-DNA and the Asp-Glu-Val-Asp (DEVD)-containing peptide sequence into alkyne and nitrile-coded Au nanoparticles (NPs). In general, during cell apoptosis, caspase-3 is generated, leading to the cleavage of the tetra-peptide sequence DEVD (Figure 2d) and removing the negatively charged protective moiety from the peptide on Au NPs. This process allows two different triple bond-labeled AuNPs to be connected through DNA hybridization to form a SERS 'hotspot', which leads to triple-bond Raman signals that are simultaneously enlarged. The SERS intensity is positively related to the concentration of caspase-3, which possesses a wide linear range (from 0.1 ng/mL to 10 μg/mL) and a low detection limit (7.18×10^{-2} ng/mL). This 'Click' SERS strategy was proposed for the precise imaging of the caspase-3 enzyme and could provide a logical judgment that can be mutually confirmed on the basis of two spliced SERS emissions, thanks mainly to their relative intensity.

Famously, SERS became a popular technique in biological applications due to its high sensitivity and ability to penetrate deep tissues. Typically, SERS nanoprobes with fluorophore attachments have Raman signals within the 1400–1700 cm^{-1} range. Conscious of the noteworthy advantages offered by the technique, Tang et al. [75] proposed a new series of SERS nanoprobes that were anchored to alkyne moieties in the biologically Raman-silent region via Au-C bond formation. To achieve target-specific recognition, two nucleic acid aptamers (namely MUC1 and AS1411) and two control oligonucleotides (T-con and C-con) carrying the same alkyne moiety were also attached to the Au surface. Both aptamer-bearing SERS nanoprobes successfully targeted MCF-7 cancer cells and were able to cross-check the target cells, potentially overcoming false-positive issues. The LOD was as low as five cancer cells, indicating a great potential for detecting circulating tumor cells. Subsequently, in vivo studies revealed that both SERS nanoprobes were successful in tumor targeting in living mice after tail intravenous injection, with a distinct signal ($\approx$2205 cm^{-1}) observed in the Raman-silent region for the first time.

SERS probes using ODNs were also used in the research on the SARS-CoV-2 virus, the causative agent of Coronavirus Disease 19 (COVID-19) [76–80], whose evolution has resulted in the emergence of various mutations that affect the virus' characteristics, such as its ability to spread and its antigenicity, which may be in response to changes in the human immune system. These mutations could potentially affect the effectiveness of treatments and diagnostic tests. Gartia and Pan et al. [81] developed a set of DNA probes (antisense oligonucleotides or ASOs) that can target a specific segment of the SARS-CoV-2 nucleocapsid phosphoprotein (or simply N protein) gene with high binding efficiency. This segment is known to not mutate among known variants. The complementary ASOs were shown to remain effective even in the presence of a hypothetical single-point mutation at the target RNA site, and their effectiveness was only slightly diminished in the case of hypothetical double or triple-point mutations. Interestingly, the mechanism of interaction between the ASOs and SARS-CoV-2 RNA has been explored in silico, using machine learning techniques, and experimentally by SERS. The study demonstrated that the N gene-targeting ASOs could efficiently detect all current SARS-CoV-2 variants regardless of their mutations, with high sensitivity and specificity up to a concentration of 63 copies/mL of SARS-CoV-2 RNA.

Theranostics is a clever combination of therapy and diagnosis, but traditional tools have had serious drawbacks, such as side effects and poor selectivity and sensitivity. To address these issues, Liu, Zheng, and Tian [82] developed a new multifunctional theranostic platform called CuPc@HG@BN, composed of hexagonal boron nitride nanosheets, conjugated DNA oligonucleotides, and copper (II) phthalocyanine. The CuPc molecule played a dual role in photodynamic therapy and in situ monitoring and imaging of miR-21 through SERS. By designing miRNA circle amplification and using the high SERS effects of copper (II) phthalocyanine on hexagonal boron nitride nanosheets, a miRNA-21 responsive concentration as low as 0.7 fM was achieved in live cells. In vitro and in vivo data showed that the integrated platform significantly enhanced photodynamic therapy efficiency with minimized damage to healthy tissues. The developed probe was successfully applied for early monitoring and guiding cancer therapy, which led to malignancy elimination.

It is highly desirable to create new nanomaterials with strong and distinctive Raman vibrations in the biological Raman-silent region (1800–2800 cm^{-1}) for Raman hyperspectral detection and imaging in living cells and organisms. To this aim, Jin et al. [83] tested polymeric nanoparticles as Raman active nanomaterials (denominated Raman beads) usable for bio-imaging purposes. The monomeric units of these Raman beads contained alkyne, azide, cyanide, and carbon-deuterate, which provoked intense Raman signals without any need for incorporating metals such as Au or Ag as Raman enhancers. The above-mentioned researchers developed a library of Raman beads with distinct Raman frequencies by substituting the endcaps of the monomers. These Raman beads were used for frequency multiplexing and demonstrated 5-color stimulated Raman scattering imaging of mixed nanoparticles. Targetable Raman beads were successfully used as probes for cancer targeting and Raman spectroscopic detection, including multi-color stimulated Raman scattering imaging in living tumor cells and tissues with high specificity by further surface functionalization with targeting moieties such as ODN aptamers and targeting peptides. In vivo studies showed that Raman beads anchored onto targeting moieties were successfully employed to target tumors in animal models, and Raman spectral detection of the tumor was accomplished only through spontaneous Raman signal at the biological Raman-silent region. This did not involve any signal enhancement due to the high density observed for the Raman reporters present in the Raman beads. Super-multiplex barcoding of Raman beads could be promptly achieved through further copolymerization of these monomers.

5. ODN-Nanostructure Conjugates for Dual Fluorescence–Raman Bio-Imaging and Detection

From the above discussion, it appears clear that detecting and imaging specific biological targets, including miRNAs, is of crucial importance for analyzing cancer cells. Establishing accurate and sensitive analytical assays for realizing it in single living cells remains challenging, especially due to the complex intracellular environment and the similarity of miRNA sequences with respect to RNA tracts of no anticancer relevance. To address these issues, Ye et al. [84] designed a dual-signal twinkling probe (DSTP) with a triplex-stem structure. This probe employed a fluorescence-SERS signal reciprocal switch to monitor the spatiotemporal dynamics of the intracellular uptake of the probe. The absolute value coupling of reciprocal signals was employed to achieve the real-time detection of miRNA targets using the SERS signals of the probe. This device effectively reduced background effects and showed favorable characteristics of sensitivity and reproducibility. The DSTP also demonstrated very high specificity and reversibility in the quantitative detection of intracellular miRNA. miRNA-203 was successfully monitored in MCF-7 cells, which was consistent with the results obtained in cell lysates as well as in vitro. This new dual-signal twinkling and dual-spectrum switch method was envisaged to be useful for detecting and imaging different types of biomolecular targets in living cells.

Even though simultaneous imaging, diagnosis, and therapy can be an effective strategy for cancer treatment, there are several challenges that must be overcome apart from those mentioned above, including a typically complex probe design and poor efficiency in anticancer drug release, as well as the well-known issue of multidrug resistance. Ye et al. [85] introduced a new one-two-three system that aimed at addressing these challenges and enhanced the imaging and detection performance of miRNA-21, providing an efficacious chemogene therapy. The term 'one-two-three' referred to the one type of miRNA-21-triggered endogenous substance accelerated cyclic reaction, two modes of signal switch, and three functions, including enhanced detection, imaging, and comprehensive treatment. The system consisted of dual-mode DNA robot nanoprobes that were assembled using two types of HDNAs and three-way branch DNAs that were modified on AuNPs, with doxorubicin being intercalated into GC base pairs of the DNA duplex moiety. Within the system, an intracellular cyclic reaction, accelerated by ATP, was triggered by miRNA-21, which led to SERS and fluorescence signals alternated with DNA structure switches. This nanosystem enabled precise SERS detection of miRNA and fluorescence imaging which facilitated on-demand release of two types of anticancer drugs, doxorubicin and anti-miRNA-21. The same system was endowed with several notable merits, such as the usage of ATP as an endogenous substance able to promote DNA structure switching and consequently accelerate the cyclic reaction. Additionally, the dual-mode signal switch is a more accurate and reliable treatment, providing more abundant information when compared to a single-mode treatment strategy. Overall, this nanosystem enabled not only the detection and imaging of intracellular miRNA targets but also an efficacious comprehensive therapy. In vivo studies were also conducted, and the results of these investigations confirmed the potential of the same system for the diagnosis and therapy of cancer.

Ye et al. [86] designed and constructed two smart binary star ratio probes (abbreviated as BSR) that connected in the presence of miRNA, leading to reciprocal changes in dual signals in living cells. This multifunctional probe integrated SERS imaging and fluorescence and provided enzyme-free numerator signal amplification for dual-signal quantitative analysis and dual-mode imaging of miRNA targets. Compared to single-mode ratio imaging methods, using fluorescence-SERS complementary ratio imaging enabled more accurate imaging contrast for direct visualization of signal changes in living cells, providing multiscale information about the dynamic behavior of miRNA and of the probe. Additionally, by using an enzyme-free numerator signal amplification and SERS reverse signal ratio response, the authors achieved amplified signals and reduced black values for the quantification of miRNA. Usefully, the BSR probes demonstrated good stability in cells and successfully traced and quantified miR-203 from MCF-7 cancer cells. Therefore, the reported BSR probe was presented as a potential tool for reliable monitoring of biomolecule dynamics in living cells.

Zhang et al. [87] developed a smart nanodevice that integrates Au@Cu_{2-x}S@polydopamine nanoparticles (ACSPs) and fuel DNA-conjugated tetrahedral DNA nanostructures (fTDNs) to achieve both antitumor therapy and in situ monitoring of miRNAs in cancer cells. The ACSP nanoprobe was used as a high-efficiency detection substrate and offered excellent SERS enhancement and high fluorescence (FL) quenching performance, which made it particularly useful for the analysis. The SERS-FL dual-spectrum biosensor led to an ultralow background signal and excellent sensitivity characteristics, with detection limits of 0.11 pM and 4.95 aM by FL and SERS, respectively, by using the ACSPs and fTDN-assisted DNA walking nanomachines as the strategy for DNA amplification. A dual-signal ratio strategy was employed for rapid FL imaging and precise SERS quantitative detection of miRNA in cancer cells, leading to improved diagnostic accuracy. Additionally, the nanodevices worked as all-in-one nanoagents in multimodal collaborative tumor therapy due to their high reactive oxygen species generation ability and excellent photothermal conversion efficiency. Both in vivo and in vitro experiments showed that the ACSPs were safe and endowed with strong anticancer activity, suggesting that these nanodevices have the potential for theranostic applications.

The measurement of telomerase activity in its natural location is crucial for identifying cancer. In this frame, a new type of nanosensor able to function in two ways, i.e., by fluorescence and SERS, was suggested by Zhang, Yang, Qu et al. [88], which proved to be useful for monitoring the activity of the telomerase enzyme. The nanosensor was created using AuNPs that were Cy5-functionalized as dual-functional nanoprobes. The telomerase substrate primer was lengthened by telomerase, forming the telomere repeats, and the hairpin DNA loop was opened upon hybridization. This caused a shift in the distance between the signaling molecule Cy5 and Au NPs, resulting in a decrease in the SERS signal and a recovery of the fluorescence one. Both the fluorescence and SERS intensities were related to the telomerase activity and could be used for sensing. In this approach, using fluorescence combined with Raman imaging, changes in human cell telomerase activity could be analyzed in real-time, allowing for the detection of anti-cancer drugs.

Ma et al. [89] proposed the detection of Ca^{2+} in a sensitive and selective manner by a dual-mode nanoprobe that was able to switch between "turn on" fluorescence and "turn off" SERS. This nanoprobe made use of gold nanostars and DNAzyme as the recognition element, where the former was utilized to quench fluorescence and enhance Raman signals. When a Ca^{2+}-specific DNAzyme was formed between the substrate chain modified with Cy5 dye and the enzyme chain on the surface of AuNSs, the fluorescence signal was quenched while the detected SERS signal was strong. Upon the cleavage of the Cy5-labeled substrate chain with Ca^{2+}, a decreased SERS signal alongside a recovered fluorescence signal was observed. This technique was successfully used to monitor Ca^{2+} in the process of T-2 toxin-induced apoptosis. This report was particularly interesting as it reported an original example of DNAzyme used with simultaneous fluorescence-SERS imaging to detect Ca^{2+} in cells. Moreover, this nanoprobe combined the benefits of SERS and fluorescence and could be applied to various cell lines, with clear benefits in a better comprehension of the role of Ca^{2+} in cellular pathways.

Due to a particular theranostic interest in the detection and visualization of miRNA and in the reported sensitivity limits depending on the strategy used, we have summarized the main investigated literature reports cited in this review and presented the corresponding results in Table 1.

Table 1. Comparison of the analytical performances (LOD) of some of the main fluorescent and/or Raman methods reported in this work.

Detected miRNA	Technique	Strategy	LOD	Ref.
miR-21	SERS	Boron nitride nanosheets-conjugated DNA oligonucleotide-copper(II) phthalocyanine.	0.7 fM	[82]
miR-21	SERS	DNA tetrahedron-mediated branched catalytic hairpin assembly reaction on AuNPs.	0.33 fM	[70]
miR-21	SERS	Two types of hairpin DNA-three-way branch DNA on AuNP.	3.26 fM	[85]
miR-21	SERS	DNA hairpins on plasmonic copper-sulfide-polydopamine AuNP.	4.95 aM	[87]
miR-21	Fluorescence	DNA hairpins on plasmonic copper-sulfide-polydopamine AuNP.	0.11 pM	[87]
miR-21	Fluorescence	Protein scaffolded DNA tetrad.	6 pM	[55]
miR-21	Fluorescence	Spherical nucleic acids with fluorescent π-conjugated polymer.	1.7 pM	[61]
miR-21	Fluorescence	Fluorescent nucleic acid probe and polymer-modified MnO_2 nanosheets.	30 pM	[54]
miR-21	Fluorescence	Zirconium metal–organic frameworks @ gold architecture functionalized with fluorophore-labeled DNA.	4.51 zmol/ng_{RNA}	[45]
miR-21 miR-10a	SERS	Gold nanopillar with self-assembling DNA probe-conjugated AuNP.	~10 aM	[72]
miR-21-D	Fluorescence	Hairpin-conjugated core-shell $NaYF_4$:Er/Gd/Yb@$NaGdF_4$ nanoparticles.	1.02 nM	[52]
miR-20a	Fluorescence	Hairpin-conjugated Fe_3O_4@C nanoparticles.	0.491 pM	[51]
miR-155	Fluorescence	Self-assembled nucleic acid nanosphere.	0.031 fM	[59]
miR-144-3p	SERS	Gold nanocage functionalized with DNA hairpins.	13.6 aM	[68]
miR-203	Fluorescence-SERS	DNA hairpins on AuNP.	≤0.13 nM *	[86]
miR-203	SERS	Triple helix structures immobilized on AuNP'.	0.63 pM	[84]
miR-1246	SERS	Core-satellite nanostructure: DNA-modified gold nanodumbells and DNA-modified AuNPs.	0.85 aM	[71]

* 0.13 nM of miR-203 detected in MCF-10A cells.

6. Conclusions and Perspectives

In summary, here we discussed the recent advancements in ODN-based nanoparticles and nanostructures for fluorescent and Raman imaging, which show promise as effective tools for live-cell analysis. These structures have the ability to provide insight into processes occurring at various levels, including those had in organelles, cells, tissues, and whole organisms. This has led to significant progress in understanding the role of certain analytes in diseases and has opened up new possibilities for diagnosing illnesses. The technological implications of these studies could lead to the development of innovative diagnostic tools that can identify diseases like cancer based on intracellular markers or guide surgical procedures using fluorescent or Raman imaging, or both. The field has focused on developing more complex probe structures, especially over the past five years, creating a versatile toolbox for live-cell analysis, with each tool possessing its own strengths and limitations for specific studies.

Due to the importance of safety, efficiency, and control in biomedical applications, it would be ideal for the design and synthesis of bio-imaging systems to be as simple as possible. However, some of the currently reviewed systems are overly complicated. In addition, it is crucial for the probes to be reproducible and easily scalable, but their synthetic procedures can often be delicate and tedious. It is also important to study the behavior of both fluorescent and Raman probes in the human body, including biodistribution, potential degradation, clearance, and biocompatibility in vivo, to ensure their efficiency.

Complex bio-imaging systems can be used as theranostics to diagnose and monitor therapeutic processes. However, caution should be taken to separate imaging and therapy to prevent unnecessary damage to normal tissues. It is also important to investigate the cytotoxicity of fluorescence and Raman probes to make them practical for use in clinical applications. With efforts to scale up preparation, probes with unique properties may hold new opportunities in personalized medicine for individual patients. Overall, based on our review of the recently reported systems, we are confident that ODN self-assembled nanostructures, including those based on nucleopeptides [90–92], and mixed ODN-functionalized inorganic nanoparticle bio-imaging systems have the potential to provide an advanced platform for personalized therapy and clinical translation.

Author Contributions: All authors contributed to all the steps of this manuscript preparation. All authors have read and agreed to the published version of the manuscript.

Funding: This project has received funding from the H2020-MSCA-RISE-2019 under grant agreement No. 872331 (acronym: NoBiasFluors) funded by the European Union.

Institutional Review Board Statement: Not applicable.

Informed Consent Statement: Not applicable.

Data Availability Statement: No data available.

Conflicts of Interest: The authors declare no conflict of interest.

References

1. Zhang, Y.; Du, Y.; Zhuo, Y.; Qiu, L. Functional Nucleic Acid-Based Live-Cell Fluorescence Imaging. *Front. Chem.* **2020**, *8*, 598013. [CrossRef] [PubMed]
2. Okamoto, A. Next-Generation Fluorescent Nucleic Acids Probes for Microscopic Analysis of Intracellular Nucleic Acids. *Appl. Microsc.* **2019**, *49*, 14. [CrossRef]
3. Cruz Da Silva, E.; Foppolo, S.; Lhermitte, B.; Ingremeau, M.; Justiniano, H.; Klein, L.; Chenard, M.-P.; Vauchelles, R.; Abdallah, B.; Lehmann, M.; et al. Bioimaging Nucleic-Acid Aptamers with Different Specificities in Human Glioblastoma Tissues Highlights Tumoral Heterogeneity. *Pharmaceutics* **2022**, *14*, 1980. [CrossRef]
4. Ebrahimi, S.B.; Samanta, D.; Mirkin, C.A. DNA-Based Nanostructures for Live-Cell Analysis. *J. Am. Chem. Soc.* **2020**, *142*, 11343–11356. [CrossRef] [PubMed]
5. Su, M.; Mei, J.; Pan, S.; Xu, J.; Gu, T.; Li, Q.; Fan, X.; Li, Z. Chapter 7—Raman Spectroscopy to Study Biomolecules, Their Structure, and Dynamics. In *Advanced Spectroscopic Methods to Study Biomolecular Structure and Dynamics*; Saudagar, P., Tripathi, T., Eds.; Academic Press: Cambridge, MA, USA, 2023; pp. 173–210. ISBN 978-0-323-99127-8.
6. Samuel, A.Z.; Sugiyama, K.; Takeyama, H. Direct Intracellular Detection of Biomolecule Specific Bound-Water with Raman Spectroscopy. *Spectrochim. Acta Part A Mol. Biomol. Spectrosc.* **2023**, *285*, 121870. [CrossRef] [PubMed]
7. Mukhopadhyay, S.; Avni, A.; Joshi, A.; Walimbe, A.; Pattanashetty, S.G. A Deep Dive into Biomolecular Condensates Using Single-Droplet Surface-Enhanced Raman Spectroscopy. *Biophys. J.* **2023**, *122*, 60a. [CrossRef]

8. Wang, X.; Wang, Y.; He, Y.; Liu, L.; Wang, X.; Jiang, S.; Yang, N.; Shi, N.; Li, Y. A Versatile Technique for Indiscriminate Detection of Unlabeled Biomolecules via Double-Enhanced Raman Scattering. *Int. J. Biol. Macromol.* **2023**, *228*, 615–623. [CrossRef] [PubMed]
9. Wang, S.; Pan, R.; He, W.; Li, L.; Yang, Y.; Du, Z.; Luan, Z.; Zhang, X. In Situ Surface-Enhanced Raman Scattering Detection of Biomolecules in the Deep Ocean. *Appl. Surf. Sci.* **2023**, *620*, 156854. [CrossRef]
10. Qiu, C.; Zhang, W.; Zhou, Y.; Cui, H.; Xing, Y.; Yu, F.; Wang, R. Highly Sensitive Surface-Enhanced Raman Scattering (SERS) Imaging for Phenotypic Diagnosis and Therapeutic Evaluation of Breast Cancer. *Chem. Eng. J.* **2023**, *459*, 141502. [CrossRef]
11. Chen, M.; Solarska, R.; Li, M. Additional Important Considerations in Surface-Enhanced Raman Scattering Enhancement Factor Measurements. *J. Phys. Chem. C* **2023**, *127*, 2728–2734. [CrossRef]
12. Nicholson, T.A.; Sagmeister, M.; Wijesinghe, S.N.; Farah, H.; Hardy, R.S.; Jones, S.W. Oligonucleotide Therapeutics for Age-Related Musculoskeletal Disorders: Successes and Challenges. *Pharmaceutics* **2023**, *15*, 237. [CrossRef] [PubMed]
13. Pan, Y.; Zhou, X.; Yang, Z. Progress of Oligonucleotide Therapeutics Target to Rna. In *Nucleic Acids in Medicinal Chemistry and Chemical Biology*; John Wiley & Sons, Ltd.: Hoboken, NJ, USA, 2023; pp. 373–427. ISBN 978-1-119-69279-9.
14. Hall, J.; Hill, A. The MOE Modification of RNA: Origins and Widescale Impact on the Oligonucleotide Therapeutics Field. *Helv. Chim. Acta* **2023**, *106*, e202200169. [CrossRef]
15. Amulya, E.; Sikder, A.; Vambhurkar, G.; Shah, S.; Khatri, D.K.; Raghuvanshi, R.S.; Singh, S.B.; Srivastava, S. Nanomedicine Based Strategies for Oligonucleotide Traversion across the Blood–Brain Barrier. *J. Control. Release* **2023**, *354*, 554–571. [CrossRef] [PubMed]
16. Fàbrega, C.; Aviñó, A.; Navarro, N.; Jorge, A.F.; Grijalvo, S.; Eritja, R. Lipid and Peptide-Oligonucleotide Conjugates for Therapeutic Purposes: From Simple Hybrids to Complex Multifunctional Assemblies. *Pharmaceutics* **2023**, *15*, 320. [CrossRef]
17. Paul, D.; Miller, M.H.; Born, J.; Samaddar, S.; Ni, H.; Avila, H.; Krishnamurthy, V.R.; Thirunavukkarasu, K. The Promising Therapeutic Potential of Oligonucleotides for Pulmonary Fibrotic Diseases. *Expert Opin. Drug. Discov.* **2023**, *18*, 193–206. [CrossRef]
18. Goyenvalle, A.; Jimenez-Mallebrera, C.; van Roon, W.; Sewing, S.; Krieg, A.M.; Arechavala-Gomeza, V.; Andersson, P. Considerations in the Preclinical Assessment of the Safety of Antisense Oligonucleotides. *Nucleic Acid Ther.* **2023**, *33*, 1–16. [CrossRef] [PubMed]
19. Lee, Y.; Ha, J.; Kim, M.; Kang, S.; Kang, M.; Lee, M. Antisense-Oligonucleotide Co-Micelles with Tumor Targeting Peptides Elicit Therapeutic Effects by Inhibiting MicroRNA-21 in the Glioblastoma Animal Models. *J. Adv. Res.* **2023**; in press. [CrossRef]
20. Xiang, Y.; Lu, Y. Using Personal Glucose Meters and Functional DNA Sensors to Quantify a Variety of Analytical Targets. *Nat. Chem.* **2011**, *3*, 697–703. [CrossRef]
21. Huang, N.-H.; Li, R.-T.; Fan, C.; Wu, K.-Y.; Zhang, Z.; Chen, J.-X. Rapid Sequential Detection of Hg^{2+} and Biothiols by a Probe DNA-MOF Hybrid Sensory System. *J. Inorg. Biochem.* **2019**, *197*, 110690. [CrossRef]
22. Bai, Y.; Shu, T.; Su, L.; Zhang, X. Functional Nucleic Acid-Based Fluorescence Polarization/Anisotropy Biosensors for Detection of Biomarkers. *Anal. Bioanal. Chem.* **2020**, *412*, 6655–6665. [CrossRef]
23. Huang, G.; Su, C.; Wang, L.; Fei, Y.; Yang, J. The Application of Nucleic Acid Probe-Based Fluorescent Sensing and Imaging in Cancer Diagnosis and Therapy. *Front. Chem.* **2021**, *9*, 705458. [CrossRef] [PubMed]
24. Saha, K.; Agasti, S.S.; Kim, C.; Li, X.; Rotello, V.M. Gold Nanoparticles in Chemical and Biological Sensing. *Chem. Rev.* **2012**, *112*, 2739–2779. [CrossRef] [PubMed]
25. Zhou, W.; Gao, X.; Liu, D.; Chen, X. Gold Nanoparticles for In Vitro Diagnostics. *Chem. Rev.* **2015**, *115*, 10575–10636. [CrossRef] [PubMed]
26. Lane, L.A.; Qian, X.; Nie, S. SERS Nanoparticles in Medicine: From Label-Free Detection to Spectroscopic Tagging. *Chem. Rev.* **2015**, *115*, 10489–10529. [CrossRef] [PubMed]
27. Yin, B.; Ho, W.K.H.; Xia, X.; Chan, C.K.W.; Zhang, Q.; Ng, Y.M.; Lam, C.Y.K.; Cheung, J.C.W.; Wang, J.; Yang, M.; et al. A Multilayered Mesoporous Gold Nanoarchitecture for Ultraeffective Near-Infrared Light-Controlled Chemo/Photothermal Therapy for Cancer Guided by SERS Imaging. *Small* **2023**, *19*, 2206762. [CrossRef] [PubMed]
28. Kumar, A.; Das, N.; Rayavarapu, R.G. Role of Tunable Gold Nanostructures in Cancer Nanotheranostics: Implications on Synthesis, Toxicity, Clinical Applications and Their Associated Opportunities and Challenges. *J. Nanotheranostics* **2023**, *4*, 1–34. [CrossRef]
29. Li, W.; Zhou, T.; Sun, W.; Liu, M.; Wang, X.; Wang, F.; Zhang, G.; Zhang, Z. A Conjugated Aptamer and Oligonucleotides-Stabilized Gold Nanoclusters Nanoplatform for Targeted Fluorescent Imaging and Efficient Drug Delivery. *Colloids. Surf. A Physicochem. Eng. Asp.* **2023**, *657*, 130521. [CrossRef]
30. Sondhi, P.; Lingden, D.; Bhattarai, J.K.; Demchenko, A.V.; Stine, K.J. Applications of Nanoporous Gold in Therapy, Drug Delivery, and Diagnostics. *Metals* **2023**, *13*, 78. [CrossRef]
31. Cândido, M.; Vieira, P.; Campos, A.; Soares, C.; Raniero, L. Gold-Coated Superparamagnetic Iron Oxide Nanoparticles Functionalized to EGF and Ce6 Complexes for Breast Cancer Diagnoses and Therapy. *Pharmaceutics* **2022**, *15*, 100. [CrossRef]
32. Fujita, H.; Ohta, S.; Nakamura, N.; Somiya, M.; Horie, M. Progress of Endogenous and Exogenous Nanoparticles for Cancer Therapy and Diagnostics. *Genes* **2023**, *14*, 259. [CrossRef]
33. Mirkin, C.A.; Letsinger, R.L.; Mucic, R.C.; Storhoff, J.J. A DNA-Based Method for Rationally Assembling Nanoparticles into Macroscopic Materials. *Nature* **1996**, *382*, 607–609. [CrossRef] [PubMed]
34. Cutler, J.I.; Auyeung, E.; Mirkin, C.A. Spherical Nucleic Acids. *J. Am. Chem. Soc.* **2012**, *134*, 1376–1391. [CrossRef]

55. Liu, Z.; Zhao, J.; Zhang, R.; Han, G.; Zhang, C.; Liu, B.; Zhang, Z.; Han, M.-Y.; Gao, X. Cross-Platform Cancer Cell Identification Using Telomerase-Specific Spherical Nucleic Acids. *ACS Nano* **2018**, *12*, 3629–3637. [CrossRef]
56. Greider, C.W.; Blackburn, E.H. Identification of a Specific Telomere Terminal Transferase Activity in Tetrahymena Extracts. *Cell* **1985**, *43*, 405–413. [CrossRef]
57. Greider, C.W.; Blackburn, E.H. A Telomeric Sequence in the RNA of Tetrahymena Telomerase Required for Telomere Repeat Synthesis. *Nature* **1989**, *337*, 331–337. [CrossRef]
58. Pan, W.; Zhang, T.; Yang, H.; Diao, W.; Li, N.; Tang, B. Multiplexed Detection and Imaging of Intracellular MRNAs Using a Four-Color Nanoprobe. *Anal. Chem.* **2013**, *85*, 10581–10588. [CrossRef] [PubMed]
59. Yang, X.-J.; Cui, M.-R.; Li, X.-L.; Chen, H.-Y.; Xu, J.-J. A Self-Powered 3D DNA Walker with Programmability and Signal-Amplification for Illuminating MicroRNA in Living Cells. *Chem. Commun.* **2020**, *56*, 2135–2138. [CrossRef] [PubMed]
60. Liang, C.-P.; Ma, P.-Q.; Liu, H.; Guo, X.; Yin, B.-C.; Ye, B.-C. Rational Engineering of a Dynamic, Entropy-Driven DNA Nanomachine for Intracellular MicroRNA Imaging. *Angew. Chem. Int. Ed. Engl.* **2017**, *56*, 9077–9081. [CrossRef] [PubMed]
61. Gao, P.; Liu, B.; Pan, W.; Li, N.; Tang, B. A Spherical Nucleic Acid Probe Based on the Au-Se Bond. *Anal. Chem.* **2020**, *92*, 8459–8463. [CrossRef] [PubMed]
62. Li, S.; Jiang, Q.; Liu, Y.; Wang, W.; Yu, W.; Wang, F.; Liu, X. Precision Spherical Nucleic Acids Enable Sensitive FEN1 Imaging and Controllable Drug Delivery for Cancer-Specific Therapy. *Anal. Chem.* **2021**, *93*, 11275–11283. [CrossRef] [PubMed]
63. Zhu, D.; Zhao, D.; Huang, J.; Zhu, Y.; Chao, J.; Su, S.; Li, J.; Wang, L.; Shi, J.; Zuo, X.; et al. Poly-Adenine-Mediated Fluorescent Spherical Nucleic Acid Probes for Live-Cell Imaging of Endogenous Tumor-Related MRNA. *Nanomedicine* **2018**, *14*, 1797–1807. [CrossRef]
64. Cui, M.-R.; Li, X.-L.; Xu, J.-J.; Chen, H.-Y. Acid-Switchable DNAzyme Nanodevice for Imaging Multiple Metal Ions in Living Cells. *ACS Appl. Mater. Interfaces.* **2020**, *12*, 13005–13012. [CrossRef] [PubMed]
65. Liang, Z.; Hao, C.; Chen, C.; Ma, W.; Sun, M.; Xu, L.; Xu, C.; Kuang, H. Ratiometric FRET Encoded Hierarchical ZrMOF @ Au Cluster for Ultrasensitive Quantifying MicroRNA In Vivo. *Adv. Mater.* **2022**, *34*, 2107449. [CrossRef]
66. Qu, A.; Sun, M.; Kim, J.-Y.; Xu, L.; Hao, C.; Ma, W.; Wu, X.; Liu, X.; Kuang, H.; Kotov, N.A.; et al. Stimulation of Neural Stem Cell Differentiation by Circularly Polarized Light Transduced by Chiral Nanoassemblies. *Nat. Biomed. Eng.* **2021**, *5*, 103–113. [CrossRef] [PubMed]
67. Zhang, Y.; Chen, W.; Zhang, Y.; Zhang, X.; Liu, Y.; Ju, H. A Near-Infrared Photo-Switched MicroRNA Amplifier for Precise Photodynamic Therapy of Early-Stage Cancers. *Angew. Chem. Int. Ed. Engl.* **2020**, *59*, 21454–21459. [CrossRef] [PubMed]
68. Li, C.; Chen, P.; Ma, X.; Lin, X.; Xu, S.; Niazi, S.; Wang, Z. Real-Time in Situ Observation of P53-Mediated Cascade Activation of Apoptotic Pathways with Nucleic Acid Multicolor Fluorescent Probes Based on Symmetrical Gold Nanostars. *Nano. Res.* **2022**, 1–10. [CrossRef]
69. Ma, Y.; Mao, G.; Wu, G.; Fan, J.; He, Z.; Huang, W. A Novel Nano-Beacon Based on DNA Functionalized QDs for Intracellular Telomerase Activity Monitoring. *Sens. Actuators B Chem.* **2020**, *304*, 127385. [CrossRef]
70. Zheng, Y.; He, S.; Jin, P.; Gao, Y.; Di, Y.; Gao, L.; Wang, J. Construction of Multifunctional Carboxymethyl Cellulose Nanohydrogel Carriers Based on Near-Infrared DNA-Templated Quantum Dots for Tumor Theranostics. *RSC Adv.* **2022**, *12*, 31869–31877. [CrossRef] [PubMed]
71. Fan, Y.; Liu, Y.; Zhou, Q.; Du, H.; Zhao, X.; Ye, F.; Zhao, H. Catalytic Hairpin Assembly Indirectly Covalent on Fe_3O_4@C Nanoparticles with Signal Amplification for Intracellular Detection of MiRNA. *Talanta* **2021**, *223*, 121675. [CrossRef]
72. Huo, M.; Li, S.; Zhang, P.; Feng, Y.; Liu, Y.; Wu, N.; Ju, H.; Ding, L. Nanoamplicon Comparator for Live-Cell MicroRNA Imaging. *Anal. Chem.* **2019**, *91*, 3374–3381. [CrossRef] [PubMed]
73. Yu, X.; Hu, L.; He, H.; Zhang, F.; Wang, M.; Wei, W.; Xia, Z. Y-Shaped DNA-Mediated Hybrid Nanoflowers as Efficient Gene Carriers for Fluorescence Imaging of Tumor-Related MRNA in Living Cells. *Anal. Chim. Acta* **2019**, *1057*, 114–122. [CrossRef] [PubMed]
74. Yan, N.; Lin, L.; Xu, C.; Tian, H.; Chen, X. A GSH-Gated DNA Nanodevice for Tumor-Specific Signal Amplification of MicroRNA and MR Imaging–Guided Theranostics. *Small* **2019**, *15*, 1903016. [CrossRef] [PubMed]
75. Huang, D.-J.; Huang, Z.-M.; Xiao, H.-Y.; Wu, Z.-K.; Tang, L.-J.; Jiang, J.-H. Protein Scaffolded DNA Tetrads Enable Efficient Delivery and Ultrasensitive Imaging of MiRNA through Crosslinking Hybridization Chain Reaction. *Chem. Sci.* **2018**, *9*, 4892–4897. [CrossRef] [PubMed]
76. Huang, Z.-M.; Lin, M.-Y.; Zhang, C.-H.; Wu, Z.; Yu, R.-Q.; Jiang, J.-H. Recombinant Fusion Streptavidin as a Scaffold for DNA Nanotetrads for Nucleic Acid Delivery and Telomerase Activity Imaging in Living Cells. *Anal. Chem.* **2019**, *91*, 9361–9365. [CrossRef]
77. Andersen, V.L.; Vinther, M.; Kumar, R.; Ries, A.; Wengel, J.; Nielsen, J.S.; Kjems, J. A Self-Assembled, Modular Nucleic Acid-Based Nanoscaffold for Multivalent Theranostic Medicine. *Theranostics* **2019**, *9*, 2662–2677. [CrossRef] [PubMed]
78. Xiao, L.; Liao, R.; Guo, J. Highly Multiplexed Single-Cell In Situ RNA and DNA Analysis by Consecutive Hybridization. *Molecules* **2020**, *25*, 4900. [CrossRef]
79. Song, J.; Mou, H.-Z.; Li, X.-Q.; Liu, Y.; Yang, X.-J.; Chen, H.-Y.; Xu, J.-J. Self-Assembled DNA/RNA Nanospheres with Cascade Signal Amplification for Intracellular MicroRNA Imaging. *Sens. Actuators B Chem.* **2022**, *360*, 131644. [CrossRef]

60. Xiao, F.; Lin, L.; Chao, Z.; Shao, C.; Chen, Z.; Wei, Z.; Lu, J.; Huang, Y.; Li, L.; Liu, Q.; et al. Organic Spherical Nucleic Acids for the Transport of a NIR-II-Emitting Dye Across the Blood-Brain Barrier. *Angew. Chem. Int. Ed. Engl.* **2020**, *59*, 9702–9710. [CrossRef] [PubMed]
61. Xiao, F.; Fang, X.; Li, H.; Xue, H.; Wei, Z.; Zhang, W.; Zhu, Y.; Lin, L.; Zhao, Y.; Wu, C.; et al. Light-Harvesting Fluorescent Spherical Nucleic Acids Self-Assembled from a DNA-Grafted Conjugated Polymer for Amplified Detection of Nucleic Acids. *Angew. Chem. Int. Ed.* **2022**, *61*, e202115812. [CrossRef]
62. Abramczyk, H.; Brozek-Pluska, B. Raman Imaging in Biochemical and Biomedical Applications. Diagnosis and Treatment of Breast Cancer. *Chem. Rev.* **2013**, *113*, 5766–5781. [CrossRef] [PubMed]
63. Smith, B.R.; Gambhir, S.S. Nanomaterials for In Vivo Imaging. *Chem. Rev.* **2017**, *117*, 901–986. [CrossRef]
64. Sheikhzadeh, E.; Beni, V.; Zourob, M. Nanomaterial Application in Bio/Sensors for the Detection of Infectious Diseases. *Talanta* **2021**, *230*, 122026. [CrossRef] [PubMed]
65. Peng, T.; Li, X.; Li, K.; Nie, Z.; Tan, W. DNA-Modulated Plasmon Resonance: Methods and Optical Applications. *ACS Appl. Mater. Interfaces.* **2020**, *12*, 14741–14760. [CrossRef]
66. Si, Y.; Bai, Y.; Qin, X.; Li, J.; Zhong, W.; Xiao, Z.; Li, J.; Yin, Y. Alkyne-DNA-Functionalized Alloyed Au/Ag Nanospheres for Ratiometric Surface-Enhanced Raman Scattering Imaging Assay of Endonuclease Activity in Live Cells. *Anal. Chem.* **2018**, *90*, 3898–3905. [CrossRef] [PubMed]
67. Sun, Y.; Wang, Y.; Lu, W.; Liu, C.; Ge, S.; Zhou, X.; Bi, C.; Cao, X. A Novel Surface-Enhanced Raman Scattering Probe Based on Au Nanoboxes for Dynamic Monitoring of Caspase-3 during Cervical Cancer Cell Apoptosis. *J. Mater. Chem. B* **2021**, *9*, 381–391. [CrossRef] [PubMed]
68. Cao, X.; Wang, Z.; Bi, L.; Bi, C.; Du, Q. Gold Nanocage-Based Surface-Enhanced Raman Scattering Probes for Long-Term Monitoring of Intracellular MicroRNA during Bone Marrow Stem Cell Differentiation. *Nanoscale* **2020**, *12*, 1513–1527. [CrossRef] [PubMed]
69. Guo, X.; Wu, X.; Sun, M.; Xu, L.; Kuang, H.; Xu, C. Tetrahedron Probes for Ultrasensitive In Situ Detection of Telomerase and Surface Glycoprotein Activity in Living Cells. *Anal. Chem.* **2020**, *92*, 2310–2315. [CrossRef] [PubMed]
70. Yue, S.; Qiao, Z.; Wang, X.; Bi, S. Enzyme-Free Catalyzed Self-Assembly of Three-Dimensional Hyperbranched DNA Structures for in Situ SERS Imaging and Molecular Logic Operations. *Chem. Eng. J.* **2022**, *446*, 136838. [CrossRef]
71. Liu, C.; Chen, C.; Li, S.; Dong, H.; Dai, W.; Xu, T.; Liu, Y.; Yang, F.; Zhang, X. Target-Triggered Catalytic Hairpin Assembly-Induced Core-Satellite Nanostructures for High-Sensitive "Off-to-On" SERS Detection of Intracellular MicroRNA. *Anal. Chem.* **2018**, *90*, 10591–10599. [CrossRef] [PubMed]
72. Kim, W.; Lee, J.U.; Jeon, M.; Park, K.; Sim, S. Three-Dimensional Hierarchical Plasmonic Nano-Architecture Based Label-Free Surface-Enhanced Raman Spectroscopy Detection of Urinary Exosomal MiRNA for Clinical Diagnosis of Prostate Cancer. *Biosens. Bioelectron.* **2022**, *205*, 114116. [CrossRef] [PubMed]
73. Zeng, Y.; Ren, J.-Q.; Shen, A.-G.; Hu, J.-M. Splicing Nanoparticles-Based "Click" SERS Could Aid Multiplex Liquid Biopsy and Accurate Cellular Imaging. *J. Am. Chem. Soc.* **2018**, *140*, 10649–10652. [CrossRef]
74. Zhu, W.; Wang, C.-Y.; Hu, J.-M.; Shen, A.-G. Promoted "Click" SERS Detection for Precise Intracellular Imaging of Caspase-3. *Anal. Chem.* **2021**, *93*, 4876–4883. [CrossRef] [PubMed]
75. Liang, D.; Jin, Q.; Yan, N.; Feng, J.; Wang, J.; Tang, X. SERS Nanoprobes in Biologically Raman Silent Region for Tumor Cell Imaging and In Vivo Tumor Spectral Detection in Mice. *Adv. Biosyst.* **2018**, *2*, 1800100. [CrossRef]
76. Wolf, J.M.; Wolf, L.M.; Bello, G.L.; Maccari, J.G.; Nasi, L.A. Molecular Evolution of SARS-CoV-2 from December 2019 to August 2022. *J. Med. Virol.* **2023**, *95*, e28366. [CrossRef] [PubMed]
77. Roviello, V.; Gilhen-Baker, M.; Vicidomini, C.; Roviello, G.N. Forest-Bathing and Physical Activity as Weapons against COVID-19: A Review. *Env. Chem. Lett.* **2022**, *20*, 131–140. [CrossRef]
78. Borbone, N.; Piccialli, I.; Falanga, A.P.; Piccialli, V.; Roviello, G.N.; Oliviero, G. Nucleic Acids as Biotools at the Interface between Chemistry and Nanomedicine in the COVID-19 Era. *Int. J. Mol. Sci.* **2022**, *23*, 4359. [CrossRef] [PubMed]
79. Roviello, V.; Roviello, G.N. Less COVID-19 Deaths in Southern and Insular Italy Explained by Forest Bathing, Mediterranean Environment, and Antiviral Plant Volatile Organic Compounds. *Env. Chem. Lett.* **2022**, *20*, 7–17. [CrossRef]
80. Vicidomini, C.; Roviello, V.; Roviello, G.N. In Silico Investigation on the Interaction of Chiral Phytochemicals from Opuntia Ficus-Indica with SARS-CoV-2 Mpro. *Symmetry* **2021**, *13*, 1041. [CrossRef]
81. Moitra, P.; Chaichi, A.; Abid Hasan, S.M.; Dighe, K.; Alafeef, M.; Prasad, A.; Gartia, M.R.; Pan, D. Probing the Mutation Independent Interaction of DNA Probes with SARS-CoV-2 Variants through a Combination of Surface-Enhanced Raman Scattering and Machine Learning. *Biosens. Bioelectron.* **2022**, *208*, 114200. [CrossRef]
82. Liu, J.; Zheng, T.; Tian, Y. Functionalized H-BN Nanosheets as a Theranostic Platform for SERS Real-Time Monitoring of MicroRNA and Photodynamic Therapy. *Angew. Chem. Int. Ed.* **2019**, *58*, 7757–7761. [CrossRef]
83. Jin, Q.; Fan, X.; Chen, C.; Huang, L.; Wang, J.; Tang, X. Multicolor Raman Beads for Multiplexed Tumor Cell and Tissue Imaging and in Vivo Tumor Spectral Detection. *Anal. Chem.* **2019**, *91*, 3784–3789. [CrossRef]
84. Zhang, N.; Ye, S.; Wang, Z.; Li, R.; Wang, M. A Dual-Signal Twinkling Probe for Fluorescence-SERS Dual Spectrum Imaging and Detection of MiRNA in Single Living Cell via Absolute Value Coupling of Reciprocal Signals. *ACS Sens.* **2019**, *4*, 924–930. [CrossRef]

35. Liu, X.; Wang, X.; Ye, S.; Li, R.; Li, H. A One-Two-Three Multifunctional System for Enhanced Imaging and Detection of Intracellular MicroRNA and Chemogene Therapy. *ACS Appl. Mater. Interfaces.* **2021**, *13*, 27825–27835. [CrossRef]
36. Zhang, J.; Zhang, H.; Ye, S.; Wang, X.; Ma, L. Fluorescent-Raman Binary Star Ratio Probe for MicroRNA Detection and Imaging in Living Cells. *Anal. Chem.* **2021**, *93*, 1466–1471. [CrossRef] [PubMed]
37. He, P.; Han, W.; Bi, C.; Song, W.; Niu, S.; Zhou, H.; Zhang, X. Many Birds, One Stone: A Smart Nanodevice for Ratiometric Dual-Spectrum Assay of Intracellular MicroRNA and Multimodal Synergetic Cancer Therapy. *ACS Nano* **2021**, *15*, 6961–6976. [CrossRef] [PubMed]
38. Luo, S.; Ma, L.; Tian, F.; Gu, Y.; Li, J.; Zhang, P.; Yang, G.; Li, H.; Qu, L.-L. Fluorescence and Surface-Enhanced Raman Scattering Dual-Mode Nanoprobe for Monitoring Telomerase Activity in Living Cells. *Microchem. J.* **2022**, *175*, 107171. [CrossRef]
39. Li, C.; Chen, P.; Wang, Z.; Ma, X. A DNAzyme-Gold Nanostar Probe for SERS-Fluorescence Dual-Mode Detection and Imaging of Calcium Ions in Living Cells. *Sens. Actuators B Chem.* **2021**, *347*, 130596. [CrossRef]
40. Roviello, G.N.; Musumeci, D.; Pedone, C.; Bucci, E.M. Synthesis, characterization and hybridization studies of an alternate nucleo-ε/γ-peptide: Complexes formation with natural nucleic acids. *Amino. Acids* **2010**, *38*, 103–111. [CrossRef]
41. Roviello, G.N.; Musumeci, D.; Moccia, M.; Castiglione, M.; Sapio, R.; Valente, M.; Bucci, E.M.; Perretta, G.; Pedone, C. dab Pna: Design, Synthesis, And Dna Binding Studies. *Nucleosides Nucleotides Nucleic Acids* **2007**, *26*, 1307–1310. [CrossRef]
42. Scognamiglio, P.L.; Riccardi, C.; Palumbo, R.; Gale, T.F.; Musumeci, D.; Roviello, G.N. Self-assembly of thyminyl l-tryptophanamide (TrpT) building blocks for the potential development of drug delivery nanosystems. *J. Nanostruct. Chem.* **2023**, 1–19. [CrossRef]

Disclaimer/Publisher's Note: The statements, opinions and data contained in all publications are solely those of the individual author(s) and contributor(s) and not of MDPI and/or the editor(s). MDPI and/or the editor(s) disclaim responsibility for any injury to people or property resulting from any ideas, methods, instructions or products referred to in the content.

Review

Recent Advances in Copper-Based Organic Complexes and Nanoparticles for Tumor Theranostics

Sergey Tsymbal [1], Ge Li [2,3], Nikol Agadzhanian [1], Yuhao Sun [4], Jiazhennan Zhang [5], Marina Dukhinova [1], Viacheslav Fedorov [6] and Maxim Shevtsov [6,7,8,*]

1. International Institute of Solution Chemistry of Advanced Materials and Technologies, ITMO University, 197101 Saint Petersburg, Russia
2. Cancer Center & Department of Breast and Thyroid Surgery, Xiang'an Hospital of Xiamen University, School of Medicine, Xiamen University, 2000 Xiang'an Road East, Xiamen 361101, China
3. Xiamen Key Laboratory for Endocrine-Related Cancer Precision Medicine, Xiang'an Hospital of Xiamen University, Xiamen 361101, China
4. Guangxi University of Chinese Medicine, Nanning 530200, China
5. Day-Care Department, Xinjiang Medical University, Urumqi 830011, China
6. Laboratory of Biomedical Nanotechnologies, Institute of Cytology of the Russian Academy of Sciences, 194064 Saint Petersburg, Russia
7. Personalized Medicine Centre, Almazov National Medical Research Centre, 2 Akkuratova Str., 197341 Saint Petersburg, Russia
8. Department of Radiation Oncology, Klinikum Rechts der Isar, Technical University of Munich, 81675 Munich, Germany
* Correspondence: maxim.shevtsov@tum.de; Tel.: +49-1731488882

Abstract: Treatment of drug-resistant forms of cancer requires consideration of their hallmark features, such as abnormal cell death mechanisms or mutations in drug-responding molecular pathways. Malignant cells differ from their normal counterparts in numerous aspects, including copper metabolism. Intracellular copper levels are elevated in various cancer types, and this phenomenon could be employed for the development of novel oncotherapeutic approaches. Copper maintains the cell oxidation levels, regulates the protein activity and metabolism, and is involved in inflammation. Various copper-based compounds, such as nanoparticles or metal-based organic complexes, show specific activity against cancer cells according to preclinical studies. Herein, we summarize the major principles of copper metabolism in cancer cells and its potential in cancer theranostics.

Keywords: copper; organic complexes; nanoparticles; tumor theranostics

1. Introduction

Copper is a transition metal that plays several important roles crucial for maintenance of cell homeostasis, regulation of cell growth and proliferation, and iron metabolism [1]. Various roles of copper are explained by its ability to act as either a recipient or a donor of electrons depending on the oxidation state: Cu^{1+} (cuprous ion) and Cu^{2+} (cupric ion). The oxidation state also affects the copper interaction with organic compounds. Thus, Cu^{1+} preferentially binds to the thiol group in cysteine or the thioether group in methionine, while Cu^{2+} exhibits a high affinity for the secondary carboxyl group in aspartic/glutamic acid or the imidazole nitrogen group in histidine. As a result, copper ions readily form complexes with biomolecules containing these amino acid residues. Copper atoms are involved in a functioning of a wide spectrum of proteins, such as copper/zinc superoxide dismutase (Cu/Zn SOD or SOD1) [2], cytochrome c oxidase (COX) [3], lysyl oxidase (LOX) [4], mitogen-activated protein kinase MEK1 [5], and cAMP-degrading phosphodiesterase PDE3B [6]. In these proteins, copper ions participate in diverse biochemical reactions (especially redox reactions) of donating or accepting of electrons and maintain specific protein structures by coordinating with the abovementioned groups.

Despite its important physiological role, free copper ions are able to damage DNA and protein molecules via generation of reactive oxygen species (ROS) and interaction with cysteine and methionine residues [7]. That is why each cell and whole organisms have distinct mechanisms for the regulation of copper absorbance, distribution, accumulation, and excretion. With the development and propagation of copper-based pharmaceuticals, it is crucial to consider these metabolic and regulatory pathways to improve biocompatibility and efficacy of such compounds. For now, only a small number of studies dedicated to the design of novel copper-containing compounds consider underlying molecular mechanisms of intracellular copper regulation. The present work aims to provide a holistic view of the problem to help researchers boost their work and realize rational approaches in drug development.

2. Copper Intake, Distribution, and Efflux in Normal and Tumor Cells

The major proteins involved in copper maintenance include: CTR1 (copper transport protein), which is responsible for copper intake either from the intestine or blood; metallochaperones and metallothioneins, including ceruloplasmin, which are responsible for metal sequestration, distribution in organisms, and transport to various proteins; ATP7A and ATP7B (ATP-ase copper transporter alpha) responsible for copper excretion via membrane efflux or Golgi apparatus [8]. All these proteins have cysteine- or methionine-rich domains responsible for the binding. A precise description of proteins involved in copper homeostasis and a comparison of copper metabolism in normal and cancer cells are given below.

As it has previously been mentioned, copper intracellular metabolism is precisely regulated by specific protein machinery, which prevents the generation of free copper ions in the cytoplasm or extracellular space and ion-mediated toxicity (Figure 1). CTR1 is a major protein responsible for copper uptake in eukaryotes. CTR1 transporter acts as a pump that facilitates copper import without ATP consumption [9]. The rate of the copper intracellular transport depends on the copper concentration, the presence of other ions (Fe^{3+}, Zn^{2+}, Ag^+) and organic compounds (e.g., ascorbate), cell type, and pH. The structure of homotrimeric CTR1 protein contains methionine gates for selective bypass of monovalent copper ions exclusively. However, isoelectric silver ions can compete with copper decreasing its intracellular content [10]. As only monovalent copper can be transported by the CTR1 protein, bivalent copper should first be restored to the monovalent state. This process is facilitated by the reductase proteins, such as STEAP, which are also reported to be overexpressed in several types of cancers and involved in tumorigenesis [11].

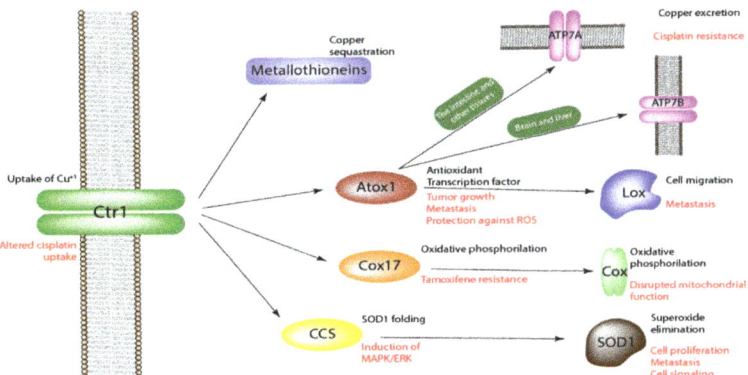

Figure 1. Proteins of copper metabolism. The arrows show how copper is metabolized. In general, CTR1 is responsible for copper intake; numerous metallothioneins and chaperones store the metal and deliver it to the active sites of functional proteins. The black font indicates the role of a protein in normal cells, and the red font indicates the protein function in cancer.

After crossing the plasma membrane, copper ions are readily sequestered by the numerous intracellular metallothioneins, metallochaperones, albumins, glutathione, and ceruloplasmin [12]. Some of these proteins can store the metal for further use, while others serve for intracellular transportation of copper. For example, metallochaperones transfer copper to the active centers of the certain proteins or buffer the metal for further use. Copper chaperon for superoxide dismutase (CCS) delivers copper to superoxide dismutase (SOD1) enzyme, which converts superoxide radical into hydrogen peroxide and oxygen [13,14]. COX17 is another metallochaperone responsible for copper transportation to COX, an important protein involved in oxidative phosphorylation [15].

Atox-1 is a transport protein that delivers copper to ATP7A and ATP7B, which are responsible for copper release into the blood or bile, respectively [16]. Additionally, ATOX1 can migrate to the cell nucleus and act as a transcription factor facilitating cell growth, proliferation, and migration. Another common localization of ATOX1 is in proximity to the plasma membrane, where copper can be transferred to the membrane-associated proteins, such as lysyl oxidase, which is involved in cell migration [16].

3. Copper Regulation in Cancer

Many proteins required for copper metabolism are known to be overexpressed or malfunctioned in cancer cell metabolism. The most known example is participation of these proteins in chemotherapeutic response to conventional drug cisplatin. There is much evidence about CTR1 involvement in the transportation of cisplatin [17,18]. Meta-analysis of gene expression in various cancer types revealed that the reduced expression of the CTR1 gene is associated with the development of cisplatin resistance [19]. The knockout of CTR1 and DMT1 (divalent metal transporter 1) in human H1299 non-small cell lung cancer cells leads to pronounced cisplatin resistance. Moreover, the CTR1 loss decreases expression of COMMD1, XIAP, and NF-κB, which have a distinct influence on the intracellular homeostasis and signaling [20]. Several works of various research groups also proved a hypothesis about involvement of CTR1 and ATOX1 in cisplatin transport and sequestration [21–23]. However, another study on HEK-293T cells provided evidence about the modest participation of copper-binding proteins (i.e., CTR1, CTR2, ATOX1, and CCS) in cisplatin uptake and distribution [24].

At the same time, a connection between high ATOX1 expression level and survival rate in primary tumor biopsies has been found. Analysis of transcription profiling of 1904 breast cancer patients on METABRIC data set suggests that overexpression of Atox1 may serve as a marker for breast cancer prognosis [25] but only in the hormone receptor-positive tumors. Considering copper involvement in the functioning of the LOX protein [26] which is responsible for cell migration, ATOX1 may facilitate the function of LOX enhancing tumor ability for metastasis [27]. Moreover, Atox1 is also involved in transcription regulation of several genes, as was mentioned earlier. First, upon copper binding ATOX1 can migrate to the cell nucleus and bind the cis element of the cyclin D1 promoter, thus stimulating cell growth and proliferation [21]. Furthermore, a more complex interplay between ATOX1 and p53 has been found [28]. Authors observed increased copper amounts in cell nuclei for HCT116 $p53^{+/+}$ cells compared to $p53^{-/-}$ cells. These facts suggest that Atox1 may play a significant role in cell signaling and regulation of gene expression which should be determined in future studies.

Cytochrome c oxidase copper chaperone (COX17) is also involved in cancer. Inhibiting COX17 in acute leukemia cells results in decreased adenosylhomocysteinase activity leading to disruption of DNA methylation and changes in cell epigenetics [29]. The link between COX17 and cisplatin distribution to mitochondria has been found [30]. The involvement of copper-binding proteins in cisplatin uptake and distribution is probably connected to the similarities in binding affinity of platinum and copper ions. Moreover, glutathione (GSH) seems to attenuate this effect. It was found that 90% of cisplatin bound to GSH is readily transferred to COX17 [31]. This suggests probable involvement of thiol-containing molecules and not only proteins in intracellular cisplatin distribution. It would be interesting to investigate the effects of combining treatment with cisplatin and thiols or cisplatin-thiol complexes or nanostructures. COX17 was also studied as a prognostic marker for prediction of tamoxifen resistance in breast cancer patients [32]. The authors reported that this protein could be employed as a predictive marker for tumor recurrence and metastasis. These features are also observed for COX5B which is a subunit of COX itself [33]. This correlates with the prognostic value of ATOX1 which was found to possess similar properties in the breast cancer. Another COX nuclear-encoded subunit, COX4, is also shown to be a valuable prognostic and therapeutic marker for medullary thyroid cancer treatment [34]. The role of the COX protein in cancer development and progression as well as its influence on altered signaling and metabolic pathways needs to be further explored.

CCS, a protein involved in copper delivery to SOD1, is also involved in tumorigenesis. SOD1 could serve as a prognostic marker which contributes to worsened prognosis and higher risk of gastric [35] and prostate [36] cancer. Another study indicates SOD1 involvement in cell proliferation and metastasis in non-small cell lung cancer [37]. At the same time, knockdown of CCS leads to decreased cell proliferation and migration of MDA-MB-231 cells but does not affect the MCF-7 cell line [38]. In addition, the MAPK/ERK pathway was inhibited upon loss of CCS activity in MDA-MB-231 cells which also correlated to the increased ROS formation. Inhibition of CCS and Atox1 with specifically designed small molecules is a promising treatment strategy with reduced side effects [39]. The expression of CCS was found to be decreased in human hepatocellular carcinoma (HCC) which is distinct from breast cancer [40]. Despite a statistical significance not being achieved, the study concluded that a low expression level of CCS is a negative prognostic marker for HCC patients. Presumably, copper trafficking in various tissues could be different, as well as the involvement of copper-binding proteins in cancer development, progression, and metastasis. This provides a foundation for further investigation on a wide panel of cancer cell lines.

Copper efflux proteins, ATP7A and ATP7B, are also involved in cancer progression. ATP7A correlates with a poor survival rate and is overexpressed in several tumor types, such as breast, lung, prostate, ovarian, and colon cancer [41]. Another study shows that ATP7A is associated with cisplatin resistance in ovarian cancer and influence effectiveness of treatment with tetrathiomolybdate, which inhibits ATP7A activity [42]. Decreased sequestration of platinum leads to its accumulation in the cell nucleus with subsequent DNA damage. Moreover, the application of tetrathiomolybdate can also result in Ctr1 high expression increasing cisplatin uptake that may be used as a solution for treatment of drug resistance tumors [43]. Another study suggests a greater impact from inhibiting ATP7B compared to ATP7A [44]. A detailed analysis of the ATP7A and ATP7B roles in ovarian cancer are discussed in the review [45]. A study in the breast cancer model reveals the opposite effects of ATP7A and ATP7B in contribution to the cisplatin resistance [46]. ATP7A seems to be more involved in this process, whereas the analysis of ATP7B did not reach statistical significance. To summarize, the above-mentioned ATP7A and ATP7B influence the cisplatin efflux leading to decreased effectiveness of this drug; however, the precise role of each protein should be determined for distinct types of cancer.

Copper takes an active part in the proangiogenic pathways via several mechanisms. First, copper stimulates endothelial cells proliferation and migration. Next, copper is involved in the expression of certain proangiogenic factors (for example, vascular endothelial growth factor VEGF) [47], particularly as a response to hypoxia-inducible factor (HIF-1) signaling [48]. When elevated, copper becomes toxic and may induce side effects leading to genetic disorders (e.g., Wilson's disease) and various types of oncological diseases. However, the exact molecular mechanisms underlying the connection between excessive copper levels and malignant cells are still unknown. It can only be hypothesized, particularly in the early stages, after considering the role copper plays in tumor angiogenesis. Malignant tissues have higher Cu accumulation levels, thus increasing the expression of human copper transporter (hCTR1). hCTR1 regulates the activation of cell-signaling pathways in embryogenesis, which leads to the development and progression of cancers [49].

The above-mentioned impact of copper ions and copper-binding proteins on cell growth, migration, and metabolism suggests that cancer cells require high copper levels to facilitate cell survival and disease progression. Indeed, tumor tissues are enriched with copper suggesting that this metal is one of the diagnostic tools for various oncological disorders [50]. Moreover, copper or copper-binding proteins are essential for the function of important signaling pathways, such as BRAF [51], NF-kB [52], MAPK [53], and EGFR/Src/VEGF [54]. Hence, the significant role of copper in cancer appearance and progression is starting to emerge in front of researchers. The accumulated data uncover the possibility to improve the efficiency of diagnostic approaches and increase treatment efficacy.

4. Therapeutic Effects of Copper-Based Compounds and Nanocarriers

The disparity in tumor cell and normal cell responses to copper have paved the way for copper complexes to evolve as anticancer agents. Copper-based compounds nowadays are receiving attention due to their target-specific therapeutic properties. Copper compounds influence the activities of several crucial cell organelles, such as the mitochondria and endoplasmic reticulum, leading to the loss of their functions and eventually resulting in cell death (Figure 2).

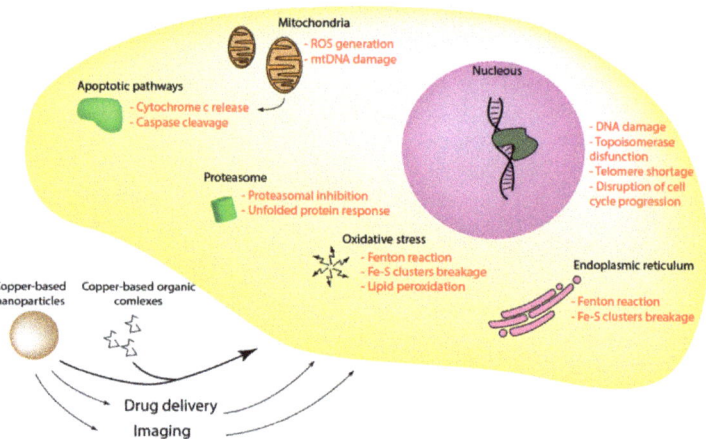

Figure 2. The main effect of copper nanoformulations on cell metabolism. The major impact of copper on cell metabolism is a result of ROS generation and DNA damage. Proteasome, endoplasmic reticulum, and mitochondria also suffer from copper excess.

Nowadays, the increasing number of metal-based compounds and nanoparticles are being investigated due to their promising potential in theranostics, and various iron, zinc, copper-based and other agents are under development and testing for these purposes. For example, superparamagnetic iron oxide nanoparticles (SPIONs) are being actively used as a contrast agent for MRI procedures and in therapy. Currently, there are several running translational studies which explore SPIONs' toxicity and biomedical applications, and ferumoxytol was FDA-approved for clinics [55]. Copper is also attracting the attention of researchers as a possible component for nanocompounds for theranostics and drug delivery. For example, copper is used in PET scanning as a radiotracer agent in cancer diagnostics, and $^{64}CuCl_2$ has successfully passed clinical studies demonstrating its diagnostic potential [56]. Several studies successfully implemented copper for efficient bone regeneration [57] and anti-inflammatory therapy [58]. Copper-based nanoparticles also found their place in chemodynamic [59] and photothermal therapy [60].

The radiotracer biodistribution has shown that the liver has the highest uptake, followed by the intestine and pancreas, with urinary excretion being insignificant. It is the first biodistribution and radiation dosimetry trial with healthy volunteers. The estimated absorbance and effective doses were higher than the ones from another report with participants suffering from prostate cancer. The measurement methodology and assumptions used in dose calculation as well as the difference between the biodistribution in cancer patients and healthy volunteers are the main reasons for that disparity [61]. An interesting combination of SPIONs and Cu (II) ions were used as a cell labeling MRI/PET agent. Contrast agents showed good cellular uptake and cell-labeling ability [62]. Furthermore, gold nanoparticles alloyed with copper-64 demonstrate higher sensitivity and stability compared to non-modified gold nanoparticles [63]. Thus, copper presence could improve the effectiveness of the iron or gold nanoparticles, which opened new opportunities for further research in the field of cancer imaging. However, the major limitation and risk factor for wide implication of copper is toxicity of copper ions for cells [64].

Extrinsic and mitochondrial pathways of apoptosis are important in the control of tumor development and could be exploited for therapy [65]. The anticancer properties of Schiff base copper (II) complexes are well-studied and known in the scientific community. For instance, [Cu(sal-5-met-L-glu)(H_2O)]·H_2O, [Cu(ethanol)2(imidazole)4][Cu2(sal-D, L-glu)2(imidazole)2] and [Cu2(sal-D,L-glu)(2-methylimidazole)] complexes activate the intrinsic pathway, while [Cu2(sal-D, L-glu)2(isoquinoline)2]·2C_2H_5OH initiates the extrinsic pathway in human HT-29 colon carcinoma cells, respectively. All these complexes also induce a cytotoxic effect on the HT-20 cell line, and as a result, prove that they might become potential anticancer agents [66]. Structural formulas of the complexes can be found in recent publications [67–69]. Another study shows that accumulation of copper ions inside the cells leads to oxidative stress and apoptosis [70]. Moreover, the usage of 2,2′-dithiodipyridine strongly enhances this effect which is bound to its ability to transport copper through the plasma membrane.

Topoisomerases play an essential role in DNA replication and are relevant in cancer research as a target for novel therapies. There are currently several drugs approved by the FDA targeting topoisomerases (e.g., irinotecan, etoposide, etc.). Thiosemicarbazones are a group of complexes proved to have anticancer activity. "Triapine" (thiosemicarbazone) has been successfully tested for uterine cervix and vaginal cancers in clinical trials phase I and II and is presently under clinical trials phase III [71]. Thiosemicarbazones copper (II) complex [Cu(PyCT4BrPh)Cl] was investigated and demonstrated a cytotoxic effect on a leukemia cell line (THP-1) and human breast cancer cell line (MCF-7). It had stronger topoisomerase inhibitor activity and generally more impact on these cell lines than its analogue without copper, which proves how transition metals can increase the effectiveness of the known compound [72].

Copper complexes are shown to influence the endoplasmic reticulum leading to immunogenic cell death in breast cancer stem cells [73–75]. In a recent study, cuprous oxide nanoparticles affect calcium transport leading to its accumulation in intracellular space resulting in oxidative stress, activation of caspases, and apoptosis. Copper complexes are also able to inhibit proteasome function [76]. Other structures allow G-quadruplex telomeric DNA reduction [77]. These effects lead to disturbances in cell cycle, activation of apoptotic pathways, and cancer cell death. One article reports copper complexes are able to accumulate inside mitochondria leading to cytotoxicity by damaging mtDNA [78]. A great variety of induced effects allows copper compounds to be used for various applications in a precisely determined manner of action.

5. Copper Nanoparticles for Cancer Imaging and Drug Delivery

Due to the recent developments in imaging technologies and biology, molecular imaging provides not only the possibility to visualize the tumor, but also to assess the expression and activity of specific molecules (e.g., protein kinases, enzymes, proteases, etc.) and various processes (including metastasis, tumor cell apoptotic death, angiogenesis, etc.) involved in cancer progression, response to therapy, and recurrence [79]. Furthermore, molecular imaging based on CuS NPs enables repetitive assessment of particles biodistribution and biokinetic properties employing positron emission tomography (PET) and photoacoustic imaging (PAI) [80,81].

Photoacoustic (PA) imaging, developed rapidly in the recent decade, represents a noninvasive biomedical imaging method which can be employed for visualization of deeply located tissues tumors, analysis of vasculature [82], or evaluation of neoangiogenesis [83]. Upon the in vivo absorbance of a short-pulse laser by various molecules (e.g., water, melanin, RNA, DNA, hemoglobin, cytochromes, lipids, etc.) ultrasonic signals are generated via the mechanism of photothermal conversion [84–86]. Up-to-date gold nanostructures (GNPs) were widely applied as contrast agents for photoacoustic imaging [87]. However, GNPs were reported to have several limitations as contrast agents, including dependence of optical properties on shape, geometry, and size of particles as well as their susceptibility to tumor microenvironmental factors. On the contrary, compared to the maximum absorption between 560 and 840 nm of GNPs, the absorption of copper nanoparticles could be tuned to peak at wavelengths greater than 900 nm, thus providing the improved sensitivity in the NIR region (i.e., stronger PA signal, higher signal-to-noise ratio, greater field-of-view) [88]. Indeed, in the study by Zhou [89] et al., it was shown that polyethylene glycol (PEG)-coated copper(II) sulfide nanoparticles (PEG-CuS NPs) (peak absorption of 1064 nm) could be successfully employed both as a contrast agent for in vivo imaging of 4T1 breast tumor vasculature and as a mediator for photothermolysis of cancer cells. However, due to the intrinsic dipole–dipole interactions among Cu-based particles, synthesis of size-tunable, biocompatible, and colloidally stable suspension of particles remains a challenge. To overcome this problem Ding [90] et al. proposed the aqueous synthesis of PEGylated copper sulfide particles with controllable size between 3 and 7 nm. Subsequent preclinical studies demonstrated that particles, particularly of less than 5 nm, had a higher tumor-imaging potential. Another approach could be based on application of tumor microenvironment-sensitive nanoparticles as was proposed in the work of Wang et al. [91]. The authors developed iron-copper co-doped polyaniline nanoparticles (Fe-Cu@PANI) which upon glutathione (GSH) redox reaction could shift in the absorption spectrum from the visible to the NIR. The etching of Fe-Cu@PANI resulted both in photoacoustic imaging of tumors and efficient photothermal therapy. In recent research by Bindra [92] et al., the authors synthesized a self-assembled nanosystem (SCP-CS) which consisted of a semiconducting polymer (SCP) and encapsulated ultrasmall CuS (CS) nanoparticles. This nanosystem demonstrated not only an improved PA-imaging ability but also significant tumor growth inhibition due to the enhanced production of ROS.

In PET apart from traditionally employed positron emitters [64Cu]-based NPs were also shown as an efficient radiotracer for tumor diagnostics [93,94]. Thus, Zhou [94] et al. in the U87 human glioblastoma xenograft model demonstrated that a novel class of chelator-free [64Cu]CuS nanoparticles (NPs) (PEG-[64Cu]CuS NPs) could effectively target the tumor cells providing a potential for image-guided PTA therapy. In a more recent study, more complex indium- and copper-based metal-phenolic nanoparticles (MPNs) (labeled with 111In and 64Cu) were proposed for in vivo multimodal PET/SPECT/CT imaging [95].

Among other applications of Cu-based NPs is their use as a chemotherapeutic drug delivery system. Recently, Zhang [96] et al. proposed hybrid hollow mesoporous organosilica nanoparticles (HMONs) that consisted of ultrasmall photothermal CuS particles and disulfiram (DSF). Upon near-infrared (NIR) irradiation, released Cu^{2+} ions from nanoparticles converted the nontoxic DSF into a highly cytotoxic diethyldithiocarbamate (DTC)-copper complex that inhibited tumor growth. In another study, thermo-responsive copper sulfide (CuS) was employed to deliver CRISPR-Cas9 ribonucleoprotein (RNP) and doxorubicin for tumor combination therapy consisting of chemotherapy, gene therapy, and photothermal therapy [97].

6. Clinical Application of Copper-Based Nanoparticles in Oncology

Although some breakthroughs have been made in the treatment of malignant tumors [98,99], therapies, such as chemotherapy and radiotherapy, have become the most commonly used clinical treatments for tumors. However, the recurrence rate, drug resistance, quality of life, and other issues of cancer patients are still a global challenge [100]. In recent years, nanomaterials can effectively deliver drugs to specific targets, protect blood circulation drugs from endogenous enzymes, extend the half-life of drugs, and have shown great potential in tumor treatment [101,102].

Breast cancer (BC) is the second most common female cancer in the world, second only to lung cancer [99]. Studies have shown that copper-based nanomaterials have broad application prospects in the treatment of BC. For example, Ahamed et al. [103] found that copper ferrite ($CuFe_2O_4$) nanoparticles (NPs) added to the culture of human breast cancer MCF-7 cells can cause intracellular oxidation stress response, exerting anti-cancer effects, specifically manifested in the production of ROS and the consumption of glutathione (GSH) (Figure 3). Furthermore, Rajagopal et al. [104] found that copper nanoparticles (Wt-CuNPs) have obvious cytotoxic effects on MCF-7 cells. The specific mechanism is mainly due to the release of copper ions from the nanoparticles and the binding of copper ions to tumor cell DNA, causing DNA damage and the resulting apoptotic cell death.

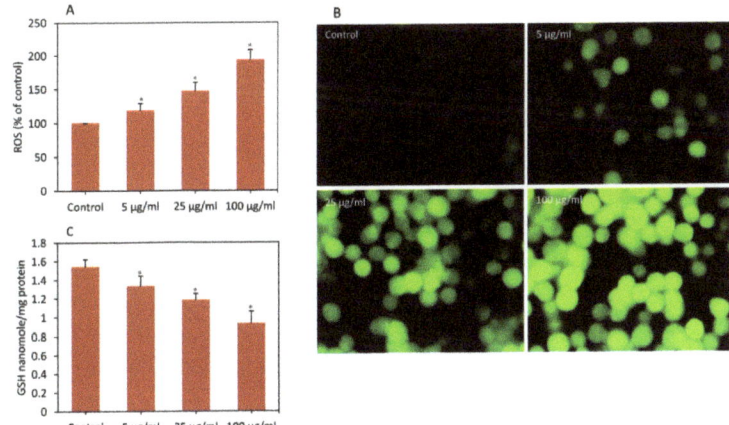

Figure 3. Copper ferrite NP-induced oxidative stress in MCF-7. Cells were exposed to copper ferrite NPs at the dosages of 0, 5, 25, and 100 g/mL for 24 h. At the end of exposure, ROS and GSH levels were determined, as described in materials and methods. (**A**) Percentage change in ROS level. (**B**) Fluorescence microscopy image of ROS generation. (**C**) GSH level. Data represented are mean ± SD of three identical experiments made in three replicates. * Significant difference as compared to control ($p < 0.05$).

Copper-based nanomaterials have also achieved good results in the treatment of esophageal cancer. Wang et al. [105] covered the silica coating on the Cu9S5 nanoparticles to form Cu9S5@MS core-shell nanostructures and added Cu9S5@MS core-shell nanostructures to human esophageal squamous carcinoma Eca109 and TE8 cells. After co-cultivation and treatment with NIR, it was found that Cu9S5@MS + NIR performs active anticancer activity against the EC109 and TE8 cancer cell lines by cell cycle arrest (Figure 4).

Furthermore, Xu et al. [106] optimized the concentration of disulfiram and Cu^{2+} ion for inhibiting esophageal cancer cells and loaded them in hyaluronic acid (HA)/polyethyleneimine (PEI) nanoparticles with specific scales to obtain NP-HPDCu^{2+} nanoparticles to improve the effectiveness and targeting of the drug. In vitro experiments proved that NP-HPDCu^{2+} nanomaterials can significantly promote the occurrence of Eca109 cell apoptosis and inhibit the migration and invasion of Eca109 (Figure 5). At the same time, the nude mouse tumor model proves that NP-HPDCu^{2+} nanomaterials can reduce the tumor volume and keep the weight of nude mice stable. The results of tumor tissue immunohistochemistry, immunofluorescence staining, and western blotting also showed that NP-HPDCu^{2+}

nanomaterials can promote apoptosis and inhibit proliferation of esophageal squamous cell carcinoma.

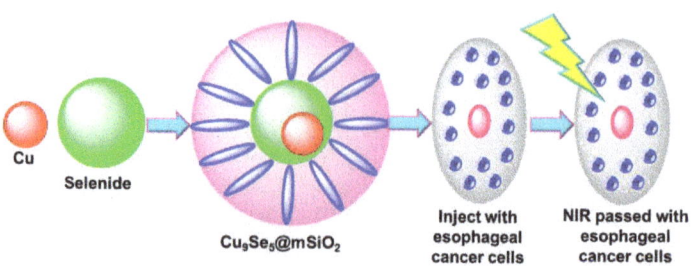

Figure 4. Portrayal of the Cu9S5@MS nanoparticles synthesis and application as a dual functional treatment stage for esophageal squamous carcinoma treatment.

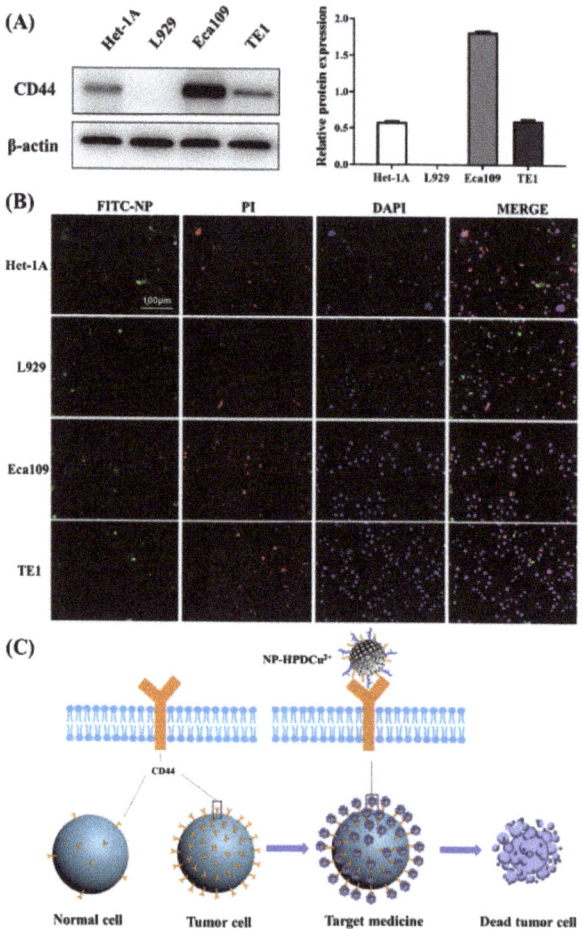

Figure 5. (**A**) Western blot of CD44 expressed on the Het-1A, L929, Eca109, and TE1 (mean ± SD, n = 3); (**B**) Fluorescence images of Het-1A, L929, Eca109, and TE1 stained with FITC-labeled NP-HPDCu^{2+} (FITC-NP, green color), PI (apoptosis marker, red color) and DAPI (nucleus marker, blue color); (**C**) mechanism diagram of targeted killing tumor cells by NP-HPDCu^{2+} nanoparticle.

Lung cancer is the malignant tumor with the highest mortality rate in the world, and non-small cell lung cancer is the most common pathological type in clinic [107,108]. Some researchers have found that copper-based nanomaterials have shown great potential in the treatment of NSCLC. Naatz et al. [109] constructed a new type of nanomaterial, Fe-doped CuO nanomaterial, which can use doped Fe to control the dissolution kinetics of copper-based nanomaterials. Using mouse lung squamous cell KLN-205 to construct a tumor-bearing nude mouse model by regulating the release of Cu^{2+}, the local long-term drug concentration can be maintained, and the occurrence of drug resistance can be reduced. Additionally, these particles can also trigger a systemic anti-cancer immune response, promote the generation of ROS, and increase the rate of tumor cell death, which shows that CuO nanomaterials also have broad prospects for anti-cancer applications (Figure 6). In addition, Kalaiarasi et al. [110] reported that in A549 cells, the anti-cancer effect of CuO copper-based nanomaterials is related to the inhibition of histone deacetylase (HDACs) expression. Specifically, CuO copper-based nanomaterials have a strong inhibitory effect on different types of HDACs, can down-regulate the expression of oncogenes and up-regulate the expression of tumor suppressor genes, and induce apoptosis of cancer cells by activating the caspase cascade pathway to exert anti-cancer effects.

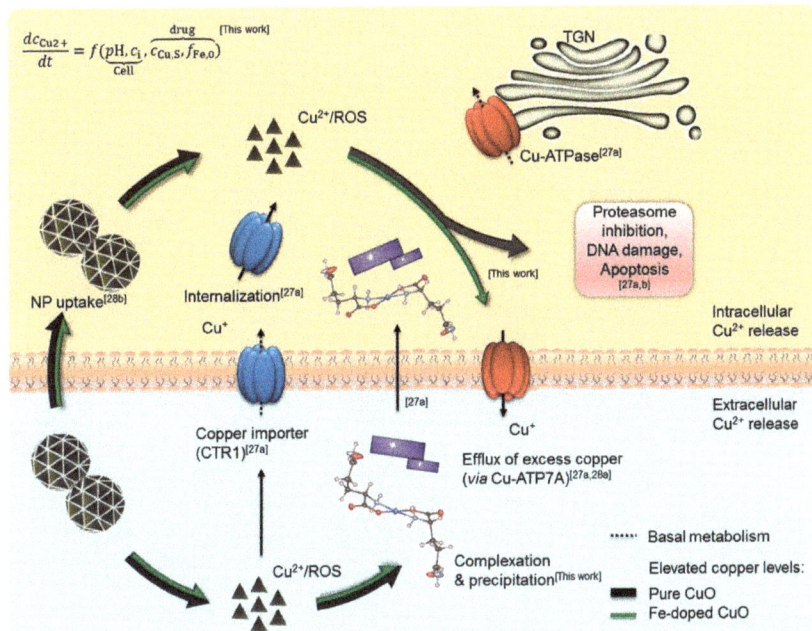

Figure 6. Copper homeostasis and regulatory mechanisms, including extra- and intracellular dissolution of pure and Fe-doped CuO NPs.

In recent years, with the continuous in-depth research of nanomaterials compared with traditional antitumor treatments, nanomaterials have been used in more and more clinical anticancer applications, showing great development potential [111]. For example, in our previous research, we found that some nanoparticles, such as superparamagnetic iron oxide nanoparticles (SPIONs), high-Z gold nanoparticles following intratumoral injection can provide a high local concentration of the agent, reduction of the particle clearance (i.e., renal or hepatic clearance) that increases the bioavailability of nanoparticles and has the effect of radiosensitizer in cancer radiotherapy, which can be used for long-term local anti-tumor therapy [112,113]. As an ideal anti-tumor drug candidate, copper-based nanomaterials have the following advantages: (i) compared with other metals, copper is cheap and rich in content [114], (ii) copper can induce reactive oxygen species (ROS)-mediated oxidative stress and promote tumor cell apoptosis [115–117], (iii) it has good biocompatibility, biodegradability, antibacterial properties, and selective cytotoxicity to cancer cells [118], and (iv) copper-based nanomaterials have less toxic effects on normal cells, fewer side effects, and are safer and more reliable [119]. Thus, copper-based nanomaterials have attracted more and more attention and have become the current research hotspot. At the same time, the emergence of copper-based nanomaterials has brought dawn to the treatment of various tumors [120].

However, copper-based nanomaterials also have limitations. For example, the production process of copper-based nanomaterials uses physical and chemical methods that are harmful to the environment and the human body [121]. Additionally, the instability and susceptibility to oxidation of copper-based nanomaterials under physiological conditions may also hinder its anti-tumor effect and reliability [122]. Furthermore, the biological safety of copper-based nanomaterials still requires further cell and molecular studies to avoid any impact on health, since Fahmy et al. [123] found that copper/copper oxide nanoparticles showed cytotoxicity to normal human lung WI-38 cells, resulting in the production of reactive oxygen species and DNA damage and inhibiting the growth and proliferation of WI-38 cells. The stability of copper nanoparticles is also one of the major concerns as copper tends to aggregate to the proteins, specifically cysteine and methionine residues. One work also found a dependency between stability and pH value [124]. However, using green synthesis, the authors successfully designed NPs which are mostly stable at various pH levels.

In short, copper-based nanomaterials are currently ideal anti-tumor drug candidates. With the continuous development of nanomaterials research, it will help provide better cancer treatment strategies in the future.

7. The Combination of Nanoparticles with Other Treatment Modalities

Based on the biological effects of copper and the physical and chemical properties of copper nanoparticles, their applications in the biomedical field mainly include externally triggered nanotherapies (photothermal therapy), drug delivery, antimicrobial applications, tissue regeneration, bioimaging, and bioeffects/biosafety. Therefore, it is reasonable to be expected that the construction of Cu-based biomaterials will have a unique integrated diagnosis and treatment function in clinical medicine. However, due to the complexity of tumors, such as the specific microenvironment and tumor metastasis, it is difficult to eradicate tumors completely through monotherapy alone. Therefore, the development of unique treatment modalities with multiple synergistic therapeutic performance has high prospects for improving therapeutic efficacy. Therefore, rational design of optimal drug combinations is important to achieve optimal synergistic therapeutic effects. Based on this, several unique multifunctional nanosystems involving copper have been constructed to jointly generate multiple nanotherapeutics [125].

Copper chalcogenides ($Cu_{2-x}E$, E:S, Se, Te, $0 \leq x \leq 1$) have been widely explored in photon-triggered disease therapy, such as photoacoustic imaging and photothermal hyperthermia. With stoichiometric ratios ($Cu_{2-x}S$), deficient cuprous sulfide exhibits stoichiometric-dependent localized surface plasmon resonance (LSPR) absorption in the near-infrared range and photothermal conversion [126]. The integration of magnetic Fe_3O_4 nanoparticles exerted a magnetic targeting function to enhance tumor accumulation. Importantly, the photonic response of these Fe_3O_4@CuS composite nanoparticles in the second NIR biological window (1064 nm) achieves higher tissue penetration ability compared to the laser activation of the first NIR biological window. Thus, a higher tumor suppression rate was achieved with no further recurrence (808 nm). In addition to the photothermal conversion efficiency (25.7%) of hydrophilic plate-like Cu_9S_5 nanocrystals at 980 nm [127], the CuS superstructure was exemplified to respond to external 980 nm laser activation for photothermal conversion and subsequent cancer ablation [128]. The cysteine-coated CuS nanoparticles were also irradiated with a 980 nm laser with a high photothermal conversion efficiency of 38.0%, efficiently inhibiting tumor growth [129]. Furthermore, encapsulation of CuS nanoparticles into zeolite imidazole framework 8 (ZIF-8) resulted in NIR-induced dissociation of ZIF-8 to release loaded chemotherapeutics, aiming to achieve synergistic photothermal ablation and NIR-triggered chemotherapy [130]. Doping iron (Fe^{3+}) can tune the vacancies of $Cu_{2-x}Se$ nanoparticles to control NIR absorption, which also enables these semiconductors to have MR-imaging properties [131].

To improve the photothermal conversion efficiency, $Cu_{2-x}S$ and Ag_2S were integrated into one system by producing Cu-Ag2S/PVP nanoparticles with a high photothermal conversion efficiency of 58.2% under 808 nm laser irradiation, which is much higher than that of $Cu_{2-x}S$/PVP nanoparticles (27.1%) [132]. The rational integration of plasmonic Au nanoparticles and plasmonic $Cu_{2-x}S$ semiconductors into one matrix can enhance the photothermal properties of Au or $Cu_{2-x}S$ components. The coupled LSPR properties of Au and $Cu_{2-x}S$ can be maximized by designing Au@$Cu_{2-x}S$ core/shell nanoparticles to enhance the PTT efficacy. Ji et al. synthesized Au@CuS nanoparticles and performed the following cation exchange between Cu^+ and CdS shells, resulting in Au@$Cu_{2-x}S$ nanostructures [133], which can be formed as nanoparticles or nanorods. The corresponding photothermal conversion efficiencies are calculated to be 59% at 808 nm and 43% at 1064 nm, which rapidly increases the ambient temperature of the Au@$Cu_{2-x}S$ nanorod aqueous solution. In particular, the design of core/shell Au@$Cu_{2-x}S$ is more favorable compared to the simple mixture of Au nanorods and $Cu_{2-x}S$ nanoparticles for photothermal conversion. This core/shell design with improved photothermal performance also induced more HeLa cell death compared to the same concentration of $Cu_{2-x}S$. The Au-Cu9S5 plasmonic hybrid nanosystem was established, which enhanced the LSPR of Cu9S5 through the coupling effect of LSPR based on the collective vibration of electrons and holes [134]. This Au-Cu9S5 hybrid nanosystem exhibits an absorption cross-section enhancement of 1.3×10^8 m^{-1} cm^{-1} and a high photothermal conduction efficiency of 37% for photothermal ablation of tumor tissue. According to the plasmonic coupling effect between core and shell, spherical Au@$Cu_{2-x}S$, Au@$Cu_{2-x}S$, and rod-shaped Au@$Cu_{2-x}S$ superparticles were synthesized for photothermal ablation of tumors (4T1 tumor model). It has X-ray-computed, tomography-imaging capabilities because of the presence of Au composition with a large atomic number and an X-ray attenuation coefficient (5.16 cm^{-2} kg^{-1}) [135].

Photothermal therapy exposes materials with the photothermal conversion ability to near-infrared light. These materials can convert the absorbed light energy into thermal energy to kill tumors, showing excellent local tumor treatment effects, but they are less effective for metastatic tumors. The combination of photothermal therapy and radiotherapy in tumor treatment can achieve a synergistic effect. Thus, Zhou et al. [89] synthesized PEG-[64Cu]CuS NPs based on a single radioactive copper sulfide nanoparticle. The study demonstrated that inhibition of tumor growth was significantly high when both methods, radiotherapy and hyperthermia, were employed.

Photothermal therapy (PTT) mainly uses photothermal materials accumulated at the tumor site, which can convert the absorbed light energy into heat energy (above 45 °C) under near-infrared irradiation. Combining tumor photothermal therapy and immunotherapy could further improve the therapeutic potency of PTT [136]. Another approach could be based on the combination of PTT with chemotherapy. Thus, Wu et al. [137] demonstrated that encapsulation of CuS nanoparticles into the zeolite imidazole framework 8 (ZIF-8) resulted in NIR-induced dissociation of ZIF-8 to release loaded chemotherapeutics, which in turn provided synergistic photothermal ablation and NIR-triggered chemotherapy.

The tumor microenvironment is usually characterized by low pH [138], altered redox states [139], hypoxia [140], and expression of particular enzymes that could be employed for the development of stimuli-responsive nanoparticles. Based on the fact that the hydrogen sulfide (H_2S)-producing enzyme of cystathionine-β-synthase (CBS) is upregulated in colon cancer, H_2S concentrations in tumors reach approximately 0.3 to 3.4 mmol·L^{-1}. Therefore, using this overexpressed endogenous H_2S to convert cuprous oxide (Cu_2O) to copper sulfide in situ can activate PA imaging and photothermal tumor ablation [141]. It is exemplified that the use of S-adenosyl-1-methionine (SAM) as an allosteric CBS activator accelerates the in situ reaction between H_2S and Cu_2O, resulting in significantly enhanced PA-imaging signal and photothermal effect. In contrast, the use of aminooxyacetic acid (AOAA) as a CBS inhibitor reduced the production of H_2S and subsequently the conversion of Cu_2O to copper sulfide, showing no significant PA signal and negligible temperature change in tumors. However, the photothermal conversion efficiency after high-dose copper sulfide conversion is low, and the ideal photon therapy effect cannot be obtained. To address this critical issue, based on the LSPR-coupling effect between noble metals and plasmonic semiconductors, Tao et al. constructed Au@Cu2O plasmonic hybrids to enhance in situ H_2S-triggered post-conversion photothermal performance [142]. Similar to the conversion of Cu_2O to Cu_9S_8, tumor-accumulated Au@Cu2O nanoparticles were also converted into Au@Cu9S8 nanoagents to achieve PA-enhanced contrast agents and photothermal tumor ablation by increasing tumor temperature. The LSPR-coupling effect induces nearly 2.1-fold stronger NIR absorption and 1.2-fold higher photothermal conversion efficiency, enabling the utilization of low nanoparticle doses with desirable therapeutic properties. These two paradigms provide another strategy for realizing photothermal hyperthermia involving copper-based nanoagents by in situ generation of copper-based nanoagents with unique photothermal properties. Cheng Y. et al. [143] took advantage of the ordered large-pore structure and easily chemically modified the property of DLMSNs, the copper sulfide (CuS) nanoparticles with high photothermal conversion efficiency. A homogenous cancer cell membrane was coated on the surfaces of these DLMSNs, followed by conjugation with the anti-PD-1 peptide. The thus-obtained AM@DLMSN@CuS/R848 was applied to holistically treat metastatic TNBC in vitro and in vivo. The data showed that AM@DLMSN@CuS/R848 had a high TNBC-targeting ability and induced efficient photothermal ablation on primary TNBC tumors under 980 nm laser irradiation. Tumor antigens thus generated and increasingly released R848 by response to the photothermal effect, combined with AUNP-12 detached from AM@DLMSN@CuS/R848 in the weakly acidic tumor microenvironment and synergistically exerted an anti-tumor effect, thus preventing TNBC recurrence and metastasis.

Table 1 summarizes the above information presenting major classes of therapeutics and some examples for detail consideration. The unique features of copper allow to create a wide spectrum of various nanostructures with great diversity of their applications.

Table 1. Copper-based compounds and nanoparticles with various applications and mechanisms of action.

	Copper-Based Compound	Mechanism of Action
Diagnostic tool	64-$CuCl_2$ [64] Combination of SPIONs and Cu(II) [62] Gold-copper alloyed NPs [63]	Contrast agent in PET/MRI scanning
Therapeutic agent	Schiff base copper (II) complexes [66]	Activation of extrinsic or intrinsic apoptotic pathways
	Copper-based nanoparticles [96,103]	Copper ions release, oxidative stress, DNA damage
	Thiosemicarbazones copper (II) complex [72]	Topoisomerase inhibition
	Polypyridyl-Schiff-base copper complex [74]	Targets endoplasmic reticulum leading to immunogenic cell death
	G-quadruplex-targeting copper complex [77]	Rapid reduction of telomeres in cancer cells
	Ferrocenyl terpyridine copper complexes [78]	Targets mitochondria, causes mtDNA damage

Table 1. *Cont.*

	Copper-Based Compound	Mechanism of Action
Combined approach	Copper chalcogenides [126] Alloyed CuAg or CuAu NPs [132,133]	Photothermal ablation and NIR-triggered chemotherapy
	PEG-[64Cu]CuS NPs [94]	Combined radiotherapy and hyperthermia against metastatic tumor cells
	Copper-doped iron NPs [109,131]	Magnetic guidance and copper release with subsequent oxidative stress

8. Conclusions

Copper is an essential trace element in cell metabolism with distinct features. Participation of copper in oxidation–reduction reactions has an important impact on cell metabolism, survival, and growth. Free copper ions could exert a cytotoxic effect; however, most of the copper is bound to the enzymes, metallochaperones, and metallothioneins. These proteins, despite their direct function, could influence functionality of other proteins affecting cell signaling and gene expression, interfering in the anti-cancer chemotherapies. Recent studies demonstrate that copper-based nanocarriers due to their unique physio-chemical properties could be efficiently employed for tumor theranostics as a monotherapeutic approach or in combination with other treatment modalities. Constant development and modification of existing systems have great potential in clinic. Some limitations, which include ROS generation and free ion emergence, should be considered. However, an understanding of the underlying molecular regulation of copper intracellular distribution and metabolism will help to improve the current development of copper-based therapeutics and nanostructures for further efficient clinical application.

Author Contributions: Conceptualization, S.T., M.D. and M.S.; methodology, S.T., G.L., N.A., Y.S., J.Z.; M.D.; V.F. and M.S.; resources, S.T., M.D. and M.S.; data curation, S.T. and M.S.; writing—original draft preparation, S.T., G.L., N.A., Y.S., J.Z., M.D., V.F. and M.S.; writing—review and editing, S.T., G.L., N.A., Y.S., J.Z., M.D., V.F. and M.S.; visualization, S.T., G.L., N.A., Y.S., J.Z., M.D., V.F. and M.S.; supervision, S.T., M.D. and M.S.; project administration, S.T., M.D. and M.S.; funding acquisition, M.S. All authors have read and agreed to the published version of the manuscript.

Funding: This research was funded by the Russian Science Foundation (Grant Number 22-15-00240).

Institutional Review Board Statement: Not applicable.

Informed Consent Statement: Not applicable.

Data Availability Statement: Not applicable.

Acknowledgments: Authors are grateful to Alexander Shtil for his supervision and support.

Conflicts of Interest: The authors declare no conflict of interest.

References

1. Shanbhag, V.C.; Gudekar, N.; Jasmer, K.; Papageorgiou, C.; Singh, K.; Petris, M.J. Copper metabolism as a unique vulnerability in cancer. *Biochim. Biophys. Acta Mol. Cell Res.* **2021**, *1868*, 118893. [CrossRef] [PubMed]
2. Skopp, A.; Boyd, S.D.; Ullrich, M.S.; Liu, L.; Winkler, D.D. Copper–zinc superoxide dismutase (Sod1) activation terminates interaction between its copper chaperone (Ccs) and the cytosolic metal-binding domain of the copper importer Ctr1. *Biometals* **2019**, *32*, 695–705. [CrossRef] [PubMed]
3. Nývltová, E.; Dietz, J.V.; Seravalli, J.; Khalimonchuk, O.; Barrientos, A. Coordination of metal center biogenesis in human cytochrome c oxidase. *Nat. Commun.* **2022**, *13*, 3615. [CrossRef] [PubMed]
4. Postma, G.C.; Nicastro, C.N.; Valdez, L.B.; Mikusic, I.A.R.; Grecco, A.; Minatel, L. Decrease lysyl oxidase activity in hearts of copper-deficient bovines. *J. Trace Elem. Med. Biol.* **2021**, *65*, 126715. [CrossRef]
5. Grasso, M.; Bond, G.J.; Kim, Y.J.; Boyd, S.; Dzebo, M.M.; Valenzuela, S.; Tsang, T.; Schibrowsky, N.A.; Alwan, K.B.; Blackburn, N.J.; et al. The copper chaperone CCS facilitates copper binding to MEK1/2 to promote kinase activation. *JBC* **2021**, *297*, 101314. [CrossRef] [PubMed]

6. Krishnamoorthy, L.; Cotruvo, J.A., Jr.; Chan, J.; Kaluarachchi, H.; Muchenditsi, A.; Pendyala, V.S.; Jia, S.; Aron, A.T.; Ackerman, C.M.; Vander Wal, M.N.; et al. Copper regulates cyclic-AMP-dependent lipolysis. *Nat. Chem. Biol.* **2016**, *12*, 586–592. [CrossRef]
7. Guo, H.; Li, K.; Wang, W.; Wang, C.; Shen, Y. Effects of copper on hemocyte apoptosis, ROS production, and gene expression in white shrimp Litopenaeus vannamei. *Biol. Trace Elem. Res.* **2017**, *179*, 318–326. [CrossRef]
8. Chen, J.; Jiang, Y.; Shi, H.; Peng, Y.; Fan, X.; Li, C. The molecular mechanisms of copper metabolism and its roles in human diseases. *Pflug Arch. Eur. J. Phy.* **2020**, *472*, 1415–1429. [CrossRef]
9. Lee, J.; Petris, M.J.; Thiele, D.J. Biochemical characterization of the human copper transporter Ctr1. *J. Biol. Chem.* **2002**, *277*, 4380–4387. [CrossRef]
10. Ren, F.; Logeman B., L.; Zhang, X.; Liu, Y.; Thiele D., J.; Yuan, P. X-ray structures of the high-affinity copper transporter Ctr1. *Nat. Commun.* **2019**, *10*, 1386. [CrossRef]
11. Chen, H.; Xu, C.; Yu, Q.; Zhong, C.; Peng, Y.; Chen, J.; Chen, G. Comprehensive landscape of STEAP family functions and prognostic prediction value in glioblastoma. *J. Cell. Physiol.* **2021**, *236*, 2988–3000. [CrossRef] [PubMed]
12. Inesi, G. Molecular features of copper binding proteins involved in copper homeostasis. *IUBMB Life* **2017**, *69*, 211–217. [CrossRef] [PubMed]
13. Luchinat, E.; Barbieri, L.; Banci, L. A molecular chaperone activity of CCS restores the maturation of SOD1 fALS mutants. *Sci. Rep.* **2017**, *7*, 17433. [CrossRef]
14. Banks, C.J.; Andersen, J.L. Mechanisms of SOD1 regulation by post-translational modifications. *Redox Biol.* **2019**, *26*, 101270. [CrossRef] [PubMed]
15. Vanišová, M.; Burská, D.; Křížová, J.; Daňhelovská, T.; Dosoudilová, Ž.; Zeman, J.; Stibůrek, L.; Hansíková, H. Stable COX17 downregulation leads to alterations in mitochondrial ultrastructure, decreased copper con-tent and impaired cytochrome c oxidase biogenesis in HEK293 cells. *Folia Biol.* **2019**, *65*, 181–187.
16. La Fontaine, S.; Mercer, J.F. Trafficking of the copper-ATPases, ATP7A and ATP7B: Role in copper homeostasis. *Arch. Biochem. Biophys.* **2007**, *463*, 149–167. [CrossRef]
17. Zhang, W.; Shi, H.; Chen, C.; Ren, K.; Xu, Y.; Liu, X.; He, L. 2 Curcumin enhances cisplatin sensitivity of human NSCLC cell lines through influencing Cu-Sp1-CTR1 regulatory loop. *Phytomedicine* **2018**, *48*, 51–61. [CrossRef]
18. Sinani, D.; Adle, D.J.; Kim, H.; Lee, J. Distinct mechanisms for Ctr1-mediated copper and cisplatin transport. *J. Biol. Chem.* **2007**, *282*, 26775–26785. [CrossRef]
19. Sun, S.; Cai, J.; Yang, Q.; Zhao, S.; Wang, Z. The association between copper transporters and the prognosis of cancer patients undergoing chemotherapy: A meta-analysis of literatures and datasets. *Oncotarget* **2017**, *8*, 16036. [CrossRef]
20. Ilyechova, E.Y.; Bonaldi, E.; Orlov, I.A.; Skomorokhova, E.A.; Puchkova, L.V.; Broggini, M. CRISP-R/Cas9 mediated deletion of copper transport genes CTR1 and DMT1 in NSCLC cell line H1299. Biological and pharmacological consequences. *Cells* **2019**, *8*, 322. [CrossRef]
21. Akerfeldt, M.C.; Tran, C.M.N.; Shen, C.; Hambley, T.W.; New, E.J. Interactions of cisplatin and the copper transporter CTR1 in human colon cancer cells. *J. Biol. Inorg. Chem.* **2017**, *22*, 765–774. [CrossRef] [PubMed]
22. Ishida, S.; Lee, J.; Thiele, D.J.; Herskowitz, I. Uptake of the anticancer drug cisplatin mediated by the copper transporter Ctr1 in yeast and mammals. *PNAS* **2002**, *99*, 14298–14302. [CrossRef] [PubMed]
23. Safaei, R.; Maktabi, M.H.; Blair, B.G.; Larson, C.A.; Howell, S. Effects of the loss of Atox1 on the cellular pharmacology of cisplatin. *J. Inorg. Biochem.* **2009**, *103*, 333–341. [CrossRef]
24. Bompiani, K.M.; Tsai, C.Y.; Achatz, F.P.; Liebig, J.K.; Howell, S.B. Copper transporters and chaperones CTR1, CTR2, ATOX1, and CCS as determinants of cisplatin sensitivity. *Metallomics* **2016**, *8*, 951–962. [CrossRef] [PubMed]
25. Blockhuys, S.; Brady, D.C.; Wittung-Stafshede, P. Evaluation of copper chaperone ATOX1 as prognostic biomarker in breast cancer. *Breast Cancer* **2020**, *27*, 505–509. [CrossRef]
26. Blockhuys, S.; Wittung-Stafshede, P. Copper chaperone Atox1 plays role in breast cancer cell migration. *Biochem. Biophys. Res. Commun.* **2017**, *483*, 301–304. [CrossRef]
27. Itoh, S.; Kim, H.W.; Nakagawa, O.; Ozumi, K.; Lessner, S.M.; Aoki, H.; Akram, K.; McKinney, R.D.; Ushio-Fukai, M.; Fukai, T. Novel role of antioxidant-1 (Atox1) as a copper-dependent transcription factor involved in cell prolifera-tion. *J. Biol. Chem.* **2008**, *283*, 9157–9167. [CrossRef]
28. Beaino, W.; Guo, Y.; Chang, A.J.; Anderson, C.J. Roles of Atox1 and p53 in the trafficking of copper-64 to tumor cell nuclei: Implications for cancer ther-apy. *J. Biol. Inorg. Chem.* **2014**, *19*, 427–438. [CrossRef]
29. Singh, R.P.; Jeyaraju, D.V.; Voisin, V.; Xu, C.; Barghout, S.H.; Khan, D.H.; Hurren, R.; Gronda, M.; Wang, X.; Jitkova, Y.; et al. Targeting the Mitochondrial Metallochaperone Cox17 Reduces DNA Methylation and Promotes AML Differentiation through a Copper Dependent Mechanism. *Blood* **2018**, *132*, 1339. [CrossRef]
30. Zhao, L.; Cheng, Q.; Wang, Z.; Xi, Z.; Xu, D.; Liu, Y. Cisplatin binds to human copper chaperone Cox17: The mechanistic implication of drug delivery to mitochondria. *Chem. Commun.* **2014**, *50*, 2667–2669. [CrossRef]
31. Zhao, L.; Wang, Z.; Wu, H.; Xi, Z.; Liu, Y. Glutathione selectively modulates the binding of platinum drugs to human copper chaperone Cox17. *Biochem. J.* **2015**, *472*, 217–223. [CrossRef] [PubMed]
32. Sotgia, F.; Fiorillo, M.; Lisanti, M.P. Mitochondrial markers predict recurrence, metastasis and tamoxifen-resistance in breast cancer patients: Early detection of treatment failure with companion diagnostics. *Oncotarget* **2017**, *8*, 68730–68745. [CrossRef] [PubMed]

33. Gao, S.P.; Sun, H.F.; Fu, W.Y.; Li, L.D.; Zhao, Y.; Chen, M.T.; Jin, W. High expression of COX5B is associated with poor prognosis in breast cancer. *Future Oncol.* **2017**, *13*, 1711–1719. [CrossRef] [PubMed]
34. Bikas, A.; Jensen, K.; Patel, A.; Costello, J.; Reynolds, S.M.; Mendonca-Torres, M.C.; Thakur, S.; Klubo-Gwiezdzinska, J.; Ylli, D.; Wartofsky, L.; et al. Cytochrome C Oxidase Subunit 4 (COX4): A Potential Therapeutic Target for the Treatment of Medullary Thyroid Cancer. *Cancers* **2020**, *12*, 2548. [CrossRef]
35. Yi, J.F.; Li, Y.-M.; Liu, T.; He, W.-T.; Li, X.; Zhou, W.-C.; Kang, S.-L.; Zeng, X.-T.; Zhang, J.-Q. Mn-SOD and CuZn-SOD polymorphisms and interactions with risk factors in gastric cancer. *World J. Gastroenterol.* **2010**, *16*, 4738–4746. [CrossRef]
36. Ahmed, A.S.; Eryilmaz, R.; Demir, H.; Aykan, S.; Demir, C. Determination of oxidative stress levels and some antioxidant en-zyme activities in prostate cancer. *Aging Male* **2019**, *22*, 198–206. [CrossRef]
37. Liu, S.; Li, B.; Xu, J.; Hu, S.; Zhan, N.; Wang, H.; Gao, C.; Li, J.; Xu, X. SOD1 Promotes Cell Proliferation and Metastasis in Non-small Cell Lung Cancer via an miR-409-3p/SOD1/SETDB1 Epigenetic Regulatory Feedforward Loop. *Front. Cell Dev. Biol.* **2020**, *8*, 213. [CrossRef]
38. Li, Y.; Liang, R.; Zhang, X.; Wang, J.; Shan, C.; Liu, S.; Li, L.; Zhang, S. Copper Chaperone for Superoxide Dismutase Promotes Breast Cancer Cell Proliferation and Migration via ROS-Mediated MAPK/ERK Signaling. *Front. Pharmacol.* **2019**, *10*, 356. [CrossRef]
39. Wang, J.; Luo, C.; Shan, C.; You, Q.; Lu, J.; Elf, S.; Zhou, Y.; Wen, Y.; Vinkenborg, J.L.; Fan, J.; et al. Inhibition of human copper trafficking by a small molecule significantly attenuates cancer cell proliferation. *Nat. Chem.* **2015**, *7*, 968–979. [CrossRef]
40. Wen, C.; Shan, C.L.; Sun, W.J.; Wan, Y.; Lin, R.; Chen, B.; Dai, H.-T.; Tang, K.-Y.; Xiang, X.-H.; Yang, J.-Y.; et al. Copper chaperone for superoxide dismutase expression is down-regulated and correlated with more malignant tumoral features and poor prognosis in human hepatocellular carcinoma. 2021, preprint. [CrossRef]
41. Samimi, G.; Varki, N.M.; Wilczynski, S.; Safaei, R.; Alberts, D.S.; Howell, S.B. Increase in expression of the copper transporter ATP7A during platinum drug-based treatment is associated with poor survival in ovarian cancer patients. *Clin. Cancer Res.* **2003**, *96*, 5853–5859.
42. Chisholm, C.L.; Wang, H.; Wong, A.H.; Vazquez-Ortiz, G.; Chen, W.; Xu, X.; Deng, C.X. Ammonium tetrathiomolybdate treatment tar-gets the copper transporter ATP7A and enhances sensitivity of breast cancer to cisplatin. *Oncotarget* **2016**, *7*, 84439. [CrossRef] [PubMed]
43. Seiko, I.; Frank, M.; Karen, S.; Douglas, H. Enhancing Tumor-Specific Uptake of the Anticancer Drug Cisplatin with a Copper Chelator. *Cancer Cell* **2010**, *17*, 574–583. [CrossRef]
44. Mangala, L.S.; Zuzel, V.; Schmandt, R.; Leshane, E.S.; Halder, J.B.; Armaiz-Pena, G.N.; Spannuth, W.A.; Tanaka, T.; Shahzad, M.M.K.; Lin, Y.G.; et al. Therapeutic Targeting of ATP7B in Ovarian Carcinoma. *Clin. Cancer Res.* **2009**, *15*, 3770–3780. [CrossRef] [PubMed]
45. David, L.; Maruša, H.; Borut, K.; Katarina, Č. The contribution of copper efflux transporters ATP7A and ATP7B to chemo-resistance and personalized medicine in ovarian cancer. *Biomed. Pharmacother.* **2020**, *129*, 110401. [CrossRef]
46. Yu, Z.; Cao, W.; Ren, Y.; Zhang, Q.; Liu, J. ATPase copper transporter A, negatively regulated by miR-148a-3p, contributes to cisplatin resistance in breast cancer cells. *Clin. Transl. Med.* **2020**, *10*, 57–73. [CrossRef]
47. Sen, C.K.; Khanna, S.; Venojarvi, M.; Trikha, P.; Ellison, E.C.; Hunt, T.K.; Roy, S. Copper-induced vascular endothelial growth factor expression and wound healing. *Am. J. Physiol. Cell Physiol.* **2002**, *282*, 9157–9167. [CrossRef]
48. Wu, Z.; Zhang, W.; Kang, Y.J. Copper affects the binding of HIF-1alpha to the critical motifs of its target genes. *Metallomics* **2019**, *11*, 429–438. [CrossRef]
49. Wee, N.K.; Weinstein, D.C.; Fraser, S.T.; Assinder, S.J. The mammalian copper transporters CTR1 and CTR2 and their roles in development and disease. *Int. J. Biochem. Cell Biol.* **2013**, *45*, 960–963. [CrossRef]
50. Wang, W.; Wang, X.; Luo, J.; Chen, X.; Ma, K.; He, H.; Li, W.; Cui, J. Serum Copper Level and the Copper-to-Zinc Ratio Could Be Useful in the Prediction of Lung Cancer and Its Prognosis: A Case-Control Study in Northeast China. *Nutr. Cancer* **2021**, *73*, 1908–1915. [CrossRef]
51. Brady, D.C.; Crowe, M.S.; Turski, M.L.; Hobbs, G.A.; Yao, X.; Chaikuad, A.; Knapp, S.; Xiao, K.; Campbell, S.L.; Thiele, D.J.; et al. Copper is required for oncogenic BRAF signal-ling and tumorigenesis. *Nature* **2014**, *509*, 492–496. [CrossRef] [PubMed]
52. Kim, D.W.; Shin, M.J.; Choi, Y.J.; Kwon, H.J.; Lee, S.H.; Lee, S.; Lee, S.; Park, J.; Han, K.H.; Eum, W.S.; et al. Tat-ATOX1 inhibits inflammatory responses via regulation of MAPK and NF-kappaB pathways. *BMB Rep.* **2018**, *51*, 654–659. [CrossRef] [PubMed]
53. Tsai, C.Y.; Finley, J.C.; Ali, S.S.; Patel, H.H.; Howell, S.B. Copper influx transporter 1 is required for FGF, PDGF and EGF-induced MAPK signaling. *Biochem. Pharmacol.* **2012**, *84*, 1007–1013. [CrossRef]
54. Li, Y.; Fu, S.Y.; Wang, L.H.; Wang, F.Y.; Wang, N.N.; Cao, Q.; Wang, Y.-T.; Yangab, J.-Y.; Wu, C.-F. Copper improves the anti-angiogenic activity of disulfiram through the EGFR/Src/VEGF pathway in gliomas. *Cancer Lett.* **2015**, *369*, 86–96. [CrossRef] [PubMed]
55. Thakor, A.S.; Jokerst, J.V.; Ghanouni, P.; Campbell, J.L.; Mittra, E.; Gambhir, S.S. Clinically Approved Nanoparticle Imaging Agents. *J. Nucl. Med.* **2016**, *57*, 1833–1837. [CrossRef]
56. Gutfilen, B.; Souza, S.A.; Valentini, G. Copper-64: A real theranostic agent. *Drug Des. Devel. Ther.* **2018**, *12*, 3235–3245. [CrossRef]
57. Xue, X.; Zhang, H.; Liu, H.; Wang, S.; Li, J.; Zhou, Q.; Chen, X.; Ren, X.; Jing, Y.; Deng, Y.; et al. Rational Design of Multifunctional CuS Nanoparticle-PEG Composite Soft Hydrogel-Coated 3D Hard Polycaprolactone Scaffolds for Efficient Bone Regeneration. *Adv. Funct. Mater.* **2022**, *32*, 2202470. [CrossRef]

58. Xue, X.; Liu, H.; Wang, S.; Hu, Y.; Huang, B.; Li, M.; Chen, X.; Ren, X.; Jing, Y.; Deng, Y.; et al. Neutrophil-erythrocyte hybrid membrane-coated hollow copper sulfide nanoparticles for targeted and photothermal/anti-inflammatory therapy of osteoarthritis. *Compos. B Eng.* **2022**, *237*, 109855. [CrossRef]
59. Lu, H.; Xu, S.; Ge, G.; Guo, Z.; Zhao, M.; Liu, Z. Boosting Chemodynamic Therapy by Tumor-Targeting and Cellular Redox Homeostasis-Disrupting Nanoparticles. *ACS Appl. Mater.* **2022**, *14*, 44098–44110. [CrossRef]
60. Fanizza, E.; Mastrogiacomo, R.; Pugliese, O.; Guglielmelli, A.; De Sio, L.; Castaldo, R.; Scavo, M.P.; Giancaspro, M.; Rizzi, F.; Gentile, G.; et al. NIR-Absorbing Mesoporous Silica-Coated Copper Sulphide Nanostructures for Light-to-Thermal Energy Conversion. *Nanomaterials* **2022**, *12*, 2545. [CrossRef]
61. Avila-Rodriguez, M.A.; Rios, C.; Carrasco-Hernandez, J.; Manrique-Arias, J.C.; Martinez-Hernandez, R.; Garcia-Perez, F.O.; Martinez-Rodriguez, E.; Romero-Piña, M.E.; Diaz-Ruiz, A. Biodistribution and radiation dosimetry of [(64)Cu]copper dichloride: First-in-human study in healthy volunteers. *EJNMMI Res.* **2017**, *7*, 98. [CrossRef] [PubMed]
62. Patel, D.; Kell, A.; Simard, B.; Xiang, B.; Lin, H.Y.; Tian, G. The cell labeling efficacy, cytotoxicity and relaxivity of cop-per-activated MRI/PET imaging contrast agents. *Biomaterials* **2011**, *32*, 1167–1176. [CrossRef] [PubMed]
63. Zhao, Y.; Sultan, D.; Detering, L.; Cho, S.; Sun, G.; Pierce, R.; Wooley, K.L.; Liu, Y. Copper-64-alloyed gold nanoparticles for cancer imaging: Im-proved radiolabel stability and diagnostic accuracy. *Angew. Chem. Int. Ed. Engl.* **2014**, *53*, 156–159. [CrossRef]
64. Zhou, M.; Tian, M.; Li, C. Copper-Based Nanomaterials for Cancer Imaging and Therapy. *Bioconjug. Chem.* **2016**, *27*, 1188–1199. [CrossRef]
65. Fulda, S.; Debatin, K.M. Extrinsic versus intrinsic apoptosis pathways in anticancer chemotherapy. *Oncogene* **2006**, *25*, 4798–4811. [CrossRef] [PubMed]
66. Konarikova, K.; Perdikaris, G.A.; Gbelcova, H.; Andrezalova, L.; Sveda, M.; Ruml, T.; Laubertová, L.; Žitňanová, I. Effect of Schiff base Cu(II) complexes on signaling pathways in HT-29 cells. *Mol. Med. Rep.* **2016**, *14*, 4436–4444. [CrossRef] [PubMed]
67. Langer, V.; Gyepesová, D.; Scholtzová, E.; Mach, P.; Kohútová, M.; Valent, A.; Smrčok, L.U. Crystal and electronic structure of aqua (N-salicylidene-methylester-L-glutamato) Cu (II) monohydrate. *Z. Kristallogr. Cryst. Mater.* **2004**, *219*, 112–116. [CrossRef]
68. Nakao, Y.; Sakurai, K.I.; Nakahara, A. Copper (II) chelates of Schiff bases derived from salicylaldehyde and various α-amino acids. *Bull. Chem. Soc. Jpn.* **1967**, *40*, 1536–1538. [CrossRef]
69. Krätsmár-Šmogrovič, J.; Pavelčík, F.; Soldánová, J.; Sivy, J.; Seressová, V.; Žemlička, M. The Crystal and Molecular Structure and Properties of Diaqua [N-salicylidene-(S)-(+)-glutamato] copper (II) Monohydrate. *Z. Nat. B* **1991**, *46*, 1323–1327. [CrossRef]
70. Zhang, J.; Duan, D.; Xu, J.; Fang, J. Redox-Dependent Copper Carrier Promotes Cellular Copper Uptake and Oxidative Stress-Mediated Apoptosis of Cancer Cells. *ACS Appl. Mater. Interfaces* **2018**, *10*, 33010–33021. [CrossRef]
71. Kunos, C.A.; Andrews, S.J.; Moore, K.N.; Chon, H.S.; Ivy, S.P. Randomized Phase II Trial of Triapine-Cisplatin-Radiotherapy for Locally Advanced Stage Uterine Cervix or Vaginal Cancers. *Front. Oncol.* **2019**, *9*, 1067. [CrossRef] [PubMed]
72. Vutey, V.; Castelli, S.; D'Annessa, I.; Samia, L.B.; Souza-Fagundes, E.M.; Beraldo, H.; Desideria, A. Human topoisomerase IB is a target of a thiosemicarbazone copper(II) complex. *Arch. Biochem. Biophys.* **2016**, *606*, 34–40. [CrossRef]
73. Kaur, P.; Johnson, A.; Northcote-Smith, J.; Lu, C.; Suntharalingam, K. Immunogenic Cell Death of Breast Cancer Stem Cells Induced by an Endoplasmic Reticulum-Targeting Copper(II) Complex. *Chembiochem* **2020**, *21*, 3618–3624. [CrossRef]
74. Tardito, S.; Isella, C.; Medico, E.; Marchio, L.; Bevilacqua, E.; Hatzoglou, M.; Bussolati, O.; Franchi-Gazzola, R. The thioxotriazole copper(II) complex A0 induces endoplasmic reticulum stress and paraptotic death in human cancer cells. *J. Biol. Chem.* **2009**, *284*, 24306–24319. [CrossRef] [PubMed]
75. Passeri, G.; Northcote-Smith, J.; Suntharalingam, K. Delivery of an immunogenic cell death-inducing copper complex to cancer stem cells using polymeric nanoparticles. *RSC Adv.* **2022**, *12*, 5290–5299. [CrossRef] [PubMed]
76. Li, D.D.; Yague, E.; Wang, L.Y.; Dai, L.L.; Yang, Z.B.; Zhi, S.; Zhang, N.; Zhao, X.-M.; Hu, Y.-H. Novel Copper Complexes That Inhibit the Proteasome and Trig-ger Apoptosis in Triple-Negative Breast Cancer Cells. *ACS Med. Chem. Lett.* **2019**, *10*, 1328–1335. [CrossRef] [PubMed]
77. Yu, Z.; Fenk, K.D.; Huang, D.; Sen, S.; Cowan, J.A. Rapid Telomere Reduction in Cancer Cells Induced by G-Quadruplex-Targeting Copper Complexes. *J. Med. Chem.* **2019**, *62*, 5040–5048. [CrossRef] [PubMed]
78. Deka, B.; Sarkar, T.; Banerjee, S.; Kumar, A.; Mukherjee, S.; Deka, S.; Saikia, K.K.; Hussain, A. Novel mitochondria targeted copper(ii) complexes of ferrocenyl terpyridine and anticancer active 8-hydroxyquinolines showing remarkable cytotoxicity, DNA and protein binding affinity. *Dalton Trans.* **2017**, *46*, 396–409. [CrossRef]
79. Weissleder, R. Molecular imaging in cancer. *Science* **2006**, *312*, 1168–1171. [CrossRef]
80. Louie, A. Multimodality imaging probes: Design and challenges. *Chem. Rev.* **2010**, *110*, 3146–3195. [CrossRef]
81. Ku, G.; Chen, J.; Vittal, J.J. Copper sulfide nanoparticles as a new class of photoacoustic contrast agent for deep tissue imaging at 1064 nm. *ACS Nano.* **2012**, *6*, 7489–7496. [CrossRef] [PubMed]
82. Zhang, H.F.; Maslov, K.; Stoica, G.; Wang, L.V. Functional photoacoustic microscopy for high-resolution and noninvasive in vivo imaging. *Nat. Biotechnol.* **2006**, *24*, 848–851. [CrossRef] [PubMed]
83. Siphanto, R.; Thumma, K.K.; Kolkman, R.G.M.; van Leeuwen, T.G.; de Mul, F.F.M.; van Neck, J.W.; van Adrichem, L.N.A.; Steenbergen, W. Serial noninvasive photoacoustic imaging of neovascularization in tumor angiogenesis. *Opt. Express* **2005**, *13*, 89–95. [CrossRef] [PubMed]
84. Yao, J.; Wang, L.V. Photoacoustic tomography: Fundamentals, advances and prospects. *Contrast Media Mol. Imaging* **2011**, *6*, 332–345. [CrossRef]

85. Zha, Z.; Deng, Z.; Li, Y.; Li, C.; Wang, J.; Wang, S.; Que, E.; Dai, Z. Biocompatible polypyrrole nanoparticles as a novel organic photoacoustic contrast agent for deep tissue imaging. *Nanoscale* **2013**, *5*, 4462–4467. [CrossRef] [PubMed]
86. Bao, B.; Yang, Z.; Liu, Y.; Xu, Y.; Gu, B.; Chen, J.; Su, P.; Tong, L.; Wang, L. Two-photon semiconducting polymer nanoparticles as a new platform for imaging of intracellular pH variation. *Biosens. Bioelectron.* **2019**, *126*, 129–135. [CrossRef] [PubMed]
87. Manohar, S.; Ungureanu, C.; van Leeuwen, T.G. Gold nanorods as molecular contrast agents in photoacoustic imaging: The promises and the caveats. *Contrast Media Mol. Imaging* **2011**, *6*, 389–400. [CrossRef]
88. Cherukula, K.; Manickavasagam Lekshmi, K.; Uthaman, S.; Cho, K.; Cho, C.S.; Park, I.K. Multifunctional inorganic nanoparticles: Recent progress in thermal therapy and imaging. *Nanomaterials* **2016**, *6*, 76. [CrossRef]
89. Zhou, M.; Ku, G.; Pageon, L.; Li, C. Theranostic probe for simultaneous in vivo photoacoustic imaging and confined photothermolysis by pulsed laser at 1064 nm in 4T1 breast cancer model. *Nanoscale* **2014**, *6*, 15228–15235. [CrossRef]
90. Ding, K.; Zeng, J.; Jing, L.; Qiao, R.; Liu, C.; Jiao, M.; Libc, Z.; Gao, M. Aqueous synthesis of PEGylated copper sulfide nanoparticles for photoacoustic imaging of tumors. *Nanoscale* **2015**, *7*, 11075–11081. [CrossRef]
91. Wang, S.; Zhang, L.; Zhao, J.; He, M.; Huang, Y.; Zhao, S. A tumor microenvironment–induced absorption red-shifted polymer nanoparticle for simultaneously activated photoacoustic imaging and photothermal therapy. *Sci. Adv.* **2021**, *7*, eabe3588. [CrossRef] [PubMed]
92. Bindra, A.K.; Wang, D.; Zheng, Z.; Jana, D.; Zhou, W.; Yan, S.; Wuac, H.; Zheng, Y.; Zhao, Y. Self-assembled semiconducting polymer based hybrid nanoagents for synergistic tumor treatment. *Biomaterials* **2021**, *279*, 121188. [CrossRef] [PubMed]
93. Phelps, M.E. Positron emission tomography provides molecular imaging of biological processes. *PNAS* **2000**, *97*, 9226–9233. [CrossRef]
94. Zhou, M.; Zhang, R.; Huang, M.; Lu, W.; Song, S.; Melancon, M.P.; Tian, M.; Liang, D.; Li, C. A chelator-free multifunctional [64Cu] CuS nanoparticle platform for simultaneous micro-PET/CT imaging and photothermal ablation therapy. *J. Am. Chem. Soc.* **2010**, *132*, 15351–15358. [CrossRef] [PubMed]
95. Suárez-García, S.; Esposito, T.V.; Neufeld-Peters, J.; Bergamo, M.; Yang, H.; Saatchi, K.; Schaffer, P.; Häfeli, U.O.; Ruiz-Molina, D.; Rodríguez-Rodríguez, C. Hybrid Metal–Phenol Nanoparticles with Polydopamine-like Coating for PET/SPECT/CT Imaging. *ACS Appl. Mater. Interfaces* **2021**, *13*, 10705–10718. [CrossRef]
96. Zhang, H.; Song, F.; Dong, C.; Yu, L.; Chang, C.; Chen, Y. Co-delivery of nanoparticle and molecular drug by hollow mesoporous organosilica for tumor-activated and photothermal-augmented chemotherapy of breast cancer. *J. Nanobiotechnology* **2021**, *19*, 290. [CrossRef]
97. Chen, C.; Ma, Y.; Du, S.; Wu, Y.; Shen, P.; Yan, T.; Li, X.; Song, Y.; Zha, Z.; Han, X. Controlled CRISPR-Cas9 Ribonucleoprotein Delivery for Sensitized Photothermal Therapy. *Small* **2021**, *17*, 2101155. [CrossRef]
98. Druzhkova, I.N.; Shirmanova, M.V.; Kuznetsova, D.S.; Lukina, M.M.; Zagaynova, E.V. Modern Approaches to Testing Drug Sensitivity of Patients' Tumors (Review). *Sovrem Tekhnologii Med.* **2021**, *12*, 91–102. [CrossRef]
99. Ganesan, K.; Wang, Y.; Gao, F.; Liu, Q.; Zhang, C.; Li, P.; Zhang, L.; Chen, J. Targeting Engineered Nanoparticles for Breast Cancer Therapy. *Pharmaceutics* **2021**, *13*, 1829. [CrossRef]
100. Es-Haghi, A.; Taghavizadeh, Y.M.; Sharifalhoseini, M.; Baghani, M.; Yousefi, E.; Rahdar, A.; Baino, F. Application of Response Sur-face Methodology for Optimizing the Therapeutic Activity of ZnO Nanoparticles Biosynthesized from Aspergillus niger. *Biomimetics* **2021**, *6*, 34. [CrossRef]
101. Sanaei, M.J.; Pourbagheri-Sigaroodi, A.; Kaveh, V.; Sheikholeslami, S.A.; Salari, S.; Bashash, D. The application of nano-medicine to overcome the challenges related to immune checkpoint blockades in cancer immunotherapy: Recent advances and opportunities. *Crit. Rev. Oncol. Hematol.* **2021**, *157*, 103160. [CrossRef] [PubMed]
102. Park, W.; Heo, Y.J.; Han, D.K. New opportunities for nanoparticles in cancer immunotherapy. *Biomater. Res.* **2018**, *22*, 24. [CrossRef] [PubMed]
103. Ahamed, M.; Akhtar, M.J.; Alhadlaq, H.A.; Alshamsan, A. Copper ferrite nanoparticle-induced cytotoxicity and oxidative stress in human breast cancer MCF-7 cells. *Colloids Surf. B. Biointerfaces* **2016**, *142*, 46–54. [CrossRef]
104. Rajagopal, G.; Nivetha, A.; Sundar, M.; Panneerselvam, T.; Murugesan, S.; Parasuraman, P.; Kumar, S.; Ilangoa, S.; Kunjiappanh, S. Mixed phytochemicals medi-ated synthesis of copper nanoparticles for anticancer and larvicidal applications. *Heliyon* **2021**, *7*, e7360. [CrossRef] [PubMed]
105. Wang, S.; Liu, J.; Qiu, S.; Yu, J. Facile fabrication of Cu9-S5 loaded core-shell nanoparticles for near infrared radiation mediated tumor therapeutic strategy in human esophageal squamous carcinoma cells nursing care of esophageal cancer pa-tients. *J. Photochem. Photobiol. B.* **2019**, *199*, 111583. [CrossRef] [PubMed]
106. Xu, R.; Zhang, K.; Liang, J.; Gao, F.; Li, J.; Guan, F. Hyaluronic acid/polyethyleneimine nanoparticles loaded with copper ion and disulfiram for esophageal cancer. *Carbohydr. Polym.* **2021**, *261*, 117846. [CrossRef]
107. Imyanitov, E.N.; Iyevleva, A.G.; Levchenko, E.V. Molecular testing and targeted therapy for non-small cell lung cancer: Cur-rent status and perspectives. *Crit. Rev. Oncol. Hematol.* **2021**, *157*, 103194. [CrossRef]
108. Herbst, R.S.; Morgenszstern, D.; Boshoff, C. The biology and management of non-small cell lung cancer. *Nature* **2018**, *553*, 446–454. [CrossRef]
109. Naatz, H.; Manshian, B.B.; Rios, L.C.; Tsikourkitoudi, V.; Deligiannakis, Y.; Birkenstock, J.; Pokhrel, S.; Mädler, L. Model-Based Nanoengineered Pharmacokinetics of Iron-Doped Copper Oxide for Nanomedical Applications. *Angew. Chem. Int. Ed. Engl.* **2020**, *59*, 1828–1836. [CrossRef]

110. Kalaiarasi, A.; Sankar, R.; Anusha, C.; Saravanan, K.; Aarthy, K.; Karthic, S.; Ravikumar, V. Copper oxide nanoparticles induce anticancer activity in A549 lung cancer cells by inhibition of histone deacetylase. *Biotechnol. Lett.* **2018**, *40*, 249–256. [CrossRef]
111. Giri, R.K.; Chaki, S.; Khimani, A.J.; Vaidya, Y.H.; Thakor, P.; Thakkar, A.B.; Pandya, S.J.; Deshpande, M.P. Biocompatible CuInS2 Nanoparticles as Poten-tial Antimicrobial, Antioxidant, and Cytotoxic Agents. *ACS Omega* **2021**, *6*, 26533–26544. [CrossRef] [PubMed]
112. Li, W.B.; Stangl, S.; Klapproth, A.; Shevtsov, M.; Hernandez, A.; Kimm, M.A.; Schuemann, J.; Qiu, R.; Michalke, B.; Bernal, M.A.; et al. Application of High-Z Gold Nanoparticles in Targeted Cancer Radiotherapy-Pharmacokinetic Modeling, Monte Carlo Simulation and Radiobiological Effect Model-ing. *Cancers* **2021**, *13*, 5370. [CrossRef] [PubMed]
113. Klapproth, A.P.; Shevtsov, M.; Stangl, S.; Li, W.B.; Multhoff, G. A New Pharmacokinetic Model Describing the Biodistribution of Intravenously and Intratumorally Administered Superparamagnetic Iron Oxide Nanoparticles (SPIONs) in a GL261 Xenograft Glioblastoma Model. *Int. J. Nanomed.* **2020**, *15*, 4677–4689. [CrossRef] [PubMed]
114. Wang, L.; Hu, C.; Shao, L. The antimicrobial activity of nanoparticles: Present situation and prospects for the future. *Int. J. Nanomed.* **2017**, *12*, 1227–1249. [CrossRef] [PubMed]
115. Zheng, R.; Cheng, Y.; Qi, F.; Wu, Y.; Han, X.; Yan, J.; Zhang, H. Biodegradable Copper-Based Nanoparticles Augmented Chemody-namic Therapy through Deep Penetration and Suppressing Antioxidant Activity in Tumors. *Adv. Healthc. Mater.* **2021**, *10*, e2100412. [CrossRef]
116. Koh, J.Y.; Lee, S.J. Metallothionein-3 as a multifunctional player in the control of cellular processes and diseases. *Mol. Brain* **2020**, *13*, 116. [CrossRef]
117. Lelievre, P.; Sancey, L.; Coll, J.L.; Deniaud, A.; Busser, B. The Multifaceted Roles of Copper in Cancer: A Trace Metal Element with Dysregulated Metabolism, but Also a Target or a Bullet for Therapy. *Cancers* **2020**, *12*, 3594. [CrossRef]
118. Camats, M.; Pla, D.; Gomez, M. Copper nanocatalysts applied in coupling reactions: A mechanistic insight. *Nanoscale* **2021**, *13*, 18817–18838. [CrossRef]
119. Mehdizadeh, T.; Zamani, A.; Abtahi, F.S. Preparation of Cu nanoparticles fixed on cellulosic walnut shell material and in-vestigation of its antibacterial, antioxidant and anticancer effects. *Heliyon* **2020**, *6*, e3528. [CrossRef]
120. Naikoo, G.; Al-Mashali, F.; Arshad, F.; Al-Maashani, N.; Hassan, I.U.; Al-Baraami, Z.; Faruck, L.H.; Qurashi, A.; Ahmed, W.; Asiri, A.M. An Overview of Copper Nanoparti-cles: Synthesis, Characterisation and Anticancer Activity. *Curr. Pharm. Des.* **2021**, *27*, 4416–4432. [CrossRef]
121. Akter, M.; Sikder, M.T.; Rahman, M.M.; Ullah, A.; Hossain, K.; Banik, S.; Hosokawa, T.; Saito, T.; Kurasaki, M. A systematic review on silver nanoparti-cles-induced cytotoxicity: Physicochemical properties and perspectives. *J. Adv. Res.* **2018**, *9*, 1–16. [CrossRef] [PubMed]
122. Da, S.D.; De Luca, A.; Squitti, R.; Rongioletti, M.; Rossi, L.; Machado, C.; Cerchiaro, G. Copper in tumors and the use of copper-based compounds in cancer treatment. *J. Inorg. Biochem.* **2022**, *226*, 111634. [CrossRef]
123. Fahmy, H.M.; Ebrahim, N.M.; Gaber, M.H. In-vitro evaluation of copper/copper oxide nanoparticles cytotoxicity and geno-toxicity in normal and cancer lung cell lines. *J. Trace. Elem. Med. Biol.* **2020**, *60*, 126481. [CrossRef] [PubMed]
124. Prasad, P.R.; Kanchi, S.; Naidoo, E.B. In-vitro evaluation of copper nanoparticles cytotoxicity on prostate cancer cell lines and their antioxidant, sensing and catalytic activity: One-pot green approach. *J. Photochem. Photobiol. B Biol.* **2016**, *161*, 375–382. [CrossRef] [PubMed]
125. Dong, C.; Feng, W.; Xu, W.; Yu, L.; Xiang, H.; Chen, Y.; Zhou, J. The Coppery Age: Copper (Cu)-Involved Nanotheranostics. *Adv. Sci.* **2020**, *21*, 2001549. [CrossRef]
126. Li, W.; Zamani, R.; Rivera Gil, P.; Pelaz, B.; Ibáñez, M.; Cadavid, D.; Shavel, A.; Alvarez-Puebla, R.A.; Parak, W.J.; Arbiol, J.; et al. CuTe Nanocrystals: Shape and Size Control, Plasmonic Properties, and Use as SERS Probes and Photothermal Agents. *J. Am. Chem. Soc.* **2013**, *135*, 7098–7101. [CrossRef] [PubMed]
127. Tian, Q.; Jiang, F.; Zou, R.; Liu, Q.; Chen, Z.; Zhu, M.; Yang, S.; Wang, J.; Wang, J.; Hu, J. Hydrophilic Cu9S5 Nanocrystals: A Photothermal Agent with a 25.7% Heat Conversion Effi-ciency for Photothermal Ablation of Cancer Cells in Vivo. *ACS Nano.* **2011**, *5*, 9761–9771. [CrossRef]
128. Tian, Q.; Tang, M.; Sun, Y.; Zou, R.J.; Chen, Z.G.; Zhu, M.F.; Yang, S.P.; Wang, J.L.; Wang, J.H.; Hu, J.Q. Hydrophilic Flower-Like CuS Superstructures as an Efficient 980 nm Laser-Driven Photother-mal Agent for Ablation of Cancer Cells. *Adv. Mater.* **2011**, *23*, 3542–3547. [CrossRef]
129. Liu, X.; Li, B.; Fu, F.; Xu, K.; Zou, R.; Wang, Q.; Zhang, B.; Chena, Z.; Hu, J. Facile synthesis of biocompatible cysteine-coated CuS nanoparticles with high photothermal conver-sion efficiency for cancer therapy. *Dalton Trans.* **2014**, *43*, 11709–11715. [CrossRef]
130. Wang, Z.; Tang, X.; Wang, X.; Yang, D.; Yang, C.; Lou, Y.; Chen, J.; He, N. Near-infrared light-induced dissociation of zeolitic imidazole framework-8 (ZIF-8) with encapsulated CuS nanoparticles and their application as a therapeutic nanoplatform. *Chem. Comm.* **2016**, *52*, 12210–12213. [CrossRef]
131. Zhang, S.; Huang, Q.; Zhang, L.; Zhang, H.; Han, Y.; Sun, Q.; Cheng, Z.; Qin, H.; Doub, S.; Li, Z. Vacancy engineering of Cu2−xSe nanoparticles with tunable LSPR and magnetism for dual-modal imaging guided photothermal therapy of cancer. *Nanoscale* **2018**, *10*, 3130–3143. [CrossRef] [PubMed]
132. Dong, L.; Ji, G.; Liu, Y.; Xu, X.; Lei, P.; Du, K.; Song, S.; Feng, J.; Zhang, H. Multifunctional Cu–Ag2S nanoparticles with high photothermal conversion efficiency for photoa-coustic imaging-guided photothermal therapy in vivo. *Nanoscale* **2018**, *10*, 825–831. [CrossRef] [PubMed]

133. Ji, M.; Xu, M.; Zhang, W.; Yang, Z.; Huang, L.; Liu, J.; Zhang, Y.; Gu, L.; Yu, Y.; Hao, W.; et al. aIStructurally Well-Defined Au@Cu2−xS Core–Shell Nanocrystals for Improved Cancer Treatment Based on Enhanced Photothermal Efficiency. *Adv. Mater.* **2016**, *28*, 3094–3101. [CrossRef] [PubMed]
134. Ding, X.; Liow, C.H.; Zhang, M.; Huang, R.; Li, C.; Shen, H.; Liu, X.; Zou, Y.; Gao, N.; Zhang, Z.; et al. Surface Plasmon Resonance Enhanced Light Absorption and Photothermal Therapy in the Second Near-Infrared Window. *J. Am. Chem. Soc.* **2014**, *136*, 15684–15693. [CrossRef]
135. Zhu, H.; Wang, Y.; Chen, C.; Ma, M.; Zeng, J.; Li, S.; Xia, Y.; Gao, M. Monodisperse Dual Plasmonic Au@Cu2−xE (E= S, Se) Core@Shell Supraparticles: Aqueous Fab-rication, Multimodal Imaging, and Tumor Therapy at in Vivo Level. *ACS Nano.* **2017**, *11*, 8273–8281. [CrossRef]
136. Chen, W.; Qin, M.; Chen, X.; Wang, Q.; Zhang, Z.; Sun, X. Combining photothermal therapy and immunotherapy against melanoma by polydopamine-coated Al2O3nanoparticles. *Theranostics* **2018**, *8*, 2229–2241. [CrossRef]
137. Wu, Z.-C.; Li, W.-P.; Luo, C.-H.; Su, C.H.; Yeh, C.S. Rattle-Type Fe3O4@CuS Developed to Conduct Magnetically Guided Photoinduced Hyper-thermia at First and Second NIR Biological Windows. *Adv. Funct. Mater.* **2015**, *25*, 6527–6537. [CrossRef]
138. Webb, B.A.; Chimenti, M.; Jacobson, M.P.; Barber, D.L. Dysregulated pH: A perfect storm for cancer progression. *Nat. Rev. Cancer* **2011**, *11*, 671–677. [CrossRef]
139. Estrela, J.M.; Ortega, A.; Obrador, E. Glutathione in Cancer Biology and Therapy. *Crit. Rev. Clin. Lab. Sci.* **2006**, *43*, 143–181. [CrossRef]
140. Harris, A.L. Hypoxia—a key regulatory factor in tumour growth. *Nat. Rev. Cancer* **2002**, *2*, 38–47. [CrossRef]
141. An, L.; Wang, X.; Rui, X.; Lin, J.; Yang, H.; Tian, Q.; Tao, C.; Yang, S. The In Situ Sulfidation of Cu2O by Endogenous H2S for Colon Cancer Theranostics. *Angew. Chem. Int. Ed* **2018**, *57*, 15782–15786. [CrossRef] [PubMed]
142. Tao, C.; An, L.; Lin, J.; Tian, Q.; Yang, S. Surface Plasmon Resonance–Enhanced Photoacoustic Imaging and Photothermal Therapy of Endog-enous H2S-Triggered Au@Cu2O. *Small* **2019**, *15*, 1903473. [CrossRef] [PubMed]
143. Cheng, Y.; Chen, Q.; Guo, Z.; Li, M.; Yang, X.; Wan, G.; Chen, H.; Zhang, Q.; Wang, Y. An Intelligent Biomimetic Nanoplatform for Holistic Treatment of Metastatic Tri-ple-Negative Breast Cancer via Photothermal Ablation and Immune Remodeling. *ACS Nano.* **2020**, *14*, 15161–15181. [CrossRef] [PubMed]

Article

Poly-α, β-D, L-Aspartyl-Arg-Gly-Asp-Ser-Based Urokinase Nanoparticles for Thrombolysis Therapy

Shuangling Chen, Meng Liang, Chengli Wu, Xiaoyi Zhang, Yuji Wang * and Ming Zhao *

School of Pharmaceutical Sciences, Capital Medical University, Beijing 100069, China
* Correspondence: wangyuji@ccmu.edu.cn (Y.W.); mingzhao@bjmu.edu.cn (M.Z.)

Abstract: The most concerning adverse effects of thrombolytic agents are major bleeding and intracranial hemorrhage due to their short half-life, low fibrin specificity, and high dosage. To alleviate bleeding side effects during thrombolytic therapy which would bring about the risk of aggravation, we try to find a novel biodegradable delivery nanosystem to carry drugs to target the thrombus, reduce the dosage of the drug, and system side effects. A novel urokinase/poly-α, β-D, L-aspartyl-Arg-Gly-Asp-Ser complex (UK/PD-RGDS) was synthesized and simply prepared. Its thrombolytic potency was assayed by the bubble-rising method and in vitro thrombolytic activity by the thrombus clot lysis assay separately. The in vivo thrombolytic activity and bleeding complication were evaluated by a rat model of carotid arteriovenous bypass thrombolysis. The thrombolytic potency (1288.19 ± 155.20 U/mg) of the UK/PD-RGDS complex nano-globule (18–130 nm) was 1.3 times that of commercial UK (966.77 ± 148.08 U/mg). In vivo, the UK/PD-RGDS complex (2000 IU/kg) could reduce the dose of UK by 90% while achieving the equivalent thrombolysis effect as the free UK (20,000 IU/kg). Additionally, the UK/PD-RGDS complex decreased the tail bleeding time compared with UK. The organ distribution of the FITC-UK/PD-RGDS complex was explored in the rat model. The UK/PD-RGDS complex could provide a promising platform to enhance thrombolytic efficacy significantly and reduce the major bleeding degree.

Keywords: urokinase; thrombolytic; targeted; nano delivery system; RGDS

1. Introduction

Thrombotic diseases seriously endanger the health and life of patients. It is believed that the best way to improve the survival rate and reduce the mortality rate of patients is immediate, early detection, and effective thrombolytic therapy [1]. In recent years a variety of technical methods have emerged, including drug thrombolytic therapy, interventional thrombolytic therapy, and interventional thrombolysis. However, thrombolytic therapy is still an important, irreplaceable, and fundamental measure for the treatment of thromboembolism [2]. Thrombolytic agents in clinical use are Plasminogen activators (PAs) to treat thromboembolism, including streptokinase (SK), urokinase (UK), alteplase (RT-PA), tissue plasminogen activator (tPA) and tenepase (TNK-TPA), etc. UK is an effective thrombolytic drug that has a low price and is widely used in primary hospitals [3]. However, due to the short half-life (2–20 min) [4], PAs need more doses to be administered. More doses of PAs could cause more serious system hemorrhage because the low thrombus-specificity of Pas activates both fibrin-bound and circulating plasminogen creating a serious risk of hemorrhage. These factors could cause a side effect risk of harmful bleeding complications and lead to the aggravation of the disease. Bleeding complications of Pas bring difficulties and risks to the clinical application of Pas including r-tPA with fibrin specificity [5,6]. On the whole, decreasing the bleeding complications of PAs could be very significant and is urgently needed for the clinical application of PAs.

Targeted delivery systems of PAs could target PAs to the site of the thrombus and at the same time reduce the dose of PAs. So, the systematic generation of broad matrix-specific fibrinolytic enzymes associated with bleeding complications could decrease and targeted delivery systems of PAs could be a better way to avoid or resolve dose-induced side effects [4]. Many targeted delivery systems have been performed to target PAs to the site of the thrombus, release PAs and perform effective thrombolytic treatment. Targeted delivery systems were generally prepared by directly bonding the thrombus-targeted ligands to PAs or the surface of drug carriers, such as liposomes [7,8], magnetic targeting delivery systems [9,10], microbubbles [11–13], polymer nanoparticles [2,9,14–16] and new inorganic-organic hybrid nanoparticles for deeper or continuous release in recent years [9,10]. The polymers in local delivery systems of PAs include PLGA Polymers, polyglutamic acid peptide dendrimer, chitosan derivatives, PEG, etc. [9,16–18]. These thrombus-targeted ligands include fibrin-specific anti-fibrin antibodies [19,20], vMF factor-specific alkaline gelatin [21], p-selectin-specific fucoidan [2,4], activated platelets-specific RGD sequence peptide, etc. [8,9,12,13,16]. Then, by wrapping or bonding PAs, these targeted delivery systems were constructed into thrombolytic targeting delivery systems. In addition, thrombin-sensitive peptides and pH-sensitive phenyl imine bonds also were used to develop PAs delivery systems and could be ruptured at the thrombus of the stroke to perform targeted thrombolytic therapy [3,4,18,21].

RGD sequence peptides exist in the α-chain of the fibrinogen and could specifically recognize Glycoprotein (GP) IIb/IIIa receptor on the activated platelet membrane. RGD sequence peptides have drawn much attention from researchers in the diagnosis of thrombosis and targeted thrombolytic therapy [22]. The cyclic RGD (cRGD) functionalized liposome has been used to carry urokinase to target thrombus in vivo thrombolysis study [8]. However, intravascular liposomes are limited due to stability problems. The major challenges in the development of liposomal PAs-targeting delivery systems are cost and still ineffective treatment [7,8]. In other research, it was shown that thrombolytic therapy by targeted microbubbles containing RGD sequence peptides under ultrasound could destroy the fibrillary network structure of the thrombus [13] and enhance the dissolution of the thrombus, while targeted nano-bubbles have a higher thrombolytic rate and penetrate deeper into thrombus than targeted micron bubbles [12]. Additionally, polymer nanoparticles are relatively stable carriers. Targeted nanoparticles, such as mesoporous carbon nanomaterials [10], poly(lactic-co-glycolic acid) magnetic nanoparticles [6], and chitosan nanoparticles [16]. Furthermore, non-bonding complex PAs-targeting delivery systems, such as the PAs complex [23,24], were used to target thrombus.

In our previous work, based on the specificity of RGD sequence peptides on activated platelets, multiple RGDS molecules have been bonded to the highly biodegradable poly-α, β-D, L-aspartic acid (PD) [25–27]. PD-RGDS with a high grafting rate of 46% was prepared to have a specific affinity for activating platelets [26]. In this study, we based on the interaction between proteins and constructed a UK/PD-RGDS complex delivery system (Scheme 1). We characterized the UK/PD-RGDS complex delivery system by Zeta Sizer and TEM. The thrombolytic potency of the UK/PD-RGDS complex was measured by the bubble-rising method. The in vivo thrombolytic activity, the bleeding complications, and organ distribution of UK/PD-RGDS were evaluated to investigate the thrombolysis efficacy and the side effects via male Wistar rats. Our study provided a foundation for the development of novel delivery systems for thrombolytic therapy.

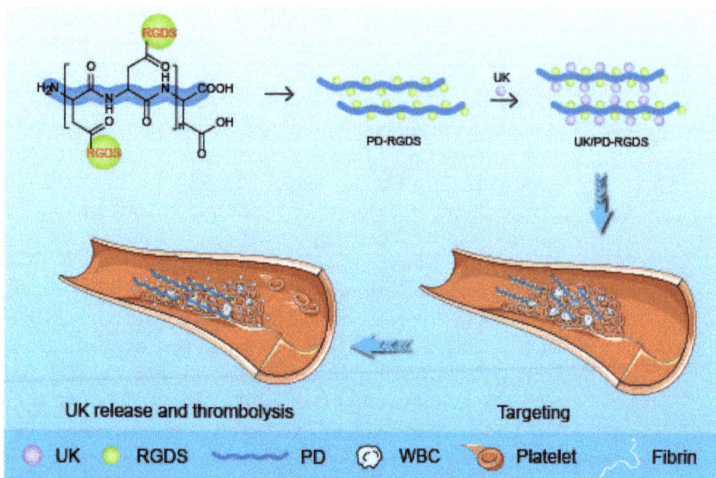

Scheme 1. The formation of the UK/PD-RGDS complex.

2. Results

2.1. Preparation of UK/PD-RGDS Nanosystem

PD-RGDS was synthesized by the condensation reaction of PD and HCl·Arg(Tos)-Gly-Asp(OBzl)-Ser(Bzl)-OBzl and the removal of protective groups at low temperatures. Its structure was characterized by ^{1}H NMR spectra, IR, and amino acid analysis. Its purity was determined by HPLC (7 mg/mL, Ultrahydrogel 120 columns, 7.8 × 30, 35) with a refractive index detector, eluted with 0.1 N NaNO$_3$ with a flow rate of 0.5 mL/min. The retention time of PD-RGDS was 11.5 min. The results of the amino acid analysis gave Asp:Arg:Gly:Ser = 3.2:1.0:1.2:1.2. It showed that 46 aspartic acids among 100 aspartic acid units were connected to RGDS.

The UK/PD-RGDS complex was simply prepared by mixing UK and PD-RGDS at 4 °C for 1 h. UK/PD-RGDS complexes with 1:5, 1:3, and 1:1 (w/w) in 10 mM PBS buffer (pH 7.4) were obtained to explore the effect of the complexation ratio on the system of the UK/PD-RGDS complex. These UK/PD-RGDS complexes were opalescent, uniform, and stable. Their z-average sizes, the size distribution of particles, and Zeta potential were measured and shown in Table 1. On the whole, the z-average sizes of all UK/PD-RGDS complexes (270–277 nm) were smaller than that of PD-RGDS (278 ± 3.772 nm). Their size distribution was narrow after the complexation of UK and PD-RGDS. The smaller sizes and narrow size distribution of the UK/PD-RGDS complexes implied that UK and PD-RGDS have a close interaction. Besides, the zeta potential of the UK/PD-RGDS complex (−15.2 ± 0.830 mV) was between that of the UK (−8.80 ± 0.285 mV) and PD-RGDS (−17.1 ± 0.993 mV) and closer to the zeta potential of PD-RGDS. The zeta potential suggested that both PD-RGDS and UK exposed on the surface of UK/PD-RGDS complexes and the ratio of PD-RGDS exposed on the surface should be more than UK.

Table 1. Z-average sizes, size distribution, and Zeta potential of UK/PD-RGDS complexes with different ratios.

Samples	Z-Average Size (nm)	PDI	Zeta Potential (mv)
PD-RGDS	278.2 ± 3.772	0.23 ± 0.010	−17.1 ± 0.993
UK	140.0 ± 0.7500	0.49 ± 0.040	−8.80 ± 0.285
UK/PD-RGDS (1:5 w/w)	270.2 ± 11.60	0.24 ± 0.029	−15.4 ± 1.10
UK/PD-RGDS (1:3 w/w)	277.0 ± 8.330	0.23 ± 0.018	−15.3 ± 0.540
UK/PD-RGDS (1:1 w/w)	272.0 ± 4.66	0.24 ± 0.013	−15.2 ± 0.830

The z-average size, size distribution, and Zeta potential of UK/PD-RGDS complexes in proportions of UK/PD-RGDS (1:5, 1:3, 1:1 w/w) were measured for 48 h to explore the different component ratios on the stability of complexes as shown in Figure 1A–C. The z-average sizes therein showed that the UK/PD-RGDS complex (1:1 w/w) (272.0 ± 4.66 nm) has a smaller range of variation over 48 h compared to the other two UK/PD-RGDS complexes ratio. So is the distribution and Zeta potential. Thus, we think that the UK/PD-RGDS complex (1:1 w/w) was the most stable according to the z-average size, size distribution, and Zeta potential. Transmission electron microscopy (TEM) revealed that the UK/PD-RGDS complex (1:1 w/w) appeared in the solid near circular spheres (18–131 nm) with a narrow size distribution, as shown in Figure 1D. Then, the complex of UK and PD-RGDS (1:1 w/w) was selected for the following experiments.

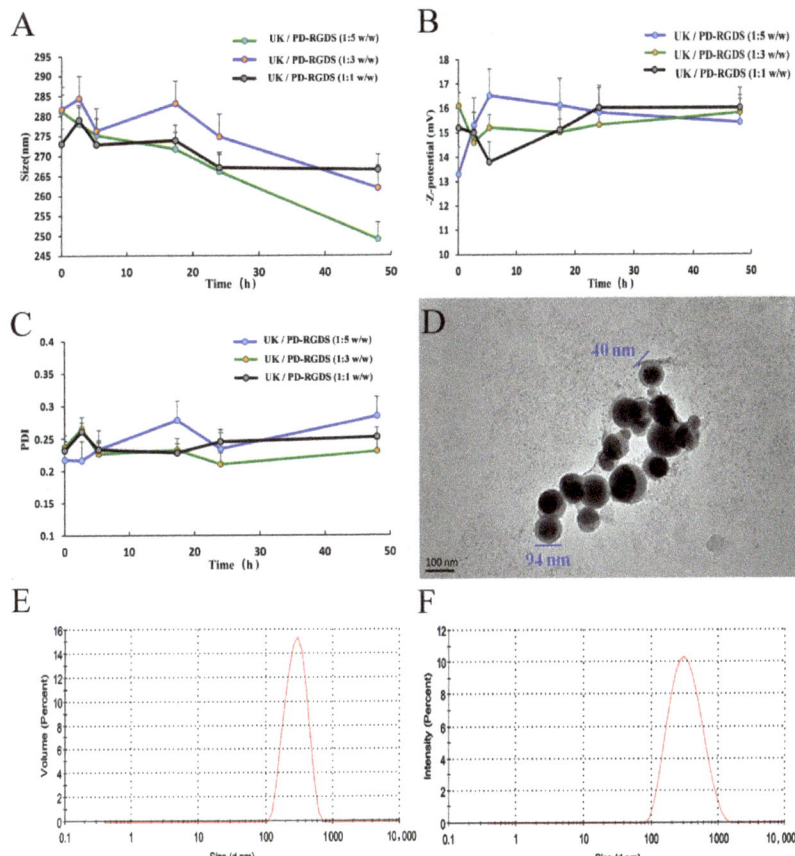

Figure 1. Physical characterization of the UK/PD-RGDS complex. (**A**) Effect of complexation ratio on z-average sizes of UK/PD-RGDS complexes in consecutive 48 h. (**B**) Effect of complexation ratio on Zeta potentials of UK/PD-RGDS complexes in consecutive 48 h. (**C**) Effect of complexation ratio on the size distribution of UK/PD-RGDS complexes in consecutive 48 h. (**D**) TEM image of the UK/PD-RGDS complex (1:1 w/w). (**E**) The size distribution of the UK/PD-RGDS complex (1:1) was measured by dynamic light scattering method. (**F**) The size distribution of the PD-RGDS measured by dynamic light scattering method.

2.2. Thrombolytic Potency of UK/PD-RGDS Complex

The bubble-rising method was adopted to determine the thrombolytic potency of UK after complexation according to the Pharmacopoeia of the People's Republic of China (2015).

The fibrin clot formed under the action of thrombin and the time of the small bubbles rising to half the volume of the reaction system from the bottom of the fibrin clot is related to the concentration of UK in Figure 2A. The logarithm of urokinase concentration showed a linear relationship with the logarithm of reaction time during the concentration of 3.53–14.12 U/mL ($y = -0.2184x + 2.8094$, $r^2 = 0.9930$). The thrombolytic potency results showed that the thrombolytic potency of the UK/PD-RGDS complex was 1288.19 ± 155.20 U/mg. Its thrombolytic potency increased and was 1.33 times of UK (966.77 ± 148.08 U/mg) at the same dose, which could not be explained.

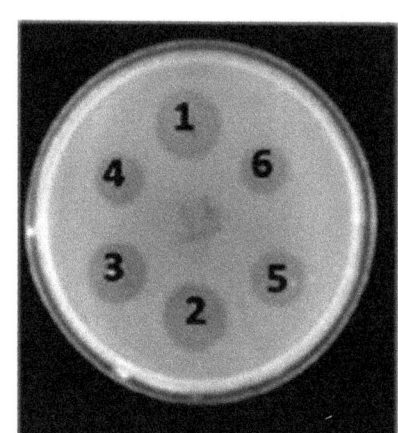

Figure 2. Bioassays of thrombolytic potency. (**A**) The thrombolytic potency was evaluated by the Bubble-rising method. (**B**) The thrombolytic potency was evaluated by the Agarose-fibrin plate method (n = 3). 1. 800 U/mL UK standard; 2. 600 U/mL UK standard; 3. 400 U/mL UK standard; 4. 200 U/mL UK standard; 5. UK/PD-RGDS complex; 6. The UK.

To further confirm the increasing thrombolytic potency of the UK/PD-RGDS complex, it was evaluated again by the Agarose-fibrin plate method. In the Agarose-fibrin plate method, agarose was used as a support and a fibrin plate was formed under the action of thrombin. Fibrin was decomposed by urokinase to generate a transparent circle at a certain temperature in Figure 2B. The reaction temperature was optimized by comparing at room temperature for 18 h, room temperature for 24 h, 37 °C for 3 h, and 37 °C for 24 h. The size of the transparent circle obtained at room temperature for 24 h is moderate, not easy to overlap and the error is small. Thus, room temperature for 24 h was selected in the following tests.

The areas of the transparent circles showed a linear relationship with the logarithm of UK concentration during the concentration of 200–800 U/mL ($y = 175.07x - 270.4$, $r^2 = 0.9971$). The thrombolytic potency of UK/PD-RGDS was 1290.80 ± 85.78 U/mg and 1.28 times of UK (1005.48 ± 66.68 U/mg) by the Agarose-fibrin plate method. Due to the thrombolytic potency results by two methods, the thrombolytic potency of UK after loading on the nanosystem of PD-RGDS increased and was consistent with each other.

Thus, the increasing thrombolytic potency of the UK/PD-RGDS complex was confirmed.

2.3. In Vivo Thrombolytic Activity of UK/PD-RGDS Complex

The in vivo thrombolytic activity of the UK/PD-RGDS complex by injection administration was evaluated in a rat model of carotid arteriovenous bypass thrombolysis (Figure 3A) [28]. The thrombus was prepared first (Figure 3B) and put into the carotid arteriovenous bypass by assembling. Via operation, the in vivo rat model of carotid arteriovenous bypass thrombolysis was established. After UK or the UK/PD-RGDS complex was injected, the blood began to circulate via carotid arteriovenous bypass. The weight of the thrombus in the bypass would decrease under the thrombolytic action of UK or the UK/PD-RGDS complex. Then the reduction in the thrombus weights after 60 min was used to evaluate the thrombolytic activity. In the evaluation, the rats were divided into three groups. NS (3 mL/kg) was used as the blank control group. The commercial UK (20,000 IU/kg) at a clinical dose (20,000 IU/Kg) was used as the positive control and the UK/PD-RGDS complex (2000 IU/kg) was used as the test group.

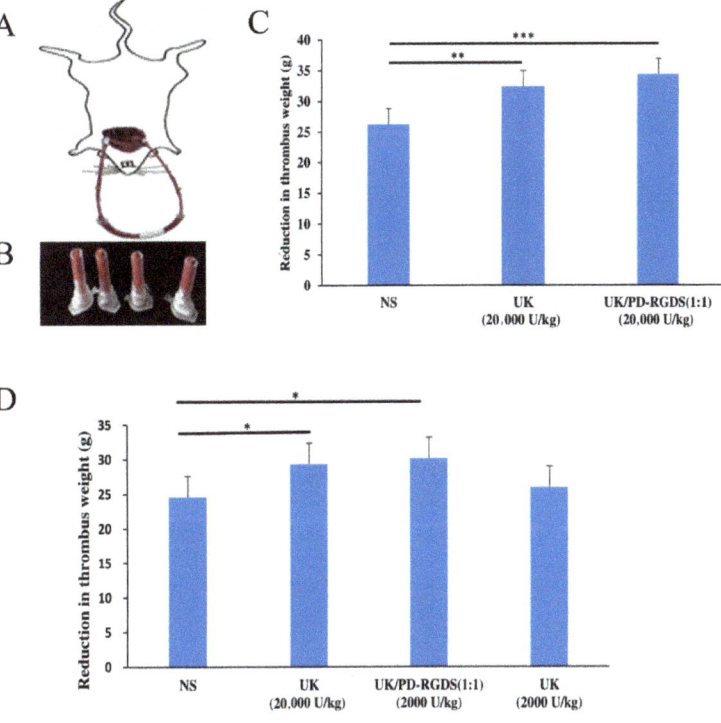

Figure 3. The assay on in vivo thrombolytic activity. (**A**) The in vivo rat model of carotid arteriovenous bypass thrombolysis. (**B**) The thrombus was prepared first. (**C,D**) In vivo thrombolytic activity of commercial UK and UK/PD-RGDS complex at high dose of 20,000 U/kg (**C**), and at dose of 2000 U/kg (**D**) (n = 6). Notes: * $p < 0.05$; ** $p < 0.01$; *** $p < 0.001$.

Figure 3C showed the in vivo thrombolytic activity results of both the positive control and our sample. Initially, we used a dosage of 20,000 IU/Kg (the clinical dose of UK) in the in vivo assay to compare the thrombosis activity between our sample (the UK/PD-RGDS complex at 1:1 ratio) and the commercial UK. However, the result showed that both of positive control (UK) and our sample (UK/PD-RGDS complex) are active and have similar in vivo thrombolytic potency (Figure 3C). We thought that both UK and our sample may reach their maximal thrombolytic activity at 20,000 IU/kg. To assess whether our sample is more potent than commercial UK, we lowered the dose of our sample to 2000 IU/kg and evaluated them in a second in vivo assay. Compared with the NS group, there was no significant difference in thrombus weight reduction in the UK (2000 IU/kg) group ($p > 0.05$). The results showed that a low dose of our sample was as effective as high doses, whereas a low dose of UK was inactive (Figure 3D). Therefore, we claimed that the thrombolytic activity of the UK/PD-RGDS complex was enhanced compared to UK.

2.4. Tail Bleeding Time of UK/PD-RGDS Complex

In the evaluation of the tail bleeding time of the UK/PD-RGDS complex, the rats were divided into three groups. The tail bleeding time of rats before administration was used as a blank control group. The UK group (20,000 IU/kg) was used as a reference control. The UK/PD-RGDS complex (2000 IU/kg) group was used as the test group. The tail bleeding time results (Figure 4) showed that the tail bleeding time of the UK/PD-RGDS complex group (428 ± 137 s) was significantly lower than in the UK group (692 ± 141 s) ($p < 0.05$, n = 5). The UK/PD-RGDS complex could reduce the side effect of bleeding.

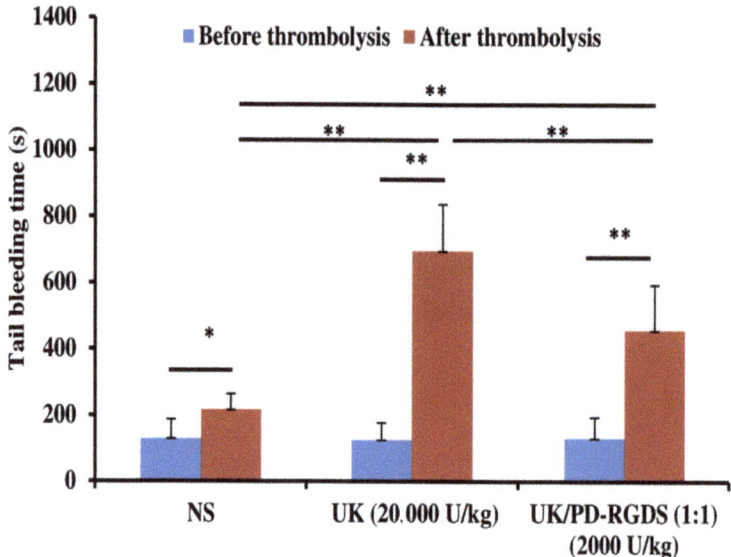

Figure 4. Tail bleeding times of with UK/PD-RGDS complexes (n = 5) * $p < 0.05$; ** $p < 0.01$.

2.5. In Vitro Thrombus Clot Lysis Assay

Thrombus clots were first prepared from rat arterial blood (65.0–75.0 mg). Thrombus clots were divided into three groups, the NS group, the group of (100 IU/mL UK), and the group of the UK/PD-RGDS complex (100 IU/mL) at 37 °C for 3 h. The reduced weights of thrombus clots present the lysis activity to thrombus clots of these compounds. Figure 5 showed the in vitro "thrombolytic activity" results of UK and the UK/PD-RGDS complex at a lower dose (n = 9). Not surprisingly, both groups showed similar potency in this assay, since they contained the same dose of functional agent UK. Our results indicated that PD-RGDS did not directly enhance the activity of UK by forming nanoparticles with UK. The UK/PD-RGDS complex is designed to improve the pharmacokinetic profile of commercial UK by increasing its stability and delivering it to the thrombus.

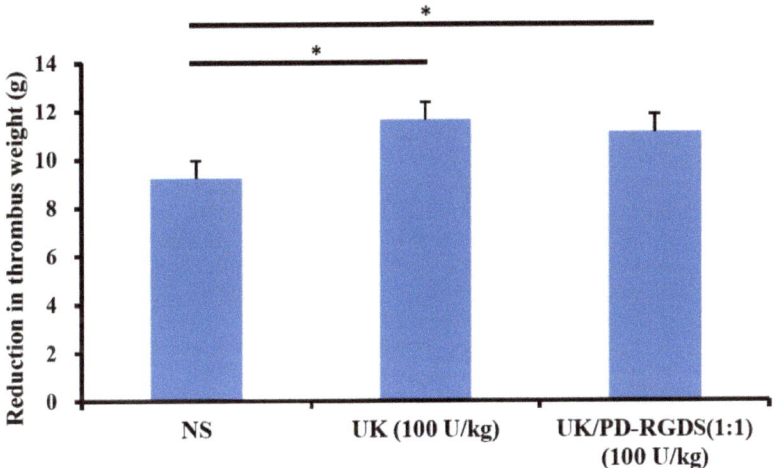

Figure 5. The thrombus clot lysis activity of UK/PD-RGDS complex (n = 9). * $p < 0.05$.

2.6. Organ Distribution Study

FITC-UK (F/P = 4.73) was performed in the organ distribution study. In the organ distribution study, the rats were divided into three groups. The NS group was used as a blank control. FITC-UK (8000 U/kg) was used as a reference control. FITC-UK/PD-RGDS complex (8000 U/kg) was used as the test group. Results in Figure 6 showed that FITC-UK in both FITC-UK or FITC-UK/PD-RGDS complex was mainly distributed in the liver and blood, while ratios of FITC-UK were very low in the spleen and lungs. The FITC-UK ratio of the FITC-UK/PD-RGDS complex group in the thrombus was significantly improved ($p < 0.05$, n = 5) compared with the FITC-UK group and was 2.3 times than the FITC-UK group. It can be seen that FITC-UK in the complex group was significantly accumulated at the site of the thrombus compared with the FITC-UK group, which suggested the thrombus targeting effect of the FITC-UK/PD-RGDS complex. It was because the RGDS in the FITC-UK/PD-RGDS complex highly binds to the activated platelet membrane GPIIb/IIIa in thrombus [26]. The ratio of the FITC-UK/PD-RGDS complex in the blood was also significantly increased ($p < 0.01$, n = 5) and 1.6 times of FITC-UK.

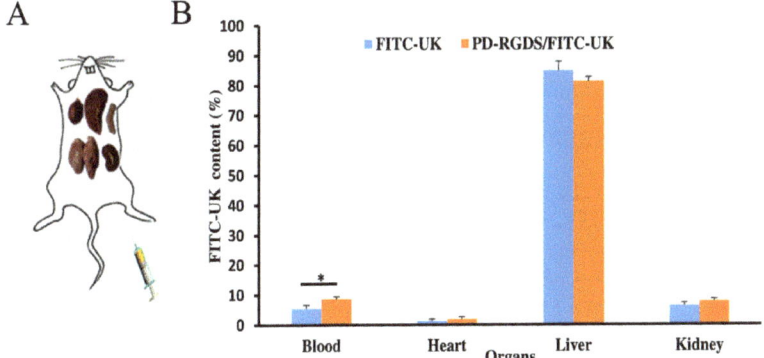

Figure 6. The assay on organ distribution. (**A**) The model of organ distribution study. (**B**) Organ distribution of FITC-UK/PD-RGDS complex and FITC-UK in vivo (n = 5). * $p < 0.01$.

3. Discussion

Thrombolytic therapy is still the most basic treatment method and the fundamental measure to treat thrombosis. Plasminogen activator is currently an effective thrombolytic drug in clinical practice. Plasminogen activators (PAs) systematically activate plasminogen to become plasmin, then the produced plasmins degrade fibrinogen and fibrin in the clots, and decompose the thrombus. When excessive plasmins are produced, bleeding of different degrees occurs. Because PAs was a group of proteases and would be degraded quickly in the blood, PAs have short half-life periods (2–20 min). To achieve the effect of thrombolytic therapy, large doses of drugs were used to reach the high blood concentration for treatment, which increases the risks of bleeding side effects and non-specific toxicity. Reducing bleeding side effects could improve the safety of the medication and reduce pressure for doctors and the risk of bleeding for patients, which was urgent and meaningful.

A targeted delivery system could target thrombolytic drugs to the thrombus site and release thrombolytic drugs for effective thrombolysis. So, targeted delivery systems could reduce the dose to achieve the treatment effect of thrombolytic therapy. Meanwhile, a targeted delivery system will also reduce the bleeding side effects caused by the increasing dose of PAs, and is the best way to avoid or solve the side effects caused by the dose [1].

The RGDS sequence peptide exists at 572–575 on the alpha chain of fibrinogen. It can specifically recognize the glycoprotein (GP) IIb/IIIa receptor on activated platelet membranes, which has been widely studied for the diagnosis of thrombosis and targeted thrombolytic therapy in recent years. RGDS can competitively bind activated platelets, thus preventing the fibrin bridging, and has the effect of inhibiting platelet aggregation. Here, PD-RGDS was a safe and biodegradable polyamino acid carrier; the main chain is biodegradable poly-α, β-D, L-aspartic acid, and the side chain consists of amino-group of poly-α, β-D, L-aspartic acid bonded with the carboxyl group of RGDS. So PD-RGDS also could specifically recognize the glycoprotein (GP) IIb/IIIa receptor on the activated platelet membrane. Moreover, in this study, the PD-RGDS carried plenty of RGDS motifs (grafting ratio 46%). The Effect of PD-RGDS on GPIIb/IIIa expression results also confirmed that PD-RGDS (10^{-5} M) could reduce to one-thousandth of the concentration of RGDS while achieving the equivalent binding to the GPIIb/IIIa receptors on the platelet surface as RGDS (2.5×10^{-2} M) [26]. Because plenty of RGDS motifs are endowed with high specific binding to activated platelets, PD-RGDS could target thrombus better. In addition, the transmission electron microscopy results showed that PD-RGDS existed as nanoparticles of 60–108 nm which was more beneficial to construct a nano-sized drug delivery system than a micron-sized carrier.

The preparation method of the UK/PD-RGDS complex was simple, green, and has a short mixing time, which better protected the activity of UK than the UK conjugation delivery system. The z-coverage sizes and zeta potentials results implied that the UK/PD-RGDS complex could exist stably for 3 days. Moreover, in UK/PD-RGDS complexes UK and PD-RGDS had a close interaction because the UK/PD-RGDS complex has smaller sizes and narrow size distribution than UK and PD-RGDS. The zeta potential results showed both UK and PD-RGDS are exposed on the surface of the UK/PD-RGDS complex. We think the complexation and closer binding of UK and PD-RGDS could cause the changing of UK and PD-RGDS conformation and be in favor of the increase in UK potency. This may be the reason for the increasing thrombolytic potency of the UK/PD-RGDS complex. In addition, the transmission electron microscopy results showed that UK/PD-RGDS existed as nanoparticles of 18–131 nm. Fibrin clots highly inhibit the penetration of particles of 1 µm or larger into fibrin clots [29], which has allowed the nanosized system to accelerate thrombolytic therapy without causing microbubbles and holes. The nanosized UK/PD-RGDS was more beneficial to penetrate the thrombus for thrombolysis than the micron-sized carrier.

Platelets are the main targets of thrombus [7]. RGD sequence peptides could specifically bind to activated platelets by targeting GPIIb/IIIa on the surface of platelets [30,31]. PD-RGDS containing 46% RGDS has specifically adhered to activated platelets [26]. Then

PD-RGDS containing these RGDS motifs loaded UK specifically to the thrombus site and then UK was concentrated to dissolve the local thrombus. Therefore, less dose of the UK/PD-RGDS complex (2000 IU/kg) could show significant thrombolytic activity as the free urokinase at the dose of 20,000 IU/kg in a rat model of carotid arteriovenous bypass thrombolysis. It could be explained by the increasing thrombolytic potency, high ratio targeting factors, and nano-sized particles. Firstly, the thrombolytic potency of the UK/PD-RGDS complex increased compared with the UK by the Bubble-rising method. That is to say, the structure of the UK/PD-RGDS complex could help to increase the thrombolytic potency of the UK and cause the stronger thrombolytic activity of the UK/PD-RGDS complex group in vivo to a certain extent. Moreover, the grafting rate of RGDS in PD-RGDS is very high and reaches up to 46%. The high grafting rate of RGDS in PD-RGDS could help UK better to concentrate on the thrombus and increase the thrombolytic activity. After complexation with PD-RGDS, UK at the dose of 2000 IU/kg had significant thrombolytic activity. However, the in vivo UK group (2000 IU/kg) has no significant thrombolytic activity compared with the NS group. It suggested that the thrombolytic activity of the UK/PD-RGDS complex group (2000 IU/kg) was 10 times of the UK group. In addition, the nanoscale UK/PD-RGDS complex system could be beneficial to dissolve thrombus. It showed that polymer conjugation by grafting with targeted motifs is a good method as the drug carrier when we construct a targeted nano-delivery system.

The tail bleeding time results indicate that the UK/PD-RGDS complex could reduce the side effect of bleeding. Meanwhile, the UK/PD-RGDS complex could improve thrombolytic activity. Therefore, this study achieved the purpose of our expected research design. The tail bleeding times after NS administration was significantly higher than that before NS administration ($p < 0.05$, n = 5) (Figure 4), indicating that the tail bleeding time of the blank control group was increased. It was perhaps caused by heparin (140 U/kg) added to the carotid arteriovenous bypass of rats.

The interaction between UK and PDRGDS could not be analyzed because the determination of protein interactions by isothermal calorimetric titration requires the unavailable UK sample with a single molecular weight; our UK is a mixture of high molecular weight UK (Mw 54,000) and low molecular weight UK (Mw 33,000).

4. Materials and Methods

4.1. Materials

All amino acids were purchased from Sichuan Sangao Biochemical Co., Ltd. (Chengdu, China). Urokinase for injection (100,000 units) was purchased from Peking University Gaoke Huatai Pharmaceutical. Bovine fibrinogen Standard (Lot 140607-201841), Bovine thrombin Standard (Lot 140605-201526), Bovine fibrinogen Standard (Lot 140606-201826), bovine fibrinogen Standard (Lot 140606-201826), Bovine fibrinogen standard (Lot 140606-201826), Bovine fibrinogen standard (Lot 140606-201826), Bovine fibrinogen standard (Lot 140606-201826) and Urokinase standard (Batch No. 140604-201224) were purchased from China National Institute for Food and Drug. Bovine thrombin(SLBV3604) and Fluorescent isothiocyanate yellow (FITC)were purchased from Sigma Company (Shanghai, China). Agarose (Batch No. 424G056) was acquired from Beijing Solebo Technology Co., Ltd. (Beijing, China). Barbiturate-sodium chloride buffer (pH 7.8, DZ331) was purchased from Xi'an Hutt Biological Company (Xian, China). Trimethylol aminomethane buffer (pH 9.0) was purchased from Beijing Regen Biotechnology Co., Ltd. (Beijing, China). Other reagents were purchased from Sinopharm Chemical Reagent Co. Ltd. (Shanghai, China).

4.2. Preparation of Poly-α, β-D, L-Aspartyl-Arg-Gly-Asp-Ser (PD-RGDS)

The preparation of poly-α, β-D, L-aspartyl-Arg-Gly-Asp-Ser was carried out according to the method in the literature [26]. In short, 82.8 mg of PD was dissolved in 1 mL of anhydrous DMF, then 97 mg of HoBt and 360 mg of HCl·EDC were added to an ice bath. After 0.5 h, HCl·Arg(Tos)-Gly-Asp(OBzl)-Ser(Bzl)-OBzl was added and the pH value was adjusted to 9. After 24 h, the reaction solution was dried, extracted by ether, and washed separately with 5% KHSO$_4$, and water three times. The solid was dried at 37 °C under reduced pressure for 48 h and provided 197 mg of the yellowish powder. The yellowish powder was dissolved and mixed in 8 mL of CF$_3$CO$_2$H:CF$_3$SO$_3$H (3:1) at 0 °C for 75–90 min. Then it was triturated with 150 mL of ether and the residue was mixed with water and dissolved until the pH value was adjusted to 7. After centrifugation for 30 min, the supernatant was dialyzed for 3 days with ultrapure water and lyophilized to provide 197 mg (72%) of the title compound as a white powder.

4.3. Preparation of UK/PD-RGDS Complex

An amount of 8 mg of PD-RGDS and 8 mg of UK were mixed in 3 mL of pH 7.4 PBS buffer (10 mM) and stirred at 4 °C for 1 h to prepare the solution of 6.667 IU/mL UK/PD-RGDS complex.

4.4. Morphology of UK/PD-RGDS Complex

The sizes and Zeta potentials of samples (2.67 mg/mL)in pH 7.4 PBS buffer (10 mM) were determined in the automatic measurement mode on Malvern's Zeta Sizer (Nano-ZS90). The sizes and morphology of UK/PD-RGDS complex particles were observed by transmission electron microscopy (JEM-2100, Japan). The solutions of the UK/PD-RGDS complex (10^3, 10^2, 10^{-2}, 10^{-5}, 10^{-7}, 10^{-9} nM) were prepared and dropped onto a formvar-coated copper grid as TEM samples. Then a drop of ethanol was added. The copper grid is first placed in the air to dry completely. The samples were observed by transmission electron microscopy (JSM-6360 LV, JEOL, Tokyo, Japan) with an electron beam acceleration voltage of 120 kV. All samples were prepared in three copies.

4.5. Bioassays of UK/PD-RGDS Nanosystem

4.5.1. Bubble-Rising Method

The test of the Bubble-rising method was performed according to Pharmacopoeia of the People's Republic of China, Part II (2015 Edition); 6.67 mg/mL bovine fibrinogen standard in barbiturate-sodium chloride buffer (pH 7.8), 6.0 bp/mL bovine thrombin standard in barbiturate-sodium chloride buffer (pH 7.8), and 1 casein unit/mL bovine plasminogen in Tris (Hydroxymethyl) aminomethane buffer solution (pH 9.0) was prepared first. Bovine thrombin and bovine plasminogen were mixed with equal volume and the mixed solution was obtained. A 60 units/mL standard solution of urokinase in a barbiturate-sodium chloride buffer (pH 7.8) was also prepared.

The UK/PD-RGDS complex prepared by method 2.2 was quantitatively diluted with barbiturate-sodium chloride buffer (pH 7.8) to the concentration of 60 units/mL UK. The sample of UK control was also prepared according to method 2.2; 0.3 mL of bovine fibrinogen was added to each tube and put at 37 ± 0.5 °C in a water bath. Then 0.9 mL, 0.8 mL, 0.7 mL, and 0.6 mL barbiturate-sodium chloride buffer (pH 7.8) were added, respectively; 0.1 mL, 0.2 mL, 0.3 mL, and 0.4 mL of UK standard solution were added successively. After that 0.4 mL of the mixed solution was added, and each tube fully oscillated until the reaction system was full of bubbles and the bubbles stay in the system. The reaction system usually condenses in 30~40 s. The end of timing was recorded when the small bubbles rose to half the volume of the reaction system in the clot. All the samples were carried out in triplicate. The potency of the UK samples and UK/PD-RGDS complex samples were measured and determined by converting measurements to the thrombolytic potency through a standard curve between the logarithm of time versus the logarithm of the concentration of urokinase.

4.5.2. Agarose-Fibrin Plate Method

A concentration of 8 mg/mL bovine fibrinogen standard, 1 mg/mL bovine thrombin standard, and 8 mg/mL agarose standard in 10 mM PBS buffer (pH 7.4) were prepared separately. Then 8 mg/mL agarose was heated in a microwave oven until boiled. After the agarose was completely dissolved, the agarose was placed in a hot water bath at 52 °C for later use; 800 U/mL, 600 U/mL, 400 U/mL, and 200 U/mL UK standard in 10 mM PBS buffer solution (pH 7.4) were prepared separately. UK samples and UK/PD-RGDS samples were diluted to an appropriate concentration with 10 mM PBS buffer (pH 7.4). Three Petri dishes (9 cm in diameter) were taken and numbered; 18 mL of agarose solution was added to a small beaker. Then 1 mL of bovine thrombin and 1 mL of bovine fibrinogen were added to the beaker. They were shaken thoroughly and quickly poured into a disposable Petri dish. A homemade punch was used and the mixtures stayed at room temperature for 1 h. After the agarose was completely solidified, the liquid in the well was drained; 5 µL of 800 U/mL, 600 U/mL, 400 U/mL, and 200 U/mL UK standard was added separately into the holes. After the mixture has been placed at room temperature for 24 h, transparent rings on the agarose-fibrin plate were observed. The potency of UK samples and UK/PD-RGDS complex samples were measured and determined by converting measurements to the thrombolytic potency through a standard curve between concentrations of UK versus the area of the transparent rings.

4.6. In Vitro Thrombus Clot Lysis Assay

The in vitro thrombus, clot lysis assay was carried out according to the procedure reported earlier [28]. Male Wistar rats (220 g ± 10 g) were anesthetized with pentobarbital sodium (20%, 7 mL·kg^{-1}, i.p.). The right carotid artery was isolated and the whole blood was collected in centrifuge tubes. The whole blood was injected into a flexible rubber hose (D 1.7 cm) containing a helix (L 15 mm; D 1.0 mm). After 40 min the thrombus with helix was carefully removed and suspended in the tri-distilled water for 1 h at room temperature. The surface water was removed using filter paper and the thrombus was weighed precisely. Then they were immersed into 8 mL NS, UK (100 IU/mL), or UK/PD-RGDS complex (100 IU/mL), respectively, at 37 °C at 70 rpm in a shaker for 3 h. The thrombi were removed and the surface water was gently removed by filter paper. The reduced weight of the thrombus was used to compare the degree of thrombus clot lysis.

4.7. In Vivo Thrombolytic Activity

Male Wistar rats (210–250 g) were anesthetized with pentobarbital sodium (20%, 7 mL·kg^{-1}, i.p.). The right common carotid artery and the left vein were operated on and isolated. The whole blood was collected from the right common carotid artery and used to prepare the thrombus clots for 40 min. The surface blood of the thrombus with helix was removed using filter paper and the thrombus was weighed precisely. The thrombus was put into a polyethylene tube as an external circulation pipeline between the right common carotid artery and the left vein. These pipelines were filled with heparin sodium (50 IU/mL NS solution). One end was inserted into the left internal jugular vein and after the heparin sodium (200 U/kg) was injected the other end was inserted into the right carotid artery. NS (3 mL/kg), UK (20,000 IU/kg), or UK/PD-RGDS complex (2000 IU/kg) were injected near the venous end. After the blood was circulated for 60 min, the thrombus was taken out and weighed after the surface blood was absorbed. The reduced weight of the thrombus was used to represent their thrombolytic activity in vivo.

4.8. Determination of the Tail Bleeding Time

The tail bleeding time was assayed as described previously with a few small modifications [32]. The operating method was referred to in 4.7. The difference is that the dose of heparin sodium (200 U/kg) was adjusted to 140 U/kg and only used to fill the tube instead of intravenous injection. Bleeding times were measured at 40 min before and after the thrombolytic treatment. The rat tail was cut off at 1 mm near the tail tip and placed in 25 mL of normal saline at 37 °C. The occurrence of uniform and continuous bloodlines was taken as the beginning of timing. The complete cessation of bleeding was recorded as the bleeding time. If the bleeding does not stop at 1800 s, it is classified as 1800 s.

4.9. Organ Distribution Study

Firstly, UK was labeled by FITC as in previous articles [33]; 10 mg of urokinase was dissolved in 1 mL of 10 mmol/L PBS (pH 7.1); 2.6 mg of fluorescence isothiocyanate yellow (FITC) was dissolved in 100 µL of 0.5 mol/L carbonate buffer (pH 9.5). FITC was dropped into urokinase and stirred in a shading environment at room temperature for 4 h. Then the mixture was centrifuged at 2500 r/min for 25 min. The supernatant was dialyzed in 10 mmol/L PBS buffer (pH 8.0) for 2–4 h. After dialyzation, FITC-UK was purified by Sephadex G50 column and eluted with 10 mmol/L PBS buffer (pH 7.1). The fluorescence of FITC-UK was measured by F-2500 Fluorescence Spectrophotometer. FITC-UK was diluted appropriately until its OD280 was close to 1.0. Its OD value was read at 495 nm and 280 nm. F/P value was calculated according to the following formula: $F/P = 2.87 \times OD_{495}/(OD_{280} - 0.35 \times OD_{495})$.

In vivo organ distribution was evaluated as in previous articles [34]. Male Wistar rats (250–300 g) were anesthetized with pentobarbital sodium (20%, 7 mL·kg^{-1}, i.p.). After the right common carotid artery was separated, filter paper (1.3cm wide) soaked in 25% FeCl$_3$ saturated solution and a small piece of Para membrane (1.7 cm wide) was put under the artery for 15 min to induce the formation of carotid artery thrombosis. After blood reperfusion for one hour, NS solution (0.3 mL/kg), FITC-UK/PD-RGDS complex solution (8000 U/kg), or FITC-UK solution (8000 U/kg) were injected via the femoral vein. The rats were sacrificed 1-h post-administration. The liver, spleen, kidneys, lungs, heart, and thrombus clots are separated and taken out. About 1 g of each tissue was added into 3 mL homogenizing buffer (0.32 M sucrose, 100 mM HEPES, pH 7.4), and homogenized in a glass homogenizer. After homogenization, the liquid was centrifuged (4000 rpm for 15 min) to obtain the supernatant samples of each tissue. The supernatant powder samples of each tissue were obtained by freeze-drying. The appropriate amount of supernatant powder samples of each tissue was dissolved in an appropriate solution (1% Triton X-100, 100 mM NaCl, 0.1% SDS, 0.5% Na-Deoxycholate) and cultured at 4 degrees for 30 min to obtain the final test samples of organs.

Standard solutions of 20, 10, 5, 2.5, 1, 0.5, 0.25, and 0.1 mg/L FITC-UK were prepared with 10 mM PBS buffer solution (pH 7.4). The fluorescence intensity was measured at the excitation wavelength (493.0 nm) and emission wavelength (524.0 nm). The concentration of FITC-UK in each tissue was determined by converting measurements to concentrations through a standard curve between the fluorescence intensity versus the concentration of the UK.

The described assessments were approved by the Ethics Committee of Capital Medical University. The committee assures the welfare of the animals was maintained under the requirements of the Animal Welfare Act and according to the guideline for the care and use of laboratory animals.

5. Conclusions

RGD sequence peptides could have specific affinities to activate platelets in the blood clot. In the present study that the UK/PD-RGDS complex delivery system was constructed by PD-RGDS containing multiple RGDS based on the hypothesis of increasing the thrombolytic activity and decreasing the bleeding complications. Results showed that in vivo, the UK/PD-RGDS complex group greatly improved the thrombolytic activity compared with the UK group and significantly reduced the bleeding at the same time. Nanoscale agents could help to penetrate the thrombus much deeper and loosen the fibrin network in the thrombus [12]. The UK/PD-RGDS complex nanosphere (270–277 nm) could penetrate the thrombus deeper and have better thrombolytic efficiency. Secondly, the thrombolysis efficiency of UK/PD-RGDS complex carrying RGDS peptides was higher [15]. The cRGD liposomes could significantly reduce the dose of urokinase by 75% [8]. The UK/PD-RGDS complex could decrease the dose of urokinase by 90%, which may be because of the high grafting rate of PD-RGDS (46%). An assay of thrombolytic potency showed that complexation of UK and PD-RGDS improved the thrombolytic potency of UK. The results of the organ distribution study revealed that the FITC-UK/PD-RGDS complex group has more FITC-UK levels in thrombus than the UK group. The in vitro thrombolytic activity of the UK/PD-RGDS complex group was reserved. The UK/PD-RGDS complex improved the thrombolytic effect of thrombolytic agents and could decrease the dose of UK by 90%. Thus, the UK/PD-RGDS complex is a promisingly safe and effective targeting delivery method for the UK.

The UK/PD-RGDS complex as the target material could be feasible to carry UK and carry out the thrombolytic therapy because it reduced the side effects of bleeding and the risk of bleeding complications.

Author Contributions: Conceptualization: M.Z.; methodology: S.C. and Y.W.; Investigation, S.C., M.L. and C.W.; writing—original draft preparation: S.C.; Writing—Review and Editing, Y.W. and X.Z.; Supervision, M.Z.; funding acquisition: S.C. and Y.W. All authors have read and agreed to the published version of the manuscript.

Funding: The article processing charge (APC) was funded by Ministry of Education (grant number: 3500-1205080302). This research was funded by Ministry of Science and Technology of the People's Republic of China (grant number: 0001-56105602) and Capital Medical University (grant number: PYZ2017107, XSKY202105, XSKY2015060).

Institutional Review Board Statement: The animal study protocol was approved by Institutional Animal Care and Use Committee of Capital Medical University (protocol code: AEEI-2018-174, 2018-10-11).

Informed Consent Statement: Not applicable.

Data Availability Statement: Not applicable.

Acknowledgments: The authors acknowledge Beijing Area Major Laboratory of Peptide and Small Molecular Drugs.

Conflicts of Interest: The authors declare no conflict of interest.

Sample Availability: Samples of the compounds are available from the authors.

References

1. Zenych, A.; Fournier, L.; Chauvierre, C. Nanomedicine Progress in Thrombolytic Therapy. *Biomaterials* **2020**, *258*, 120297. [CrossRef] [PubMed]
2. Zhang, H.J.; Pei, Y.M.; Gao, L.Y.; He, Q.Q.; Zhang, H.L.; Zhu, L.; Zhang, Z.Z.; Hou, L. Shear Force Responsive and Fixed-Point Separated System for Targeted Treatment of Arterial Thrombus. *Nano Today* **2021**, *38*, 101186. [CrossRef]
3. Cui, W.; Liu, R.; Jin, H.Q.; Lv, P.; Sun, Y.Y.; Men, X.; Yang, S.N.; Qu, X.Z.; Yang, Z.Z.; Huang, Y.N. PH Gradient Difference around Ischemic Brain Tissue Can Serve as a Trigger for Delivering Polyethylene Glycol-Conjugated Urokinase Nanogels. *J. Control. Release* **2016**, *225*, 53–63. [CrossRef] [PubMed]
4. Zhang, H.J.; Qu, H.Y.; He, Q.Q.; Gao, L.Y.; Zhang, H.L.; Wang, Y.F.; Zhang, Z.Z.; Hou, L. Thrombus-Targeted Nanoparticles for Thrombin-Triggered Thrombolysis and Local Inflammatory Microenvironment Regulation. *J. Control. Release* **2021**, *339*, 195–207. [CrossRef] [PubMed]
5. Landais, A.; Chaumont, H.; Dellis, R. Thrombolytic Therapy of Acute Ischemic Stroke during Early Pregnancy. *J. Stroke Cerebrovasc. Dis.* **2018**, *27*, E20–E23. [CrossRef]
6. Chen, H.-A.; Ma, Y.-H.; Hsu, T.-Y.; Chen, J.-P. Preparation of Peptide and Recombinant Tissue Plasminogen Activator Conjugated Poly(Lactic-Co-Glycolic Acid) (PLGA) Magnetic Nanoparticles for Dual Targeted Thrombolytic Therapy. *Int. J. Mol. Sci.* **2020**, *21*, 2690. [CrossRef]
7. Huang, T.; Li, N.; Gao, J. Recent Strategies on Targeted Delivery of Thrombolytics. *Asian J. Pharm. Sci.* **2019**, *14*, 233–247. [CrossRef]
8. Zhang, N.P.; Li, C.L.; Zhou, D.Y.; Ding, C.; Jin, Y.Q.; Tian, Q.M.; Meng, X.Z.; Pu, K.F.; Zhu, Y.M. Cyclic RGD Functionalized Liposomes Encapsulating Urokinase for Thrombolysis. *Acta Biomater.* **2018**, *70*, 227–236. [CrossRef]
9. Huang, M.J.; Zhang, S.F.; Lu, S.Y.; Qi, T.M.; Yan, J.; Gao, C.M.; Liu, M.Z.; Li, T.; Ji, Y.Z. Synthesis of Mesoporous Silica/Polyglutamic Acid Peptide Dendrimer with Dual Targeting and Its Application in Dissolving Thrombus. *J. Biomed. Mater. Res. A* **2019**, *107*, 1824–1831. [CrossRef]
10. Zhang, Y.; Liu, Y.; Zhang, T.; Wang, Q.; Huang, L.; Zhong, Z.; Lin, J.; Hu, K.; Xin, H.; Wang, X. Targeted Thrombolytic Therapy with Metal Organic-Framework-Derived Carbon Based Platforms with Multimodal Capabilities. *ACS Appl. Mater. Interfaces* **2021**, *13*, 24453–24462. [CrossRef]

11. Nederhoed, J.H.; Ebben, H.P.; Slikkerveer, J.; Hoksbergen, A.W.J.; Kamp, O.; Tangelder, G.-J.; Wisselink, W.; Musters, R.J.P.; Yeung, K.K. Intravenous Targeted Microbubbles Carrying Urokinase versus Urokinase Alone in Acute Peripheral Arterial Thrombosis in a Porcine Model. *Ann. Vasc. Surg.* **2017**, *44*, 400–407. [CrossRef] [PubMed]
12. Ma, L.; Wang, Y.J.; Zhang, S.M.; Qian, X.C.; Xue, N.Y.; Jiang, Z.Q.; Akakuru, O.U.; Li, J.; Xu, Y.F.; Wu, A.G. Deep Penetration of Targeted Nanobubbles Enhanced Cavitation Effect on Thrombolytic Capacity. *Bioconjug. Chem.* **2020**, *31*, 369–374. [CrossRef] [PubMed]
13. Guan, L.N.; Wang, C.M.; Yan, X.; Liu, L.Y.; Li, Y.H.; Mu, Y.M. A Thrombolytic Therapy Using Diagnostic Ultrasound Combined with RGDS-Targeted Microbubbles and Urokinase in a Rabbit Model. *Sci. Rep.* **2020**, *10*, 12511. [CrossRef] [PubMed]
14. Zhang, S.-F.; Lü, S.; Yang, J.; Huang, M.; Liu, Y.; Liu, M. Synthesis of Multiarm Peptide Dendrimers for Dual Targeted Thrombolysis. *ACS Macro Lett.* **2020**, *9*, 238–244. [CrossRef]
15. Zhong, Y.; Gong, W.J.; Gao, X.H.; Li, Y.N.; Liu, K.; Hu, Y.G.; Qi, J.S. Synthesis and Evaluation of a Novel Nanoparticle Carrying Urokinase Used in Targeted Thrombolysis. *J. Biomed. Mater. Res. A* **2020**, *108*, 193–200. [CrossRef]
16. Liao, J.; Ren, X.T.; Yang, B.W.; Li, H.; Zhang, Y.X.; Yin, Z.N. Targeted Thrombolysis by Using C-RGD-Modified N,N,N-Trimethyl Chitosan Nanoparticles Loaded with Lumbrokinase. *Drug Dev. Ind. Pharm.* **2019**, *45*, 88–95. [CrossRef]
17. Kolmakov, A.G.; Baikin, A.S.; Gudkov, S.V.; Belosludtsev, K.N.; Nasakina, E.O.; Kaplan, M.A.; Sevostyanov, M.A. Polylactide-Based Stent Coatings: Biodegradable Polymeric Coatings Capable of Maintaining Sustained Release of the Thrombolytic Enzyme Streptokinase. *Pure Appl. Chem.* **2020**, *92*, 1329–1340. [CrossRef]
18. Nan, D.; Jin, H.Q.; Yang, D.; Yu, W.W.; Jia, J.J.; Yu, Z.M.; Tan, H.; Sun, Y.G.; Hao, H.J.; Qu, X.Z.; et al. Combination of Polyethylene Glycol-Conjugated Urokinase Nanogels and Urokinase for Acute Ischemic Stroke Therapeutic Implications. *Transl. Stroke Res.* **2021**, *12*, 844–857. [CrossRef]
19. Hanaoka, S.; Saijou, S.; Matsumura, Y. A Novel and Potent Thrombolytic Fusion Protein Consisting of Anti-Insoluble Fibrin Antibody and Mutated Urokinase. *Thromb. Haemost.* **2021**, *122*, 57–66. [CrossRef]
20. Petroková, H.; Mašek, J.; Kuchař, M.; Vítečková Wünschová, A.; Štikarová, J.; Bartheldyová, E.; Kulich, P.; Hubatka, F.; Kotouček, J.; Knotigová, P.T.; et al. Targeting Human Thrombus by Liposomes Modified with Anti-Fibrin Protein Binders. *Pharmaceutics* **2019**, *11*, 642. [CrossRef]
21. Absar, S.; Kwon, Y.M.; Ahsan, F. Bio-Responsive Delivery of Tissue Plasminogen Activator for Localized Thrombolysis. *J. Control. Release* **2014**, *177*, 42–50. [CrossRef] [PubMed]
22. Zhang, L.; Li, Z.; Ye, X.; Chen, Z.; Chen, Z.-S. Mechanisms of Thrombosis and Research Progress on Targeted Antithrombotic Drugs. *Drug Discov. Today* **2021**, *26*, 2282–2302. [CrossRef] [PubMed]
23. Kawata, H.; Uesugi, Y.; Soeda, T.; Takemoto, Y.; Sung, J.H.; Umaki, K.; Kato, K.; Ogiwara, K.; Nogami, K.; Ishigami, K.; et al. A New Drug Delivery System for Intravenous Coronary Thrombolysis with Thrombus Targeting and Stealth Activity Recoverable by Ultrasound. *J. Am. Coll. Cardiol.* **2012**, *60*, 2550–2557. [CrossRef] [PubMed]
24. Liang, J.F.; Li, Y.T.; Song, H.; Park, Y.J.; Naik, S.S.; Yang, V.C. ATTEMPTS: A Heparin/Protamine-Based Delivery System for Enzyme Drugs. *J. Control. Release* **2002**, *78*, 67–79. [CrossRef] [PubMed]
25. Zhang, H.; Wang, Y.; Zhao, M.; Wu, J.; Zhang, X.; Gui, L.; Zheng, M.; Li, L.; Liu, J.; Peng, S. Synthesis and In Vivo Lead Detoxification Evaluation of Poly-α,β-Dl-Aspartyl-l-Methionine. *Chem. Res. Toxicol.* **2012**, *25*, 471–477. [CrossRef]
26. Chen, S.L.; Peng, Z.D.; Wang, Y.J.; Wu, J.H.; An, R.; Miao, R.R.; Zhao, M.; Peng, S.Q. Development and Activity Evaluation of Arg-Gly-Asp-Containing Antithrombotic Conjugate. *J. Mol. Struct.* **2019**, *1198*, 126816. [CrossRef]
27. Luo, Y.; Wang, Y.; Fu, J. Nanomaterials in Cerebrovascular Disease Diagnose and Treatment. *Part. Part. Syst. Charact.* **2021**, *38*, 2000311. [CrossRef]
28. Gui, L.; Zhao, M.; Wang, Y.; Qin, Y.; Liu, J.; Peng, S. Synthesis, Nanofeatures, in Vitro Thrombus Lysis Activity and in Vivo Thrombolytic Activity of Poly-α,β-Aspartyl-L-Alanine. *Nanomedicine* **2010**, *5*, 703–714. [CrossRef]
29. Marsh, J.N.; Senpan, A.; Hu, G.; Scott, M.J.; Gaffney, P.J.; Wickline, S.A.; Lanza, G.M. Fibrin-Targeted Perfluorocarbon Nanoparticles for Targeted Thrombolysis. *Nanomedicine* **2007**, *2*, 533–543. [CrossRef]
30. Zhao, S.; Li, Z.; Huang, F.; Wu, J.; Gui, L.; Zhang, X.; Wang, Y.; Wang, X.; Peng, S.; Zhao, M. Nano-Scaled MTCA-KKV: For Targeting Thrombus, Releasing Pharmacophores, Inhibiting Thrombosis and Dissolving Blood Clots in Vivo. *Int. J. Nanomed.* **2019**, *14*, 4817–4831. [CrossRef]
31. Pawlowski, C.L.; Li, W.; Sun, M.; Ravichandran, K.; Hickman, D.; Kos, C.; Kaur, G.; Gupta, A.S. Platelet Microparticle-Inspired Clot-Responsive Nanomedicine for Targeted Fibrinolysis. *Biomaterials* **2017**, *128*, 94–108. [CrossRef] [PubMed]
32. Bi, F.; Zhang, J.; Su, Y.; Tang, Y.-C.; Liu, J.-N. Chemical Conjugation of Urokinase to Magnetic Nanoparticles for Targeted Thrombolysis. *Biomaterials* **2009**, *30*, 5125–5130. [CrossRef]
33. Hou, X.W.L.Z. Preparation of Thrombus-Targeted Urokinase Liposomes and Its Thrombolytic Effect in Model Rats. *Acta Pharm. Sin.* **2003**, *38*, 231–235.
34. Vaidya, B.; Agrawal, G.P.; Vyas, S.P. Platelets Directed Liposomes for the Delivery of Streptokinase: Development and Characterization. *Eur. J. Pharm. Sci.* **2011**, *44*, 589–594. [CrossRef] [PubMed]

Disclaimer/Publisher's Note: The statements, opinions and data contained in all publications are solely those of the individual author(s) and contributor(s) and not of MDPI and/or the editor(s). MDPI and/or the editor(s) disclaim responsibility for any injury to people or property resulting from any ideas, methods, instructions or products referred to in the content.

Review

Hybrid Nanoplatforms Comprising Organic Nanocompartments Encapsulating Inorganic Nanoparticles for Enhanced Drug Delivery and Bioimaging Applications

Fatih Yanar [1,*], Dario Carugo [2] and Xunli Zhang [3,*]

1. Department of Molecular Biology and Genetics, Bogazici University, 34342 Istanbul, Türkiye
2. Nuffield Department of Orthopedics, Rheumatology and Musculoskeletal Sciences (NDORMS), University of Oxford, Oxford OX3 7LD, UK; dario.carugo@ndorms.ox.ac.uk
3. School of Engineering, Faculty of Engineering and Physical Sciences, University of Southampton, Southampton SO17 1BJ, UK
* Correspondence: fatih.yanar@boun.edu.tr (F.Y.); xl.zhang@soton.ac.uk (X.Z.)

Abstract: Organic and inorganic nanoparticles (NPs) have attracted significant attention due to their unique physico-chemical properties, which have paved the way for their application in numerous fields including diagnostics and therapy. Recently, hybrid nanomaterials consisting of organic nanocompartments (e.g., liposomes, micelles, poly (lactic-co-glycolic acid) NPs, dendrimers, or chitosan NPs) encapsulating inorganic NPs (quantum dots, or NPs made of gold, silver, silica, or magnetic materials) have been researched for usage in vivo as drug-delivery or theranostic agents. These classes of hybrid multi-particulate systems can enable or facilitate the use of inorganic NPs in biomedical applications. Notably, integration of inorganic NPs within organic nanocompartments results in improved NP stability, enhanced bioavailability, and reduced systemic toxicity. Moreover, these hybrid nanomaterials allow synergistic interactions between organic and inorganic NPs, leading to further improvements in therapeutic efficacy. Furthermore, these platforms can also serve as multifunctional agents capable of advanced bioimaging and targeted delivery of therapeutic agents, with great potential for clinical applications. By considering these advancements in the field of nanomedicine, this review aims to provide an overview of recent developments in the use of hybrid nanoparticulate systems that consist of organic nanocompartments encapsulating inorganic NPs for applications in drug delivery, bioimaging, and theranostics.

Keywords: organic nanoparticles; inorganic nanoparticles; encapsulation; hybrid nanoparticles; drug delivery; bioimaging

1. Introduction

Nanoparticles (NPs) are submicroscopic particles with dimensions typically ranging between 1 and 100 nanometers (nm) in diameter. The unique physico-chemical properties of NPs, including their small size, large surface-to-volume ratio, and unique optical behaviour, make them a suitable candidate system for usage in the field of nanomedicine, especially in drug delivery and bioimaging applications. Examples of commonly employed NPs in nanomedicine include organic particulate systems, such as liposomes, micelles, dendrimers, poly (lactic-co-glycolic acid) (PLGA), and chitosan NPs, as well as inorganic NPs such as quantum dots (QDs) and NPs made of gold (AuNPs), silver (AgNPs), silica (SNPs), or magnetic materials (MNPs).

NP-based drug-delivery systems (DDSs) are designed for delivering a drug (or a combination of drugs) to a specific region within the body, in order to primarily cause damage to target cells whilst reducing side-effects due to systemic drug distribution in off-target regions. For example, it has been demonstrated that drug-encapsulating liposomes can enhance targeting and treatment efficiency compared with the free form of the drug [1].

Notably, DDSs have been successful in treating cancer, as well as a wide range of other diseases and conditions. For example, they have shown potential for improving treatment of infectious diseases [2], respiratory diseases [3], hypertension [4], diabetes [5], and for targeting the brain vasculature to enable drug transport across the blood–brain barrier [6].

Building upon the achievements of liposomal products, it has become increasingly apparent that NPs hold potential for overcoming widely recognized challenges such as those associated with the delivery of poorly water-soluble drugs, the transport of drugs across tight epithelial barriers, the intracellular delivery of large molecules, and the co-delivery of two or more drugs [7]. The range of applications has also extended beyond drug delivery to include detection of proteins, tissue engineering, tumour diagnosis, purification of molecules/cells, and biomedical imaging [8]. In recent years, the use of different types of NPs in nanomedicine has increased significantly, largely due to their ability to lower toxicity and improve bioavailability of therapeutic payloads, as well as for their applicability as contrast agents in biomedical imaging.

Although organic and inorganic NPs individually are cornerstones of nanomedicine, these nanoparticulate systems can also be used in combination to achieve multifunctional features. These can be obtained via encapsulation or surface modification of the NPs, resulting in hybrid (organic–inorganic) nanoplatforms with enhanced diagnostic and/or therapeutic performance. Significant efforts have been dedicated to the development of hybrid nanoplatforms with a core–shell architecture where inorganic NPs are encapsulated within the aqueous core of organic nanocompartments, or where inorganic NPs are stabilised by an outer layer of organic compounds. These hybrid nanoplatforms have been evaluated for different applications [9–11], since the combination of organic/inorganic NPs is thought to elicit synergistic effects and to also improve the effectiveness and safety of inorganic NPs.

Inorganic NPs have characteristics of large surface area and high reactivity, which can lead to agglomeration or degradation [12]. Moreover, they generally exhibit poor solubility in biological fluids, which can potentially trigger an immune response. In this context, the protection of inorganic NPs using an organic 'shield' can provide a physical barrier that prevents direct contact of inorganic NPs with biological structures, thereby reducing unwanted toxic effects. The presence of an organic shell or coating also improves the solubility, stability, and biocompatibility of inorganic NPs, while promoting or enabling targeted delivery and cellular uptake. Additionally, surface functionalization of inorganic NPs can allow application in targeted delivery and controlled release. Overall, the presence of an organic compartment (i.e., shell or coating) can allow and extend the use of inorganic NPs in drug delivery and biomedical imaging. As a result, hybrid organic/inorganic nanoplatforms have attracted significant attention in recent years and have been extensively studied in the literature. This is evident from Figure 1, which shows the number of published research articles specifically focusing on the use of such nanoparticulate systems in the fields of drug delivery and bioimaging.

This review presents a comprehensive overview of recent developments in this field of research, with a particular focus on nanoparticulate systems for enhanced therapeutic efficacy and real-time biomedical imaging. Specifically, this review covers the use of hybrid platforms (such as core–shell structures) that employ a variety of organic nanocompartments (such as liposomes, micelles, PLGA and chitosan NPs, and dendrimers) that encapsulate inorganic NPs (such as AuNPs, AgNPs, QDs, SNPs, and MNPs). These nanoscale systems represent promising nanodevices for drug delivery and bioimaging applications within the field of nanomedicine.

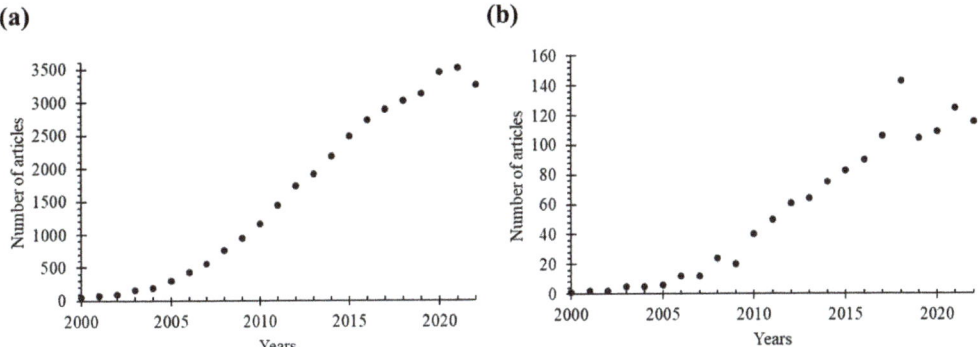

Figure 1. The number of research articles indexed in the Web of Science Core Collection (in Science Citation Index Expanded) was determined through a search using the following keywords: (**a**) "nanoparticle AND drug delivery" OR "nanoparticle AND imaging", and (**b**) combinations of "hybrid", "organic", "inorganic", and "nanoparticle" AND "drug delivery" OR "imaging" in the title or abstract.

2. Organic Nanoparticles/Nanocompartments

Organic NPs (or nanocompartments) are nanoparticulate systems composed of organic materials or compounds, and have been researched and utilised for a plethora of biomedical applications as drug nanocarriers or imaging probes. A schematic overview of organic NPs that are commonly employed in the field of nanomedicine is given in Figure 2.

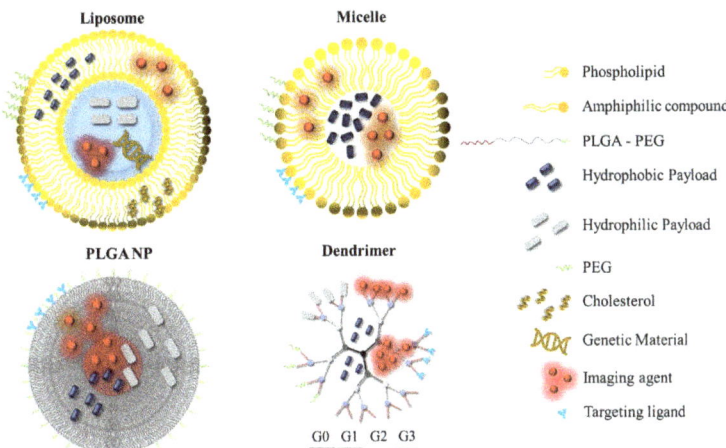

Figure 2. Organic nanoparticles that are commonly utilised in the field of nanomedicine. The nanoparticles can be modified by incorporating hydrophilic/hydrophobic drugs, PEG, targeting ligands or imaging agents for improved performance or added functionalities across different application areas. The different generations of the dendrimer are shown in different colours. PEG: polyethylene glycol; PLGA: poly (lactic-co-glycolic acid); G: generation number.

Liposomes are composed of natural or synthetic lipids (typically phospholipids), and consist of lipid bilayers encapsulating aqueous compartments. They have characteristics of high biocompatibility and biodegradability, low toxicity, and the ability to encapsulate both hydrophobic and hydrophilic molecules [13,14]. Together, those features make liposomes highly attractive as nanocarriers for drug-delivery applications. Liposomes can be synthesized in different sizes, ranging between 50 and 1000 nm, using different techniques such as thin-film hydration, reverse-phase evaporation, lipid extrusion, or microfluidics [14,15].

The size of NPs has great importance in drug-delivery applications as it determines their circulation half-life and drug encapsulation efficiency, and also influences their ability to extravasate across the tumour vasculature (i.e., through the so-called enhanced permeability and retention (EPR) effect) as well as their recognition by macrophages [16]. It has been reported that liposomes with size greater than 100 nm can be easily identified by macrophages, leading to accumulation in organs containing the mononuclear phagocyte system, such as the liver and spleen. NPs smaller than 10 nm can be rapidly cleared by the renal excretion system, considering that the size of the average renal filtration pore is of around 10 nm. NPs can also be subject to rapid clearance due to opsonization by serum proteins, whereby liposomes become recognizable by the immune system and can be digested by macrophages of the reticuloendothelial system (RES) [17], especially in the liver and spleen, resulting in lowered therapeutic efficacy overall. To overcome this challenge, a hydrophilic polymer, polyethylene glycol (PEG), has been widely employed to coat the surface of NPs to improve their stability and circulation time [18]. The steric barrier provided by PEG mitigates the extent of opsonization and recognition by macrophages, thus resulting in increased accumulation at the intended site of treatment and reduced side-effects [19].

Liposomes are generally utilised as drug-delivery vehicles, mainly due to the versatility offered by their architecture that makes them suitable for incorporation of both lipophilic and hydrophilic pharmaceutical actives within their lipid bilayer and aqueous core, respectively. Liposomal products are able to deliver drugs to a specific site within the body while minimizing the negative side-effects of systemic exposure, since the pharmaceutical agent is confined by the liposome membrane and therefore healthy tissues are not directly exposed to it. Moreover, the liposome surface can be functionalized with targeting ligands, antibodies, peptides, and/or imaging moieties for application in diagnostics, therapy, or the concurrent combination of these (often referred to as 'theranostics').

Micelles consist of closed monolayered structures typically formed by the aggregation of amphiphilic compounds into stable ordered units, by self-assembly above a critical micelle concentration (CMC) [20]. The structure of micelles is characterized by a fatty acid core with a polar surface, or by a polar core with fatty acids on the surface which is referred to as an 'inverted micelle' configuration [21]. Preparation methods of micelles include solvent evaporation [22], dialysis [23], and direct dissolution [24]. The size of micelles typically ranges between 10 and 100 nm, and various shapes can be synthesized depending on the specific application of interest [20]. Polymeric micelles are a type of micelle formed by amphiphilic block copolymers (composed of hydrophobic and hydrophilic blocks), and are typically characterized by a larger volume, lower CMC, and greater stability. The structure of micelles makes them suitable for encapsulation of hydrophobic bioactive compounds within their core, while the outer hydrophilic layer provides stability and protection from recognition by RES. Thus, micelles are a suitable candidate system for improving solubility, stability, and bioavailability of hydrophobic payloads. The surface of micelles can also be functionalized using targeting moieties such as antibodies or peptides, providing the ability to bind to specific receptors on target cells. In this context, micelles can be employed in a number of biomedical applications including drug delivery, extraction of proteins, and bioimaging [25,26].

Poly (lactic-co-glycolic acid) (PLGA) is a widely used polymer in the field of nanomedicine, especially due to its remarkable biodegradability and biocompatibility. It is composed of two monomeric units of lactic acid and glycolic acid, which can be readily metabolized with minimal toxicity [27]. The structure of PLGA can be designed by adjusting the relative amount and/or the molecular weight of these monomers, facilitating the tailored synthesis of NPs towards specific applications. The synthesis of PLGA NPs can be performed by nanoprecipitation or emulsification–evaporation techniques [28]. Depending on the application and the desired encapsulation performance, single (oil-in-water) or double (water-in-oil-in-water) emulsion templating techniques can be utilised [27]. The size of PLGA NPs can vary between 10 and 1000 nm [29], and it can be regulated by controlling experimental conditions such as stirring rate and temperature [30]. PLGA NPs are suitable for encapsulating hydrophilic and/or hydrophobic molecules, while the surface can be modified with various moieties for reduced toxicity, enhanced stability, targeted delivery, or theranostic functionalities. PLGA NPs are especially considered for sustained drug release, since their controllable properties enable the design of desirable drug-release profiles, overall making them ideal nanocarriers for treatments demanding long-term therapy with reduced dosing frequency.

Dendrimers are characterized by a well-defined tree-like structure with a central core and multiple layers of branch units on the outer shell. Branches originate from the central core and their number and branching architecture can be designed based on the target application. The architecture of dendrimers allows precise control over their size and shape, as well as the functionalization of their surface. The most commonly employed techniques for the synthesis of dendrimers are divergent and convergent methods [31]. The size of dendrimers typically ranges between 1 and 15 nm; however, high-generation dendrimers can have larger sizes [32]. Dendrimers have a high loading capacity for pharmaceutical agents and can provide sustained drug-release profiles. They can be functionalized with imaging probes or contrast agents, making them suitable for diagnostic imaging. Moreover, they can be employed in a number of other applications such as gene delivery, tissue engineering, and catalysis. Polyamidoamine (PAMAM) dendrimers represent one of the most utilised families of dendrimers due to their unique physico-chemical properties, such as high water solubility, biocompatibility, and precise structural control, overall making them suitable for therapeutic and diagnostic applications [33].

Chitosan is a deacetylated form of chitin, which is derived from crustacean shells or the cell walls of fungi. It is a widely used, FDA-approved, biodegradable, biocompatible polymer and is often utilised as a nanocarrier material in drug-delivery applications. Various methods have been proposed for the synthesis of chitosan NPs, such as microemulsification, emulsification–solvent diffusion, emulsion-based solvent evaporation, and ionotropic gelation [34]. Beyond application in drug delivery, these NPs have demonstrated potential for application in tissue engineering and as antimicrobial or antioxidant agents. As for other nanoparticulate systems, the surface of chitosan NPs can also be modified for targeting or imaging purposes. Moreover, chitosan NPs are known for their mucoadhesive and sustained drug-release properties, making them an ideal candidate system for mucosal drug-delivery applications [34].

3. Inorganic Nanoparticles

Inorganic NPs are nanoscale particulate systems composed of inorganic materials such as metals or semiconductors. They have unique optical properties due to their small size and high surface-to-volume ratio, which make them valuable tools for application in drug delivery, biomedical imaging, and theranostics [35]. Examples of inorganic NPs utilised in the field of nanomedicine are shown in Figure 3.

Figure 3. Examples of inorganic nanoparticles that are commonly employed in nanomedicine, which include: gold, silver, magnetic, and silica NPs, and quantum dots.

Gold NPs (AuNPs) and silver NPs (AgNPs) have attracted significant attention in industry, and in technological development more broadly, especially due to their remarkable optical properties. Notably, these NPs can display surface plasmon resonance (SPR) behaviour, which occurs when light excites electrons at the interface between the conductor and insulator components of the material [36]. The optical properties of these NPs are mainly shape- and size-dependent, and accordingly, the SPR also depends on the composition, shape, and size of the NPs [37,38]. When light is absorbed on the surface of metal NPs, electrons undergo a collective oscillation, which generates local heat that can be utilised in biomedical applications. In principle, when plasmonic NPs, such as gold or silver, are encapsulated inside an outer layer, they can cause this layer to phase transition when they are subjected to light irradiation as they can convert optical energy into local heat energy [39]. It is important to note that this phenomenon also enables photothermal therapy (PTT), which refers to the utilization of light energy to kill cancer cells [40]. Various physico-chemical methods for the synthesis of AuNPs and AgNPs have been reported in the literature, including laser ablation, chemical reduction, and green synthesis [41]. The size and shape of the NPs are highly dependent on the production technique employed. AuNPs and AgNPs are also attractive for drug-delivery applications, as they can be designed to function as controlled drug-release systems. This feature of NPs relies on the photothermal effect (PTE), which induces the release of encapsulated payloads through an outer layer surrounding the NP, due to a phase change in the layer caused by localized temperature increases, as described above. In addition, among metallic NPs, AgNPs are widely used in healthcare products and in the food industry due to their remarkable antibacterial properties.

Quantum dots (QDs) are semiconductor nanomaterials composed of groups II-VI, III-V, or IV-VI elements [42]. Synthesis methods of QDs are classified into two main categories: top-down and bottom-up approaches. In top-down approaches, a semiconductor is thinned down to form QDs, while in bottom-up approaches, a self-assembly process is performed to create QDs [43]. These NPs have a small size of around 10 nm and possess unique optical and electronic properties arising from quantum confinement effects. Their tuneable emission wavelength makes them ideal agents for various applications in bioimaging, detection, and tracking. Notably, QDs have an absorption wavelength higher than 650 nm in the near-infrared (NIR) region, which is advantageous in biomedical imaging applications as it presents minimal tissue absorption [44]. In drug-delivery applications, the surface of QDs can be functionalized with phospholipids or amphiphilic polymers to create a hydrophilic protective shell. This can improve the stability and biocompatibility of QDs in biological fluids, enabling their utilization in biomedical applications, e.g., for improving treatment efficacy. It can also provide steric stabilization, reduce non-specific interactions, and enhance circulation time within the body. Overall, QDs are particularly attractive for improving imaging contrast, detection sensitivity of biological targets, targeted drug delivery, and theranostics.

Silica NPs (SNPs) are composed primarily of silicon dioxide (SiO_2) and their size typically ranges between a few nm to a few hundred nm. Commonly employed techniques for the synthesis of SNPs include the microemulsion method and the Stöber process. The structure of SNPs can be adapted to achieve various architectural shapes including spheres, rods, or tubes [45]. They have characteristics of high stability and biocompatibility, making them suitable nanomaterials for a range of biomedical applications. The surface of SNPs can be functionalized with specific molecules allowing targeted delivery, controlled release, or contrast enhancement in bioimaging. The porous structure of SNPs also enables high drug-loading capacity and efficient delivery of pharmaceutical agents. SNPs can be engineered to incorporate fluorescent dyes, QDs, or other imaging agents. This property allows them to be utilised as imaging probes for fluorescence imaging, magnetic resonance imaging (MRI), or computed tomography (CT). For this reason, SNPs can serve as effective platforms in biosensing applications, including for the detection of biomarkers or for 'tracking' biological processes dynamically [46].

Magnetic NPs (MNPs) refer to particles in the nanoscale that possess magnetic properties. These particles are typically composed of a magnetic core material (e.g., iron oxide), coated with a functional shell [47]. The core of MNPs is responsible for their magnetic behaviour. Iron oxide magnetic NPs (IONPs) are commonly used as MNPs due to their high magnetization and biocompatibility. Their small size (up to a few hundred nm) provides benefits such as large surface-area-to-volume ratio, improved colloidal stability, and enhanced magnetic response. Among various methods for the synthesis of MNPs, the sol–gel method, coprecipitation, microemulsification, and thermal decomposition are commonly employed [48]. MNPs can be surface-modified with biocompatible coatings to enhance stability and biocompatibility; this makes them suitable for various biological and biomedical applications. They are commonly employed in MRI as contrast agents to enhance imaging resolution and sensitivity. Interestingly, MNPs can also be employed to enable magnetic hyperthermia, which is a therapeutic approach where NPs generate heat in response to an alternating magnetic field, leading to localized tumour ablation [49]. They can also be employed in targeted DDSs, biosensing, and magnetic cell separation. Thanks to the magnetic properties of these NPs, they can be controlled using external magnetic fields to localize in specific regions of the body, i.e., for targeted drug delivery or imaging.

4. Properties of Nanoparticles with Encapsulation

As discussed earlier, efficient delivery and 'shielding' of inorganic NPs through encapsulation can significantly enhance their performance in a range of biomedical applications. The advantages that the encapsulation process introduces include improved stability and biocompatibility, as well as controlled drug release and targeted drug-delivery capabilities. This can typically be achieved by encapsulating the NPs within a shell or modifying the surface of the NPs, as shown in Figure 4. Additionally, other types of hybrid NPs have been investigated in the literature, including hybrid nanogels or Janus particles [50]. Examples of methods for the encapsulation of inorganic NPs include the polymerization of the organic NPs around inorganic NPs, and the encapsulation via adsorption of oppositely charged particles onto inorganic NPs [35].

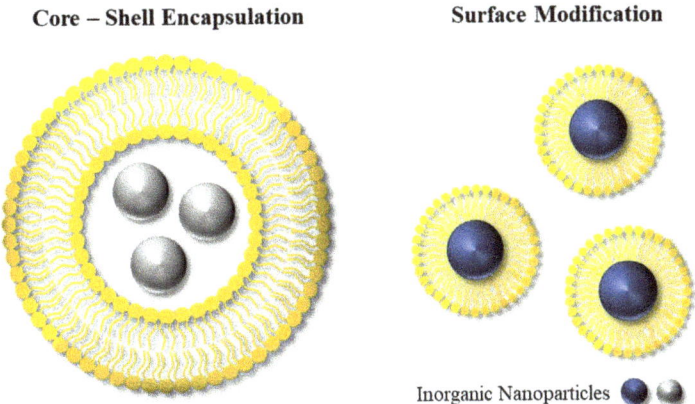

Figure 4. The design of organic–inorganic hybrid platforms can have various architectures, including those based on core–shell structures and surface modifications.

Typically, a polymeric nanocompartment or layer is created around the NPs, which can be achieved by heterogeneous polymerization techniques or through nanoprecipitation [51–53]. On the other hand, the surface modification method involves attaching organic molecules to the surface of the NPs, enhancing their functionality without creating an outer shell structure. Depending on the desired application, the surface chemistry of NPs can be tailored by selecting appropriate molecules to meet specific requirements.

The efficiency of a given encapsulation method depends on various physico-chemical properties of the NPs, including their size, surface chemistry, surface charge, and solubility, as well as the desired characteristics of the hybrid nanoplatform, such as stability, biocompatibility, and controlled release and/or targeted delivery capabilities.

5. Biomedical Applications of Hybrid Nanoparticles

Hybrid NPs have been utilised as powerful tools in biomedical applications, especially for targeted drug delivery, bioimaging, and theranostics. Concerning applications in drug delivery, hybrid NPs—like other nanomedicines—can deliver drugs by active or passive targeting. Active targeting typically involves some modification on the particle surface, e.g., using charged lipids, antibodies, or attachment of ligands that enable the NP to bind to the receptors of the target cells and to cross biological membranes more effectively (Figure 5c). Active targeting therefore reduces undesired side-effects of drugs, while providing high therapeutic efficacy by allowing high dosing at the diseased site [54].

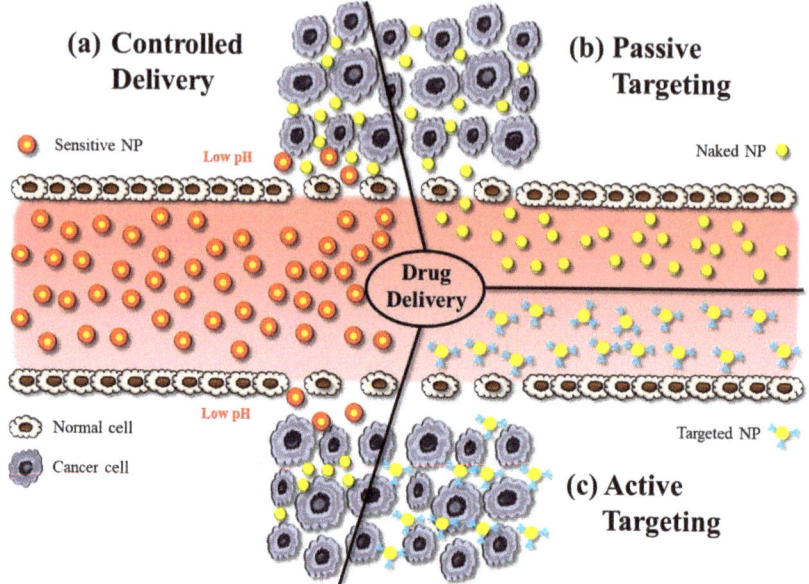

Figure 5. The illustration shows various approaches in nanoparticle-based drug delivery: (**a**) controlled delivery, (**b**) active targeting, and (**c**) passive targeting. NP: nanoparticle.

Passive targeting is the accumulation of NPs at pathological sites due to the EPR effect, whereby the increased vascular permeability enables enhanced extravasation of NPs and drugs (Figure 5b). Notably, the size of intercellular gaps in the vascular endothelium of these pathological regions increases by about 1 µm after exposure to inflammatory mediators. This helps NPs, for example, at tumour sites much more effectively than in physiological tissues [55]. The delivery of the payload to cells occurs via the interaction between the NP and the cell membrane, which can occur through different mechanisms. NPs can indirectly enter the cytoplasm by endocytosis, resulting in the delivery of the payload. An alternative process is fusion, whereby NP layers merge with the cell membrane, resulting in direct delivery of the payload. Another mechanism of interaction between NPs and cell membranes is lipid exchange, which involves the exchange of bilayer materials between the NP layers and the cell membrane. It is important to note that these interactions can trigger the immune system. Therefore, it is crucial to develop the surface chemistry of NPs in such a way to make them unrecognizable by the RES [17].

Controlled drug delivery, on the other hand, refers to the regulation of the release of a payload from a nanocarrier system (Figure 5c). These systems are designed to initiate drug release in response to an exogenous or endogenous trigger such as pH, temperature, or light. Sustained drug-delivery systems on the other hand provide prolonged release of a drug over a certain period of time. In DDSs designed for sustained release, the drug is typically encapsulated within the NP or a matrix system, and drug release occurs via diffusion [56].

Another important biomedical application of NPs is in biomedical imaging, i.e., as contrast agents. This approach has the potential to provide more accurate information about a given disease condition compared with traditional clinical imaging. Owing to the aforementioned targeting capabilities and the small size of NPs, nanoparticulate contrast agents can be directed to the targeted area, thus enabling accurate detection and imaging of a biological target tissue. This approach can be applied in fluorescence imaging, magnetic resonance imaging (MRI), computerized tomography (CT), ultrasound (US), and multimodal imaging [57].

Fluorescence imaging often involves the use of fluorophores or fluorescent dye molecules attached to NPs. These compounds can absorb light at specific wavelengths and emit light at longer wavelengths, when excited by an appropriate light source. Fluorophores can be targeted to specific cell types or cell structures for various purposes including protein analysis, gene detection, diagnostics, and real-time monitoring [58]. Fluorescence imaging can provide vital information, especially when NIR light is utilised, due to improved tissue penetration of light and reduced autofluorescence, allowing for enhanced imaging sensitivity. Fluorescence imaging can be implemented by conjugating fluorescent dyes with NPs or by encapsulating fluorescent agents within NPs.

MRI is one of the most commonly applied methods for disease diagnosis and monitoring in the clinic for various medical conditions. The fundamental principle of MRI is based on the movement of protons within a strong magnetic field. This provides high-resolution images of internal body structures in multiple planes. Commonly employed contrast agents in MRI include gadolinium-based NPs and superparamagnetic materials, due to their remarkable magnetic properties [59]. These contrast agents help enhance the visibility of targeted areas, improving the accuracy of the diagnostic process.

US imaging is also one of the most commonly employed methods for medical imaging, due to its simplicity of usage, safety, and real-time imaging capabilities. In this approach, sound waves (with a frequency >20 kHz) are generated by an extracorporeal transducer positioned in contact with the body. As these waves penetrate tissues, they encounter biological structures with different acoustical properties. These differences in properties between structures (i.e., mainly in density and compressibility) result in reflections of the ultrasound wave, which are captured by a probe and are then converted into images [60]. Different types of contrast agents are used in US imaging in order to enhance the acoustical properties mismatch between the vasculature and surrounding tissues; these include gas-filled (e.g., microbubbles with a core of a heavy gas, like perfluorocarbon), solid-based (e.g., silica NPs), and liquid-based (e.g., perfluorooctyl bromide) particles. Some of these NP systems present a core–shell design configuration. It is important to note that the aforementioned imaging approaches primarily focus on providing single imaging modalities; however, advancements have allowed the development of multifunctional NPs serving as imaging agents with multimodal imaging capabilities. These NPs can incorporate single (e.g., silicon naphthalocyanine) [61] or multiple imaging agents, enabling the simultaneous use of distinct imaging modalities [62]. This approach allows complementary information to be captured from different imaging techniques, thereby also improving accuracy and reliability of diagnosis.

Another significant area of application is photoablation therapy, which can be classified into two main modalities: photodynamic therapy (PDT) and photothermal therapy (PTT) [40,63] (Figure 6). PDT involves the use of (initially) non-toxic compounds called photosensitizers, which are exposed to light at a specific wavelength (e.g., in the NIR) resulting in the formation of toxic compounds. These toxic products can react with hydroxyl ions or water, and in turn form reactive oxygen species (ROS) that can cause cell death.

This approach is mainly employed in the context of anticancer therapy, as the production of ROS can induce suppression of tumour growth. In PTT, the targeted area is irradiated using a light source at a specific wavelength similar to PDT. This light energy is converted into heat energy by specific materials, such as AuNPs or AgNPs, resulting in hyperthermia and cell death. PDT and PTT are examples of common therapeutic approaches utilising hybrid platforms.

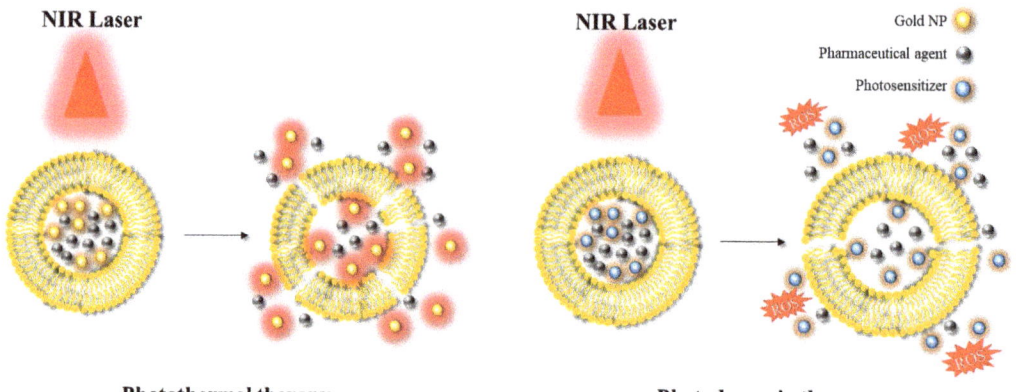

Figure 6. Illustration depicting the application of photothermal therapy and photodynamic therapy. NIR: near-infrared; ROS: reactive oxygen species.

5.1. Liposomes-Based Hybrid Platforms

Liposomes are one of the most commonly used nanocarriers for integration into hybrid platforms in nanomedicine. These platforms combine the unique advantages of liposomes and inorganic NPs, offering multi-functional features. Examples of studies that focused on hybrid platforms consisting of liposomes as the organic nanocompartment and inorganic NPs as the core are given in Table 1.

Table 1. Liposome-based hybrid platforms for biomedical applications.

Organic NPs	Inorganic NPs	Payload	Application	Target Condition	Highlights	Ref.
Liposomes	AuNPs	Dox	Controlled release/PTT	HeLa and U14 cell line/mice	The platform showed excellent anticancer effects, with up to 78.28% inhibition rate of tumour cells.	[40]
Liposomes	AuNPs	Dox	Controlled release/PTT	A549 cell line	NIR irradiation resulted in >80% drug release within 1 min.	[64]
Liposomes	AuNPs	Fish oil protein (tagged with AuNPs)	Sustained release and targeted delivery	HIG-82 cell line/Osteoarthritic rat model	The first study to report on the anti-osteoarthritic activity of fish oil protein and AuNP encapsulating liposomes.	[65]
Liposomes (conjugated with apo E)	AuNPs	miRNA inhibitors (integrated with AuNPs)	Targeted delivery	U87 cell line/mice	Greater accumulation in brain tumour tissue compared to controls.	[66]
Liposomes (cationic)	AuNPs	Carboplatin	Chemo-radiation therapy	HCT116 cell line/mice	Combination of carboplatin and the hybrid platform was found to be remarkably more efficient in terms of radiosensitization effect.	[67]

Table 1. Cont.

Organic NPs	Inorganic NPs	Payload	Application	Target Condition	Highlights	Ref.
Liposomes	Hollow AuNPs	Perfluorocarbon and Dox (stearic acid conjugated)	US-guided fluorescence imaging/PTT	MCF-7, 4T1 and HEK293 cell line/mice	Tumour accumulation of the platform was observed by in vivo fluorescence imaging and the antitumour effect was verified.	[68]
Liposomes	Au Nanorods	Ganoderic acid A	PTT, chemotherapy, antibacterial therapy	E. coli and S. aureus/MCF-7 cell line/mice	NIR irradiation exhibited broad-spectrum antibacterial effects against drug-resistant E. coli and S. aureus. Strong anticancer activity was observed against MCF-7 cells.	[69]
Liposomes	Au Nanorods	Ruthenium (II) polypyridyl complexes	Targeted release/PTT	SGC-7901 cell line/mice	NIR irradiation in combination with the nanoplatform could alter the morphology of cells in vitro and could inhibit tumour growth significantly.	[70]
Liposomes	AuNPs-aptamers	Morin	pH-sensitive targeted release	SGC-7901 cell line/mouse	The platform exhibited tumour targeting properties and could inhibit tumour growth.	[71]
Liposomes	Solid AuNPs or hollow AuNPs	Dox	Targeted release/PTT	HDF and MCF-7 cell line/mice	Hollow AuNPs presented eight-fold anticancer efficacy compared with solid AuNPs.	[72]
Liposomes (with FA)	AuNPs and graphene QDs	Dox	Targeted release/bioimaging/PTT/PDT	4T1 and MCF-7 cell line/mice	The platform exhibited in vivo tumour diagnosis capabilities through imaging along with successful PDT.	[73]
Liposomes (folate modified)	Au nanorods and MNPs	Dox	Magnetic and photothermal responsive targeted delivery	5637 and A549 cell line	The hybrid nanoplatform was synthesized using microfluidics-based production; 95% of the drug was released after 3 h.	[74]
Liposomes (with reduced graphene oxide sheets)	Carbon QDs	Dox	Stimuli-sensitive delivery/PTT	MD-MB-231 cell line/mice	Monitoring drug release was accomplished using the emission intensity of the theranostic platform.	[75]
Liposomes	CdSe QDs (modified with oleic acid) and SPIONs (Fe_3O_4)	-	Targeted delivery/fluorescence imaging	HepG2 cell line	Magnetic fluorescent liposomes could be drifted by an external magnet that could further be characterized using a fluorescence microscope.	[76]
Liposomes (with RGD peptide)	QDs	L-arginine	Fluorescence imaging-guided PTT	4T1 cell line/mice	The theranostic platform demonstrated the generation of NO, which was toxic to tumour cells in vitro. The accumulation of liposomes in the tumour tissue could be tracked in vivo.	[77]
Liposomes	Graphene QDs	-	US-triggered release	HCT116 cell line	Controlled delivery of QDs (as biomarkers) could be achieved by employing low-frequency US.	[78]

Table 1. Cont.

Organic NPs	Inorganic NPs	Payload	Application	Target Condition	Highlights	Ref.
Liposomes (cationic)	CMNPs	-	Magneto-PTT	U87 cell line	The study revealed efficient intracellular uptake of the nanoplatform and exhibited superior hyperthermia effects.	[79]
Liposomes	MNPs	Tenofovir disoproxil fumarate	Multimodal imaging/targeted release	HIV-infected microglia cell line	The platform demonstrated the capability for brain-targeted delivery with assistance of image guidance in vitro.	[62]
Liposomes (ICG loaded and FA modified)	Mesoporous SNPs (with gadolinium)	Dox	PDT/PTT/NIR fluorescence/MRI/PAI	4T1 cell line/mice	The multifunctional theranostic platform demonstrated capability for multimodal imaging, enabled effective diagnostics, and presented Dox release upon NIR irradiation.	[80]
Liposomes (as lipid coating)	Mesoporous SNPs	Berberine	Brain-targeted drug delivery	In vitro assay/mice	The hybrid nanoplatform achieved sustained release of berberine and inhibition of acetylcholine esterase, potentially contributing to the treatment of Alzheimer's disease.	[81]
Liposomes	AgNPs	-	Drug release analysis	Dialysis bag method	Greater AgNP release was observed at pH 5.5, which corresponds to the pH found in mature endosomes of tumour cells.	[82]
Liposomes	AgNPs	-	Evaluation of cytotoxicity	THP1 cell line	The nanoplatform was found to increase reactive oxygen species-independent induction of apoptosis, suggesting that the encapsulation could potentially reduce the concentration of AgNP required to exert a biological effect.	[83]

NPs: nanoparticles; AuNPs: gold nanoparticles; Dox: doxorubicin; PTT: photothermal therapy; NIR: near-infrared; RT: radiotherapy; QDs: quantum dots; PDT: photodynamic therapy; MNPs: magnetic nanoparticles; US: ultrasound; CMNPs: citric acid-coated iron oxide magnetic nanoparticles; SNPs: silica nanoparticles; MRI: magnetic resonance imaging; PAI: photoacoustic imaging; AgNPs: silver nanoparticles; SPIONs: superparamagnetic iron oxide nanoparticles; NO: nitric oxide; FA: folic acid; ICG: indocyanine green.

Xing et al. demonstrated the implementation of PTT by employing a liposomal system encapsulating doxorubicin (Dox), i.e., a chemotherapeutic compound and AuNPs as inorganic NPs [40]. When the hybrid platform was irradiated by NIR light, the liposome layers became permeable to both Dox and AuNPs, allowing the release of both payloads. It was reported that this platform resulted in effective tumour suppression with a cell growth inhibition rate of up to 78.28%. In a similar study, Koga et al. utilised AuNPs to achieve controlled drug release via PTE. Their results showed that the hybrid system was capable of releasing >80% of the drug in less than 1 min upon NIR irradiation [64]. Another approach, investigated by Lv et al., employed thermosensitive liposomes encapsulating Au nanorods, MNPs, and Dox for targeted delivery assisted by PTE [74]. This platform showed superparamagnetic properties and enabled controlled release, with approximately 95% of the drug released after 3 h of irradiation using a 980 nm laser beam. This approach was found to be highly effective in treating bladder tumour cells. Among many studies investigating liposome–AuNP hybrid platforms as DDSs, Li et al. demonstrated that incorporating hollow AuNPs in liposomes resulted in greater efficacy both in terms of the hyperthermia achieved within the targeted tissue and the drug (Dox) release profile, compared with liposomes loaded with solid AuNPs [72]. It is worth mentioning that, in addition to employing chemotherapeutic agents (such as Dox) as encapsulated payloads in liposomes for targeted and controlled drug-delivery applications, various other types of molecules have also been encapsulated in nanocompartments, including miRNA inhibitors [66] and fish oil protein [65]. Grafals-Ruiz et al. proposed a liposomal system for the treatment of glioblastoma (GBM), one of the most common types of brain tumours [66]. The system comprised liposomes encapsulating AuNPs functionalized with oligonucleotide miRNA inhibitors. In addition, the liposomes were conjugated to either apolipoprotein E (ApoE) or rabies virus glycoprotein. The hybrid nanoplatform was investigated in GBM syngeneic mice by intravenous administration. Results showed that the expression of miRNA-92b was effectively inhibited. Furthermore, compared with controls, liposomes conjugated with ApoE accumulated to a greater extent at the tumour tissue, suggesting improved targeted delivery.

Liposomal platforms have also been employed for enhanced bioimaging applications. A hybrid nanoplatform consisting of liposomes encapsulating AuNPs, perfluorocarbon, and Dox has been utilised for image-guided PTT [68], with positive outcomes in terms of both bioimaging and drug delivery. Another study conducted by Prasad et al. synthesized liposomes encapsulating AuNPs along with emissive graphene QDs for application in in vivo bioimaging and NIR-mediated cancer therapy [73]. The liposomes also encapsulated Dox as a chemotherapeutic drug, and their surface was functionalized with folic acid (FA) as a targeting ligand. The developed theranostic system demonstrated capability for in vivo bioimaging of tumour tissue using NIR light (wavelength of 750 nm). Moreover, it offered PDT and chemotherapeutic performance. Notably, NIR light exposure resulted in the generation of ROS, resulting in tumour reduction.

In a similar theranostic approach by Li et al., perfluorocarbon was encapsulated in liposomes by film dispersion, with the aim of developing US contrast agents that are more effective at penetrating into a target tissue than conventional, micrometer-sized contrast agents [68]. This hybrid system also encapsulated Dox along with hollow AuNPs, to achieve PTT upon exposure to NIR light (wavelength of 808 nm). In vivo fluorescence imaging demonstrated the accumulation of NPs at the targeted area in a 4T1 tumour model. NIR illumination also resulted in localized hyperthermia leading to significant Dox release. The nanoplatform was also found effective at performing US image-guided PTT and chemotherapy. Charest et al. developed a liposomal formulation of AuNPs with carboplatin and evaluated its radiosensitizing potential [67]. The study found that simultaneous administration of low doses of carboplatin and AuNPs through encapsulation in liposomal nanocarriers resulted in effective radiosensitization.

The investigation of liposomal encapsulation of QDs, MNPs, and SNPs was also conducted for enhanced therapeutic efficacy. Chen et al. examined liposomes encapsulating both CdSe QDs modified with oleic acid and superparamagnetic iron oxide NPs (SPIONs), for the treatment of hepatocellular carcinoma [76]. This strategy led to the synthesis of magnetic fluorescent liposomes with multifunctional properties. The platform could label and image cancer cells with high biocompatibility, suggesting that it has the potential for improved targeted drug delivery. A study by Sun et al. demonstrated the capabilities of liposomes encapsulating MSNs, for triple-modal image-guided cancer therapy as a theranostic drug-delivery platform [80]. In this research, gadolinium-doped MSNs were encapsulated in liposomes along with Dox. Liposomes were coated with FA to prevent the leakage of Dox and for achieving targeted delivery. The nanoplatform was also conjugated with indocyanine green (ICG) to enable triple-modal imaging through NIR irradiation. Results demonstrated that ICG enabled PTT and PDT, while allowing NIR fluorescence imaging and photoacoustic imaging (PAI). The addition of gadolinium also enabled MRI capabilities. In vitro and in vivo studies showed improved antitumour effects with good imaging contrast, suggesting that this theranostic platform is a candidate for image-guided phototherapy. Overall, the system was successful at providing triple-modal imaging in a single platform, capable of NIR fluorescence imaging, PAI, and MRI.

5.2. Micelle-Based Hybrid Platforms

Micelles are widely used as organic NPs in biomedical applications, acting as nanocompartments through the creation of shells or coatings, as well as for encapsulating inorganic NPs. Some example biomedical applications of such micelle-based hybrid platforms are given in Table 2.

Table 2. Micelle-based hybrid platforms for biomedical applications.

Organic NPs	Inorganic NPs	Payload	Application	Target Condition	Highlights	Ref.
Micelles (oleic acid and tetraethylene glycol)	AuNPs and IONPs	Dexa	Drug delivery/bioimaging	Dialysis bag	The micellar system was capable of encapsulating Dexa, AuNPs, IONPs, and demonstrated its potential for the delivery of multiple types of therapeutic and diagnostic agents.	[84]
Micelles (polylacticacid stereocomplex)	AuNPs (tethered in the shell)	Dox	PTT/chemotherapy	HepG2 cell line/mice	The nanoplatform was able to provide accelerated drug release via PTE, and showed improved efficacy in tumour reduction.	[85]
Micelles (PHEA-LA-PEG-FA)	Au core (with silica)/QDs shell	Dox	Drug delivery/PTT/bioimaging	MCF7 cell line	The platform was utilised as a theranostic device capable of real-time imaging.	[86]
Micelles	QDs and/or SPIONs	Single stranded DNA (p53) or avidin	Biomolecular detection/tracking	-	The hybrid platform successfully performed rapid, sensitive, and specific separation and detection of DNA and/or protein from a small sample volume.	[87]
Micelles	CuInS$_2$/ZnS QDs	-	Intracellular temperature sensing	HeLa and PC3 cell line (mice)	The nanoplatform was efficient in microscale temperature sensing/hyperthermia monitoring through NIR emission with no cytotoxic effect.	[88]
Micelles	CdSe/ZnS QDs	-	Evaluation of toxicity/biosensing	HepG2 cell line	CdSe/ZnS QDs coated with micelles showed minimal toxicity, suggesting that thicker protective polymer layers reduced cytotoxicity and were suitable for bioimaging applications.	[89]
Micelles	SPIONs	Dox	pH-sensitive delivery/bioimaging	HepG2 cell line/mice	The platform successfully achieved drug release and could be imaged through MRI.	[90]

NPs: nanoparticles; AuNPs: gold nanoparticles; Dox: doxorubicin; QDs: quantum dots; IONPs: iron oxide nanoparticles; Dexa: dexamethasone; PTE: photothermal effect; PTT: photothermal therapy; PHEA: α,β-poly(N-hydroxyethyl)-DL-aspartamide; FA: folic acid; SPIONs: superparamagnetic iron oxide nanoparticles; MRI: magnetic resonance imaging; PEG: polyethylene glycol; LA: lipoic acid.

Volsi et al. studied the design of a theranostic micellar nanoplatform for targeted cancer therapy [86]. The polymeric micellar structure consisted of α,β-poly(N-hydroxyethyl)-DL-aspartamide functionalized with lipoic acid (LA), PEG as a hydrophilic moiety, and FA as a targeting moiety which was able to self-assemble in aqueous solution. The platform also encapsulated Dox and Au core–shell QD NPs. Experiments showed that the nanocompartment was stable and efficient at targeting/delivering Dox to MCF7 cells, as well as capable of exploiting heat generation by PTE of QD-Au NPs. It was suggested that this theranostic hybrid system has potential for cancer treatment considering its enhanced drug-delivery behaviour and imaging capabilities, which can assist in diagnostics and therapy monitoring purposes.

Micellar nanocompartments were also examined by Li et al. for pH-sensitive delivery and MRI imaging [90]. Researchers synthesized poly(ethylene glycol)-b-poly(β-benzyl L-aspartate) and aminolyzed it with N,N-diisopropylamino ethylamine and N,N-dibutalamino ethylamine at different molar ratios, for the development of an amphiphilic block copolymer capable of encapsulating SPIONs and Dox. The system was designed to encapsulate its payloads, i.e., drugs or contrast agents, under neutral pH conditions, providing stability and preventing premature release of payloads during circulation or storage. The nanoplatform was also designed to provide triggered release in weak acidic environments, typically found in certain pathological conditions. Experiments demonstrated effective uptake by HepG2 cells and successful release of Dox at low pH conditions, demonstrating the nanoplatform's potential for therapeutic purposes. In addition, fluorescence and MRI studies revealed that the weak positive charge of the hybrid system contributed to longer blood circulation in vivo. The system thus exhibited successful pH-sensitive tumour targeting with efficient and non-invasive MRI visibility, allowing improved non-invasive image-guided therapy. Furthermore, it had a minimal side-effects profile, while displaying impressive anticancer outcomes. Further studies focusing on the applicability of micellar nanocompartments for enhanced tracking and biomolecular detection have been also carried out by encapsulating QDs and SPIONs [87].

5.3. PLGA-Based Hybrid Platforms

PLGA-based hybrid systems have recently emerged as promising platforms, especially for drug delivery and bioimaging applications. Table 3 shows selected publications covering PLGA-based hybrid platforms utilised in biomedical applications.

Table 3. PLGA-based hybrid platforms for biomedical applications.

Organic NPs	Inorganic NPs	Payload	Application	Target Condition	Highlights	Ref.
PLGA	Hollow Au nanoshell	Anti-PD-1 peptide	Sustained release/PTT	4T1 and CT26 cell line/mice	Efficient PD-1 blocking was achieved through sustained release (for 40 days) of anti-PD-1 peptide by NIR irradiation.	[91]
PLGA	Graphene QDs	Dox	pH-responsive delivery/bioimaging	HeLa cell line	Drug release was observed in a mild acidic environment in vitro. The platform showed its potential for bioimaging applications.	[92]
PLGA	CdSe/ZnS QDs	Chlorophyllin copper complex	PDT	NIH-3T3 cell line	The nanoplatform could generate ROS when excited at 365 nm.	[63]
PLGA (with PEG and Wy5a aptamer)	SPIONs	Docetaxel	Controlled drug delivery/MRI	PC-3 cell line/mice	In vitro investigations demonstrated high-sensitivity MRI detection and enhanced cytotoxic effects. In vivo studies showed that NPs exhibited superior antitumour efficacy while causing minimal systemic toxicity.	[93]

Table 3. Cont.

Organic NPs	Inorganic NPs	Payload	Application	Target Condition	Highlights	Ref.
PLGA (with PEG-FA)	SPIONs	Dox	US/MRI/focused US-triggered drug delivery	4T1 cell line/mice	The nanoplatform exhibited enhanced tumour targeting, effective US/MRI contrast, and focused US-triggered drug release.	[94]
PLGA	SNPs (conjugated with Cy7.5)	Docetaxel	Chemo-radiation therapy	Mice	Tracking and sustained drug release from spacers (made of PLGA and loaded with SNPs) were achieved, demonstrating the combined therapeutic efficacy of chemo-radiation therapy.	[95]
PLGA	AgNPs	IFNγ	Cancer therapy	HeLa and MCF-7 cell line	The nanoplatform induced apoptosis through the delivery of AgNPs and IFNγ.	[96]

NPs: nanoparticles; PLGA: poly (lactic-co-glycolic acid); Dox: doxorubicin; PTT: photothermal therapy; PDT: photodynamic therapy; QDs: quantum dots; FA: folic acid; PEG: polyethylene glycol; SPIONs: superparamagnetic iron oxide nanoparticles; IFNγ: recombinant interferon gamma; SNPs: silica nanoparticles; AgNPs: silver nanoparticles; US: ultrasound; MRI: magnetic resonance imaging; ROS: reactive oxygen species.

Luo et al. utilised PLGA NPs for the encapsulation of anti-PD-1 peptide and hollow Au nanoshells for improved immunotherapy achieved by PD-1 blocking combined with PTT [91]. This hybrid system demonstrated long-term activation of the immune system over 40 days, which could also be accelerated using NIR laser illumination. It was also revealed that multiple irradiations using NIR laser illumination enhanced the antitumour effect, resulting in the inhibition of primary tumours as well as distant tumours. The therapy was also capable of enhancing immune cell activation.

In another study conducted by Galliani et al., PLGA-based hybrid platforms were designed to implement PDT for enhanced cancer therapy [63]. Chlorophyllin–copper complex and CdSe/ZnS core–shell QDs were encapsulated successfully in the PLGA nanocompartments. Irradiation at 365 nm by UV resulted in the generation of ROS due to fluorescence resonance energy transfer between QDs and chlorophyllin. It was indicated by the authors that this platform has potential for PDT as it could generate ROS upon irradiation; however, further analysis is required to assess the underlying mechanisms and optimize the formulation.

As reported by Jin et al., a PLGA-based nanoplatform was designed for multimodal imaging and US-triggered drug delivery. SPIONs and Dox were successfully encapsulated in PLGA-based nanocompartments conjugated with PEG and FA [94]. In vitro experiments demonstrated the potential for US and MRI contrast imaging, as well as increased targeting ability due to FA conjugation. Focused US was utilised as a remote-control technique to trigger Dox release and induce cell membrane permeabilization. These findings highlighted the promising application potential of this system as a tool for US- and MRI-guided drug delivery in anticancer therapy.

Kumar et al. aimed to develop implantable nanoplatforms composed of PLGA-based nanocompartments encapsulating docetaxel and Cy7.5 (fluorophore) conjugated with SNPs [95]. In vivo studies demonstrated efficient sustained drug release near tissues. The docetaxel-loaded spacers exhibited suppression of tumour growth compared with the control over 16 days, demonstrating improved therapeutic efficacy. It is important to note that this system was also suitable for contrast imaging due to its fluorescent moiety (Cy7.5), with potential for use in disease monitoring.

5.4. Dendrimer-Based Hybrid Platforms

Table 4 represents a selection of recently published articles focusing on the utilization of dendrimer-based hybrid platforms in biomedical applications.

Table 4. Dendrimer-based hybrid platforms developed for use in biomedical applications.

Organic NPs	Inorganic NPs	Payload	Application	Target Condition	Highlights	Ref.
Dendrimers (PAMAM)	Carbon QDs (conjugated with RGDS peptide)	-	Targeted delivery/bioimaging	MDA-MB-231 cell line (for TNBC)	Green synthesis of carbon QDs was successfully performed. The nanoplatform showed potential as a theranostic tool for TNBC, with the capability of detecting and monitoring the presence of Cu (II) ions.	[97]
Dendrimers (conjugated with PEG and Herceptin)	AuNPs	Gadolinium	Targeted delivery/bioimaging	HER-2 overexpressing cell lines	Successful in vitro internalization was achieved with no cytotoxicity. The nanoplatform worked as a nanoimaging agent as well as a nanocarrier for targeted delivery of cytotoxic drugs.	[98]
Dendrimers (PAMAM)	Mesoporous SNPs	Curcumin	Fluorescence imaging/pH-responsive drug delivery	HeLa cell line	The study demonstrated the first-time use of PAMAM dendrimers as pH-sensitive capping and self-fluorescent agents.	[99]
Dendrimer–stabilized Au nanoflowers	Ultrasmall IONPs	-	MRI/CT/PAI-guided combination of PTT and RT	4T1 cell line/subcutaneous tumour model	The multifunctional theranostic platform presented enhanced photothermal conversion efficiency and compatibility with multiple imaging modalities.	[100]

NPs: nanoparticles; AuNPs: gold nanoparticles; QDs: quantum dots; SNPs: silica nanoparticles; IONPs: iron oxide nanoparticles; Dox: doxorubicin; MRI: magnetic resonance imaging; CT: computed tomography; PAI: photoacoustic imaging; PTT: photothermal therapy; RT: radiotherapy; PAMAM: polyamidoamine; TNBC: triple-negative breast cancer; PEG: polyethylene glycol.

In a study by Ghosh et al., a dendrimer-based hybrid platform was utilised for targeted gene delivery for the treatment of triple-negative breast cancer (TNBC) [97]. The authors successfully synthesized carbon QDs conjugated with PAMAM dendrimers of different generations. RGDS peptides were further conjugated to the nanoplatform to be able to target the $\alpha v \beta 3$ integrin, which is known to be overexpressed in TNBC. Among different conjugates, QD–PAMAM conjugate 3 showed superior capabilities for gene complexation and protection against enzymatic digestion. Furthermore, it exhibited efficient detection of Cu (II) ions, with a fluorescence quenching efficiency of 93%. It is important to note that TNBC often has higher levels of Cu (II) ions, which offers potential for the detection of the metastatic phase of TNBC.

In another approach, a pH-sensitive and self-fluorescent nanoplatform was developed using mesoporous SNPs and PAMAM dendrimers [99]. It was found that the inclusion of PAMAM dendrimers provided improved encapsulation efficiency and additional eligible reaction sites for modifications. In addition, the structure of PAMAM dendrimers affected the drug-release performance and prevented burst release. The fluorescence behaviour of this hybrid system offers the capability for potential biological tracking and bio-detection. Importantly, PAMAM dendrimers served as both pH-sensitive capping agents and self-fluorescent agents, possessing multiple functions in a single platform. This research demonstrated the versatility of dendrimer-based systems for developing multifunctional and biocompatible drug-delivery platforms for biomedical imaging, diagnosis, and simultaneous therapy.

Dendrimer-based nanocompartments have also been utilised for the application in multimodal image-guided cancer therapy [100]. A theranostic nanoplatform was developed, comprising generation 5 poly(amidoamine) dendrimer-stabilized AuNPs embedded with ultrasmall IONPs, capable of MR/CT/PAI-guided PTT and radiotherapy (RT). This multifunctional nanoplatform induced significant cell death under laser irradiation. In vivo experiments demonstrated accumulation within the tumour tissue along with MRI, CT, and PAI imaging enhancement capabilities. These findings suggest that this nanoplatform has the potential for image-guided cancer therapy, leading to improved diagnosis and treatment while minimizing side-effects.

5.5. Chitosan-Based Hybrid Platforms

Chitosan has also been used as a constitutive material for hybrid platforms in biomedical applications. Examples of recently published studies focusing on chitosan-based hybrid platforms are given in Table 5.

Table 5. Chitosan-based hybrid platforms developed for usage in biomedical applications.

Organic NPs	Inorganic NPs	Payload	Application	Target Condition	Highlights	Ref.
Chitosan/tripolyphosphate nanogels	Cysteine-functionalized AuNPs	Dox	CT imaging/targeted delivery	OSCC cell line/mice	The hybrid nanogel exhibited high drug-loading capacity (87%) and controlled drug release at acidic pH. AuNPs enabled the monitoring of drug delivery and accumulation in tumours.	[101]
Thermosensitive hydrogel with chitosan	Multiwalled carbon nanotubes	Dox and rhodamine B	Sustained drug delivery/fluorescence imaging	BEL-7402 cell line/mice	Dual drug delivery could be successfully monitored using fluorescence imaging.	[102]
Chitosan	SPION	Dox	Drug delivery/bioimaging	C6 glioma cell line	The nanoplatform could be used as an MRI contrast agent as well as a theranostic tool for glioblastoma.	[103]
Chitosan/alginate	Fe_3O_4	Lutein	Magnetic targeting delivery	MDA-MB-231 and MCF-7 cell line	The platform showed enhanced cytotoxicity upon exposure to a magnetic field.	[104]
Magnetic chitosan	Aptamer-modified graphene QDs	Dox	Photothermal chemotherapy	Hepatoma cell line H22/mice	There was no evidence of substantial biological toxicity or adverse effects in either in vivo or in vitro experiments.	[105]
Chitosan	Mesoporous SNP	Dox and indocyanine green	Chemotherapy/PDT	HepG2 cell line	The platform could successfully target and kill cells via chemotherapy combined with PDT.	[106]

NPs: nanoparticles; AuNPs: gold nanoparticles; Dox: doxorubicin; CT: computed tomography; SPIONs: superparamagnetic iron oxide nanoparticles; MRI: magnetic resonance imaging; QDs: quantum dots; SNPs: silica nanoparticles; PDT: photodynamic therapy.

Liu et al. developed a theranostic platform by incorporating cysteine functionalized AuNPs into chitosan/tripolyphosphate NPs (modified with polyacrylic acid), aiming for improved cellular uptake, high loading capacity, controlled release, and efficient bioimaging [101]. This hybrid platform was also loaded with Dox as the chemotherapeutic agent. In vitro experiments showed sustained drug release for up to 48 h under acidic conditions; however, drug release was accelerated at higher pH values. The hybrid platform also showed greater cellular uptake compared to free Dox. In vivo studies revealed that the drug accumulation could be tracked, which was confirmed by CT scans. Notably, significant inhibition of tumour growth compared with free Dox was observed in vivo. These findings demonstrated the potential of this hybrid platform as a theranostic tool for tumour treatment.

A study by Gholami et al. explored chitosan-based hybrid platforms loaded with SPIONs and Dox for treating glioblastoma [103]. Drug-release tests demonstrated a rapid release of Dox at pH 5.5, which resembles the pH of the tumour microenvironment, indicating pH-dependent drug-release capability. In addition, the cellular internalization of this hybrid platform was confirmed through fluorescence microscopy. Overall, the study demonstrated the potential of this formulation for both diagnosis and treatment of glioblastoma.

Chen et al. investigated a nanoplatform comprising aptamer-modified graphene QDs and magnetic chitosan for the treatment of hepatocellular carcinoma [105]. It was designed to utilise an aptamer for active targeting and graphene QDs for PTT. In vitro experiments demonstrated the internalization of this hybrid platform in cancer cells and subsequent NIR-triggered drug release. Additionally, the platform showed low cytotoxicity profiles and enhanced accumulation at the tumour site in vivo, which was further validated by imaging. Overall, this system has potential for combined photothermal chemotherapy in cancer treatment.

6. Conclusions

This review provides a description and analysis of recent findings in drug delivery and bioimaging applications using hybrid nanoplatforms composed of organic NPs encapsulating inorganic NPs. Findings from these recent investigations suggest that the development of hybrid nanoplatforms holds great potential for targeted cancer therapy and imaging applications. Previous studies have focused on various types of nanocompartments functionalized with different agents for specific biomedical applications. Hybrid nanoplatforms have demonstrated enhanced biocompatibility, stability, efficient drug delivery, improved bioimaging performance, and applicability in different treatment approaches such as PTT, PDT, and US-based. These unique characteristics of hybrid nanoplatforms have enabled successful tumour targeting along with remarkable anticancer outcomes with low side-effect profiles. This demonstrates the potential of hybrid nanoplatforms in improving diagnosis, monitoring, and treatment, allowing new avenues of opportunity for the development of novel therapeutic modalities.

On the other hand, there is a clear need for the development of effective formulation processes for these hybrid nanostructures. Ongoing challenges relate to (i) the design and optimization of suitable nanostructures, (ii) the controllable production of NPs with desired properties including shape, size, drug loading efficiency, and well-defined drug release or imaging performance, and (iii) scale-up of manufacturing processes for large-scale production. Addressing these challenges requires close collaboration across multiple disciplines, including nanotechnology, chemistry, engineering, and pharmaceutics. In addition, the transition from laboratory-scale research to practical applications in humans requires extensive research to optimize nanostructure formulations, elucidate the underlying mechanisms of action, identify suitable administration routes, and further assess the extent of performance improvement compared with more conventional drug delivery and imaging approaches.

Author Contributions: Conceptualization, F.Y., D.C. and X.Z.; writing—original draft preparation, F.Y.; writing—review and editing, supervision, D.C. and X.Z. All authors have read and agreed to the published version of the manuscript.

Funding: This research received no external funding.

Institutional Review Board Statement: Not applicable.

Informed Consent Statement: Not applicable.

Data Availability Statement: Not applicable.

Conflicts of Interest: The authors declare no conflict of interest.

References

1. Gabizon, A.; Peretz, T.; Sulkes, A.; Amselem, S.; Ben-Yosef, R.; Ben-Baruch, N.; Catane, R.; Biran, S.; Barenholz, Y. Systemic administration of doxorubicin-containing liposomes in cancer patients: A phase I study. *Eur. J. Cancer Clin. Oncol.* **1989**, *25*, 1795–1803. [CrossRef] [PubMed]
2. Tom, R.T.; Suryanarayanan, V.; Reddy, P.G.; Baskaran, S.; Pradeep, T. Ciprofloxacin-protected gold nanoparticles. *Langmuir* **2004**, *20*, 1909–1914. [CrossRef] [PubMed]
3. Rudokas, M.; Najlah, M.; Alhnan, M.A.; Elhissi, A. Liposome Delivery Systems for Inhalation: A Critical Review Highlighting Formulation Issues and Anticancer Applications. *Med. Princ. Pract.* **2016**, *25* (Suppl 2), 60–72. [CrossRef] [PubMed]
4. Hajos, F.; Stark, B.; Hensler, S.; Prassl, R.; Mosgoeller, W. Inhalable liposomal formulation for vasoactive intestinal peptide. *Int. J. Pharm.* **2008**, *357*, 286–294. [CrossRef]
5. Liu, J.; Gong, T.; Fu, H.; Wang, C.; Wang, X.; Chen, Q.; Zhang, Q.; He, Q.; Zhang, Z. Solid lipid nanoparticles for pulmonary delivery of insulin. *Int. J. Pharm.* **2008**, *356*, 333–344. [CrossRef]
6. Vieira, D.B.; Gamarra, L.F. Getting into the brain: Liposome-based strategies for effective drug delivery across the blood–brain barrier. *Int. J. Nanomed.* **2016**, *11*, 5381–5414. [CrossRef]
7. Farokhzad, O.C.; Langer, R. Impact of nanotechnology on drug delivery. *ACS Nano* **2009**, *3*, 16–20. [CrossRef]
8. Salata, O. Applications of nanoparticles in biology and medicine. *J. Nanobiotechnology* **2004**, *2*, 3. [CrossRef]
9. Jiang, Y.; Krishnan, N.; Heo, J.; Fang, R.H.; Zhang, L. Nanoparticle–hydrogel superstructures for biomedical applications. *J. Control. Release* **2020**, *324*, 505–521. [CrossRef]
10. Park, W.; Shin, H.; Choi, B.; Rhim, W.-K.; Na, K.; Han, D.K. Advanced hybrid nanomaterials for biomedical applications. *Prog. Mater. Sci.* **2020**, *114*, 100686. [CrossRef]
11. Sun, Y.; Zheng, L.; Yang, Y.; Qian, X.; Fu, T.; Li, X.; Yang, Z.; Yan, H.; Cui, C.; Tan, W. Metal–organic framework nanocarriers for drug delivery in biomedical applications. *Nano-Micro Lett.* **2020**, *12*, 103. [CrossRef]
12. Zhou, H.; Ge, J.; Miao, Q.; Zhu, R.; Wen, L.; Zeng, J.; Gao, M. Biodegradable Inorganic Nanoparticles for Cancer Theranostics: Insights into the Degradation Behavior. *Bioconjugate Chem.* **2020**, *31*, 315–331. [CrossRef] [PubMed]
13. Barenholz, Y. (Chezy) Doxil®—The first FDA-approved nano-drug: Lessons learned. *J. Control. Release* **2012**, *160*, 117–134. [CrossRef]
14. Yanar, F.; Mosayyebi, A.; Nastruzzi, C.; Carugo, D.; Zhang, X. Continuous-Flow Production of Liposomes with a Millireactor under Varying Fluidic Conditions. *Pharmaceutics* **2020**, *12*, 1001. [CrossRef]
15. Carugo, D.; Bottaro, E.; Owen, J.; Stride, E.; Nastruzzi, C. Liposome production by microfluidics: Potential and limiting factors. *Sci. Rep.* **2016**, *6*, 25876. [CrossRef]
16. Kibria, G.; Hatakeyama, H.; Ohga, N.; Hida, K.; Harashima, H. The effect of liposomal size on the targeted delivery of doxorubicin to Integrin $\alpha v \beta 3$-expressing tumor endothelial cells. *Biomaterials* **2013**, *34*, 5617–5627. [CrossRef]
17. Wang, X.; Wang, Y.; Chen, Z.G.; Shin, D.M. Advances of Cancer Therapy by Nanotechnology. *Cancer Res. Treat.* **2009**, *41*, 1–11. [CrossRef] [PubMed]
18. Ishida, T.; Harashima, H.; Kiwada, H. Liposome clearance. *Biosci. Rep.* **2002**, *22*, 197–224. [CrossRef] [PubMed]
19. Abuchowski, A.; McCoy, J.R.; Palczuk, N.C.; van Es, T.; Davis, F.F. Effect of covalent attachment of polyethylene glycol on immunogenicity and circulating life of bovine liver catalase. *J. Biol. Chem.* **1977**, *252*, 3582–3586. [CrossRef] [PubMed]
20. Perumal, S.; Atchudan, R.; Lee, W. A Review of Polymeric Micelles and Their Applications. *Polymers* **2022**, *14*, 2510. [CrossRef]
21. Daza, E.A.; Schwartz-Duval, A.S.; Volkman, K.; Pan, D. Facile Chemical Strategy to Hydrophobically Modify Solid Nanoparticles Using Inverted Micelle-Based Multicapsule for Efficient Intracellular Delivery. *ACS Biomater. Sci. Eng.* **2018**, *4*, 1357–1367. [CrossRef] [PubMed]
22. Liu, J.; Zeng, F.; Allen, C. Influence of serum protein on polycarbonate-based copolymer micelles as a delivery system for a hydrophobic anti-cancer agent. *J. Control. Release* **2005**, *103*, 481–497. [CrossRef] [PubMed]
23. Letchford, K.; Zastre, J.; Liggins, R.; Burt, H. Synthesis and micellar characterization of short block length methoxy poly(ethylene glycol)-block-poly(caprolactone) diblock copolymers. *Colloids Surf. B Biointerfaces* **2004**, *35*, 81–91. [CrossRef] [PubMed]
24. Yang, L.; Wu, X.; Liu, F.; Duan, Y.; Li, S. Novel Biodegradable Polylactide/poly(ethylene glycol) Micelles Prepared by Direct Dissolution Method for Controlled Delivery of Anticancer Drugs. *Pharm. Res.* **2009**, *26*, 2332–2342. [CrossRef] [PubMed]

25. Sun, X.; Bandara, N. Applications of reverse micelles technique in food science: A comprehensive review. *Trends Food Sci. Technol.* **2019**, *91*, 106–115. [CrossRef]
26. Lee, R.-S.; Lin, C.-H.; Aljuffali, I.A.; Hu, K.-Y.; Fang, J.-Y. Passive targeting of thermosensitive diblock copolymer micelles to the lungs: Synthesis and characterization of poly(N-isopropylacrylamide)-block-poly(ε-caprolactone). *J. Nanobiotechnology* **2015**, *13*, 42. [CrossRef]
27. Tabatabaei Mirakabad, F.S.; Nejati-Koshki, K.; Akbarzadeh, A.; Yamchi, M.R.; Milani, M.; Zarghami, N.; Zeighamian, V.; Rahimzadeh, A.; Alimohammadi, S.; Hanifehpour, Y.; et al. PLGA-Based Nanoparticles as Cancer Drug Delivery Systems. *Asian Pac. J. Cancer Prev.* **2014**, *15*, 517–535. [CrossRef]
28. Hernández-Giottonini, K.Y.; Rodríguez-Córdova, R.J.; Gutiérrez-Valenzuela, C.A.; Peñuñuri-Miranda, O.; Zavala-Rivera, P.; Guerrero-Germán, P.; Lucero-Acuña, A. PLGA nanoparticle preparations by emulsification and nanoprecipitation techniques: Effects of formulation parameters. *RSC Adv.* **2020**, *10*, 4218–4231. [CrossRef]
29. Huang, W.; Zhang, C. Tuning the size of poly(lactic-co-glycolic acid) (PLGA) nanoparticles fabricated by nanoprecipitation. *Biotechnol. J.* **2018**, *13*, 1700203. [CrossRef]
30. Reis, C.P.; Neufeld, R.J.; Ribeiro, A.J.; Veiga, F.; Nanoencapsulation, I. Methods for preparation of drug-loaded polymeric nanoparticles. *Nanomedicine* **2006**, *2*, 8–21. [CrossRef]
31. Hodge, P. Polymer science branches out. *Nature* **1993**, *362*, 18–19. [CrossRef]
32. Mittal, P.; Saharan, A.; Verma, R.; Altalbawy, F.M.A.; Alfaidi, M.A.; Batiha, G.E.-S.; Akter, W.; Gautam, R.K.; Uddin, M.S.; Rahman, M.S. Dendrimers: A New Race of Pharmaceutical Nanocarriers. *BioMed Res. Int.* **2021**, *2021*, 8844030. [CrossRef]
33. Nanjwade, B.K.; Bechra, H.M.; Derkar, G.K.; Manvi, F.V.; Nanjwade, V.K. Dendrimers: Emerging polymers for drug-delivery systems. *Eur. J. Pharm. Sci.* **2009**, *38*, 185–196. [CrossRef] [PubMed]
34. Mohammed, M.A.; Syeda, J.T.M.; Wasan, K.M.; Wasan, E.K. An Overview of Chitosan Nanoparticles and Its Application in Non-Parenteral Drug Delivery. *Pharmaceutics* **2017**, *9*, 53. [CrossRef]
35. Ladj, R.; Bitar, A.; Eissa, M.M.; Fessi, H.; Mugnier, Y.; Le Dantec, R.; Elaissari, A. Polymer encapsulation of inorganic nanoparticles for biomedical applications. *Int. J. Pharm.* **2013**, *458*, 230–241. [CrossRef] [PubMed]
36. Mahmudin, L.; Suharyadi, E.; Utomo, A.B.S.; Abraha, K. Optical Properties of Silver Nanoparticles for Surface Plasmon Resonance (SPR)-Based Biosensor Applications. *J. Mod. Phys.* **2015**, *06*, 1071. [CrossRef]
37. Link, S.; Wang, Z.L.; El-Sayed, M.A. Alloy Formation of Gold–Silver Nanoparticles and the Dependence of the Plasmon Absorption on Their Composition. *J. Phys. Chem. B* **1999**, *103*, 3529–3533. [CrossRef]
38. Cobley, C.M.; Skrabalak, S.E.; Campbell, D.J.; Xia, Y. Shape-Controlled Synthesis of Silver Nanoparticles for Plasmonic and Sensing Applications. *Plasmonics* **2009**, *4*, 171–179. [CrossRef]
39. Guerrero, A.R.; Hassan, N.; Escobar, C.A.; Albericio, F.; Kogan, M.J.; Araya, E. Gold nanoparticles for photothermally controlled drug release. *Nanomedicine* **2014**, *9*, 2023–2039. [CrossRef] [PubMed]
40. Xing, S.; Zhang, X.; Luo, L.; Cao, W.; Li, L.; He, Y.; An, J.; Gao, D. Doxorubicin/gold nanoparticles coated with liposomes for chemo-photothermal synergetic antitumor therapy. *Nanotechnology* **2018**, *29*, 405101. [CrossRef] [PubMed]
41. Slepička, P.; Slepičková Kasálková, N.; Siegel, J.; Kolská, Z.; Švorčík, V. Methods of Gold and Silver Nanoparticles Preparation. *Materials* **2019**, *13*, 1. [CrossRef] [PubMed]
42. Iga, A.M.; Robertson, J.H.P.; Winslet, M.C.; Seifalian, A.M. Clinical Potential of Quantum Dots. *BioMed Res. Int.* **2008**, *2007*, e76087. [CrossRef] [PubMed]
43. Bera, D.; Qian, L.; Tseng, T.-K.; Holloway, P.H. Quantum Dots and Their Multimodal Applications: A Review. *Materials* **2010**, *3*, 2260–2345. [CrossRef]
44. Matea, C.T.; Mocan, T.; Tabaran, F.; Pop, T.; Mosteanu, O.; Puia, C.; Iancu, C.; Mocan, L. Quantum dots in imaging, drug delivery and sensor applications. *Int. J. Nanomed.* **2017**, *12*, 5421–5431. [CrossRef]
45. Selvarajan, V.; Obuobi, S.; Ee, P.L.R. Silica Nanoparticles—A Versatile Tool for the Treatment of Bacterial Infections. *Front. Chem.* **2020**, *8*, 602. [CrossRef]
46. Parra, M.; Gil, S.; Gaviña, P.; Costero, A.M. Mesoporous Silica Nanoparticles in Chemical Detection: From Small Species to Large Bio-Molecules. *Sensors* **2021**, *22*, 261. [CrossRef]
47. Gul, S.; Khan, S.B.; Rehman, I.U.; Khan, M.A.; Khan, M.I. A Comprehensive Review of Magnetic Nanomaterials Modern Day Theranostics. *Front. Mater.* **2019**, *6*, 179. [CrossRef]
48. Ali, A.; Shah, T.; Ullah, R.; Zhou, P.; Guo, M.; Ovais, M.; Tan, Z.; Rui, Y. Review on Recent Progress in Magnetic Nanoparticles: Synthesis, Characterization, and Diverse Applications. *Front. Chem.* **2021**, *9*, 629054. [CrossRef]
49. Vilas-Boas, V.; Carvalho, F.; Espiña, B. Magnetic Hyperthermia for Cancer Treatment: Main Parameters Affecting the Outcome of In Vitro and In Vivo Studies. *Molecules* **2020**, *25*, 2874. [CrossRef]
50. Macchione, M.A.; Biglione, C.; Strumia, M. Design, Synthesis and Architectures of Hybrid Nanomaterials for Therapy and Diagnosis Applications. *Polymers* **2018**, *10*, 527. [CrossRef]
51. Ding, X.; Zhao, J.; Liu, Y.; Zhang, H.; Wang, Z. Silica nanoparticles encapsulated by polystyrene via surface grafting and in situ emulsion polymerization. *Mater. Lett.* **2004**, *58*, 3126–3130. [CrossRef]
52. Deng, Y.; Wang, L.; Yang, W.; Fu, S.; Elaıssari, A. Preparation of magnetic polymeric particles via inverse microemulsion polymerization process. *J. Magn. Magn. Mater.* **2003**, *257*, 69–78. [CrossRef]

53. Landfester, K.; Weiss, C. Encapsulation by Miniemulsion Polymerization. *Mod. Tech. Nano-Microreactors/-React.* **2010**, *229*, 1–49. [CrossRef]
54. Noble, G.T.; Stefanick, J.F.; Ashley, J.D.; Kiziltepe, T.; Bilgicer, B. Ligand-targeted liposome design: Challenges and fundamental considerations. *Trends Biotechnol.* **2014**, *32*, 32–45. [CrossRef]
55. Hua, S. Targeting sites of inflammation: Intercellular adhesion molecule-1 as a target for novel inflammatory therapies. *Front. Pharmacol.* **2013**, *4*, 127. [CrossRef] [PubMed]
56. Adepu, S.; Ramakrishna, S. Controlled Drug Delivery Systems: Current Status and Future Directions. *Molecules* **2021**, *26*, 5905. [CrossRef]
57. Ryvolova, M.; Chomoucka, J.; Drbohlavova, J.; Kopel, P.; Babula, P.; Hynek, D.; Adam, V.; Eckschlager, T.; Hubalek, J.; Stiborova, M.; et al. Modern Micro and Nanoparticle-Based Imaging Techniques. *Sensors* **2012**, *12*, 14792–14820. [CrossRef]
58. Han, X.; Xu, K.; Taratula, O.; Farsad, K. Applications of Nanoparticles in Biomedical Imaging. *Nanoscale* **2019**, *11*, 799–819. [CrossRef]
59. Caspani, S.; Magalhães, R.; Araújo, J.P.; Sousa, C.T. Magnetic Nanomaterials as Contrast Agents for MRI. *Materials* **2020**, *13*, 2586. [CrossRef]
60. Li, L.; Guan, Y.; Xiong, H.; Deng, T.; Ji, Q.; Xu, Z.; Kang, Y.; Pang, J. Fundamentals and applications of nanoparticles for ultrasound-based imaging and therapy. *Nano Sel.* **2020**, *1*, 263–284. [CrossRef]
61. Taratula, O.; Doddapaneni, B.S.; Schumann, C.; Li, X.; Bracha, S.; Milovancev, M.; Alani, A.W.G.; Taratula, O. Naphthalocyanine-Based Biodegradable Polymeric Nanoparticles for Image-Guided Combinatorial Phototherapy. *Chem. Mater.* **2015**, *27*, 6155–6165. [CrossRef]
62. Tomitaka, A.; Arami, H.; Huang, Z.; Raymond, A.; Rodriguez, E.; Cai, Y.; Febo, M.; Takemura, Y.; Nair, M. Hybrid magneto-plasmonic liposomes for multimodal image-guided and brain-targeted HIV treatment. *Nanoscale* **2018**, *10*, 184–194. [CrossRef] [PubMed]
63. Galliani, M.; Signore, G. Poly(Lactide-Co-Glycolide) Nanoparticles Co-Loaded with Chlorophyllin and Quantum Dots as Photodynamic Therapy Agents. *ChemPlusChem* **2019**, *84*, 1653–1658. [CrossRef] [PubMed]
64. Koga, K.; Tagami, T.; Ozeki, T. Gold nanoparticle-coated thermosensitive liposomes for the triggered release of doxorubicin, and photothermal therapy using a near-infrared laser. *Colloids Surf. A Physicochem. Eng. Asp.* **2021**, *626*, 127038. [CrossRef]
65. Sarkar, A.; Carvalho, E.; D'souza, A.A.; Banerjee, R. Liposome-encapsulated fish oil protein-tagged gold nanoparticles for intra-articular therapy in osteoarthritis. *Nanomedicine* **2019**, *14*, 871–887. [CrossRef]
66. Grafals-Ruiz, N.; Rios-Vicil, C.I.; Lozada-Delgado, E.L.; Quiñones-Díaz, B.I.; Noriega-Rivera, R.A.; Martínez-Zayas, G.; Santana-Rivera, Y.; Santiago-Sánchez, G.S.; Valiyeva, F.; Vivas-Mejía, P.E. Brain Targeted Gold Liposomes Improve RNAi Delivery for Glioblastoma. *Int. J. Nanomed.* **2020**, *15*, 2809–2828. [CrossRef] [PubMed]
67. Charest, G.; Tippayamontri, T.; Shi, M.; Wehbe, M.; Anantha, M.; Bally, M.; Sanche, L. Concomitant Chemoradiation Therapy with Gold Nanoparticles and Platinum Drugs Co-Encapsulated in Liposomes. *Int. J. Mol. Sci.* **2020**, *21*, 4848. [CrossRef]
68. Li, W.; Hou, W.; Guo, X.; Luo, L.; Li, Q.; Zhu, C.; Yang, J.; Zhu, J.; Du, Y.; You, J. Temperature-controlled, phase-transition ultrasound imaging-guided photothermal-chemotherapy triggered by NIR light. *Theranostics* **2018**, *8*, 3059–3073. [CrossRef]
69. Zhang, W.; Yu, W.; Ding, X.; Yin, C.; Yan, J.; Yang, E.; Guo, F.; Sun, D.; Wang, W. Self-assembled thermal gold nanorod-loaded thermosensitive liposome-encapsulated ganoderic acid for antibacterial and cancer photochemotherapy. *Artif. Cells Nanomed. Biotechnol.* **2019**, *47*, 406–419. [CrossRef]
70. Zhu, L.; Kuang, Z.; Song, P.; Li, W.; Gui, L.; Yang, K.; Ge, F.; Tao, Y.; Zhang, W. Gold nanorod-loaded thermosensitive liposomes facilitate the targeted release of ruthenium(II) polypyridyl complexes with anti-tumor activity. *Nanotechnology* **2021**, *32*, 455103. [CrossRef]
71. Ding, X.; Yin, C.; Zhang, W.; Sun, Y.; Zhang, Z.; Yang, E.; Sun, D.; Wang, W. Designing Aptamer-Gold Nanoparticle-Loaded pH-Sensitive Liposomes Encapsulate Morin for Treating Cancer. *Nanoscale Res. Lett.* **2020**, *15*, 68. [CrossRef] [PubMed]
72. Li, Y.; He, D.; Tu, J.; Wang, R.; Zu, C.; Chen, Y.; Yang, W.; Shi, D.; Webster, T.J.; Shen, Y. The comparative effect of wrapping solid gold nanoparticles and hollow gold nanoparticles with doxorubicin-loaded thermosensitive liposomes for cancer thermo-chemotherapy. *Nanoscale* **2018**, *10*, 8628–8641. [CrossRef] [PubMed]
73. Prasad, R.; Jain, N.K.; Yadav, A.S.; Chauhan, D.S.; Devrukhkar, J.; Kumawat, M.K.; Shinde, S.; Gorain, M.; Thakor, A.S.; Kundu, G.C.; et al. Liposomal nanotheranostics for multimode targeted in vivo bioimaging and near-infrared light mediated cancer therapy. *Commun. Biol.* **2020**, *3*, 284. [CrossRef] [PubMed]
74. Lv, S.; Jing, R.; Liu, X.; Shi, H.; Shi, Y.; Wang, X.; Zhao, X.; Cao, K.; Lv, Z. One-Step Microfluidic Fabrication of Multi-Responsive Liposomes for Targeted Delivery of Doxorubicin Synergism with Photothermal Effect. *Int. J. Nanomed.* **2021**, *16*, 7759–7772. [CrossRef] [PubMed]
75. Hashemi, M.; Mohammadi, J.; Omidi, M.; Smyth, H.D.C.; Muralidharan, B.; Milner, T.E.; Yadegari, A.; Ahmadvand, D.; Shalbaf, M.; Tayebi, L. Self-assembling of graphene oxide on carbon quantum dot loaded liposomes. *Mater. Sci. Eng. C Mater. Biol. Appl.* **2019**, *103*, 109860. [CrossRef] [PubMed]
76. Chen, M.; Huang, H.; Pan, Y.; Li, Z.; Ouyang, S.; Ren, C.; Zhao, Q. Preparation of layering-structured magnetic fluorescent liposomes and labeling of HepG2 cells. *Biomed. Mater. Eng.* **2022**, *33*, 147–158. [CrossRef]
77. Tang, T.; Huang, B.; Liu, F.; Cui, R.; Zhang, M.; Sun, T. Enhanced delivery of theranostic liposomes through NO-mediated tumor microenvironment remodeling. *Nanoscale* **2022**, *14*, 7473–7479. [CrossRef]

78. Awad, N.S.; Haider, M.; Paul, V.; AlSawaftah, N.M.; Jagal, J.; Pasricha, R.; Husseini, G.A. Ultrasound-Triggered Liposomes Encapsulating Quantum Dots as Safe Fluorescent Markers for Colorectal Cancer. *Pharmaceutics* **2021**, *13*, 2073. [CrossRef]
79. Anilkumar, T.S.; Lu, Y.-J.; Chen, J.-P. Optimization of the Preparation of Magnetic Liposomes for the Combined Use of Magnetic Hyperthermia and Photothermia in Dual Magneto-Photothermal Cancer Therapy. *Int. J. Mol. Sci.* **2020**, *21*, 5187. [CrossRef]
80. Sun, Q.; You, Q.; Wang, J.; Liu, L.; Wang, Y.; Song, Y.; Cheng, Y.; Wang, S.; Tan, F.; Li, N. Theranostic Nanoplatform: Triple-Modal Imaging-Guided Synergistic Cancer Therapy Based on Liposome-Conjugated Mesoporous Silica Nanoparticles. *ACS Appl. Mater. Interfaces* **2018**, *10*, 1963–1975. [CrossRef]
81. Singh, A.K.; Singh, S.S.; Rathore, A.S.; Singh, S.P.; Mishra, G.; Awasthi, R.; Mishra, S.K.; Gautam, V.; Singh, S.K. Lipid-Coated MCM-41 Mesoporous Silica Nanoparticles Loaded with Berberine Improved Inhibition of Acetylcholine Esterase and Amyloid Formation. *ACS Biomater. Sci. Eng.* **2021**, *7*, 3737–3753. [CrossRef]
82. Jayachandran, P.; Ilango, S.; Suseela, V.; Nirmaladevi, R.; Shaik, M.R.; Khan, M.; Khan, M.; Shaik, B. Green Synthesized Silver Nanoparticle-Loaded Liposome-Based Nanoarchitectonics for Cancer Management: In Vitro Drug Release Analysis. *Biomedicines* **2023**, *11*, 217. [CrossRef] [PubMed]
83. Yusuf, A.; Brophy, A.; Gorey, B.; Casey, A. Liposomal encapsulation of silver nanoparticles enhances cytotoxicity and causes induction of reactive oxygen species-independent apoptosis. *J. Appl. Toxicol.* **2018**, *38*, 616–627. [CrossRef]
84. Valdivia, V.; Gimeno-Ferrero, R.; Pernia Leal, M.; Paggiaro, C.; Fernández-Romero, A.M.; González-Rodríguez, M.L.; Fernández, I. Biologically Relevant Micellar Nanocarrier Systems for Drug Encapsulation and Functionalization of Metallic Nanoparticles. *Nanomaterials* **2022**, *12*, 1753. [CrossRef]
85. Fan, X.; Luo, Z.; Ye, E.; You, M.; Liu, M.; Yun, Y.; Loh, X.J.; Wu, Y.-L.; Li, Z. AuNPs Decorated PLA Stereocomplex Micelles for Synergetic Photothermal and Chemotherapy. *Macromol. Biosci.* **2021**, *21*, 2100062. [CrossRef] [PubMed]
86. Li Volsi, A.; Fiorica, C.; D'Amico, M.; Scialabba, C.; Palumbo, F.S.; Giammona, G.; Licciardi, M. Hybrid Gold/Silica/Quantum-Dots supramolecular-nanostructures encapsulated in polymeric micelles as potential theranostic tool for targeted cancer therapy. *Eur. Polym. J.* **2018**, *105*, 38–47. [CrossRef]
87. Mahajan, K.D.; Ruan, G.; Vieira, G.; Porter, T.; Chalmers, J.J.; Sooryakumar, R.; Winter, J.O. Biomolecular detection, tracking, and manipulation using a magnetic nanoparticle-quantum dot platform. *J. Mater. Chem. B* **2020**, *8*, 3534–3541. [CrossRef] [PubMed]
88. Zhang, H.; Wu, Y.; Gan, Z.; Yang, Y.; Liu, Y.; Tang, P.; Wu, D. Accurate intracellular and in vivo temperature sensing based on CuInS2/ZnS QD micelles. *J. Mater. Chem. B* **2019**, *7*, 2835–2844. [CrossRef]
89. Fan, Q.; Dehankar, A.; Porter, T.K.; Winter, J.O. Effect of Micelle Encapsulation on Toxicity of CdSe/ZnS and Mn-Doped ZnSe Quantum Dots. *Coatings* **2021**, *11*, 895. [CrossRef]
90. Li, B.; Cai, M.; Lin, L.; Sun, W.; Zhou, Z.; Wang, S.; Wang, Y.; Zhu, K.; Shuai, X. MRI-visible and pH-sensitive micelles loaded with doxorubicin for hepatoma treatment. *Biomater. Sci.* **2019**, *7*, 1529–1542. [CrossRef] [PubMed]
91. Luo, L.; Yang, J.; Zhu, C.; Jiang, M.; Guo, X.; Li, W.; Yin, X.; Yin, H.; Qin, B.; Yuan, X.; et al. Sustained release of anti-PD-1 peptide for perdurable immunotherapy together with photothermal ablation against primary and distant tumors. *J. Control. Release* **2018**, *278*, 87–99. [CrossRef]
92. Liang, J.; Huang, Q.; Hua, C.; Hu, J.; Chen, B.; Wan, J.; Hu, Z.; Wang, B. pH-Responsive Nanoparticles Loaded with Graphene Quantum Dots and Doxorubicin for Intracellular Imaging, Drug Delivery and Efficient Cancer Therapy. *ChemistrySelect* **2019**, *4*, 6004–6012. [CrossRef]
93. Fang, Y.; Lin, S.; Yang, F.; Situ, J.; Lin, S.; Luo, Y. Aptamer-Conjugated Multifunctional Polymeric Nanoparticles as Cancer-Targeted, MRI-Ultrasensitive Drug Delivery Systems for Treatment of Castration-Resistant Prostate Cancer. *BioMed Res. Int.* **2020**, *2020*, e9186583. [CrossRef]
94. Jin, Z.; Chang, J.; Dou, P.; Jin, S.; Jiao, M.; Tang, H.; Jiang, W.; Ren, W.; Zheng, S. Tumor Targeted Multifunctional Magnetic Nanobubbles for MR/US Dual Imaging and Focused Ultrasound Triggered Drug Delivery. *Front. Bioeng. Biotechnol.* **2020**, *8*, 586874. [CrossRef]
95. Kumar, R.; Belz, J.; Markovic, S.; Jadhav, T.; Fowle, W.; Niedre, M.; Cormack, R.; Makrigiorgos, M.G.; Sridhar, S. Nanoparticle-Based Brachytherapy Spacers for Delivery of Localized Combined Chemoradiation Therapy. *Int. J. Radiat. Oncol. *Biol. *Phys.* **2015**, *91*, 393–400. [CrossRef]
96. Chaubey, N.; Sahoo, A.K.; Chattopadhyay, A.; Ghosh, S.S. Silver nanoparticle loaded PLGA composite nanoparticles for improving therapeutic efficacy of recombinant IFNγ by targeting the cell surface. *Biomater. Sci.* **2014**, *2*, 1080–1089. [CrossRef] [PubMed]
97. Ghosh, S.; Ghosal, K.; Mohammad, S.A.; Sarkar, K. Dendrimer functionalized carbon quantum dot for selective detection of breast cancer and gene therapy. *Chem. Eng. J.* **2019**, *373*, 468–484. [CrossRef]
98. Otis, J.B.; Zong, H.; Kotylar, A.; Yin, A.; Bhattacharjee, S.; Wang, H.; Jr, J.R.B.; Wang, S.H. Dendrimer antibody conjugate to target and image HER-2 overexpressing cancer cells. *Oncotarget* **2016**, *7*, 36002–36013. [CrossRef] [PubMed]
99. Xu, X.; Lü, S.; Gao, C.; Wang, X.; Bai, X.; Gao, N.; Liu, M. Facile preparation of pH-sensitive and self-fluorescent mesoporous silica nanoparticles modified with PAMAM dendrimers for label-free imaging and drug delivery. *Chem. Eng. J.* **2015**, *266*, 171–178. [CrossRef]
100. Lu, S.; Li, X.; Zhang, J.; Peng, C.; Shen, M.; Shi, X. Dendrimer-Stabilized Gold Nanoflowers Embedded with Ultrasmall Iron Oxide Nanoparticles for Multimode Imaging–Guided Combination Therapy of Tumors. *Adv Sci (Weinh)* **2018**, *5*, 1801612. [CrossRef]
101. Liu, Z.; Zhou, D.; Yan, X.; Xiao, L.; Wang, P.; Wei, J.; Liao, L. Gold Nanoparticle-Incorporated Chitosan Nanogels as a Theranostic Nanoplatform for CT Imaging and Tumour Chemotherapy. *IJN* **2022**, *17*, 4757–4772. [CrossRef]

102. Wei, C.; Dong, X.; Zhang, Y.; Liang, J.; Yang, A.; Zhu, D.; Liu, T.; Kong, D.; Lv, F. Simultaneous fluorescence imaging monitoring of the programmed release of dual drugs from a hydrogel-carbon nanotube delivery system. *Sens. Actuators B Chem.* **2018**, *273*, 264–275. [CrossRef]
103. Gholami, L.; Tafaghodi, M.; Abbasi, B.; Daroudi, M.; Kazemi Oskuee, R. Preparation of superparamagnetic iron oxide/doxorubicin loaded chitosan nanoparticles as a promising glioblastoma theranostic tool. *J. Cell. Physiol.* **2019**, *234*, 1547–1559. [CrossRef]
104. Bulatao, B.P.; Nalinratana, N.; Jantaratana, P.; Vajragupta, O.; Rojsitthisak, P.; Rojsitthisak, P. Lutein-loaded chitosan/alginate-coated Fe_3O_4 nanoparticles as effective targeted carriers for breast cancer treatment. *Int. J. Biol. Macromol.* **2023**, *242*, 124673. [CrossRef] [PubMed]
105. Chen, L.; Hong, W.; Duan, S.; Li, Y.; Wang, J.; Zhu, J. Graphene quantum dots mediated magnetic chitosan drug delivery nanosystems for targeting synergistic photothermal-chemotherapy of hepatocellular carcinoma. *Cancer Biol. Ther.* **2022**, *23*, 281–293. [CrossRef] [PubMed]
106. Chen, Y.; Wang, X.; Lu, Z.; Chang, C.; Zhang, Y.; Lu, B. Dual-responsive targeted hollow mesoporous silica nanoparticles for cancer photodynamic therapy and chemotherapy. *J. Macromol. Sci. Part. A* **2023**, *60*, 474–483. [CrossRef]

Disclaimer/Publisher's Note: The statements, opinions and data contained in all publications are solely those of the individual author(s) and contributor(s) and not of MDPI and/or the editor(s). MDPI and/or the editor(s) disclaim responsibility for any injury to people or property resulting from any ideas, methods, instructions or products referred to in the content.

Review

The Power of Field-Flow Fractionation in Characterization of Nanoparticles in Drug Delivery

Juan Bian [1], Nemal Gobalasingham [2], Anatolii Purchel [2] and Jessica Lin [1,*]

[1] Genentech Research and Early Development, Genentech Inc., 1 DNA Way, South San Francisco, CA 94080, USA
[2] Wyatt Technology Corporation, 6330 Hollister Ave, Santa Barbara, CA 93117, USA
* Correspondence: linz20@gene.com; Tel.: +1-650-491-7249

Abstract: Asymmetric-flow field-flow fractionation (AF4) is a gentle, flexible, and powerful separation technique that is widely utilized for fractionating nanometer-sized analytes, which extend to many emerging nanocarriers for drug delivery, including lipid-, virus-, and polymer-based nanoparticles. To ascertain quality attributes and suitability of these nanostructures as drug delivery systems, including particle size distributions, shape, morphology, composition, and stability, it is imperative that comprehensive analytical tools be used to characterize the native properties of these nanoparticles. The capacity for AF4 to be readily coupled to multiple online detectors (MD-AF4) or non-destructively fractionated and analyzed offline make this technique broadly compatible with a multitude of characterization strategies, which can provide insight on size, mass, shape, dispersity, and many other critical quality attributes. This review will critically investigate MD-AF4 reports for characterizing nanoparticles in drug delivery, especially those reported in the last 10–15 years that characterize multiple attributes simultaneously downstream from fractionation.

Keywords: nanoparticle drug delivery systems; asymmetrical flow field-flow fractionation; light scattering detection; multi-attribute characterization

1. Introduction

The application of nanotechnology for medical purposes has defined nanomedicine. Nowadays, nanomedicines such as nanoparticles (NPs) for drug and gene delivery have become an emerging field of medicine. Nanomedicines have significant potential to improve human health for prevention and treatment of diseases. Nanoparticles for drug delivery are revolutionizing the nanomedicine field, especially most recently, as several approved COVID-19 vaccines use nanoparticles to carry messenger RNA [1,2]. Additionally, nanoparticle-based vaccines and therapeutics in preclinical or clinical studies play an increasingly significant role against the COVID-19 pandemic [3]. The growing interest in applying NPs for drug delivery can be attributed to the appealing features such as improved stability and biocompatibility, enhanced permeability, and retention effect, as well as precise targeting [4–6].

Numerous NPs for drug delivery have been developed recently, including lipid-based NPs [7–9], polymer-based NPs [10–12], virus-like NPs [13,14], extracellular vesicles (EVs) [15,16], and inorganic NPs [17,18]. The chemical composition, stability, particle size distribution, nanoparticle shape, and morphology, as well as drug encapsulation and distribution, are critical parameters to characterize for nanoparticles [19]. In addition, regulatory agencies, including the U.S. Food and Drug Administration (FDA) and the European Medical Agency (EMA), have published multiple regulatory guidance documents to define the quality expectation for premarket submission [19–21]. Therefore, thorough understanding and characterization of nanoparticle drug delivery systems are critical for the identification of the critical quality attributes (CQAs) and successful development of nanomedicines.

To advance the development of the nanoparticles in drug delivery applications, gentle analytical techniques with full preservation of their native properties are in urgent demand. Field flow fractionation (FFF) is uniquely suited for the analysis of delicate nanoparticles because of the use of an external force field for gentle separation. FFF, which was invented and patented in 1966 by J. Calvin Giddings, applies an external field to an open channel to achieve separations [22]. The open channel design is highly conducive for separating fragile species with a wide particle size range, and offers the flexibility in carrier liquid selection [23]. Compared to commonly used size exclusion chromatography (SEC), the fundamental difference is the absence of a stationary phase in the FFF channel, which significantly reduces the system backpressure and shear force. This feature makes FFF noticeably gentler than SEC. The absence of a stationary phase as in liquid chromatography (LC) makes FFF a "soft" fractionation technique with minimal shear or mechanical stress towards analytes [24]. Another feature of FFF is the applicability for particles across a wide size range (1 nm to 10 μm) [23].

FFF is a group of distinctive techniques that use different types of separation fields perpendicular to the open channel. Thermal FFF (ThFFF) utilizes temperature difference across the channel to create the thermal gradient necessary to induce the separation [25,26]. Sedimentation FFF (SdFFF) uses sedimentation induced by gravity or a centrifugal force to separate particles [27,28]. Magnetic FFF (MgFFF) separates analytes according to their difference in magnetic properties [29]. Electrical FFF (ElFFF) introduces a transverse electrical current to create an electric field [30]. Electrical asymmetrical flow (EAF4) was developed afterwards as a variation of ElFFF to combine the electric field with the crossflow field [31]. Flow FFF (FlFFF) applies a crossflow field to facilitate separation and is the most versatile sub-technique of FFF. FlFFF has various formats, namely symmetric flow FFF (SF4), asymmetric flow FFF (AF4), or hollow fiber flow FFF (HF5), which differ in the geometrical channel shape and the way the crossflow is applied [32–34]. Table 1 summarizes the main FFF sub-techniques, the corresponding separation field, and the critical physicochemical properties of analyzed samples as the basis of separation.

Table 1. Summary of FFF sub-techniques.

Sub-Techniques of FFF	External Field	Physicochemical Property
Thermal FFF (ThFFF)	Thermal gradient	Soret coefficient
Sedimentation FFF (SeFFF)	Gravity/Centrifugal force	Effective mass (density)
Electrical FFF (ElFFF)	Electric field	Electrophoretic mobility
Magnetic FFF (MgFFF)	Magnetic field	Magnetic properties
Flow FFF (FlFFF)	Cross flow	Diffusion coefficient

Among these FFF sub-techniques, AF4 is the most commonly used because of its broad separation range, great versatility, and wide commercial availability, indicated by the dominance in scientific publications. The principle of AF4 has been reviewed previously [23,24,32,35,36]. In an AF4 setup, separation of analytes of different hydrodynamic sizes is achieved by an applied crossflow perpendicular to the separation channel, which is built from two blocks, one of which contains a semi-permeable membrane supported by a porous frit. The main separation zone is defined either with a spacer of defined thickness or can be built into the top block and influences the parabolic flow profile within the channel perpendicular to the crossflow. The semi-permeable membrane retains the analytes, while allowing the mobile phase to traverse. Then, the particles migrate along the parabolic laminar flow of liquid carrier and a dynamic equilibrium is established, where smaller particles (with a higher diffusion coefficient) equilibrate toward the middle of the AF4 channel (with higher velocity) and elute earlier than larger particles, as illustrated in Figure 1. Based on the sample relaxation prior to separation, two types of channels are currently available in the market for AF4 technology: focusing and hydrodynamic relaxation [37]. In the conventional AF4 channel, the sample is relaxed close to the membrane during the focusing step (Figure 1A), while the frit inlet channel or dispersion inlet channel uses hydrodynamic relaxation where no focusing is needed [38] (Figure 1B). The separation in AF4 can theoretically benefit from "focusing", which reduces band broadening and improves resolution and efficiency. However, the focusing step might cause sample loss in some cases due to the adsorption on the membrane or aggregation of the sample, while sample loss on the membrane is typically negligible once the elution mode starts [39]. Therefore, frit-inlet FFF can be a preferred approach for particles that suffer from sample aggregation or undesirable membrane interaction. For example, the colloidal particle stability and particle size alterations were noted during the focusing step using a conventional AF4 channel, which could be circumvented by using a frit-inlet channel [40].

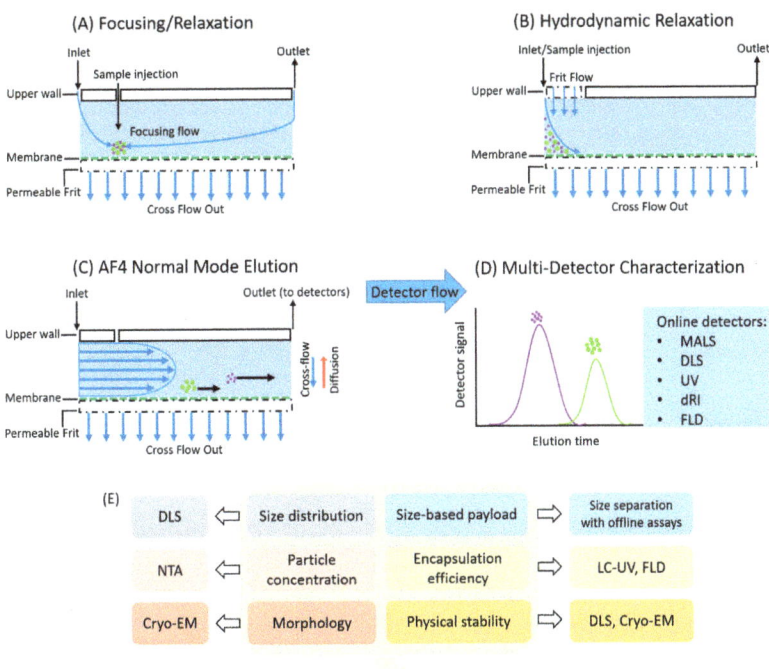

Figure 1. Representative diagram of relaxation and elution process in AF4: (**A**) focusing relaxation in conventional AF4; (**B**) hydrodynamic relaxation in frit inlet or dispersion AF4; (**C**) elution in the channel, showing the sample migration along with the parabolic flow, where smaller species travel faster than larger species; (**D**) following AF4 separation, the multi-detector system that is coupled with AF4 enables online characterization towards size-resolved fractions (Purple: population of particles with smaller size, green: population of particles with larger size); (**E**) critical quality attributes of the nanoparticle drug delivery systems and corresponding analytical techniques; multi-attribute characterization can also be achieved by the multi-detector AF4 (MD-AF4) system. MALS, multi-angle light scattering; DLS, dynamic light scattering; UV, ultraviolet; RI, refractive index; FLD, fluorescence detector; LC, liquid chromatography; NTA, nanoparticle tracking analysis; Cryo-EM, cryogenic electron microscopy.

The critical quality attributes of nanoparticle drug delivery systems are usually determined through individual detection techniques (Figure 1E). As an elution technique, AF4 can be coupled to multiple detectors to enable multi-attribute characterization of the size-resolved fractions. Multi-angle light scattering (MALS) is capable of determining molar mass and root-mean-square radius or radius of gyration (R_g). The range of R_g measured by MALS is ~10 nm to 500 nm, and even up to 1000 nm assuming shape-specific models and sufficient angular coverage [41]. In addition, particle concentration can be derived from the scattered intensity from MALS, and the lower limit for particle concentration is 10^7 particles per milliliter for MALS analysis, below which it could result in noisy MALS peaks due to particle fluctuation [42]. Dynamic light scattering enables the measurement of Stokes radius or hydrodynamic radius (R_h). R_h can be measured accurately using batch DLS from 0.2 to 1000–5000 nm depending on the instrument. However, the upper limit of online DLS coupling with FFF depends on the flow rate and detection angle, and assuming a flow rate of 0.5 mL/min and appropriate configuration, R_h can be determined accurately from 0.5 nm to ~300 nm [41]. With both online MALS and DLS, particle structure and morphology can be evaluated by the shape factor, which is the ratio of the R_g and R_h [43]. Ultraviolet (UV) and refractive index (RI) detectors, as well as fluorescence detectors (FLD), serve as concentration detectors depending on the analyte properties. The combination of dual-concentration detectors with MALS further expands the analysis. For example, UV and RI detector together with MALS enables the determination of molar mass for each component and the composition ratio of the complex nanoparticle drug delivery systems. This approach has been applied to analyze drug-loaded liposomes [44] and virus-like particles with nucleic acids [45]. Additionally, nanoparticle tracking analysis (NTA) and cryogenic electron microscopy (cryo-EM) are often used offline with AF4 for complementary particle quantitation and structure/morphology visualization.

Applications of FFF in environmental matrices [46], food macromolecules [47], pharmaceutics and biopharmaceutics [48], and nanomedicines [24,49] were reviewed previously. This article briefly revisits different FFF sub-techniques and the basic principles, and the following sections will focus on discussion of the recent applications of FFF, especially AF4 in the characterization of nanoparticles used in the drug delivery. Challenges and opportunities of FFF will also be outlined in this review.

2. Applications of FFF in Nanoparticle Drug Delivery Systems

One of the main challenges in developing nanoparticle drug delivery systems is particle characterization. Analytical scientists always need to consider the tradeoff between the level of details that the sample can be characterized with and the complexity of the analytical tools. For example, batch DLS offers a very quick and simple way to characterize particle size but lacks resolution and sensitivity to dispersity. On the other hand, cryo-EM provides an unprecedented level of detail but at a cost of much lower throughput and a limited sampling amount for statistical significance. Recently, FFF has gained increasing popularity for various nanoparticles owing to its separation capability towards a wide particle size range. FFF, especially AF4, has found plentiful use in the fields of nanomedicines and nanomaterials [32,49]. Coupling AF4 with flow-detectors, such as light scattering detectors (MALS, DLS) and concentration detectors (UV, RI, FLD), allows both in-depth understanding of structural information and quantitative measurement of the size-based distributions. To reveal the complexity, nanoparticles are separated by AF4 followed by online or offline detectors. For example, nanoparticle samples can be separated using AF4 and sent into a set of flow-through detectors for simultaneous online characterization or collected into fractions for offline characterization. Coupling with online detections has apparent advantages, as it does not cause agglomeration or breakage, nor alter NPs structure during the characterization. Among the online flow detectors used in combination with AF4, MALS and DLS are the most popular choices because they are size-based techniques that couple easily with a size-based separation [50]. In addition, spectroscopy detection (UV/FLD) is informative beyond quantitation, revealing information such as size-dependent plasmon shift and the location of labelled species [51,52]. Lastly, a combination of online and offline approaches can be used in parallel to provide orthogonal measurement.

This section is intended to provide a detailed overview of the recent applications using FFF, especially AF4, in the most common nanoparticle drug delivery systems. Our goal is to provide insights in the nanoparticle characterization when the FFF technique is combined with various detectors. Most importantly, the application sections highlight the use of multi-detector AF4 (MD-AF4) towards the multiple attribute characterization, including size distribution, shape or morphology, stability, drug encapsulation, and drug release. While the emphasis is mostly on AF4, which is the most widely used FFF technique, it is noteworthy to mention alternative FFF techniques, such as frit-inlet AF4 or HF5, for the characterization of different nanoparticles. Table 2 summarizes the FFF applications in different types of drug delivery nanoparticles discussed in this review, including the FFF technique, additional characterization techniques, and critical results.

Table 2. Example of FFF applications in drug delivery nanoparticles.

Nanoparticles	FFF Technique and Applications	Key Results	Ref
Lipid-based nanoparticles	AF4 with offline NTA and LC-MS for doxorubicin liposome formulations	Particle size distribution of the liposome and drug-to-lipid ratios were analyzed and compared across different doxorubicin formulations.	[53]
	AF4-MALS-DLS for peptide-liposome interaction	Selectivity of the peptide, quantity of the bound peptide, and size distribution and morphology of liposomes were revealed for understanding of structure-activity relationship.	[54]
	AF4-MALS-Gamma ray detector for liposome loaded with high energy alpha emitter (^{212}Bi)	Liposome particle size and stability of encapsulation in the serum were studied.	[55]
	AF4-MALS for drug transfer assay to quantify retention of lipophilic model compounds	Transfer kinetics of lipophilic model compounds from donor liposomes to acceptor liposomes were elucidated at different lipid mass ratios, and with different vesicle morphology and lamellarity.	[56]
	AF4-MALS-RI for stability evaluation of liposomes against the intestinal bile salts in oral delivery application	Different mechanisms of entrapped calcein leakages were revealed.	[57]
	AF4-MALS-DLS-dRI-UV for liposome-plasma protein interaction (from albumin HDL and LDL)	Liposomes were separated from albumin, and HDL, LHL, and the size were determined. The effect of the biolayer composition on liposome stability was also observed.	[58]
	AF4 with frit-inlet channel coupled with MALS to analyze LNP for RNA delivery	Frit-inlet channel enabled size and physical stability of LNP-RNA with great reliability and recovery.	[43]
Polymer-based nanoparticles	AF4-DLS-UV-FLD for enrofloxacin in PLGA nanoparticles	Comprehensive analysis of nanoparticle concentration (via UV), drug concentration (via FLD), and particle size distribution (via DLS). Unentrapped drug was easily removed via crossflow.	[59]
	AF4-MALS for protein-conjugated polysaccharides	Complementary analysis by SEC-MALS and AF4-MALS revealed heterogeneity in conformation and aggregation of the conjugates from molar mass and size determination.	[60]
	AF4-RI-FLD-DLS for polymer micelles in vitro stability	AF4 enabled the separation of polymer micelles from plasma protein and can be used to study the in vitro instability of drug-loaded nanoparticles.	[61]
	AF4-MALS-DLS for PEG-PDLLA polymersomes	Insights in size and shape of polymersomes via combination of MALS and DLS, and whether they are empty or loaded.	[62]
	AF4-RI for PAMAM dendrimers	Separate impurities (i.e., missing arm) and aggregates from PAMAM main populations and monitor interactions of PAMAM dendrimers with BSA.	[63]
	AF4-MALS-RI-UV for PEI-Mal dendrimers	Characterization of crossflow pathway enabled quantification of free, unencapsulated dye in addition to molar mass distributions.	[64]

Table 2. Cont.

Nanoparticles	FFF Technique and Applications	Key Results	Ref
Viral vectors and Virus-like Nanoparticles	AF4-MALS-DLS-UV-FLD-RI for VLPs derived from human polyoma JC virus	Comprehensive analysis of VLP molar mass and radius (via MALS), hydrodynamic radius (via DLS), concentration (via RI), sample composition and concentration (via UV), and improved small molecule limit of detection (via FLD).	[49]
	AF4-MALS-UV & ElFFF-MALS-UV of bacteriophage-like VLPs	Complementary analysis by ElFFF and AF4 obtained size and electrophoretic mobility of three VLPs.	[65]
Extracellular vesicles	AF4-UV-MALS for characterization of EVs from urine and comparison with ultrafiltration combined with SEC method	AF4-UV-MALS was demonstrated to be a straightforward and reproducible method for determining size, amount, and purity of isolated urinary EVs.	[66]
	AF4-UV-MALS combined with batch DLS and NTA for size separation, characterization and quantification of exosomes	Fractionation quality of exosomes was significantly influenced by crossflow conditions and channel thickness where focusing time has less impact. AF4-UV-MALS and DLS both showed the presence of two particle subpopulations. Compared to DLS and AF4-MALS, NTA overestimated the size and number density for the larger exosome population.	[67]
	AF4-UV-DLS with EM imaging for identification of subsets of EVs	Two exosome subpopulations and one non-membrane NPs exomere were discovered and identified	[68]
	AF4 and nanoflow-LC-ESI-MS/MS for size dependent lipidomic analysis of urinary exosomes	AF4 enabled the fractionation of exosomes with different sizes that originated from different types of cells. Degree of lipid increase was more significant in the smaller fractions, indicating that AF4 is capable of screening of urinary exosomes in cancer patients.	[69]
	Offline coupling of AF4 and CE for separation of EVs	EVs could be resolved from free proteins and high-density lipoproteins by AF4 and further separated from the low-density lipoproteins co-eluted in AF4 by offline CE.	[70]
	Orthogonal approach of ultracentrifugation and HF5-MALS-UV-FLD for purification and mapping of EV subtypes	Size, abundance, and DNA/protein content of the large and small EVs were characterized by HF5-MALS-UV-FLD as the second dimension, showing potential in sorting particles with different sizes and contents.	[71]
	EAF4 hyphenated with MALS and NTA for fast and purification-free characterization of NPs	EAF4 provided online sample purification and simultaneous access to size and Zeta-potential; high resolution size and number concentration was achieved by hyphenation of EAF4 with MALS and NTA.	[72]
Inorganic nanoparticles	AF4-MALS-DLS-ICPMS for quantitative characterization of GNPs	Mixtures of three GNPs were separated by AF4 and then each fraction was quantified by ICPMS. Both geometric diameters and hydrodynamic diameters were determined online by MALS and DLS.	[73]
	AF4 for characterization of elution behavior of non-spherical GNPs	Elution behavior of the GNPs with three different morphologies was studied by AF4 and particle size was compared with DLS and TEM.	[74]
	AF4-MALS-UV-RI for characterization and stability evaluation of drug-loaded metal-organic framework (MOF) NPs	Empty and drug-loaded nanoMOFs were studied in terms of particle size distribution and stability. Detection of aggregate formation and monitoring of nanoMOF morphological changes indicates their interaction with the drug molecules.	[75]

2.1. Lipid-Based Nanoparticles

Liposomes are small artificial vesicles of spherical shape that are comprised of one or more concentric bilayers encapsulating an aqueous core. Due to their amphiphilicity, biocompatibility, and appropriate particle size, liposomes have been widely used in the past 40 years as membrane modeling, drug delivery vehicles, and nanoreactor vessels [76–78]. The first liposomal drug formulation ever approved by the FDA in 1995 was PEGylated liposome-encapsulated doxorubicin (Doxil®) to treat Kaposi's sarcoma [79]. Nowadays, a large variety of liposomal-formulated drugs are approved by the FDA for clinical use [80]. Critical quality attributes of liposomal drug products include particle size distribution, charge, and payload amount. For instance, it has been shown that while smaller liposomes can decrease recognition by the complement system and innate immunity, thus enhancing the bioavailability, larger liposomes can increase drug payload [81,82].

These critical quality attributes of liposomes can be characterized using FFF as a separation technique followed by either online or offline detectors like discussed earlier. For example, Ansar et al., used AF4 for size-based fractionation of doxorubicin liposomal formulations, followed by offline NTA for particle sizing of the collected fractions, and online LC-MS for lipid and payload quantification [53]. Interestingly, this study concluded that the formulated liposomes had a narrow size distribution without any significant variation in D10, D50, and D90 values, and the drug to lipid ratios remained constant as a function of particle size, indicating that the drug loading to the liposomal particles is size independent. These findings are summarized in Figure 2, where the amount of DOX drug stays constant relative to the nanoparticle size.

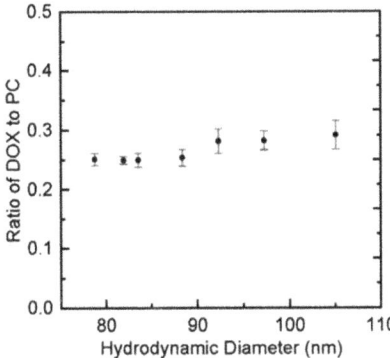

Figure 2. The mass ratio of DOX and total lipids as a function of the number averaged hydrodynamic diameter of DLF-1. Reprinted from Ref. [53] with permission from Elsevier.

Lavicoli and coworkers used AF4 coupled with MALS/DLS to study the peptide-liposome interaction [54]. They were able to characterize, in a single analysis, the selectivity of the peptides, the amount of peptides bonded to each liposome, and the induced change in the size distribution and morphology of the liposomes. By adding MALS and DLS downstream of the separation, AF4 provided information on particle shape and morphology by measuring their shape factor, which allowed them to identify subtle differences in complexes between positively charged F-AmP peptides with both the negatively charged POs (palmitoyloleoylphosphatidylcholine) and DL-AUVs (dilaurylphosphatidylcholine-anionic unilamellar vesicles). Huclier-Markai et al., used AF4 with MALS and a gamma ray detector to monitor the liposome size together with the incorporation of the high energy alpha emitter (^{212}Bi) [55]. Considering 212Bi's short half-life, it can only be delivered using labelled carrier particles (most notably, liposomes) that would rapidly accumulate in the target tumor. Animal studies suggested that the in vivo biologic period for the alpha-emitter is around 14 h, during which the metal must stay encapsulated. The AF4-gamma ray analysis has proven that more than 85% of radionuclides were retained in liposomes after incubating for 24 h at 37 °C in human serum. These results confirm that liposomes with a diameter of 100 nm represent a good vesicle to transport radionuclides for applications in targeted alpha therapy.

AF4 conditions have been investigated to study their influence on liposome fractionation. For instance, Hupfeld et al., studied the effect of ionic strength and osmolality of the carrier fluid in AF4 [76]. It was discovered that the liposomes eluted at different times when the ionic strength in the carrier fluid was changed. This was explained by osmotic stress-induced changes in vesicle size. The osmotic stress-induced size change in the liposome was found to be size dependent. Larger liposomes appeared to shrink or swell when exposed to hyper- or hypo-osmotic media, respectively. Smaller liposomes tend to shrink but not to swell under the same conditions. This study confirms the necessity to adjust the ionic strength of the carrier fluid to reduce inter-liposomal repulsion and interaction between liposomes and the FFF channel accumulation wall. Additionally, the osmotic pressure of the carrier liquid should be adjusted to match the pressure inside of liposomes using non-ionic additives. Kuntsche et al., evaluated the effect of fractionation conditions (flow profiles, injection volume, buffer composition) on the liposome and payload recovery [83]. The importance of osmolality match between liposome inner solution and carrier fluid was also confirmed. However, hydrophobic drug recovery had a strong dependence on its octanol-water partition coefficient. Because the sample is highly diluted during the fractionation, an alteration in the sample composition has to be studied and taken into consideration.

These newly discovered effects of AF4 elution conditions on sample composition have been utilized in developing drug release assays in several research studies [56–58]. Hinna et al., demonstrated that AF4 could be used for the drug transfer assay to quantify the retention of lipophilic compounds within liposomal carriers in the presence of lipophilic biological sinks [56]. The approach was extended for stability assessment of liposomes against the intestinal bile salts in applications for oral drug delivery, and it was found that the addition of taurocholate to egg-PC liposomes led to the formation of mixed-micelles and leakage of the calcein drug encapsulated inside liposomes [57]. Additionally, Holzschuh et al., introduced a novel approach to measure liposome-plasma protein interactions based on size by employing AF4 coupling with online detectors that enabled a simultaneous analysis of the sample (e.g., size determination). The authors obtained a good separation profile for liposomes and three main acceptor domains (albumin, HDL, and LDL) [58]. This study confirmed that rigid liposomes and PEGylated fluid liposomes showed higher stability in human plasma when compared to non-PEGylated fluid liposomes. This means that the bilayer composition of a liposomal formulation plays a significant role in stability and drug release in biological media. Hence, separation of the plasma-liposome sample by AF4 seems to be a potential alternative to already established methods.

Lipid nanoparticles (LNPs) are typically spherical nanoparticles with a solid lipid core that act as a novel pharmaceutical drug delivery system [84,85]. LNPs were first approved as a drug delivery vehicle in 2018 for the drug Onpattro to treat polyneuropathy in patients with hereditary transthyretin-mediated amyloidosis [86]. It became more widely known in late 2020 because some COVID-19 vaccines, notably mRNA-1273 [87] and BNT162b [88], used PEGylated-lipid nanoparticles for mRNA delivery [89]. Similar to liposomal drug formulations, determination of payload content relative to LNP size can be important to understand the efficacy and safety. AF4 has been successfully used for the separation and characterization of lipid-based drug delivery systems; however, electrostatically interacting LNP complexes with the relatively labile lipid-monolayer coating are more prone to destabilization during the focusing step in the conventional AF4 channel [90].

Non-focusing AF4 channels (frit-inlet or dispersion channel) can circumvent the instability of the LNPs during conventional AF4 separation. In a recent study, Mildner et al., demonstrated AF4 with the frit-inlet channel was well-suited for the analysis of lipid-based nanoparticles for RNA delivery with satisfactory reproducibility and sample recovery [39].

Downstream characterization by multi-detectors would benefit from the high sample recovery from AF4 separation, which makes AF4 become compliant with ISO/TS 21362:2018 (nanotechnologies-analysis of nano-objects, using asymmetrical-flow and centrifugal field-flow fractionation), resulting in better alignment with orthogonal techniques. As shown in Figure 3, particle concentration from MALS following the AF4 separation agrees well with batch NTA characterization. Nevertheless, AF4-MALS provides data across a wider radius distribution compared to batch characterization techniques.

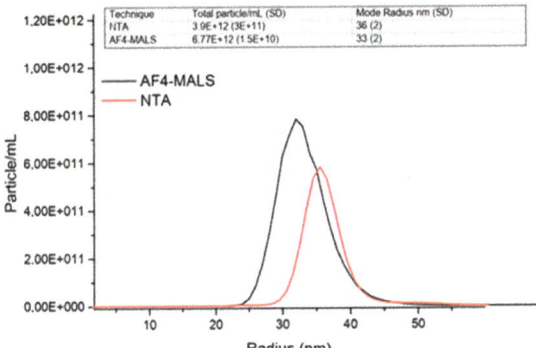

Figure 3. Number-based particle size distribution obtained by NTA and AF4-MALS. Total particle/mL concentration and radius are reported in the table in the graph. Reprinted from Ref. [39] with permission from Elsevier.

It has been shown that the in vivo potency and tissue-penetration ability of LNPs are related to particle size [91]. Researchers from Merck & CO. enabled the online determination of the size-dependent RNA content in LNPs, which was validated through RNA quantitation using a reserved-phase liquid chromatography (RPLC) assay performed on individual size fractions [92]. This study involved an optimized MD-AF4 analysis with a patented UV scattering correction approach, which eliminates overestimation of UV absorption at 260 nm caused by the scattering of LNPs. Figure 4A demonstrates the significant contribution of UV scattering to the apparent absorption of 260 nm light for unloaded LNPs with no chromophore. Interestingly, after UV scattering was removed using the correction algorithm, the calculated RNA weight percentages for four different LNP formulations were found to be in excellent agreement with the data obtained by offline RPLC analysis of collected AF4 fractions. Figure 4B demonstrated the distribution of RNA content for one LNP formulation (LNP-2) using both online and offline approaches. The authors envisioned the potential for this application in QC environment to evaluate the total RNA content in LNPs within a specified size range, which is one of the critical quality attributes for RNA-LNP products.

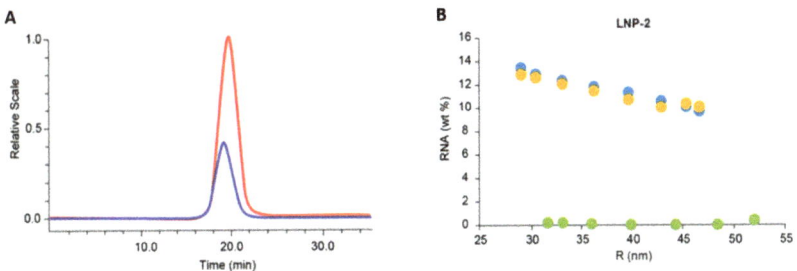

Figure 4. (A) UV chromatogram of empty LNP-2E (blue) and RNA-filled LNP-2F (red) showing

the significant UV signal from LNP-2E due to the scattering phenomenon in the UV detector. (**B**) Size-dependent RNA distribution in LNPs. Average of duplicate fractionation and offline RPLC analyses (yellow) vs. data from online analysis (blue for RNA-LNP and green for empty LNP). Reprinted from Ref. [92] with permission from Elsevier.

2.2. Polymer-Based Nanoparticles

Polymer-based nanoparticles for drug delivery have many advantages because of their versatility, customizability, and broad variety of structure-function relationships [93]. These structures can include nanoparticle capsules, micelles, polymersomes, dendrimers, and many other polymer nanoparticle complexes, which can be tuned to achieve tailored functions, such as targeted delivery, improved solubility, or desired biodegradability. Delivery mechanisms can vary from direct conjugation of the drug to polymer (either covalently or ionically), physical adsorption to the carrier, or encapsulation [94]. The consequence of this versatility is that detailed characterization can be quite challenging because polymers are generally heterogeneous, which impacts their physiochemical properties, and their behavior in solution may vary from the solid state. Additionally, structure modification may inadvertently impact loading capacity, release rate, or efficacy [95]. This only further highlights the need for robust, reliable, and comprehensive characterization.

Polylactic-co-glycolic acid (PLGA) has been approved by the FDA for drug formulations and various therapeutic devices. PLGA NPs are biodegradable, biocompatible, and readily tunable by composition or by molar mass [96,97]. PLGA NPs can entrap drugs for drug delivery, and the particle size and shape can influence the drug loading. Shakiba et al., explored AF4 coupled with UV, FLD, and DLS to measure the release profiles for enrofloxacin entrapped in PLGA nanoparticles [59]. The AF4 methodology in comparison with the dialysis approach is provided in Figure 5. The combination of UV (for nanoparticle concentration), FLD (for drug concentration), and DLS (for size distributions) with AF4 provided comprehensive characterization, which was more streamlined and convenient than the traditional dialysis approach. Polysaccharides are another class of biodegradable polymers that are explored in drug delivery, which can load drugs by either covalent binding or entrapment [98–100]. The significant particle size distribution (25–150 nm radius or higher) and potential high molar mass (1–10 MDa) present analytical challenges during traditional SEC separation due to shear degradation [101,102]. Deng et al. explored AF4-MALS for the characterization of ultra-high molecular weight polysaccharides, highlighting the advantages of AF4 as a "soft" separation to ensure the integrity of the complex [60].

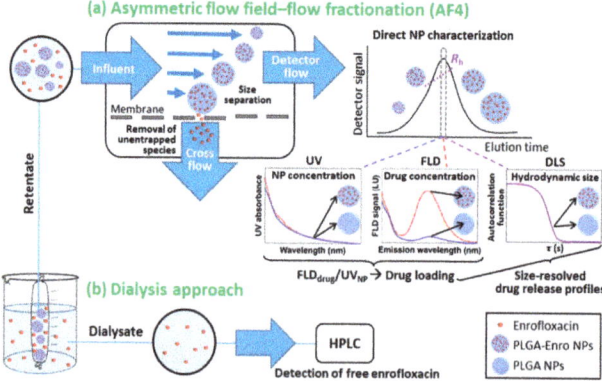

Figure 5. A scheme showing the convenience of AF4 for separating the unentrapped drug from the entrapped drug for determining drug loading. The free drug is removed via semi-permeable membrane via crossflow, while the entrapped drug elutes and is quantified. Reprinted from Ref. [59] with permission from Elsevier.

Polymer micelles are self-assembled colloids by amphiphilic polymers based on thermodynamic favorability at critical micellar concentrations, thus in aqueous media would present a hydrophobic core and an external hydrophilic shell. Challenges in the characterization of polymeric micelles are remarkable, as their stability is directly correlated to concentration, and the micelles may disassemble upon dilution (whether in the bloodstream or in analytical methods). Environmental factors like pH, temperature, and micelle composition can also affect their shape and morphology, leading to varying delivery efficiency [103,104]. Ideal fractionation methods should provide the distinct capability of separating intact micelles from disassembled micelles, unimers, and unencapsulated nanomedicine. However, traditional size separation methods like SEC have several limitations, such as micelles' disassembly on the column, interacting with or adsorbing to the stationary phase, and likely not eluting out from the column [12]. In this case, AF4 is very well suited for micelle fractionation. For example, Liu et al., employed AF4 to investigate the in vitro stability of micelles in human plasma using both empty micelles and those loaded with tetra(hydroxyphenyl)chlorin (mTHPC) [61]. They explored both covalently crosslinked and non-crosslinked micelles based on amphiphilic block copolymers with poly(ε-caprolactone), poly(1,2-dithiolane-carbonate), and/or poly(ethylene glycol). Micelles were prepared with and without mTHPC, and release was studied by incubating loaded micelles and taking samples at various time points and running AF4 coupled to RI, FLD, and DLS referenced against empty micelles. Size distributions, achieved with inline DLS, helped to elucidate stability, and the results indicated covalently crosslinked micelles had much better stability than non-crosslinked micelles. A representative hydrodynamic radius distribution is provided in Figure 6.

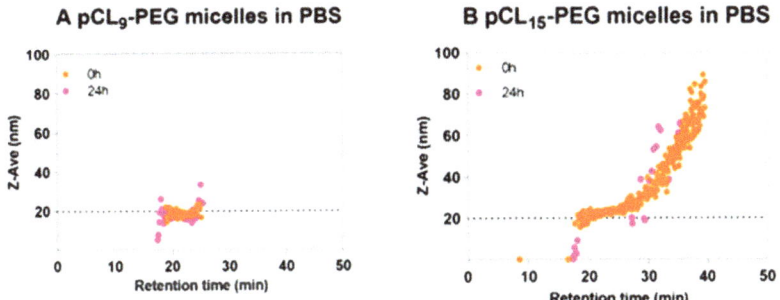

Figure 6. Hydrodynamic radius from in-line DLS after AF4 fractionation comparing micelle size distributions at different incubation periods. Reprinted from Ref. [61] with permission from Elsevier.

Polymersomes are self-assembled hollow nanostructures that are analogous to liposomes and capable of forming spherical or non-spherical shapes [105]. Compared to micelles, polymersomes are in the form of a bilayer with a solvent core for drug encapsulation (i.e., aqueous core for water-based assemblies). For aqueous systems, amphiphilic block copolymers generally form micelles when the hydrophilic polymer fraction is greater than 50%, and form polymersomes when the hydrophilic polymer fraction is between 25 and 45%. Because of the many unique shapes and structures that polymersomes can take, including rod-like assemblies, spheroids, discocyte, and stomatocyte structures, characterizing their size and morphology can be quite challenging [105].

It has been demonstrated that the relationship between R_g and R_h can provide insight on the conformation of macromolecules. The ratio of R_g/R_h defines the shape factor ($\rho = R_g/R_h$), which can be plotted as a Burchard-Stockmayer plot [106,107]. As a result, the ratio of R_g/R_h approaches $\rho = 1$ for a theoretical hollow sphere with a thin shell, or $\rho = 0.77$ for a solid sphere. Wauters et al., explored polymersomes made of amphiphilic block copolymers based on polyethylene glycol and poly(D,L-lactide) (PEG-PDLLA), which were polymerized with various polymer chain lengths and block ratios to achieve spherical and cylindrical assemblies [62]. By analyzing the Burchard-Stockmayer plots from online MALS and DLS data, they were able to not only determine the size of the empty polymersomes, but also provide insights for the shape: whether the polymersomes were spherical or cylindrical. They also investigated polymersomes loaded with BSA or DiD (a far-red fluorescent small molecule) and used the shape factor derived from the Burchard-Stockmayer plot to evaluate if the polymersome was empty or filled. As plotted in Figure 7, BSA-loaded polymersomes showed a reduction of the R_g values, yielding an average ρ of 0.77 ± 0.09, indicating the presence of BSA inside polymersomes [62].

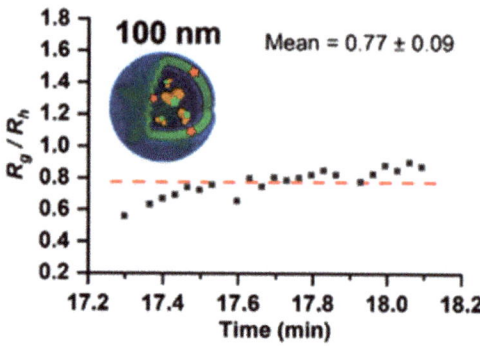

Figure 7. Ratio of R_g and R_h for spherical BSA-loaded polymersomes extruded using a 100 nm filter. The red line represents the mean value of these ratios. Reprinted from Ref. [62] with permission from American Chemical Society.

Dendrimers are branched polymers with a defined structure, typically hyperbranched polymers emanating from a central core [108]. Dendrimers are capable of forming scaffolds and cavities that lead to advanced complexes, allowing further functionalization to improve critical quality attributes like biocompatibility. Examples of dendrimers include polyamidoamine (PAMAM), polypropyleneimine (PPI), and several other amine- or ether-derivatives [109]. When characterizing dendrimers, it is critical to separate dendrimers from the impurities, including dendrimer defects with missing arms, entangled or aggregated dendrimers, and other suboptimal structures. Lee et al., explored the structural changes of PAMAM with AF4, including generational dispersity (inter-molecularly coupled dendrimers), skeletal dispersity (missing arms and molecular loops), and other structural defects that occur during synthesis [63]. Various analytical techniques have been tested for characterization of dendrimers, such as SEC, infrared spectroscopy, capillary electrophoresis (CE), nuclear magnetic resonance (NMR), and mass spectrometry, while their performance tends to be worse as the size of the dendrimer increases. However, AF4 was able to separate PAMAM dendrimers with optimized conditions (flow rate, pH, and salt concentration), including separation of four different dendrimer structures with a wide range of molecular weights of 14 to 467 kDa [63].

One of the more creative studies for evaluating the drug encapsulation comes from the work of Boye and coworkers, who installed a UV detector on the crossflow pathway of the AF4 to measure the free drug [64]. In this case, they studied hyperbranched PEI with a maltose shell (PEI-Mal) dendrimers complexed with a dye, Rose Bengal (RB). Traditionally, multiple detectors are installed downstream of the channel outlet, while the authors included a UV detector on the crossflow outlet. This innovative setup allows for measuring the concentration of small molecules that traverse the membrane and exit via what is normally the crossflow waste pathway, as illustrated in Figure 8. The authors investigated parameters like membrane material and molecular weight cutoff with pure RB and RB-PEI-Mal complex, and the free RB was quantified using a calibration curve established by the UV detector at the crossflow pathway. This work demonstrated the power of AF4 in fractionation and purification of a mixture of nanoparticles and small drug molecules, as well as free drug quantification by exploiting the semi-permeable membrane.

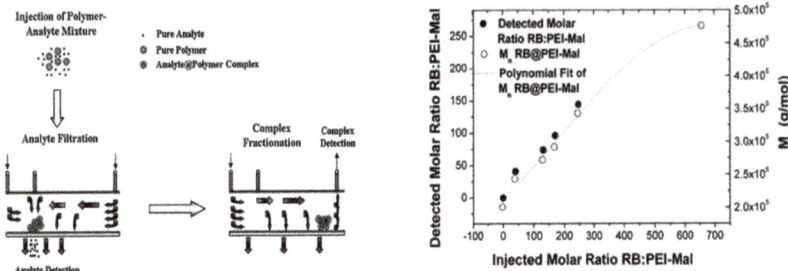

Figure 8. Conceptual diagram of executing free, small analyte detection via crossflow pathway detectors (**right**) and complex fractionation and subsequent detection (**left**). The molar ratio of RB:PEI-Mal in the complex was achieved by separation of free dye from complex. Reprinted from Ref. [64] with permission from Elsevier.

2.3. Viral Vectors and Virus-like Nanoparticles

The biotechnology sector has been investing in viral vectors for gene therapy for many years now [110,111]. As of 2023, more than 3600 gene therapy clinical trial studies are ongoing or have been approved, and more than 70% of them are based on viral vectors [112,113]. Despite all this progress, commercial-scale production remains challenging, and the final viral particles that contain drug substances are not well characterized. As aforementioned, AF4 separation with downstream flow-through detectors (MALS, DLS, UV) is a very promising methodology for viral vector analysis. For example, Cirkowitz et al., have utilized AF4 to guide the development of the scalable process to produce virus-like particles (VLP) derived from the human polyoma JC virus, and then conducted scattering detector-based analytical characterization [45]. Figure 9 shows a fractogram that demonstrates the necessity of size-based separation because VLP expression in the insect cells produced not only desired VLPs (17,000 kDa), but also VP1 aggregates with a lower molecular mass (2500 kDa). Therefore, the use of AF4-MALS as an analytical tool enabled the development of a scalable process for the production, purification, and packaging of the VLPs based on the human polyoma JC virus.

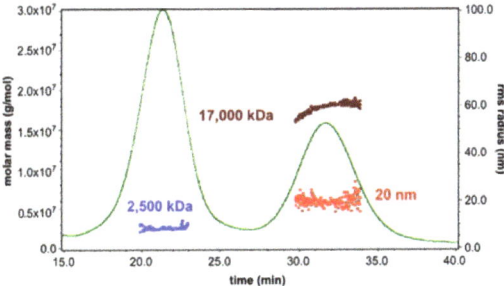

Figure 9. AF4 fractogram of the VLP with the radius and molar mass distributions of the two different size populations measured by MALS and RI signal. Reprinted from Ref. [45] with permission from Elsevier.

Additionally, a combination of various modes of FFF can be used to better understand the complicated nanoparticle samples. A research group from the University of Utah in collaboration with Pfizer used AF4 in combination with ElFFF to obtain size and electrophoretic mobility of three bacteriophage-like VLPs: a blank Q beta bacteriophage, which is denoted as VLP, and two conjugated particles with different peptides [65]. The comparison of electrical and asymmetric flow modes of FFF revealed that separation of samples with similar size but different electrical properties can be achieved to a small extent. ElFFF showed consistent shoulder peaks in fractograms, indicating the presence of particle population with different surface charge properties. Additionally, this allows for the quantification of surface charge properties of polydisperse samples with multiple species present in the mixture.

2.4. Extracellular Vesicles

In addition to VLPs, cell-secreted nanoparticles, extracellular vesicles, also attracted increasing interest in the field of target drug delivery [114], and the remarkable advance in the development of EV-based drug delivery systems has been witnessed in the last decades [115–117]. EV is a cluster of heterogeneous lipid bilayer-delimited nanoparticles of different sizes, cargos, and surface markers, including exosomes, microvesicles, and apoptotic bodies. Exosomes are a subtype of EVs and are typically 30–100 nm in diameter, which is the smallest population in EVs compared to microvesicles (50–1000 nm in diameter) and apoptotic bodies (100 nm to several micrometers in diameter) [118]. EVs enable intercellular communication by serving as delivery vehicles for a wide range of endogenous cargo molecules, like proteins and nucleic acids. For instance, Zhang et al., transfected HEK293T cells with si-RNA (small interfering RNA) and incubated the isolated exosomes with gastric cancer cell lines. They demonstrated that exosome-delivered si-RNA could reverse chemoresistance to cisplatin in gastric cancer [119]. Additionally, exosomes have been used to incorporate small drug molecules with poor bioavailability to improve the delivery efficiency. In the study led by Pascucci et al., mesenchymal stromal cells (MSCs) were incubated with a high dosage of paclitaxel (PTX), a hydrophobic mitotic inhibitor with a powerful anticancer effect. Exosomes released by MSCs contained encapsulated PTX, showing stronger anti-proliferative activity than PTX alone towards pancreatic adenocarcinoma [120].

EVs as drug delivery systems present unique advantages, namely low immunogenicity and excellent biocompatibility and biostability. Currently, only two engineered exosome therapeutic candidates, both from Codiak BioSciences, have entered clinical development (ExoIL-12™ and ExoSTING™) [116]. To expand its industrial applications, the International Society for Extracellular Vesicles (ISEV) initiated the efforts "the minimal information for studies of extracellular vesicles" (MISEV) towards EV separation and characterization techniques in 2014, which was updated in 2018 and 2021, suggesting that the size distribution, morphology, purity, and stability of EVs should be investigated [16,121,122]. As the field continues to grow, a powerful separation technique coupled with online detectors is needed to obtain a full picture, where AF4-MALS could lend itself well to such applications. Despite very few applications of AF4-MALS directly towards EV drug delivery systems, researchers have already paved the way for the EV drug delivery characterization by using AF4-MALS for the characterization of various EV subtypes from different body fluids [66,123–125]. Thereby, AF4-MALS plays a promising role in separating and characterizing EVs in various settings, including drug delivery [126,127].

Sitar et al., used AF4-MALS-UV for size-based separation, characterization, and quantitation of exosomes by varying the AF4 parameters. They found the crossflow velocity and channel thickness significantly influenced the fractionation performance, whereas the focusing time had less impact [67]. AF4-MALS also showed broad size distribution and two subpopulations present in the exosome sample, larger exosomes and smaller vesicle-like particles. Batch NTA analyses were also conducted directly for the bulk exosome, and the results showed that NTA overestimates the size and the number density for the larger exosome population [67]. This issue has been reported previously for size measurement when applying light scattering towards heterogeneous suspensions. Large particles scatter more light than small particles, and if present in polydisperse samples, could potentially dominate the scattered light fluctuations and thus shift the particle size distribution and uplift the average diameter [128]. Therefore, AF4 became a powerful tool to address this issue by separating different extracellular vesicles prior to size characterization and quantitation. Oeyen et al., described a method using AF4-MALS-UV for characterization and quantitation of urinary EVs, where R_g defined by MALS was in the range of 40–160 nm. The online UV detector allows for the determination of contaminating proteins in the sample fraction. The study also demonstrated that AF4-MALS-UV was a highly reproducible technique compared to NTA, showing its potential as a reliable quality control method for EVs. It is noteworthy that authors proposed to include AF4-MALS-UV as a standard characterization method for EVs in the ISEV guidelines to improve the quality of the EV-related research [66].

AF4 MALS/DLS-UV was also successfully used for the identification of small EV subpopulations and corresponding biophysical and molecular characterization [68]. The AF4 fractogram of B16-F10 melanoma-derived small EVs is displayed in Figure 10a. A total of two exosome subsets, including large exosome vesicles (hydrodynamic diameter 90–120 nm) and small exosome vesicles (hydrodynamic diameter 60–80 nm), as well as one abundant non-membranous nanoparticle termed "exomeres" (hydrodynamic diameter < 50 nm). Representative AF4 fractions were further analyzed by TEM, showing distinct morphology and size for each small EV subset (Figure 10c) [68]. Based on this study, Zhang et al., established a protocol for "asymmetric-flow field flow fractionation of small extracellular vesicles" consisting of four sections: I. Preparation of small extracellular vesicles (sEVs) from cell culture. II. AF4 fractionation of sEVs. III Online data collection and analysis. IV. Fraction collection, concentrations, and characterization [129]. This protocol provides general guidance for the EV separation using AF4, which makes AF4 more accessible and friendly to new users.

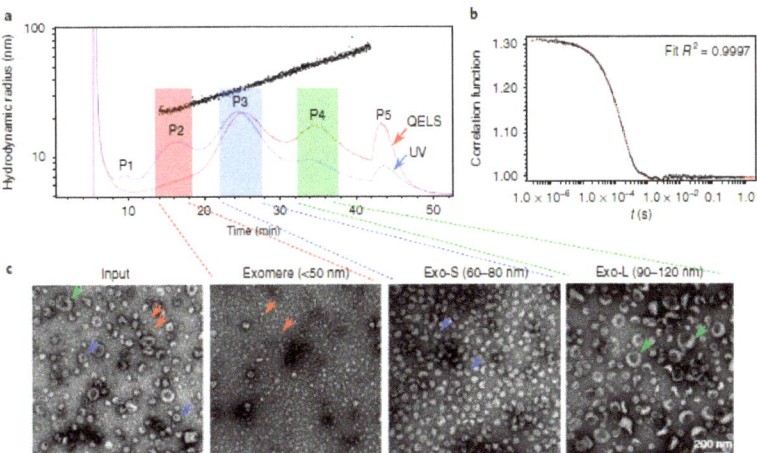

Figure 10. Separation and characterization of EVs using multi-detector asymmetrical flow field-flow fractionation (MD-AF4). (a) A representative AF4 fractionation profile of B16-F10-derived exosomes with UV and QELS (DLS) signals in blue and red separately; black dots illustrate hydrodynamic radius (Rh, nm), showing the particle size distribution over retention time. P1-P5 mark the peaks detected based on UV absorbance. Fractions were pooled for exomeres (hydrodynamic diameter < 50 nm), Exo-S (60–80 nm), and Exo-L (90–120 nm). (b) Representative correlation function in QELS for P3 (t = 25.1 min). (c) TEM imaging analysis of exosome input mixture (pre-fractionation) and fractionated exomere, Exo-S and Exo-L subpopulations. Arrows indicate exomeres (red), Exo-S (blue) and Exo-L (green). Reprint from Ref. [68] with permission from Springer Nature.

Size separation of exosomes is critical for monitoring the size changes of EV subpopulations associated with various biological statuses. Moon's group used AF4 for size sorting of exosomes, followed by exosome fraction collection and characterization by offline analytical tools [69,130]. For example, Joon-Seon et al., utilized AF4 to separate urinary exosomes by size, demonstrating a significant difference in exosome sizes between healthy controls and patients with prostate cancer [69]. Gao et al., highlighted the versatility of AF4 offline coupling with CE for EV analysis [70]. The authors demonstrated that EVs could be resolved from free proteins and high-density lipoproteins by AF4, which could be further separated from co-eluted low-density lipoproteins through CE by different surface charges (Figure 11). The AF4 fraction allowed for rapid EV quantitation in various samples in the matrix, showing the great potential of AF4 in reducing the matrix interference for the characterization of EV subpopulations produced by cell lines or present in clinical samples [70].

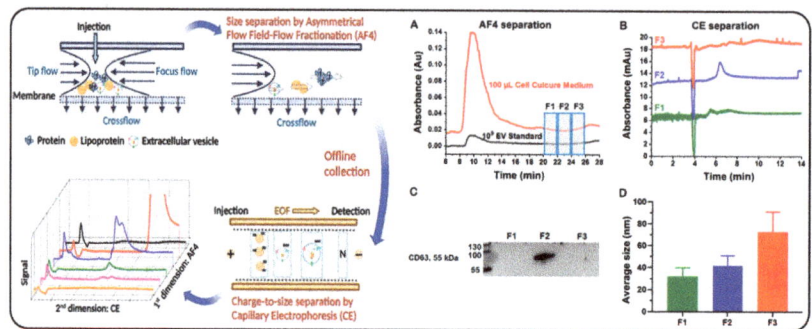

Figure 11. Representative workflow of offline coupling of AF4 and CE for separation of extracellular vesicles. (**A**) Fractograms of injection of HeLa cell medium (red trace) and standard EVs (black trace) to AF4. (**B**) CE traces of three AF4 fractions were collected from injection of 109 standard EVs, (F1: 20–22 min in green trace, F2: 22–24 min in blue trace, and F3: 24–26 min in red trace). (**C**) Western-Blot analysis of the CD63 protein, an EV marker in three AF4 fractions collected from injection of a HeLa cell medium. (**D**) Average diameter of the particles in the AF4 fractions collected from a HeLa cell medium observed in SEM. Reprint from Ref. [70] with permission from ACS publications.

In addition to conventional AF4, efforts have also been made to separate and characterize EVs using alternative flow FFF techniques. Although HF5 plays a much less significant role compared to other flow FFF techniques due to the lack of flexibility and limited sample loading, its improved resolution, sensitivity, and disposability make it suitable for nanoparticles with limited sample volume and/or require disposable separation devices [23,131]. Marassi et al., separated different EV populations derived from the C2C12 cell line using HF5 followed by MALS-UV-FLD detection, which provided insights on the content of different EV subsets in addition to size distribution; for example, DNA/RNA was observed to release from the large EV populations while protein was detected from the small EV populations [71]. Derivative AF4 techniques were also evaluated in this field. EAF4 is another variant of AF4, which combines two complementary fields for separation. Drexel et al., described a method using EAF4-MALS combined with NTA through a flow splitter for the analysis of liposomes and exosomes in the biological matrix, where EAF4 provided online sample purification while simultaneously enabling access to size and Zeta potential and MALS and NTA detection added high resolution particle size and concentration information [72]. This study highlights the benefits of the EAF4-MALS-NTA platform to study the behavior of EV-based drug delivery vesicles under in vivo-like conditions.

2.5. Inorganic Nanoparticles

Inorganic NPs attracted increasing attention in the past decades because of their potential in carrying various therapeutic agents, such as small molecule drugs, peptides, proteins, and genes. When employed as nanocarriers, inorganic NPs have shown good drug loading capacity, stability, and biocompatibility [132,133]. The finely controlled size of the inorganic nanoparticles provides a versatile platform for drug encapsulation either in the cavity of the nanoparticle structure or on the surface of the nanomaterials due to the high surface-area-to-volume ratio [134,135]. The properties of the inorganic nanoparticles, including size, shape, and composition, could affect their performance in drug delivery [18,136]. The most investigated inorganic nanocarriers include gold nanoparticles (GNPs) and mesoporous silica nanoparticles (MSNs), and some of the inorganic nanocarriers are also investigated in the clinical trials [136–140]. Kong et al., reported the use of polyethyleneimine (PEI)-entrapped GNPs modified with peptide via a polyethylene glycol (PEG) spacer as a vector for B-cell lymphoma-2 (Bcl-2) siRNA delivery to glioblastoma cells [141]. Their results revealed that the modified GNPs could deliver Bcl-2 siRNA to the target cells with excellent transfection efficiency, leading to specific gene silencing in the target cells. In another study, the anticancer drug doxorubicin (DOX) was attached to GNPs with an average diameter of 30 nm through a pH-sensitive linker, which allowed for the intracellular triggered release of DOX from the GNPs once inside acidic organelles [142].

As the drug delivery system, inorganic NPs hold structural strength compared to organic NPs. Their surface is often coated by other materials to form hybridized framework, some of which can change their size or morphology to improve the drug loading [18]. Since the drug loading may alter due to framework disintegration, size, or morphology change, thereby, a size-indicating analytical method that allows structure and composition characterization is highly needed. In this context, AF4-MALS represents an exciting opportunity for inorganic nanoparticles for size separation and characterization. Schmidt et al., developed an analytical platform coupling AF4 with MALS, DLS, and inductively coupled plasma mass spectrometry (ICP-MS) to separate GNPs by size and quantitatively measure the GNP mass concentration (Figure 12) [73]. The authors successfully separated three GNP populations, which were quantified by ICP-MS with recovery within 50–95% [73]. In this study, to ensure the stability of GNPs during separation, SDS was added as a surfactant in the aqueous carrier to ensure the NP stability during the separation in AF4. The influence of the membrane was also tested to improve the GNP recovery and results demonstrated that the polyethersulfone (PES) membrane was superior to regenerated cellulose, resulting in higher recovery for the GNPs and better peak shape of GNPs in the fractogram [73]. Indeed, utilization of a representative medium as AF4 mobile-phase is critical for the separation and characterization of GNPs and their conjugates, as the properties of mobile phase (e.g., pH and ionic strength) can influence the electrostatic property of the nanoparticle samples and membrane in the channel [143,144]. Wang and coworkers found GNPs alone aggregated or precipitated in the AF4 channel when the ionic strength of the mobile phase was increased. However, when proteins were present, they formed a corona on the GNPs' surface to increase the GNP stability, making ionic mobile phases such as phosphate buffer appropriate [143].

Lee et al., used the AF4-DLS to study the elution behavior of the GNPs with three different morphologies: gold nanospheres (GNS), gold nanotriangles (GNT), and gold nanorods (GNR) [74]. The authors found that although the diameter of the GNS was approximately similar to the length of the GNR from TEM, its elution time (3.7 min) was earlier than that of the GNS (4.5 min), which indicated that non-spherical particles move down the AF4 channel by different mechanisms compared to the spherical particles [74]. Additionally, nanosized metal-organic frameworks (nanoMOFs) were also investigated in drug delivery applications. Roda et al., used AF4-MALS-UV-RI system to study the MIL-100(Fe) nanoMOFs loaded with azidothymidine derivatives with three different degrees of phosphorylation: azidothymidine (AZT, native drug), azidothymidine monophosphate (AZT-MP), and azidothymidine triphosphate (AZT-TP) [75]. The gentle separation nature of AF4 allows for the detection of low abundance aggregation in the MOFs. The authors found that AZT-loaded nanoMOF had an identical PSD profile with the empty nanoMOF, confirmed by their similar scattering behavior, while AZT-MP and AZT-TP-loaded nanoMOFs showed increased scattering intensity and particle size compared to the empty ones [75]. Their findings through AF4-MALS also highlighted the key role of the phosphate group for improved encapsulation of AZT derivatives to nanoMOFs [145]. They successfully demonstrated the capability of AF4-MALS to provide evidence for particle size distribution and stability, as well as surface modification of the drug-loaded nanoMOFs.

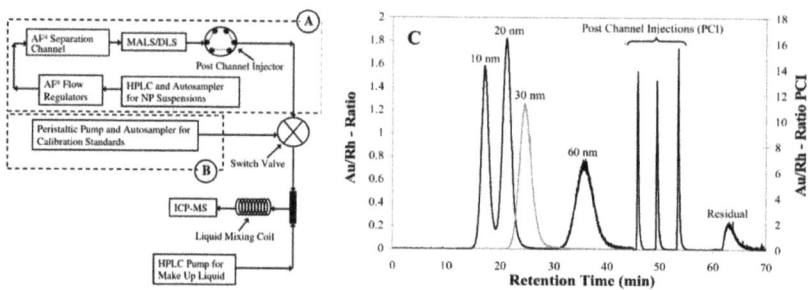

Figure 12. Representative diagram of AF4-MALS-DLS-ICP-MS platform including (**A**) the AF4-MALS-DLS system with post channel injection and (**B**) flow injection of calibrant solution. A switch valve allowed A or B to be operational and a separate HPLC pump delivered make-up liquid. (**C**) AF4-ICP MS fractogram of a mixture of 10, 20, and 60 nm Au NPs (black line) superimposed on a fractogram corresponding to 30 nm NIST Au NPs (light gray line). The signal intensities of post channel injections of 10, 20, and 60 nm Au NPs have been indicated on the secondary y-axis. Reprint from Ref. [73] with permission from ACS publications.

3. Current Challenges and Future Trends

FFF coupled with light scattering detectors (MALS and DLS) as well as concentration detectors (UV, RI, FLD) has become the present-day analytical technique to tackle the unique challenges in nanoparticle characterization, which are currently unaddressed by other size-based separation approaches. Compared to SEC, FFF separation is usually gentler and more protective for fragile particles in terms of degradation or aggregation. However, analyte-membrane interaction has been noted, which leads to sample loss and low recovery, especially when using conventional AF4 with focusing prior to separation [146]. The mitigation strategies have been described in literature to reduce the analyte-membrane interaction by careful selection of the liquid carrier, membrane type, and molecular weight cutoff (MWCO) [147,148]. The advancement in membrane manufacturing, including robustness, solvent compatibility, and surface properties, could smooth the AF4 methodology development as well. Additionally, using a non-focusing channel in AF4 (frit-inlet or dispersion channel) could potentially circumvent a membrane interaction issue for some vulnerable nanoparticles during the focusing step.

Numerous applications of FFF in the separation of nanoparticles from a complex matrix have demonstrated that FFF is a very promising purification technique where the particle integrity could be maintained during the separation process. However, the loading capacity is still a drawback for purification when a large volume is needed to yield sufficient material. This necessitates the development of instrumentation, such as preparative channels, and this combination with fraction collectors will make FFF applicable as a preparative system. Another challenge arises from the sample dilution due to the high flow rate in FFF. Efforts have been made to increase the sample concentration for the following analysis. Manufacturers have developed dilution control modules to extract up to 90% of the sample-free liquid carrier to waste and deliver the concentrated sample to the online detectors, thus improving the detector limit for the low-abundance analytes [131]. Alternatively, the enrichment step post-FFF separation can be employed to concentrate the FFF fractions offline. For example, filtration was reported to concentrate the FFF fractions prior to offline quantitative analysis by CE [70]. Online sample enrichment could be another future direction to improve the performance of FFF both quantitatively and qualitatively. A solid-phase extraction (SPE) pre-column was introduced into the online enrichment system, combined with atmospheric pressure chemical ionization-mass spectrometry (APCI-MS) [149]. The next challenge is to explore the online coupling of AF4 with complementary detection techniques, such as high-resolution MS, to obtain simultaneous structure and compositional information, or with orthogonal separation techniques, such as LC and CE, to achieve improved selectivity from multi-dimensional separation. These will benefit from the development of online enrichment systems and customized interfaces. Such efforts have been made to achieve automated online isolation and fractionation for nanosized biomacromolecules by online coupled immunoaffinity chromatography-AF4 [150], demonstrating the potential of FFF in multi-dimensional analysis. In combination with hyphenated techniques, the investigation scope of FFF could be remarkably improved towards nanoparticles with higher complexity.

Intelligent software advancement is also an essential part of the development of technology, which will make the technique more user-friendly and thus facilitate the data analysis and interpretation, especially for non-spherical particle analysis. At the same time, substantial efforts have been made to develop standard analytical methods or protocols to guide FFF development in the pharmaceutical industry and address the regulatory expectations [35,151]. Standardized FFF methods for each type of nanoparticle would be helpful to flatten the learning curve for new users and make the technique readily accessible for researchers. Therefore, challenges and opportunities co-exist in the field of FFF. The remarkable versatility of FFF makes it a highly promising analytical platform for comprehensive physicochemical characterization of nanomedicines in pre-clinical investigation, product development, and quantity control of manufacturing. Moving forward, continuous improvements in instrumentation and software are expected to create more opportunities for FFF in different fields, in addition to drug delivery.

4. Summary

Multi-detector AF4 (MD-AF4) represents a multi-attribute characterization platform, which has been demonstrated to be a promising, powerful, and versatile analytical technique for size-dependent characterization of drug delivery nanoparticles. In particular, MD-AF4 is employed to (i) measure particle size distribution of highly heterogeneous samples; (ii) evaluate the morphology through shape factor; (iii) determine size-dependent payloads or drug-nanoparticle interactions; and (iv) study drug release and stability of the nanoparticle in the formulation buffer or biological matrix. For nanoparticles used in drug delivery applications, their size, shape, protein binding, and release kinetics play a significant role in biodistribution, off-target toxicities, and ultimately safety and efficacy. Those critical quality attributes must be carefully monitored during formulation development and manufacturing control. As a gentle size-based separation technique, AF4 is positioned to be the foremost technique for such analysis where traditional SEC fails, i.e., separation and characterization of lipid NPs, liposomes, EVs, and gene vectors. As presented in this review, MD-AF4 can be used either as a single technique or in combination with other complementary analytical techniques for the physical-chemical characterization of drug delivery nanoparticles. The wide applications of AF4 and its unique separation nature make MD-AF4 an enabling technology platform to provide high resolution and size-dependent characterization for various nanoparticles. AF4 will be more widely used in the pharmaceutical industry with the advancement of instrumentation and software, as well as regulatory guidance.

Author Contributions: Conceptualization, J.B. and J.L.; writing—original draft preparation, J.B., N.G. and A.P.; writing—review and editing, J.B., N.G., A.P. and J.L.; supervision, J.L. All authors have read and agreed to the published version of the manuscript.

Funding: This research received no external funding.

Institutional Review Board Statement: Not applicable.

Informed Consent Statement: Not applicable.

Data Availability Statement: Not applicable.

Acknowledgments: Graphical abstract was partially created with BioRender.

Conflicts of Interest: Author Juan Bian and Jessica Lin were employed by the company Genentech Inc. Author Nemal Gobalasingham and Anatolii Purchel were employed by the company Wyatt Technology. The authors declare no conflict of interest.

References

1. Sharma, K.; Koirala, A.; Nicolopoulos, K.; Chiu, C.; Wood, N.; Britton, P.N. Vaccines for COVID-19: Where do we stand in 2021? *Paediatr. Respir. Rev.* **2021**, *39*, 22–31. [CrossRef] [PubMed]
2. Hussain, A.; Yang, H.; Zhang, M.; Liu, Q.; Alotaibi, G.; Irfan, M.; He, H.; Chang, J.; Liang, X.J.; Weng, Y.; et al. Mrna vaccines for COVID-19 and diverse diseases. *J. Control Release* **2022**, *345*, 314–333. [CrossRef] [PubMed]
3. Rauf, A.; Abu-Izneid, T.; Khalil, A.A.; Hafeez, N.; Olatunde, A.; Rahman, M.; Semwal, P.; Al-Awthan, Y.S.; Bahattab, O.S.; Khan, I.N.; et al. Nanoparticles in clinical trials of COVID-19: An update. *Int. J. Surg.* **2022**, *104*, 106818. [CrossRef] [PubMed]
4. Jahangirian, H.; Lemraski, E.G.; Webster, T.J.; Rafiee-Moghaddam, R.; Abdollahi, Y. A review of drug delivery systems based on nanotechnology and green chemistry: Green nanomedicine. *Int. J. Nanomed.* **2017**, *12*, 2957–2978. [CrossRef]
5. Patra, J.K.; Das, G.; Fraceto, L.F.; Campos, E.V.R.; Rodriguez-Torres, M.D.P.; Acosta-Torres, L.S.; Diaz-Torres, L.A.; Grillo, R.; Swamy, M.K.; Sharma, S.; et al. Nano based drug delivery systems: Recent developments and future prospects. *J. Nanobiotechnol.* **2018**, *16*, 71. [CrossRef]
6. Yao, Y.; Zhou, Y.; Liu, L.; Xu, Y.; Chen, Q.; Wang, Y.; Wu, S.; Deng, Y.; Zhang, J.; Shao, A. Nanoparticle-based drug delivery in cancer therapy and its role in overcoming drug resistance. *Front. Mol. Biosci.* **2020**, *7*, 193. [CrossRef]
7. Dos Santos Rodrigues, B.; Lakkadwala, S.; Kanekiyo, T.; Singh, J. Development and screening of brain-targeted lipid-based nanoparticles with enhanced cell penetration and gene delivery properties. *Int. J. Nanomed.* **2019**, *14*, 6497–6517. [CrossRef]
8. Ickenstein, L.M.; Garidel, P. Lipid-based nanoparticle formulations for small molecules and rna drugs. *Expert. Opin. Drug. Deliv.* **2019**, *16*, 1205–1226. [CrossRef]
9. Thi, T.T.H.; Suys, E.J.A.; Lee, J.S.; Nguyen, D.H.; Park, K.D.; Truong, N.P. Lipid-based nanoparticles in the clinic and clinical trials: From cancer nanomedicine to COVID-19 vaccines. *Vaccines* **2021**, *9*, 359. [CrossRef]
10. Sur, S.; Rathore, A.; Dave, V.; Reddy, K.R.; Chouhan, R.S.; Sadhu, V. Recent developments in functionalized polymer nanoparticles for efficient drug delivery system. *Nano-Struct. Nano-Objects* **2019**, *20*, 100397. [CrossRef]
11. Charelli, L.E.; de Mattos, G.C.; de Jesus Sousa-Batista, A.; Pinto, J.C.; Balbino, T.A. Polymeric nanoparticles as therapeutic agents against coronavirus disease. *J. Nanopart. Res.* **2022**, *24*, 12. [CrossRef] [PubMed]
12. Ghezzi, M.; Pescina, S.; Padula, C.; Santi, P.; Del Favero, E.; Cantu, L.; Nicoli, S. Polymeric micelles in drug delivery: An insight of the techniques for their characterization and assessment in biorelevant conditions. *J. Control Release* **2021**, *332*, 312–336. [CrossRef] [PubMed]
13. Jeevanandam, J.; Pal, K.; Danquah, M.K. Virus-like nanoparticles as a novel delivery tool in gene therapy. *Biochimie* **2019**, *157*, 38–47. [CrossRef]
14. Chung, Y.H.; Cai, H.; Steinmetz, N.F. Viral nanoparticles for drug delivery, imaging, immunotherapy, and theranostic applications. *Adv. Drug. Deliv. Rev.* **2020**, *156*, 214–235. [CrossRef] [PubMed]
15. Walker, S.; Busatto, S.; Pham, A.; Tian, M.; Suh, A.; Carson, K.; Quintero, A.; Lafrence, M.; Malik, H.; Santana, M.X.; et al. Extracellular vesicle-based drug delivery systems for cancer treatment. *Theranostics* **2019**, *9*, 8001–8017. [CrossRef]
16. Herrmann, I.K.; Wood, M.J.A.; Fuhrmann, G. Extracellular vesicles as a next-generation drug delivery platform. *Nat. Nanotechnol.* **2021**, *16*, 748–759. [CrossRef]
17. Ghosn, Y.; Kamareddine, M.H.; Tawk, A.; Elia, C.; El Mahmoud, A.; Terro, K.; El Harake, N.; El-Baba, B.; Makdessi, J.; Farhat, S. Inorganic nanoparticles as drug delivery systems and their potential role in the treatment of chronic myelogenous leukaemia. *Technol. Cancer Res. Treat.* **2019**, *18*, 1533033819853241. [CrossRef]
18. Shi, Z.; Zhou, Y.; Fan, T.; Lin, Y.; Zhang, H.; Mei, L. Inorganic nano-carriers based smart drug delivery systems for tumor therapy. *Smart Mater. Med.* **2020**, *1*, 32–47. [CrossRef]
19. *Drug Products, Including Biological Products, That Contain Nanomaterials—Guidance for Industry*; US Food and Drug Administration: Silver Spring, MD, USA, 2022. Available online: https://www.fda.gov/regulatory-information/search-fda-guidance-documents/drug-products-including-biological-products-contain-nanomaterials-guidance-industry (accessed on 20 April 2022).
20. *Liposome Drug Products Chemistry, Manufacturing, and Controls; Human Pharmacokinetics and Bioavailability; and Labeling Documentation*; US Food and Drug Administration: Silver Spring, MD, USA, 2018. Available online: https://www.fda.gov/regulatory-information/search-fda-guidance-documents/liposome-drug-products-chemistry-manufacturing-and-controls-human-pharmacokinetics-and (accessed on 4 October 2021).
21. *Reflection Paper on the Data Requirements for Intravenous Liposomal Products Developed with Reference to an Innovator Liposomal Product*; European Medicines Agency: London, UK, 2013. Available online: https://www.ema.europa.eu/en/data-requirements-intravenous-liposomal-products-developed-reference-innovator-liposomal-product-0#current-effective-version-section (accessed on 13 March 2013).

22. Giddings, J.C. A new separation concept based on a coupling of concentration and flow nonuniformities. *Sep. Sci.* **1966**, *1*, 123–125. [CrossRef]
23. Plavchak, C.L.; Smith, W.C.; Bria, C.R.M.; Williams, S.K.R. New advances and applications in field-flow fractionation. *Annu. Rev. Anal. Chem.* **2021**, *14*, 257–279. [CrossRef]
24. Zattoni, A.; Roda, B.; Borghi, F.; Marassi, V.; Reschiglian, P. Flow field-flow fractionation for the analysis of nanoparticles used in drug delivery. *J. Pharm. Biomed. Anal.* **2014**, *87*, 53–61. [CrossRef] [PubMed]
25. Hovingh, M.E.; Thompson, G.H.; Giddings, J.C. Column parameters in thermal field-flow fractionation. *Anal. Chem.* **1970**, *42*, 195–203. [CrossRef]
26. Liu, G.; Giddings, J.C. Separation of particles in aqueous suspensions by thermal field-flow fractionation—Measurement of thermal-diffusion coefficients. *Chromatographia* **1992**, *34*, 483–492. [CrossRef]
27. Giddings, J.C.; Yang, F.J.F.; Myers, M.N. Sedimentation field-flow fractionation. *Anal. Chem.* **1974**, *46*, 1917–1924. [CrossRef]
28. Chianéa, T.; Assidjo, N.E.; Cardot, P.J.P. Sedimentation field-flow-fractionation: Emergence of a new cell separation methodology. *Talanta* **2000**, *51*, 835–847. [CrossRef]
29. Williams, P.S.; Carpino, F.; Zborowski, M. Magnetic nanoparticle drug carriers and their study by quadrupole magnetic field-flow fractionation. *Mol. Pharm.* **2009**, *6*, 1290–1306. [CrossRef]
30. Caldwell, K.D.; Gao, Y.S. Electrical field-flow fractionation in particle separation. 1. Monodisperse standards. *Anal. Chem.* **1993**, *65*, 1764–1772. [CrossRef]
31. Johann, C.; Elsenberg, S.; Schuch, H.; Rosch, U. Instrument and method to determine the electrophoretic mobility of nanoparticles and proteins by combining electrical and flow field-flow fractionation. *Anal. Chem.* **2015**, *87*, 4292–4298. [CrossRef]
32. Contado, C. Field flow fractionation techniques to explore the "nano-world". *Anal. Bioanal. Chem.* **2017**, *409*, 2501–2518. [CrossRef]
33. Giddings, J.C.; Yang, F.J.; Myers, M.N. Flow-field-flow fractionation: A versatile new separation method. *Science* **1976**, *193*, 1244–1245. [CrossRef]
34. Wahlund, K.G.; Giddings, J.C. Properties of an asymmetrical flow field-flow fractionation channel having one permeable wall. *Anal. Chem.* **1987**, *59*, 1332–1339. [CrossRef] [PubMed]
35. Caputo, F.; Mehn, D.; Clogston, J.D.; Rosslein, M.; Prina-Mello, A.; Borgos, S.E.; Gioria, S.; Calzolai, L. Asymmetric-flow field-flow fractionation for measuring particle size, drug loading and (in)stability of nanopharmaceuticals. The joint view of european union nanomedicine characterization laboratory and national cancer institute—Nanotechnology characterization laboratory. *J. Chromatogr. A* **2021**, *1635*, 461767. [PubMed]
36. Wahlund, K.G. Flow field-flow fractionation: Critical overview. *J. Chromatogr. A* **2013**, *1287*, 97–112. [CrossRef]
37. Moon, M.H.; Hwang, I. Hydrodynamic vs. Focusing relaxation in asymmetrical flow field-flow fractionation. *J. Liq. Chromatogr. Relat. Technol.* **2007**, *24*, 3069–3083. [CrossRef]
38. Fuentes, C.; Choi, J.; Zielke, C.; Penarrieta, J.M.; Lee, S.; Nilsson, L. Comparison between conventional and frit-inlet channels in separation of biopolymers by asymmetric flow field-flow fractionation. *Analyst* **2019**, *144*, 4559–4568. [CrossRef] [PubMed]
39. Mildner, R.; Hak, S.; Parot, J.; Hyldbakk, A.; Borgos, S.E.; Some, D.; Johann, C.; Caputo, F. Improved multidetector asymmetrical-flow field-flow fractionation method for particle sizing and concentration measurements of lipid-based nanocarriers for rna delivery. *Eur. J. Pharm. Biopharm.* **2021**, *163*, 252–265. [CrossRef]
40. Elvang, P.A.; Stein, P.C.; Bauer-Brandl, A.; Brandl, M. Characterization of co-existing colloidal structures in fasted state simulated fluids fassif: A comparative study using af4/malls, dls and dosy. *J. Pharm. Biomed. Anal.* **2017**, *145*, 531–536. [CrossRef]
41. Champagne, J. Vlp Characterization: The Light Scattering Biophysical Toolbox. Available online: https://www.wyatt.com/library/webinars/vlp-characterization-light-scattering-biophysical-toolbox.html (accessed on 5 February 2014).
42. Gobalasingham, N. Expanding the Characterization Toolkit with fff-mals: Developments, Techniques, and Applications. Available online: https://www.wyatt.com/library/webinars/expanding-the-characterization-toolkit-with-fff-mals-developments-techniques-and-applications.html?utm_source=lcgc&utm_medium=digital-ad&utm_campaign=resource-center-08-2021 (accessed on 15 June 2021).
43. Gioria, S.; Caputo, F.; Urban, P.; Maguire, C.M.; Bremer-Hoffmann, S.; Prina-Mello, A.; Calzolai, L.; Mehn, D. Are existing standard methods suitable for the evaluation of nanomedicines: Some case studies. *Nanomedicine* **2018**, *13*, 539–554. [CrossRef]
44. Hinna, A.; Steiniger, F.; Hupfeld, S.; Brandl, M.; Kuntsche, J. Asymmetrical flow field-flow fractionation with on-line detection for drug transfer studies: A feasibility study. *Anal. Bioanal. Chem.* **2014**, *406*, 7827–7839. [CrossRef]
45. Citkowicz, A.; Petry, H.; Harkins, R.N.; Ast, O.; Cashion, L.; Goldmann, C.; Bringmann, P.; Plummer, K.; Larsen, B.R. Characterization of virus-like particle assembly for DNA delivery using asymmetrical flow field-flow fractionation and light scattering. *Anal. Biochem.* **2008**, *376*, 163–172. [CrossRef]
46. Janwitayanuchit, W.; Suwanborirux, K.; Patarapanich, C.; Pummangura, S.; Lipipun, V.; Vilaivan, T. Synthesis and anti-herpes simplex viral activity of monoglycosyl diglycerides. *Phytochemistry* **2003**, *64*, 1253–1264. [CrossRef] [PubMed]
47. Nilsson, L. Separation and characterization of food macromolecules using field-flow fractionation: A review. *Food Hydrocoll.* **2013**, *30*, 1–11. [CrossRef]
48. Fraunhofer, W.; Winter, G. The use of asymmetrical flow field-flow fractionation in pharmaceutics and biopharmaceutics. *Eur. J. Pharm. Biopharm.* **2004**, *58*, 369–383. [CrossRef] [PubMed]

49. Wagner, M.; Holzschuh, S.; Traeger, A.; Fahr, A.; Schubert, U.S. Asymmetric flow field-flow fractionation in the field of nanomedicine. *Anal. Chem.* **2014**, *86*, 5201–5210. [CrossRef]
50. Wyatt, P.J. Measurement of special nanoparticle structures by light scattering. *Anal. Chem.* **2014**, *86*, 7171–7183. [CrossRef]
51. Mogensen, K.B.; Kneipp, K. Size-dependent shifts of plasmon resonance in silver nanoparticle films using controlled dissolution: Monitoring the onset of surface screening effects. *J. Phys. Chem. C* **2014**, *118*, 28075–28083. [CrossRef]
52. Thomsen, T.; Ayoub, A.B.; Psaltis, D.; Klok, H.A. Fluorescence-based and fluorescent label-free characterization of polymer nanoparticle decorated t cells. *Biomacromolecules* **2021**, *22*, 190–200. [CrossRef]
53. Ansar, S.M.; Mudalige, T. Characterization of doxorubicin liposomal formulations for size-based distribution of drug and excipients using asymmetric-flow field-flow fractionation (af4) and liquid chromatography-mass spectrometry (lc-ms). *Int. J. Pharm.* **2019**, *574*, 118906. [CrossRef]
54. Iavicoli, P.; Urban, P.; Bella, A.; Ryadnov, M.G.; Rossi, F.; Calzolai, L. Application of asymmetric flow field-flow fractionation hyphenations for liposome-antimicrobial peptide interaction. *J. Chromatogr. A* **2015**, *1422*, 260–269. [CrossRef]
55. Huclier-Markai, S.; Grivaud-Le Du, A.; N'Tsiba, E.; Montavon, G.; Mougin-Degraef, M.; Barbet, J. Coupling a gamma-ray detector with asymmetrical flow field flow fractionation (af4): Application to a drug-delivery system for alpha-therapy. *J. Chromatogr. A* **2018**, *1573*, 107–114. [CrossRef]
56. Hinna, A.H.; Hupfeld, S.; Kuntsche, J.; Bauer-Brandl, A.; Brandl, M. Mechanism and kinetics of the loss of poorly soluble drugs from liposomal carriers studied by a novel flow field-flow fractionation-based drug release-/transfer-assay. *J. Control Release* **2016**, *232*, 228–237. [CrossRef] [PubMed]
57. Bohsen, M.S.; Tychsen, S.T.; Kadhim, A.A.H.; Grohganz, H.; Treusch, A.H.; Brandl, M. Interaction of liposomes with bile salts investigated by asymmetric flow field-flow fractionation (af4): A novel approach for stability assessment of oral drug carriers. *Eur. J. Pharm. Sci.* **2023**, *182*, 106384. [CrossRef] [PubMed]
58. Holzschuh, S.; Kaess, K.; Fahr, A.; Decker, C. Quantitative in vitro assessment of liposome stability and drug transfer employing asymmetrical flow field-flow fractionation (af4). *Pharm. Res.* **2016**, *33*, 842–855. [CrossRef]
59. Shakiba, S.; Astete, C.E.; Cueto, R.; Rodrigues, D.F.; Sabliov, C.M.; Louie, S.M. Asymmetric flow field-flow fractionation (af4) with fluorescence and multi-detector analysis for direct, real-time, size-resolved measurements of drug release from polymeric nanoparticles. *J. Control Release* **2021**, *338*, 410–421. [CrossRef] [PubMed]
60. Deng, J.Z.; Lin, J.; Chen, M.; Lancaster, C.; Zhuang, P. Characterization of high molecular weight pneumococcal conjugate by sec-mals and af4-mals. *Polymers* **2022**, *14*, 3769. [CrossRef]
61. Liu, Y.; Fens, M.; Capomaccio, R.B.; Mehn, D.; Scrivano, L.; Kok, R.J.; Oliveira, S.; Hennink, W.E.; van Nostrum, C.F. Correlation between in vitro stability and pharmacokinetics of poly(epsilon-caprolactone)-based micelles loaded with a photosensitizer. *J. Control Release* **2020**, *328*, 942–951. [CrossRef]
62. Wauters, A.C.; Pijpers, I.A.B.; Mason, A.F.; Williams, D.S.; Tel, J.; Abdelmohsen, L.; van Hest, J.C.M. Development of morphologically discrete peg-pdlla nanotubes for precision nanomedicine. *Biomacromolecules* **2019**, *20*, 177–183. [CrossRef]
63. Lee, S.; Kwen, H.D.; Lee, S.K.; Nehete, S.V. Study on elution behavior of poly(amidoamine) dendrimers and their interaction with bovine serum albumin in asymmetrical flow field-flow fractionation. *Anal. Bioanal. Chem.* **2010**, *396*, 1581–1588. [CrossRef]
64. Boye, S.; Polikarpov, N.; Appelhans, D.; Lederer, A. An alternative route to dye-polymer complexation study using asymmetrical flow field-flow fractionation. *J. Chromatogr. A* **2010**, *1217*, 4841–4849. [CrossRef]
65. Shiri, F.; Petersen, K.E.; Romanov, V.; Zou, Q.; Gale, B.K. Characterization and differential retention of q beta bacteriophage virus-like particles using cyclical electrical field-flow fractionation and asymmetrical flow field-flow fractionation. *Anal. Bioanal. Chem.* **2020**, *412*, 1563–1572. [CrossRef]
66. Oeyen, E.; Van Mol, K.; Baggerman, G.; Willems, H.; Boonen, K.; Rolfo, C.; Pauwels, P.; Jacobs, A.; Schildermans, K.; Cho, W.C.; et al. Ultrafiltration and size exclusion chromatography combined with asymmetrical-flow field-flow fractionation for the isolation and characterisation of extracellular vesicles from urine. *J. Extracell. Vesicles* **2018**, *7*, 1490143. [CrossRef] [PubMed]
67. Sitar, S.; Kejžar, A.; Pahovnik, D.; Kogej, K.; Tušek-Žnidarič, M.; Lenassi, M.; Žagar, E. Size characterization and quantification of exosomes by asymmetrical-flow field-flow fractionation. *Anal. Chem.* **2015**, *87*, 9225–9233. [CrossRef]
68. Zhang, H.; Freitas, D.; Kim, H.S.; Fabijanic, K.; Li, Z.; Chen, H.; Mark, M.T.; Molina, H.; Martin, A.B.; Bojmar, L.; et al. Identification of distinct nanoparticles and subsets of extracellular vesicles by asymmetric flow field-flow fractionation. *Nat. Cell Biol.* **2018**, *20*, 332–343. [CrossRef] [PubMed]
69. Yang, J.S.; Lee, J.C.; Byeon, S.K.; Rha, K.H.; Moon, M.H. Size dependent lipidomic analysis of urinary exosomes from patients with prostate cancer by flow field-flow fractionation and nanoflow liquid chromatography-tandem mass spectrometry. *Anal. Chem.* **2017**, *89*, 2488–2496. [CrossRef] [PubMed]
70. Gao, Z.; Hutchins, Z.; Li, Z.; Zhong, W. Offline coupling of asymmetrical flow field-flow fractionation and capillary electrophoresis for separation of extracellular vesicles. *Anal. Chem.* **2022**, *94*, 14083–14091. [CrossRef]
71. Marassi, V.; Maggio, S.; Battistelli, M.; Stocchi, V.; Zattoni, A.; Reschiglian, P.; Guescini, M.; Roda, B. An ultracentrifugation—Hollow-fiber flow field-flow fractionation orthogonal approach for the purification and mapping of extracellular vesicle subtypes. *J. Chromatogr. A* **2021**, *1638*, 461861. [CrossRef]

72. Drexel, R.; Siupa, A.; Carnell-Morris, P.; Carboni, M.; Sullivan, J.; Meier, F. Fast and purification-free characterization of bio-nanoparticles in biological media by electrical asymmetrical flow field-flow fractionation hyphenated with multi-angle light scattering and nanoparticle tracking analysis detection. *Molecules* **2020**, *25*, 4703. [CrossRef]
73. Schmidt, B.; Loeschner, K.; Hadrup, N.; Mortensen, A.; Sloth, J.J.; Koch, C.B.; Larsen, E.H. Quantitative characterization of gold nanoparticles by field-flow fractionation coupled online with light scattering detection and inductively coupled plasma mass spectrometry. *Anal. Chem.* **2011**, *83*, 2461–2468. [CrossRef]
74. Lee, J.; Goda, E.S.; Choi, J.; Park, J.; Lee, S. Synthesis and characterization of elution behavior of nonspherical gold nanoparticles in asymmetrical flow field-flow fractionation (asflfff). *J. Nanoparticle Res.* **2020**, *22*, 256. [CrossRef]
75. Roda, B.; Marassi, V.; Zattoni, A.; Borghi, F.; Anand, R.; Agostoni, V.; Gref, R.; Reschiglian, P.; Monti, S. Flow field-flow fractionation and multi-angle light scattering as a powerful tool for the characterization and stability evaluation of drug-loaded metal-organic framework nanoparticles. *Anal. Bioanal. Chem.* **2018**, *410*, 5245–5253. [CrossRef]
76. Hupfeld, S.; Moen, H.H.; Ausbacher, D.; Haas, H.; Brandl, M. Liposome fractionation and size analysis by asymmetrical flow field-flow fractionation/multi-angle light scattering: Influence of ionic strength and osmotic pressure of the carrier liquid. *Chem. Phys. Lipids* **2010**, *163*, 141–147. [CrossRef] [PubMed]
77. Akbarzadeh, A.; Rezaei-Sadabady, R.; Davaran, S.; Joo, S.W.; Zarghami, N.; Hanifehpour, Y.; Samiei, M.; Kouhi, M.; Nejati-Koshki, K. Liposome: Classification, preparation, and applications. *Nanoscale Res. Lett.* **2013**, *8*, 102. [CrossRef]
78. Cevc, G. Rational design of new product candidates: The next generation of highly deformable bilayer vesicles for noninvasive, targeted therapy. *J. Control Release* **2012**, *160*, 135–146. [CrossRef]
79. Barenholz, Y. Doxil(r)—The first fda-approved nano-drug: Lessons learned. *J. Control Release* **2012**, *160*, 117–134. [CrossRef] [PubMed]
80. Chang, H.I.; Yeh, M.K. Clinical development of liposome-based drugs: Formulation, characterization, and therapeutic efficacy. *Int. J. Nanomed.* **2012**, *7*, 49–60.
81. Hupfeld, S.; Holsaeter, A.M.; Skar, M.; Frantzen, C.B.; Brandl, M. Liposome size analysis by dynamic/static light scattering upon size exclusion-/field flow-fractionation. *J. Nanosci. Nanotechnol.* **2006**, *6*, 3025–3031. [CrossRef]
82. Yohannes, G.; Pystynen, K.-H.; Riekkola, M.-L.; Wiedmer, S.K. Stability of phospholipid vesicles studied by asymmetrical flow field-flow fractionation and capillary electrophoresis. *Anal. Chim. Acta* **2006**, *560*, 50–56. [CrossRef]
83. Kuntsche, J.; Decker, C.; Fahr, A. Analysis of liposomes using asymmetrical flow field-flow fractionation: Separation conditions and drug/lipid recovery. *J. Sep. Sci.* **2012**, *35*, 1993–2001. [CrossRef]
84. Rades, A.S.T. Solid lipid nanoparticles. In *Nanocarrier Technologies: Frontiers of Nanotherapy*; Mozafari, M.R., Ed.; Springer: Berlin/Heidelberg, Germany, 2006; pp. 41–50.
85. Jenning, V.; Thunemann, A.F.; Gohla, S.H. Characterisation of a novel solid lipid nanoparticle carrier system based on binary mixtures of liquid and solid lipids. *Int. J. Pharm.* **2000**, *199*, 167–177. [CrossRef]
86. Kristen, A.V.; Ajroud-Driss, S.; Conceição, I.; Gorevic, P.; Kyriakides, T.; Obici, L. Patisiran, an rnai therapeutic for the treatment of hereditary transthyretin-mediated amyloidosis. *Neurodegener. Dis. Manag.* **2018**, *9*, 5–23. [CrossRef]
87. Baden, L.R.; El Sahly, H.M.; Essink, B.; Kotloff, K.; Frey, S.; Novak, R.; Diemert, D.; Spector, S.A.; Rouphael, N.; Creech, C.B.; et al. Efficacy and safety of the mrna-1273 SARS-CoV-2 vaccine. *N. Engl. J. Med.* **2021**, *384*, 403–416. [CrossRef] [PubMed]
88. Polack, F.P.; Thomas, S.J.; Kitchin, N.; Absalon, J.; Gurtman, A.; Lockhart, S.; Perez, J.L.; Perez Marc, G.; Moreira, E.D.; Zerbini, C.; et al. Safety and efficacy of the bnt162b2 mrna COVID-19 vaccine. *N. Engl. J. Med.* **2020**, *383*, 2603–2615. [CrossRef] [PubMed]
89. Pardi, N.; Hogan, M.J.; Porter, F.W.; Weissman, D. Mrna vaccines—A new era in vaccinology. *Nat. Rev. Drug. Discov.* **2018**, *17*, 261–279. [CrossRef] [PubMed]
90. Parot, J.; Caputo, F.; Mehn, D.; Hackley, V.A.; Calzolai, L. Physical characterization of liposomal drug formulations using multi-detector asymmetrical-flow field flow fractionation. *J. Control Release* **2020**, *320*, 495–510. [CrossRef]
91. Chen, S.; Tam, Y.Y.C.; Lin, P.J.C.; Sung, M.M.H.; Tam, Y.K.; Cullis, P.R. Influence of particle size on the in vivo potency of lipid nanoparticle formulations of sirna. *J. Control Release* **2016**, *235*, 236–244. [CrossRef]
92. Jia, X.; Liu, Y.; Wagner, A.M.; Chen, M.; Zhao, Y.; Smith, K.J.; Some, D.; Abend, A.M.; Pennington, J. Enabling online determination of the size-dependent rna content of lipid nanoparticle-based rna formulations. *J. Chromatogr. B Anal. Technol. Biomed. Life Sci.* **2021**, *1186*, 123015. [CrossRef]
93. Quattrini, F.; Berrecoso, G.; Crecente-Campo, J.; Alonso, M.J. Asymmetric flow field-flow fractionation as a multifunctional technique for the characterization of polymeric nanocarriers. *Drug. Deliv. Transl. Res.* **2021**, *11*, 373–395. [CrossRef]
94. Zhao, H.; Lin, Z.Y.; Yildirimer, L.; Dhinakar, A.; Zhao, X.; Wu, J. Polymer-based nanoparticles for protein delivery: Design, strategies and applications. *J. Mater. Chem. B* **2016**, *4*, 4060–4071. [CrossRef]
95. Shakiba, S.; Shariati, S.; Wu, H.; Astete, C.E.; Cueto, R.; Fini, E.H.; Rodrigues, D.F.; Sabliov, C.M.; Louie, S.M. Distinguishing nanoparticle drug release mechanisms by asymmetric flow field–flow fractionation. *J. Control Release* **2022**, *352*, 485–496. [CrossRef]
96. Sadat Tabatabaei Mirakabad, F.; Nejati-Koshki, K.; Akbarzadeh, A.; Yamchi, M.R.; Milani, M.; Zarghami, N.; Zeighamian, V.; Rahimzadeh, A.; Alimohammadi, S.; Hanifehpour, Y.; et al. Plga-based nanoparticles as cancer drug delivery systems. *Asian Pac. J. Cancer Prev.* **2014**, *15*, 517–535. [CrossRef]
97. Alvi, M.; Yaqoob, A.; Rehman, K.; Shoaib, S.M.; Akash, M.S.H. Plga-based nanoparticles for the treatment of cancer: Current strategies and perspectives. *AAPS Open* **2022**, *8*, 1–7. [CrossRef]

98. Yadav, N.; Francis, A.P.; Priya, V.V.; Patil, S.; Mustaq, S.; Khan, S.S.; Alzahrani, K.J.; Banjer, H.J.; Mohan, S.K.; Mony, U.; et al. Polysaccharide-drug conjugates: A tool for enhanced cancer therapy. *Polymers* **2022**, *14*, 950. [CrossRef] [PubMed]
99. Dacoba, T.G.; Omange, R.W.; Li, H.; Crecente-Campo, J.; Luo, M.; Alonso, M.J. Polysaccharide nanoparticles can efficiently modulate the immune response against an hiv peptide antigen. *ACS Nano* **2019**, *13*, 4947–4959. [CrossRef] [PubMed]
100. Klein, M.; Menta, M.; Dacoba, T.G.; Crecente-Campo, J.; Alonso, M.J.; Dupin, D.; Loinaz, I.; Grassl, B.; Séby, F. Advanced nanomedicine characterization by dls and af4-uv-mals: Application to a hiv nanovaccine. *J. Pharm. Biomed. Anal.* **2020**, *179*, 113017. [CrossRef]
101. Biemans, R.; Micoli, F.; Romano, M.R. Glycoconjugate vaccines, production and characterization. In *Recent Trends in Carbohydrate Chemistry*; Elsevier: Amsterdam, The Netherlands, 2020; pp. 285–313.
102. Barth, H.G.; Carlin, F.J. A review of polymer shear degradation in size-exclusion chromatography. *J. Liq. Chromatogr.* **1984**, *7*, 1717–1738. [CrossRef]
103. Kuntsche, J.; Horst, J.C.; Bunjes, H. Cryogenic transmission electron microscopy (cryo-tem) for studying the morphology of colloidal drug delivery systems. *Int. J. Pharm.* **2011**, *417*, 120–137. [CrossRef]
104. Truong, N.P.; Whittaker, M.R.; Mak, C.W.; Davis, T.P. The importance of nanoparticle shape in cancer drug delivery. *Expert. Opin. Drug. Deliv.* **2015**, *12*, 129–142. [CrossRef]
105. Wong, C.K.; Stenzel, M.H.; Thordarson, P. Non-spherical polymersomes: Formation and characterization. *Chem. Soc. Rev.* **2019**, *48*, 4019–4035. [CrossRef]
106. Burchard, W.; Schmidt, M.; Stockmayer, W.H. Information on polydispersity and branching from combined quasi-elastic and intergrated scattering. *Macromolecules* **1980**, *13*, 1265–1272. [CrossRef]
107. Kok, C.M.; Rudin, A. Relationship between the hydrodynamic radius and the radius of gyration of a polymer in solution. *Die Makromol. Chem. Rapid Commun.* **1981**, *2*, 655–659. [CrossRef]
108. Tomalia, D.A.; Naylor, A.M.; Goddard, W.A. Starburst dendrimers: Molecular-level control of size, shape, surface chemistry, topology, and flexibility from atoms to macroscopic matter. *Angew. Chem. Int. Ed. Engl.* **1990**, *29*, 138–175. [CrossRef]
109. Wang, J.; Li, B.; Qiu, L.; Qiao, X.; Yang, H. Dendrimer-based drug delivery systems: History, challenges, and latest developments. *J. Biol. Eng.* **2022**, *16*, 18. [CrossRef] [PubMed]
110. Kootstra, N.A.; Verma, I.M. Gene therapy with viral vectors. *Annu. Rev. Pharmacol. Toxicol.* **2003**, *43*, 413–439. [CrossRef] [PubMed]
111. Shah, P.B.; Losordo, D.W. Non-viral vectors for gene therapy: Clinical trials in cardiovascular disease. In *Advances in Genetics*; Academic Press: Cambridge, MA, USA, 2005; Volume 54, pp. 339–361.
112. Eisenman, D. The united states' regulatory environment is evolving to accommodate a coming boom in gene therapy research. *Appl. Biosaf.* **2019**, *24*, 147–152. [CrossRef] [PubMed]
113. Lee, C.S.; Bishop, E.S.; Zhang, R.; Yu, X.; Farina, E.M.; Yan, S.; Zhao, C.; Zheng, Z.; Shu, Y.; Wu, X.; et al. Adenovirus-mediated gene delivery: Potential applications for gene and cell-based therapies in the new era of personalized medicine. *Genes. Dis.* **2017**, *4*, 43–63. [CrossRef] [PubMed]
114. Rodriguez, D.A.; Vader, P. Extracellular vesicle-based hybrid systems for advanced drug delivery. *Pharmaceutics* **2022**, *14*, 267. [CrossRef]
115. Jayasinghe, M.K.; Tan, M.; Peng, B.; Yang, Y.; Sethi, G.; Pirisinu, M.; Le, M.T.N. New approaches in extracellular vesicle engineering for improving the efficacy of anti-cancer therapies. *Semin. Cancer Biol.* **2021**, *74*, 62–78. [CrossRef]
116. Ferreira, D.; Moreira, J.N.; Rodrigues, L.R. New advances in exosome-based targeted drug delivery systems. *Crit. Rev. Oncol. Hematol.* **2022**, *172*, 103628. [CrossRef]
117. Amiri, A.; Bagherifar, R.; Ansari Dezfouli, E.; Kiaie, S.H.; Jafari, R.; Ramezani, R. Exosomes as bio-inspired nanocarriers for rna delivery: Preparation and applications. *J. Transl. Med.* **2022**, *20*, 125. [CrossRef]
118. De Jong, O.G.; Kooijmans, S.A.A.; Murphy, D.E.; Jiang, L.; Evers, M.J.W.; Sluijter, J.P.G.; Vader, P.; Schiffelers, R.M. Drug delivery with extracellular vesicles: From imagination to innovation. *Acc. Chem. Res.* **2019**, *52*, 1761–1770. [CrossRef]
119. Zhang, Q.; Zhang, H.; Ning, T.; Liu, D.; Deng, T.; Liu, R.; Bai, M.; Zhu, K.; Li, J.; Fan, Q.; et al. Exosome-delivered c-met sirna could reverse chemoresistance to cisplatin in gastric cancer. *Int. J. Nanomed.* **2020**, *15*, 2323–2335. [CrossRef] [PubMed]
120. Pascucci, L.; Cocce, V.; Bonomi, A.; Ami, D.; Ceccarelli, P.; Ciusani, E.; Vigano, L.; Locatelli, A.; Sisto, F.; Doglia, S.M.; et al. Paclitaxel is incorporated by mesenchymal stromal cells and released in exosomes that inhibit in vitro tumor growth: A new approach for drug delivery. *J. Control Release* **2014**, *192*, 262–270. [CrossRef] [PubMed]
121. Thery, C.; Witwer, K.W.; Aikawa, E.; Alcaraz, M.J.; Anderson, J.D.; Andriantsitohaina, R.; Antoniou, A.; Arab, T.; Archer, F.; Atkin-Smith, G.K.; et al. Minimal information for studies of extracellular vesicles 2018 (misev2018): A position statement of the international society for extracellular vesicles and update of the misev2014 guidelines. *J. Extracell. Vesicles* **2018**, *7*, 1535750. [CrossRef]
122. Witwer, K.W.; Goberdhan, D.C.; O'Driscoll, L.; Thery, C.; Welsh, J.A.; Blenkiron, C.; Buzas, E.I.; Di Vizio, D.; Erdbrugger, U.; Falcon-Perez, J.M.; et al. Updating misev: Evolving the minimal requirements for studies of extracellular vesicles. *J. Extracell. Vesicles* **2021**, *10*, e12182. [CrossRef]

123. Yang, J.S.; Kim, J.Y.; Lee, J.C.; Moon, M.H. Investigation of lipidomic perturbations in oxidatively stressed subcellular organelles and exosomes by asymmetrical flow field-flow fractionation and nanoflow ultrahigh performance liquid chromatography-tandem mass spectrometry. *Anal. Chim. Acta* **2019**, *1073*, 79–89. [CrossRef] [PubMed]
124. Wu, B.; Chen, X.; Wang, J.; Qing, X.; Wang, Z.; Ding, X.; Xie, Z.; Niu, L.; Guo, X.; Cai, T.; et al. Separation and characterization of extracellular vesicles from human plasma by asymmetrical flow field-flow fractionation. *Anal. Chim. Acta* **2020**, *1127*, 234–245. [CrossRef]
125. Kim, Y.B.; Yang, J.S.; Lee, G.B.; Moon, M.H. Evaluation of exosome separation from human serum by frit-inlet asymmetrical flow field-flow fractionation and multiangle light scattering. *Anal. Chim. Acta* **2020**, *1124*, 137–145. [CrossRef]
126. Li, P.; Kaslan, M.; Lee, S.H.; Yao, J.; Gao, Z. Progress in exosome isolation techniques. *Theranostics* **2017**, *7*, 789–804. [CrossRef]
127. Chia, B.S.; Low, Y.P.; Wang, Q.; Li, P.; Gao, Z. Advances in exosome quantification techniques. *TrAC Trends Anal. Chem.* **2017**, *86*, 93–106. [CrossRef]
128. Anderson, W.; Kozak, D.; Coleman, V.A.; Jamting, A.K.; Trau, M. A comparative study of submicron particle sizing platforms: Accuracy, precision and resolution analysis of polydisperse particle size distributions. *J. Colloid. Interface Sci.* **2013**, *405*, 322–330. [CrossRef]
129. Zhang, H.; Zhang, H.; Lyden, D. A protocol for asymmetric-flow field-flow fractionation (af4) of small extracellular vesicles. *Protocol Exchange* **2018**, 1–9. [CrossRef]
130. Kim, Y.B.; Lee, G.B.; Moon, M.H. Size separation of exosomes and microvesicles using flow field-flow fractionation/multiangle light scattering and lipidomic comparison. *Anal. Chem.* **2022**, *94*, 8958–8965. [CrossRef] [PubMed]
131. Podzimek, S.; Johann, C. Asymmetric flow field-flow fractionation: Current status, possibilities, analytical limitations and future trends. *Chromatographia* **2021**, *84*, 531–534. [CrossRef]
132. Jia, Y.-P.; Ma, B.-Y.; Wei, X.-W.; Qian, Z.-Y. The in vitro and in vivo toxicity of gold nanoparticles. *Chin. Chem. Lett.* **2017**, *28*, 691–702. [CrossRef]
133. Xu, Z.P.; Zeng, Q.H.; Lu, G.Q.; Yu, A.B. Inorganic nanoparticles as carriers for efficient cellular delivery. *Chem. Eng. Sci.* **2006**, *61*, 1027–1040. [CrossRef]
134. Biener, J.; Wittstock, A.; Baumann, T.; Weissmüller, J.; Bäumer, M.; Hamza, A. Surface chemistry in nanoscale materials. *Materials* **2009**, *2*, 2404–2428. [CrossRef]
135. Kong, F.Y.; Zhang, J.W.; Li, R.F.; Wang, Z.X.; Wang, W.J.; Wang, W. Unique roles of gold nanoparticles in drug delivery, targeting and imaging applications. *Molecules* **2017**, *22*, 1445. [CrossRef]
136. Wang, W.; Wang, J.; Ding, Y. Gold nanoparticle-conjugated nanomedicine: Design, construction, and structure-efficacy relationship studies. *J. Mater. Chem. B* **2020**, *8*, 4813–4830. [CrossRef]
137. Song, M.; Wang, X.; Li, J.; Zhang, R.; Chen, B.; Fu, D. Effect of surface chemistry modification of functional gold nanoparticles on the drug accumulation of cancer cells. *J. Biomed. Mater. Res. A* **2008**, *86*, 942–946. [CrossRef]
138. Zeng, X.; Liu, G.; Tao, W.; Ma, Y.; Zhang, X.; He, F.; Pan, J.; Mei, L.; Pan, G. A drug-self-gated mesoporous antitumor nanoplatform based on ph-sensitive dynamic covalent bond. *Adv. Funct. Mater.* **2017**, *27*, 1605985. [CrossRef]
139. Farjadian, F.; Roointan, A.; Mohammadi-Samani, S.; Hosseini, M. Mesoporous silica nanoparticles: Synthesis, pharmaceutical applications, biodistribution, and biosafety assessment. *Chem. Eng. J.* **2019**, *359*, 684–705. [CrossRef]
140. Anselmo, A.C.; Mitragotri, S. A review of clinical translation of inorganic nanoparticles. *AAPS J.* **2015**, *17*, 1041–1054. [CrossRef] [PubMed]
141. Kong, L.; Qiu, J.; Sun, W.; Yang, J.; Shen, M.; Wang, L.; Shi, X. Multifunctional pei-entrapped gold nanoparticles enable efficient delivery of therapeutic sirna into glioblastoma cells. *Biomater. Sci.* **2017**, *5*, 258–266. [CrossRef] [PubMed]
142. Wang, F.; Wang, Y.C.; Dou, S.; Xiong, M.H.; Sun, T.M.; Wang, J. Doxorubicin-tethered responsive gold nanoparticles facilitate intracellular drug delivery for overcoming multidrug resistance in cancer cells. *ACS Nano* **2011**, *5*, 3679–3692. [CrossRef] [PubMed]
143. Wang, J.; Giordani, S.; Marassi, V.; Roda, B.; Reschiglian, P.; Zattoni, A. Quality control and purification of ready-to-use conjugated gold nanoparticles to ensure effectiveness in biosensing. *Front. Sens.* **2022**, *3*, 1087115. [CrossRef]
144. Marassi, V.; Zanoni, I.; Ortelli, S.; Giordani, S.; Reschiglian, P.; Roda, B.; Zattoni, A.; Ravagli, C.; Cappiello, L.; Baldi, G.; et al. Native study of the behaviour of magnetite nanoparticles for hyperthermia treatment during the initial moments of intravenous administration. *Pharmaceutics* **2022**, *14*, 2810. [CrossRef]
145. Agostoni, V.; Anand, R.; Monti, S.; Hall, S.; Maurin, G.; Horcajada, P.; Serre, C.; Bouchemal, K.; Gref, R. Impact of phosphorylation on the encapsulation of nucleoside analogues within porous iron(iii) metal-organic framework mil-100(fe) nanoparticles. *J. Mater. Chem. B* **2013**, *1*, 4231–4242. [CrossRef] [PubMed]
146. Kowalkowski, T.; Sugajski, M.; Buszewski, B. Impact of ionic strength of carrier liquid on recovery in flow field-flow fractionation. *Chromatographia* **2018**, *81*, 1213–1218. [CrossRef]
147. Mudalige, T.K.; Qu, H.; Sanchez-Pomales, G.; Sisco, P.N.; Linder, S.W. Simple functionalization strategies for enhancing nanoparticle separation and recovery with asymmetric flow field flow fractionation. *Anal. Chem.* **2015**, *87*, 1764–1772. [CrossRef]
148. Gigault, J.; Pettibone, J.M.; Schmitt, C.; Hackley, V.A. Rational strategy for characterization of nanoscale particles by asymmetric-flow field flow fractionation: A tutorial. *Anal. Chim. Acta* **2014**, *809*, 9–24. [CrossRef]
149. Valto, P.; Knuutinen, J.; Alen, R. Evaluation of resin and fatty acid concentration levels by online sample enrichment followed by atmospheric pressure chemical ionization-mass spectrometry (APCI-MS). *Environ. Sci. Pollut. Res. Int.* **2009**, *16*, 287–294. [CrossRef] [PubMed]

150. Multia, E.; Liangsupree, T.; Jussila, M.; Ruiz-Jimenez, J.; Kemell, M.; Riekkola, M.L. Automated on-line isolation and fractionation system for nanosized biomacromolecules from human plasma. *Anal. Chem.* **2020**, *92*, 13058–13065. [CrossRef] [PubMed]
151. Caputo, F.; Clogston, J.; Calzolai, L.; Rosslein, M.; Prina-Mello, A. Measuring particle size distribution of nanoparticle enabled medicinal products, the joint view of euncl and nci-ncl. A step by step approach combining orthogonal measurements with increasing complexity. *J. Control Release* **2019**, *299*, 31–43. [CrossRef] [PubMed]

Disclaimer/Publisher's Note: The statements, opinions and data contained in all publications are solely those of the individual author(s) and contributor(s) and not of MDPI and/or the editor(s). MDPI and/or the editor(s) disclaim responsibility for any injury to people or property resulting from any ideas, methods, instructions or products referred to in the content.

Article

Peroxidase-like Activity of CeO$_2$ Nanozymes: Particle Size and Chemical Environment Matter

Arina D. Filippova [1], Madina M. Sozarukova [1], Alexander E. Baranchikov [1], Sergey Yu. Kottsov [1], Kirill A. Cherednichenko [2] and Vladimir K. Ivanov [1,*]

[1] Kurnakov Institute of General and Inorganic Chemistry of the Russian Academy of Sciences, 119991 Moscow, Russia
[2] Department of Physical and Colloid Chemistry, Faculty of Chemical and Environmental Engineering, National University of Oil and Gas "Gubkin University", 119991 Moscow, Russia
* Correspondence: van@igic.ras.ru

Abstract: The enzyme-like activity of metal oxide nanoparticles is governed by a number of factors, including their size, shape, surface chemistry and substrate affinity. For CeO$_2$ nanoparticles, one of the most prominent inorganic nanozymes that have diverse enzymatic activities, the size effect remains poorly understood. The low-temperature hydrothermal treatment of ceric ammonium nitrate aqueous solutions made it possible to obtain CeO$_2$ aqueous sols with different particle sizes (2.5, 2.8, 3.9 and 5.1 nm). The peroxidase-like activity of ceria nanoparticles was assessed using the chemiluminescent method in different biologically relevant buffer solutions with an identical pH value (phosphate buffer and Tris-HCl buffer, pH of 7.4). In the phosphate buffer, doubling CeO$_2$ nanoparticles' size resulted in a two-fold increase in their peroxidase-like activity. The opposite effect was observed for the enzymatic activity of CeO$_2$ nanoparticles in the phosphate-free Tris-HCl buffer. The possible reasons for the differences in CeO$_2$ enzyme-like activity are discussed.

Keywords: cerium dioxide; colloids; surface; hydroxyl species; enzyme-like activity; buffer; size effect

1. Introduction

Nanocrystalline cerium dioxide is well known as a multifunctional catalyst [1], a UV-protective material [2–4] and a component for highly sensitive gas sensors [5–7]. One of the main factors determining the functional characteristics of ceria-based materials is the size of CeO$_2$ particles. For instance, Wu et al. found that the rate of photoinduced decomposition of the herbicide, N-(phosphonomethyl)-glycine, decreases by a factor of 6 with an increase in the particle size of cerium dioxide from 2 to 5 nm [8]. Torrente-Murciano et al. demonstrated that doubling the particle size (from 5 to 10 nm) leads to a decrease in the catalytic activity of CeO$_2$ in the reaction of naphthalene oxidation to CO$_2$ by a factor of 2.5 [9]. It is important to note that the size effect is typical not only for ultrasmall particles of cerium dioxide (less than 10 nm), but also for larger particles (up to 50 nm). Lin et al. found that the rate of conversion of carbon dioxide to methane (at 548 K) decreases by a factor of 3 with an increase in the size of CeO$_2$ particles from 32 to 50 nm [10]. The dependence of the catalytic activity of cerium dioxide on particle size is generally associated with a number of factors, including changes in the surface-to-volume ratio, band gap energy, surface chemistry and electronic structure [11].

In recent years, it has been found that cerium dioxide demonstrates exceptional biological activity, exhibits antibacterial [12–14] and antiviral properties [15,16], is characterised by low cytotoxicity [17–19], and can act as a UV- and radioprotective agent [20,21]. One of the key mechanisms of the biological action of CeO$_2$ is associated with its ability to mimic the activity of a number of enzymes and exhibit peroxidase- [22], catalase- [23], oxidase- [24], superoxide dismutase- [25], lipoperoxidase-, phospholipoperoxidase- [26], phosphatase- [27], phospholipase- [28], photolipase- [29], haloperoxidase- [30] and urease-like activity [31]. In particular, Lang et al. demonstrated a direct correlation between the antiviral activity of cerium dioxide against human coronavirus OC43 and its haloperoxidase-like activity [30].

Since cerium dioxide is able to catalyse reactions involving enzyme substrates, it might be expected that the size of CeO_2 particles will affect the rate of such reactions. There is, however, a scarcity of data in the literature on the effect of particle size on the enzyme-like activity of CeO_2. Shlapa et al. showed that an increase in the size of CeO_2 particles by a factor of ~2 (from 7 to 15 nm) leads to a decrease in the oxidase-like activity of cerium dioxide by a factor of 1.2 [24]. Henych et al. found that the phospholipase-like activity of CeO_2 with a particle size of 5 nm is more than 30 times higher than the activity of 10 nm CeO_2 particles [28]. To the best of the authors' knowledge, there are virtually no data on the size effect on the peroxidase-like activity of CeO_2. At the same time, hydrogen peroxide is the most important reactive oxygen species (ROS) that causes oxidative stress in living systems, and the peroxidase activity of enzymes and their mimetics is of paramount importance in biological processes [32].

As a rule, spectrophotometric methods, based on determining the concentration of coloured products of catalytic reactions, are used to determine the ROS-scavenging ability of materials. For this purpose, TMB (3,3',5,5'-tetramethylbenzidine) assays are the most commonly used. Nevertheless, this approach has several limitations due to the complex mechanism of TMB oxidation that can proceed via either one-electron or two-electron pathways. Moreover, a charge-transfer complex between TMB and its diimine final product can form. All of these products are characterised by different absorption wavelengths, extinction coefficients and formation rate constants [33,34]. For a deeper insight into the chemical interactions of cerium dioxide with biological systems, however, selective methods for determining its enzyme-like activity are of primary importance, especially those that are specific to particular reactive oxygen species (e.g., $OH\cdot$, $HO_2\cdot$ and $O_2\cdot^{-}$). In this context, fluorescent [22] or chemiluminescent [35] methods are considered to be more accurate and informative.

Special attention should be paid to the correct choice of the medium used for the analysis of enzyme-like activity, e.g., the choice of a physiologically relevant buffer solution [36]. It is generally accepted that it is the pH of the medium used that determines the mechanism of CeO_2 interaction with hydrogen peroxide [37], while the presence of phosphate ions affects the activity of cerium dioxide, although not in a completely unambiguous way [38–41].

In this regard, the accurate analysis of the size effect on the enzyme-like activity of cerium dioxide requires the use of a single synthetic technique that will allow the production of CeO_2 with different particle sizes under the same, or similar, conditions, as well as an analysis of its enzyme-like activity within a single analytical approach, taking into account the chemical environment of the nanoparticles. In the present work, a quantitative analysis of the size effect on the peroxidase-like activity of nanocrystalline cerium dioxide was carried out using the chemiluminescent method with two different buffer solutions (namely phosphate and Tris-HCl buffer solutions). These buffer solutions had the same pH (7.4) but differed in the presence or absence of phosphate ions.

2. Results

2.1. Synthesis of Aqueous Cerium Dioxide Sols with a Given Particle Size

Since cerium dioxide is characterised by low solubility in aqueous media and there is a weak dependence of particle size on synthesis temperature, the preparation of CeO_2 sols with different sizes of nanoparticles is a complex problem. The approach used in the present study, based on the thermohydrolysis (95 °C) of ceric ammonium nitrate, makes it possible to obtain ultrasmall (up to 5 nm) CeO_2 particles possessing high surface activity [42]. In this work, solutions of $(NH_4)_2[Ce(NO_3)_6]$ with concentrations of 0.046, 0.092, 0.185, 0.277 and 0.370 M were used to obtain cerium dioxide with a high yield (Table 1). In the course of the preliminary experiments, it was found that the use of a ceric ammonium nitrate solution with a lower concentration (0.046 M) did not produce a CeO_2 sol. The resulting solution did not demonstrate the Tyndall effect, which indicated the absence of CeO_2 nanoparticles in the solution. As a result of the thermal treatment of the $(NH_4)_2[Ce(NO_3)_6]$ solutions with concentrations of 0.092–0.370 M, a series of CeO_2 sols with different particle sizes were obtained.

Table 1. Synthesis conditions and concentrations of aqueous CeO_2 sols prepared by thermohydrolysis of ceric ammonium nitrate.

Sample	$(NH_4)_2[Ce(NO_3)_6]$, M	$c(CeO_2)$, M (g/L)	Yield, %
1	0.092	0.09 (15.6)	85
2	0.185	0.15 (26.1)	89
3	0.277	0.15 (26.2)	90
4	0.370	0.15 (25.9)	86

The XRD patterns of the CeO_2 sols, which were dried at a low temperature (50 °C), are shown in Figure 1. According to the data obtained, the phase composition of the solid residues corresponds with nanocrystalline cubic cerium dioxide (sp. gr. $\overline{Fm3m}$, PDF2 00-034-0394). As can be seen from Figure 1a, the peak width decreases with an increase in $(NH_4)_2[Ce(NO_3)_6]$ concentration. CeO_2 crystallite size was evaluated using the XRD data based on the Scherrer formula. According to Figure 1b, the size of cerium dioxide crystallites increases consistently, in the range of 2–5 nm, with the changes in the concentration of $(NH_4)_2[Ce(NO_3)_6]$.

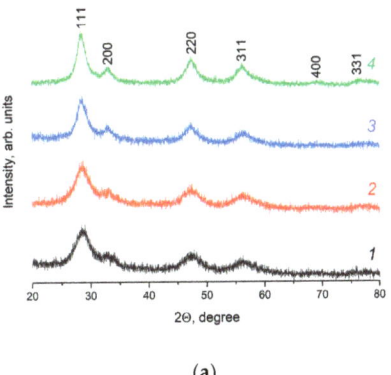

(a)

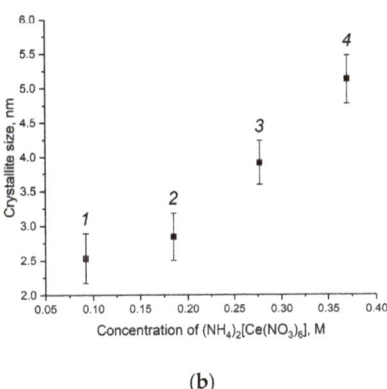

(b)

Figure 1. (a) XRD patterns of CeO_2 nanopowders prepared upon the drying (50 °C) of aqueous ceria sols. (b) Dependence of CeO_2 crystallite size on the concentration of ceric ammonium nitrate in the reaction mixture. Ceria sols were obtained from (1) 0.092 M, (2) 0.185 M, (3) 0.277 M and (4) 0.370 M solutions of $(NH_4)_2[Ce(NO_3)_6]$.

According to the generally accepted concepts of nucleation and crystal growth, the growth of solid-phase particles typically proceeds via the dissolution–crystallisation mechanism (Ostwald ripening). Conversely, for cerium dioxide and some other oxides, the mechanism of oriented attachment and growth of particles is usually observed [43,44], particularly under hydrothermal conditions [45]. Apparently, when taking into account the extremely low solubility of CeO_2, the observed increase in the size of CeO_2 particles with an increase in the initial concentration of $(NH_4)_2[Ce(NO_3)_6]$ is due to the implementation of the nonclassical particle growth mechanism.

The high-resolution transmission electron microscopic (HRTEM) images confirm the results of the X-ray diffraction. As can be seen from Figure 2 (see also ESI, Figure S1

(Supplementary Materials)), the CeO$_2$ particle size is approximately 3 nm. The HRTEM images display interplanar distances of about 1.9 Å, which can be attributed to the (220) planes in the crystal lattice of CeO$_2$ [46–48].

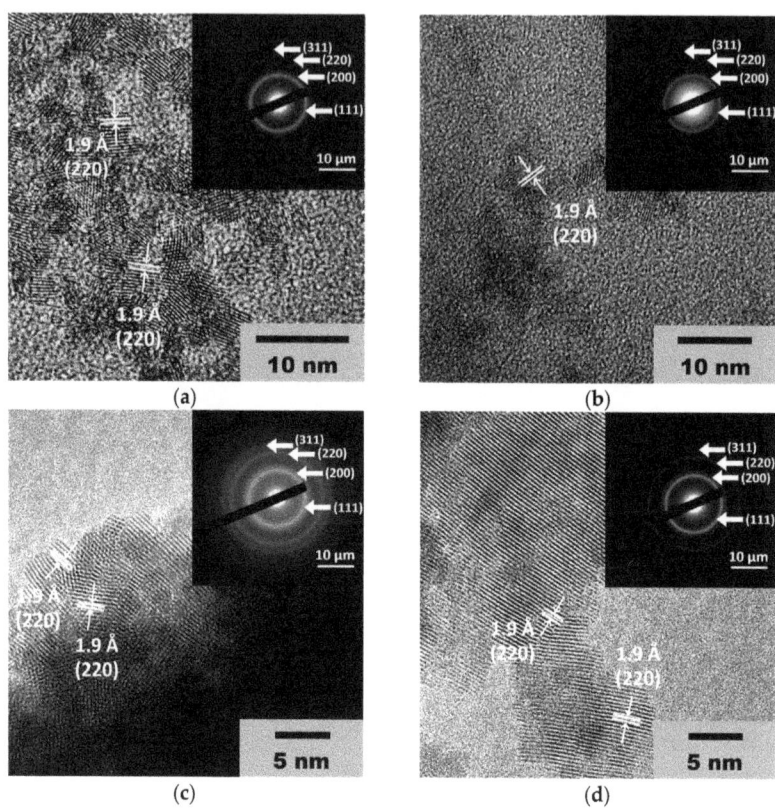

Figure 2. TEM images and SAED patterns of the CeO$_2$ sols obtained from (**a**) 0.092 M, (**b**) 0.185 M, (**c**) 0.277 M and (**d**) 0.370 M aqueous solutions of (NH$_4$)$_2$[Ce(NO$_3$)$_6$].

The dynamic light-scattering technique allowed the study of the size distribution of aggregates of individual ceria nanoparticles in the sols (Figure 3). The CeO$_2$ sol obtained from 0.092 M ceric ammonium nitrate solution is characterised by bimodal distribution of aggregates. As the concentration of (NH$_4$)$_2$[Ce(NO$_3$)$_6$] in the starting solution increases, a transition to monomodal distribution of aggregates in the sols is observed. At the same time, with an increase in (NH$_4$)$_2$[Ce(NO$_3$)$_6$] concentration, the size of aggregates increases. It should be noted that during the 3 months of storage under ambient conditions, the size of aggregates in the sol obtained from the solution with the lowest concentration of (NH$_4$)$_2$[Ce(NO$_3$)$_6$] increases by 30% (from 10 to 13 nm), and the size of agglomerates increases by 10% (from 120 to 130 nm), while the aggregate size of the sols obtained from the solutions with higher concentrations of ceric ammonium nitrate (0.185–0.37 M) changes by no more than 15% (Figure 3b).

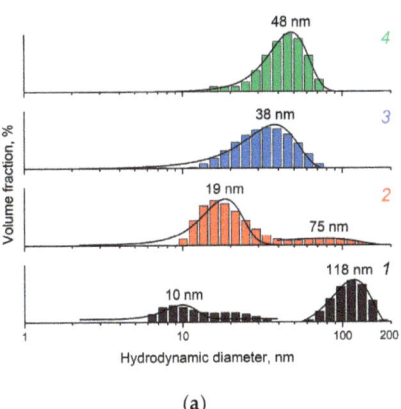 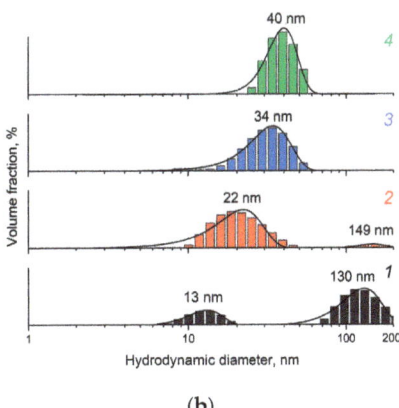

(a) (b)

Figure 3. Hydrodynamic diameter distribution for CeO_2 particles in aqueous ceria sols obtained from (1) 0.092 M, (2) 0.185 M, (3) 0.277 M and (4) 0.370 M $(NH_4)_2[Ce(NO_3)_6]$ aqueous solutions: (**a**) after synthesis and (**b**) after three months of storage under ambient conditions. The pH value of all the sols is approximately 2.4.

The results of the electrokinetic measurements are shown in Figures 4 and S2. Thermohydrolysis of ceric ammonium nitrate yields an acidic environment; thus, CeO_2 particles acquire a positive charge due to the protonation of surface hydroxyl groups [47]. The pH value of all of the sols obtained is approximately 2.4, so the values of the ζ-potentials are positive (Figure 4). As it follows from Figure 4, with an increase in the concentration of ceric ammonium nitrate in the starting solutions, the values of the ζ-potential of CeO_2 nanoparticles increase from +29.9 to +38.2 mV. High ζ-potential values ensure long-term stability of CeO_2 colloids, which demonstrate no signs of precipitation after three months of storage under ambient conditions (Figure 3).

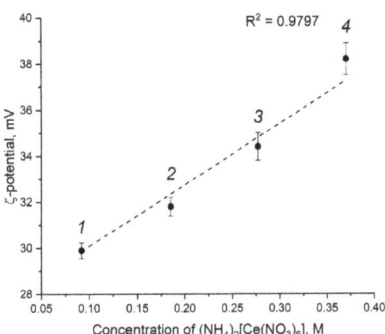

Figure 4. Dependence of the ζ-potential of CeO_2 aqueous sols on the concentration of initial ceric ammonium nitrate solution. The sols were obtained from (1) 0.092 M, (2) 0.185 M, (3) 0.370 M and (4) 0.554 M solutions of $(NH_4)_2[Ce(NO_3)_6]$. The pH value of all the sols is approximately 2.4.

The IR spectroscopy (Figure 5) shows a broad absorption band in the 3400–3000 cm^{-1} region, which can be attributed to the stretching vibrations of physically adsorbed water [49]. The absorption band at 1625 cm^{-1} corresponds to the bending vibrations of molecular water [49]. The peaks at 1033 and 450 cm^{-1} can be attributed to the bending vibrations of the Ce–OH surface groups and to the stretching vibration of the Ce–O bond [50,51]. The absorption bands in the range of 1500–1300 cm^{-1} and at 1280 cm^{-1} correspond to the bending C–OH and stretching C–O vibrations of isopropanol traces [52].

The weak absorption band at 807 cm^{-1} can be attributed to the vibrations of nitrate species adsorbed on the oxide surface [53].

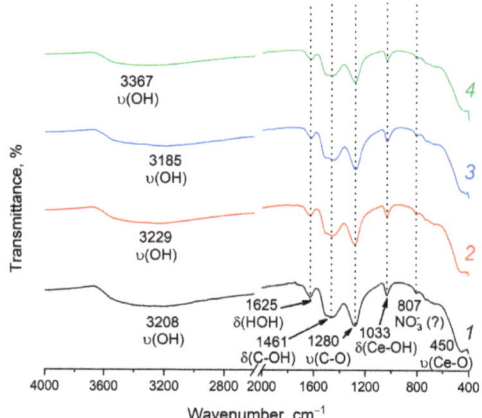

Figure 5. FTIR spectra of CeO$_2$ nanopowders prepared by drying aqueous ceria sols obtained from (1) 0.092 M, (2) 0.185 M, (3) 0.277 M and (4) 0.370 M aqueous solutions of (NH$_4$)$_2$[Ce(NO$_3$)$_6$].

The Raman spectra of the dried ceria sols are typical for nanocrystalline CeO$_2$ (Figure 6) and show an intense F$_{2g}$ mode peak at 453–448 cm^{-1}, which is characteristic of a fluorite structure and is due to the symmetric stretching vibrations of the Ce-O bond [8]. A weak band at about 600 cm^{-1} can be attributed to the defect-induced (D) mode [54,55]. The band centred at ~270 cm^{-1} can be ascribed to the second-order transverse acoustic (2TA) mode [7]. The band at 330 cm^{-1} is characteristic of the 3TL mode of O-Ce-O vibrations [56].

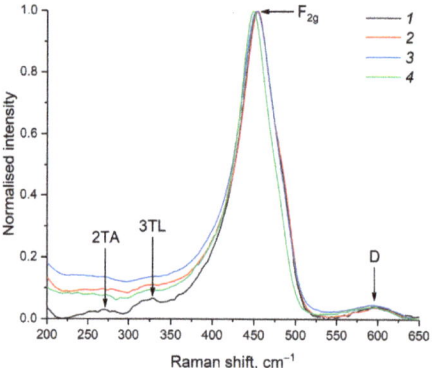

Figure 6. Raman spectra of CeO$_2$ nanopowders prepared by drying aqueous ceria sols obtained from (1) 0.092 M, (2) 0.185 M, (3) 0.277 M and (4) 0.370 M solutions of (NH$_4$)$_2$[Ce(NO$_3$)$_6$].

The main physicochemical characteristics (particle sizes according to the XRD and DLS data, and ζ-potential values) of cerium dioxide colloidal solutions are presented in Table 2.

Table 2. The main parameters of CeO$_2$ sols.

CeO$_2$ Sample	Concentration of the Initial (NH$_4$)$_2$[Ce(NO$_3$)$_6$] Aqueous Solution (M)	Particle Size Estimated from XRD Data (D$_{XRD}$, nm)	Particle Size Estimated from DLS Data (D$_{DLS}$, nm)	ζ-Potential (mV)	Specific Surface Area (SSA *, m^2/g)	Proportion of Surface Cerium Atoms, (Ce$_{surf}$ *, %)
1	0.092	2.5 ± 0.4	10	+29.9 ± 0.3	329	60
2	0.185	2.8 ± 0.3	19	+31.8 ± 0.4	293	54
3	0.277	3.9 ± 0.3	38	+34.4 ± 0.6	213	39
4	0.370	5.1 ± 0.4	48	+38.2 ± 0.7	162	30

* Calculated values.

The specific surface area (SSA) of CeO$_2$ was estimated from the particle sizes, determined according to the powder X-ray diffraction data and the literature data on the density of CeO$_2$ (ρ = 7.215 g/cm^3, PDF2 N°00-034-0394), according to Formula (1):

$$SSA = 6000/(D \cdot \rho) \quad (1)$$

where 6000 is the shape factor, D is the diameter of a spherical CeO$_2$ particle, and ρ is the density of CeO$_2$. The proportion of surface cerium atoms in spherical CeO$_2$ particles (Ce$_{surf}$) was also evaluated, according to a previously proposed method [57].

2.2. Enzyme-like Activity of CeO$_2$ Sols towards H$_2$O$_2$ Decomposition

The chemiluminescent method used for the analysis of peroxidase-like activity is based on the interaction of a chemiluminescent probe molecule (luminol) with reactive oxygen species through the formation of monoprotonated 3-aminophthalic acid in an electronically excited state [58]. The luminescence intensity of the luminol oxidation product is proportional to the concentration of free radicals, which depends on the activity of the enzyme-like inorganic material. This method of analysis makes it possible to determine the enzyme-like activity of nanomaterials with high analytical precision [59–65].

At the first stage, the analysis of the enzyme-like activity of cerium dioxide was carried out in a phosphate-buffered solution. The concentration of cerium dioxide in the reaction mixture was the same for all the sols and amounted to 250 μM. When the cerium dioxide sols were added to the reaction mixture containing a phosphate-buffered solution, luminol and hydrogen peroxide, an increase in the luminescence intensity of the luminol oxidation product was observed (Figure 7a). Thus, cerium dioxide catalysed the oxidation of luminol in the presence of H$_2$O$_2$; however, the appearance of the chemiluminescence curves differs from those obtained for horseradish peroxidase [58]. It is most likely that, in a phosphate-buffered solution, cerium dioxide exhibits prooxidant (more precisely, peroxidase-like) activity [66]. In this case, the chemiluminescence intensity increases with an increase in the size of CeO$_2$ particles. A similar effect was observed for the different concentrations of the CeO$_2$ sols at 500 μM and 6 mM.

The kinetic parameters of the luminol oxidation reactions with hydrogen peroxide in the presence of CeO$_2$ were mathematically modelled (Figure 7b). At the first stage, the interaction of cerium dioxide with H$_2$O$_2$ with the formation of hydroxyl radicals was considered [67]. The chosen kinetic model of H$_2$O$_2$-induced luminol oxidation includes the decomposition of hydrogen peroxide (2), the interaction of luminol with hydroxyl radicals (3), and the final stage of chemiluminescence (4), where P is the reaction product:

$$CeO_2 + H_2O_2 \rightarrow 2OH\cdot \quad (2)$$
$$Lum + OH\cdot \rightarrow Lum* \quad (3)$$
$$Lum* + Lum* \rightarrow P + h\nu \quad (4)$$

As can be seen from Figure 7b, the theoretical kinetic curves fit the experimental data well. The rate constants of reactions (2)–(4) obtained as a result of the mathematical modelling are shown in Table 3.

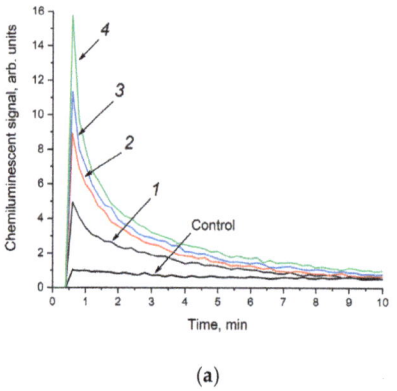
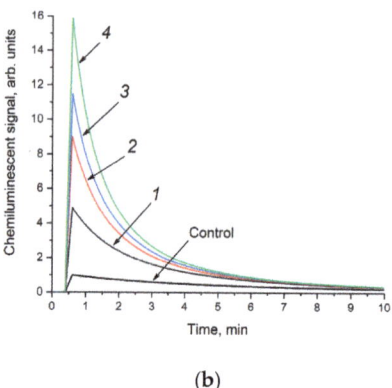

(a)　　　　　　　　　　　　　　　　　　　(b)

Figure 7. Luminol (50 µM) chemiluminescence kinetic curves in the presence of H_2O_2 (10 mM) and CeO_2 sols (250 µM) in a phosphate-buffered solution (100 mM, pH = 7.4): the experimental data were obtained via direct chemiluminescent measurements (**a**), and the calculated curves were obtained through kinetic modelling of the experimental data (**b**). The aqueous ceria sols were obtained from (1) 0.092 M, (2) 0.185 M, (3) 0.277 M and (4) 0.370 M solutions of $(NH_4)_2[Ce(NO_3)_6]$. Control: luminol and H_2O_2 in a phosphate-buffered solution without CeO_2. The chemiluminescence signal was recorded at 36 °C.

Table 3. Rate constants of H_2O_2-induced luminol oxidation reactions in the presence of aqueous sols of cerium dioxide in a phosphate-buffered solution (100 mM, pH = 7.4).

Sample	Particle Size Estimated from XRD Data (D_{XRD}, nm)	k_2 (µM/min)	k_3 (µM/min)	k_4 (µM/min)
Control	–	6.9×10^4	1.9×10^{-6}	3.7×10^{-5}
1	2.5 ± 0.4	8.9×10^4	2.9×10^{-6}	4.7×10^{-5}
2	2.8 ± 0.3	9.7×10^4	3.7×10^{-6}	5.3×10^{-5}
3	3.9 ± 0.3	9.8×10^4	3.8×10^{-6}	5.3×10^{-5}
4	5.1 ± 0.4	1.0×10^5	4.3×10^{-6}	5.5×10^{-5}

Since the chemiluminescent analysis technique is based on the registration of the luminescence of the reaction product (4), the constant k_4 was chosen as the key kinetic parameter used to compare the peroxidase-like activity of cerium dioxide with different particle sizes. The rate of the chemiluminescence reaction of luminol (4) increased by a factor of 1.2 with a two-fold increase in the size of cerium dioxide particles (see Table 3).

It is known that cerium compounds have a high chemical affinity with phosphate ions, which leads to a change in the composition of CeO_2 surface in phosphate-containing media [68–70]. The formation of surface complexes with phosphate species affects the enzyme-like activity of cerium dioxide. Thus, it was previously shown that ageing CeO_2 in a phosphate-buffered solution reduces the superoxide dismutase activity of cerium dioxide (analysis in Tris-HCl, with pH of 7.4–7.5) [71,72]. Zhao et al. found that phosphate-containing compounds increase the oxidase-like activity of CeO_2 by up to six times (acetate buffer, with pH of 4) [40].

In order to exclude the effect of phosphate species on the enzyme-like activity of cerium dioxide, an analysis was also performed on the peroxidase-like activity of CeO_2 in Tris-HCl, which has a pH identical to the phosphate-buffered solution (pH = 7.4). Previously, it was shown that the use of Tris-HCl as a dispersion medium does not affect the antioxidant properties of the dispersed CeO_2 phase [36]. The concentration of cerium dioxide in Tris-HCl was similar to that in the phosphate buffer and amounted to 250 µM. Upon the introduction of the ceria sols into the buffer solutions containing luminol and hydrogen peroxide, the resulting mixtures remained stable with no visible signs of sedimentation. Figure 8a shows that the addition of the CeO_2 aqueous sols to a mixture of Tris-HCl, hydrogen peroxide and luminol leads to an increase in chemiluminescence intensity, compared with a control without cerium dioxide. At the same time, the luminescence intensity of the luminol oxidation product decreases with increasing CeO_2 particle size.

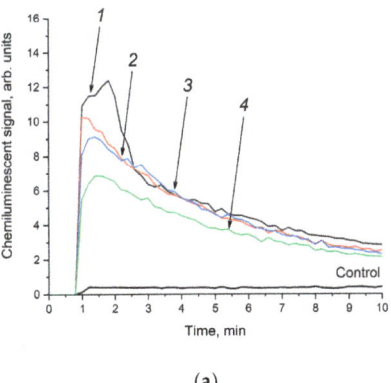
(a)

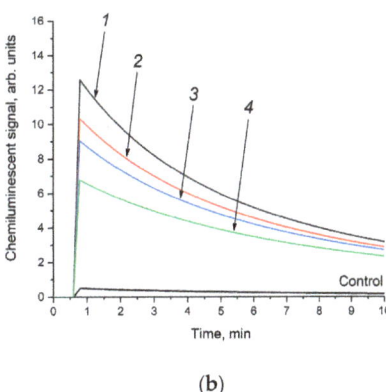
(b)

Figure 8. Luminol (50 µM) chemiluminescence kinetic curves in the presence of H_2O_2 (10 mM) and CeO_2 sols (250 µM) in a Tris-HCl buffer solution (100 mM, pH = 7.4): the experimental data were obtained via direct chemiluminescent measurements (a), and the calculated curves were obtained through kinetic modelling of the experimental data (b). The aqueous ceria sols were obtained from (1) 0.092 M, (2) 0.185 M, (3) 0.277 M and (4) 0.370 M solutions of $(NH_4)_2[Ce(NO_3)_6]$. Control: luminol and H_2O_2 in a Tris-HCl buffer solution without CeO_2. The chemiluminescence signal was recorded at 36 °C.

The appearance of the chemiluminescence curves in the Tris-HCl solution is almost similar to the shape of the curves recorded in the phosphate-buffered solution (Figures 7a and 8a). However, in the Tris-HCl solution, the chemiluminescence intensity decreases more slowly, indicating a decrease in the rate of reactions (3) and (4). As it can be seen from Table 4, the k_4 values for the experiments conducted in the Tris-HCl solution are lower than in the phosphate-buffered solution. At the same time, in the Tris-HCl solution, the k_2 values are three orders of magnitude higher than in the phosphate buffer. The latter observation supports the abovementioned crucial influence of phosphate species on the reactivity of CeO_2 nanoparticles. In the absence of phosphate species, the rate of CeO_2 interaction with hydrogen peroxide is significantly higher.

Table 4. Rate constants of H_2O_2-induced luminol oxidation in the presence of aqueous cerium dioxide sols in a Tris-HCl solution (100 mM, pH = 7.4).

Sample	Particle Size Estimated from XRD Data (D_{XRD}, nm)	k_2 (μM/min)	k_3 (μM/min)	k_4 (μM/min)
Control	–	6.9×10^4	1.5×10^{-6}	3.0×10^{-5}
1	2.5 ± 0.4	1.2×10^8	1.6×10^{-8}	2.0×10^{-6}
2	2.8 ± 0.3	1.2×10^8	1.6×10^{-8}	1.8×10^{-6}
3	3.9 ± 0.3	1.1×10^8	1.6×10^{-8}	1.7×10^{-6}
4	5.1 ± 0.4	1.1×10^8	1.5×10^{-8}	1.5×10^{-6}

It is common knowledge that the treatment of cerium dioxide sol with hydrogen peroxide can lead to the formation of peroxide and hydroperoxide species on the surface of CeO_2 [73,74]. When H_2O_2 is added to the CeO_2 sol in the Tris-HCl buffer solution (Figure 9, curve 3), a red shift of the Ce^{4+} absorption band edge (curve 1) is observed, which can be attributed to the formation of peroxo complexes on the CeO_2 surface [75,76]. The appearance of a shoulder in the absorption spectrum at 400 nm is consistent with the fact that the colloidal solution acquires a yellow colour, which is characteristic of cerium peroxo or hydroperoxo complexes [77,78]. Conversely, the solution of CeO_2 in the phosphate buffer remains colourless after hydrogen peroxide has been added (curve 2). In this case, the red shift of the absorption band upon treatment with H_2O_2 is less pronounced than for the CeO_2 solution in Tris-HCl, and no shoulder is observed for the absorption spectrum at ~400 nm. This result clearly indicates the difference in the manifestation of the enzyme-like activity of cerium dioxide in a decomposition reaction of hydrogen peroxide in a phosphate-buffered solution and in Tris-HCl.

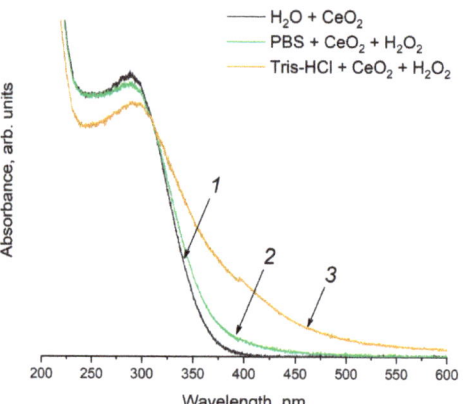

Figure 9. UV-vis absorption spectra of (1) CeO_2 sol (250 μM, particle size is 2.8 nm) in (2) phosphate-buffered solution and (3) Tris-HCl buffer solution after the addition of hydrogen peroxide (10 mM).

For the quantitative comparison of the enzyme-like activity of the cerium dioxide sols in the phosphate-buffered solution and Tris-HCl buffer solution, the value of the integrated intensity (light sum) for a selected period of time (10 min) was used. Figure 10 shows the light sum as a function of (a) CeO_2 particle size and (b) the calculated specific surface area of cerium dioxide. In the phosphate-buffered solution, the peroxidase-like activity of CeO_2 almost doubles as the particle size doubles, while in the Tris-HCl solution, it decreases by a factor of 1.4.

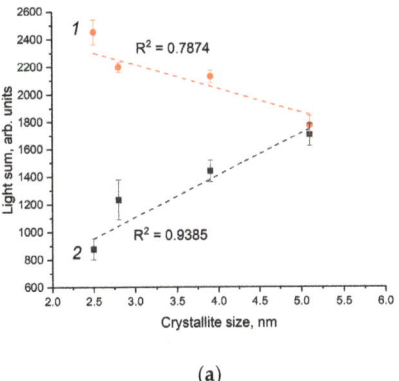 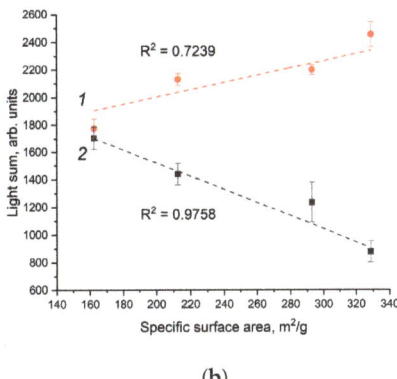

(a)   (b)

Figure 10. Dependence of light sums for the reaction of luminol oxidation in the presence of cerium dioxide, obtained by integrating the chemiluminescence kinetic curves, on (**a**) particle size and (**b**) CeO_2 specific surface area in the Tris-HCl buffer solution (1) and phosphate-buffered solution (2). The pH value of both buffer solutions is 7.4.

As follows from Figure 10, the quantitative dependence obtained for the enzyme-like activity of cerium dioxide towards H_2O_2 is opposite. In the phosphate-buffered solution, the peroxidase-like activity of CeO_2 sols increases with an increase in the particle size of the dispersed phase. This result is quite unexpected, since, with a decrease in particle size, the specific surface area available for interaction with the components of the reaction mixture, including reactive oxygen species, increases. Liu et al., however, previously reported a similar result for citrate-stabilised CeO_2 sols. The peroxidase-like activity of cerium dioxide in an acetate buffer solution (pH 4.5) increased by a factor of ~2.3 with an increase in the particle size by a factor of ~1.5 (from 3 to 4 nm) [79]. It is important to note that the redox activity of cerium dioxide depends on pH and that, according to some data, the prooxidant activity of CeO_2 increases notably in an acidic environment [37,80].

The interaction of cerium dioxide with the components of the reaction mixture, including phosphate ions, is expected to change the chemical composition of the surface of CeO_2 particles. The main functional groups on the surface of cerium dioxide are hydroxyl groups, the amount and reactivity of which directly depend on particle size [81,82]. Previously, Wang et al. showed that surface hydroxylation of ceria nanoparticles is a key factor in determining the adsorption kinetics of phosphate species from KH_2PO_4 solution. With an increase in the concentration of hydroxyl groups by 1.3 times (from 60 to 76%), the number of adsorbed phosphate species increases by 2.5 times, and the rate of their adsorption increases by 620 times [83]. Thus, phosphate ions can inhibit the interaction of surface-active centres of CeO_2 with substrates (luminol and hydrogen peroxide). This effect is apparently the main reason for the decrease in the peroxidase-like activity of CeO_2 with a decrease in particle size from 5.1 to 2.8 nm (Figure 10).

It should be noted that the enzyme-like activity of cerium dioxide with a particle size of 5.1 nm is virtually the same in both the phosphate-buffered solution and the Tris-HCl solution (Figure 10). Conversely, the peroxidase-like activity of CeO_2 with a particle size of 2.5–3.9 nm in the Tris-HCl solution noticeably exceeds the activity of cerium dioxide in the phosphate-buffered solution. Most probably, this difference can be attributed to the different chemical composition of the Tris-HCl solution and the phosphate-buffered solution. McCormack et al. found that the surface of cerium dioxide particles acquires a negative charge due to the interaction with phosphate species [84]. The surface charge significantly affects the rate of reactions involving charged substrates [34,85]. In the process of free-radical oxidation of luminol, negatively charged intermediates (luminol hydroperoxide anion) can also be formed [58,86]. It is likely that in the phosphate-buffered solution, the interaction of cerium dioxide with negatively charged luminol radical is hindered, resulting in reduced peroxidase-like activity (Figure 10). In addition, the chemiluminescence intensity of luminol depends on the chemical composition of the buffer solution and increases in the presence of halide ions, including Cl^- [87].

However, in the Tris-HCl solution, the enzyme-like activity of CeO_2 towards hydrogen peroxide naturally decreases with increasing particle size (Figure 10). Similarly, Baldim et al. showed that, in a buffer solution of Tris-HCl (pH 7.5), the superoxide dismutase-like activity of colloidal solutions of CeO_2 decreased by 5 times with an increase in the particle size of cerium dioxide from 5 to 23 nm [76]. Lee et al. found that the antioxidant activity of CeO_2 stabilised with polyacrylic acid and octylamine also depends on the particle size and decreases by 6 times with the increase in particle size from 4 to 8 nm [88]. In both cases, the decrease in the enzyme-like activity of CeO_2 was attributed to a decrease in the ratio of Ce^{3+} ions; however, the actual oxidation state of cerium in nanocrystalline CeO_2 is currently the subject of extensive debates [89]. Recently, the presence of trivalent cerium in nanoscale CeO_2 has been questioned [81,89].

Since peroxidases and nanozymes that mimic their activity use hydrogen peroxide to oxidise the substrate (including luminol) [90], the sorption of hydrogen peroxide on the CeO_2 surface can be considered the most important stage of the catalytic reaction. Using computational methods, Wang et al. proposed the following mechanism for the peroxidase-like activity of cerium dioxide: adsorption of hydrogen peroxide, dissociation of H_2O_2 with the formation of hydroxyl radicals, and $\cdot OH$ reduction [91]. This mechanism is fully consistent with the abovementioned reactions (2)–(4) used to describe the kinetics of H_2O_2-induced luminol oxidation in the presence of cerium dioxide in a Tris-HCl buffer solution. Therefore, with an increase in the specific surface of CeO_2 (Table 2) in the absence of phosphate ions, the sorption of hydrogen peroxide increases, and the process of luminol oxidation proceeds more intensively (see Table 4, constant k_4), which was observed in the Tris-HCl solution (Figure 10b).

Thus, the data obtained in the present study indicate that the determination of the enzyme-like activity of CeO_2 is a complex task that requires a detailed analysis of a number of factors. The peroxidase-like activity of cerium dioxide is affected by the pH of the medium [37] and the composition of the particle surface [88]. The reactivity of the luminol chemiluminescent probe directly depends on the pH of the buffer solution and, presumably, can be enhanced by the components of the analysed mixture containing halide species (including the buffer). It is also necessary to take into account the presence of species in the reaction mixture that can specifically interact with the enzyme-like material and significantly change its activity. The correct choice of the chemical environment is fundamentally important for obtaining reliable data on the enzyme-like activity of CeO_2, including its activity under conditions close to the intracellular environment. An analysis of the interaction of cerium dioxide with reactive oxygen species requires further thorough investigation, and this work gives impetus to the development of appropriate approaches.

3. Materials and Methods

3.1. Materials

Ammonium cerium(IV) nitrate (99.9%, Lanhit (Moscow, Russia)), isopropyl alcohol (99.9%, Chimmed (Moscow, Russia)), hydrogen peroxide (99.9%, Chimmed), luminol (Sigma-Aldrich, St. Louis, MO, USA, 123072), potassium dihydrogen phosphate (Sigma-Aldrich, P0662), potassium hydrogen phosphate (Sigma-Aldrich, P5655), tris-hydrochloride (Merck, Lowe, NJ, USA, 10812846001), and Milli-Ω Water (18.2 MΩ/cm).

3.2. Synthesis of CeO_2 Sols

Four samples of nanoceria were prepared via thermohydrolysis of ammonium cerium(IV) nitrate without using additives. $(NH_4)_2[Ce(NO_3)_6]$ solutions with different concentrations (0.092, 0.185, 0.277, 0.370 and 0.554 M) were placed in a 100 mL Synthware^{TM} autoclave (filling degree of 25%) and heated at 95 °C for 24 h. The resulting yellow precipitates were separated by centrifugation (relative centrifugal force of 20,000× g), washed three or four times with isopropyl alcohol, re-dispersed in an appropriate amount of distilled water, and boiled for 3 h to remove the remaining isopropanol. The concentration of CeO_2 in the resulting sols was determined gravimetrically. Before the analysis of enzyme-like activity, the resulting sols were diluted to the same concentration, $c(CeO_2) = 0.05$ M.

3.3. Characterisation Methods

The X-ray powder diffraction pattern analysis (XRD) of the sols (dried at 50 °C) was performed using a Bruker (Billerica, MA, USA) D8 Advance diffractometer (CuKα radiation) in the angle range of 20–80° 2θ, with a step of 0.02° 2θ and a signal acquisition time of 0.4 s per step. The full-profile analysis of diffraction patterns was performed using the Bruker (Billerica, MA, USA) TOPAS v.4.2 software, and the diffraction maxima were fitted to Voigt pseudo-functions.

CeO_2 nanoparticles were investigated using a JEOL JEM 2100 UHR (Akisima, Tokyo, Japan) transmission electron microscope (TEM) at an acceleration voltage of 200 kV. A drop of the aqueous sol was placed on the formvar/carbon Cu grid (Ted Pella Inc., Redding, CA, USA) and left to evaporate. The samples were treated in an HPT-100 plasma cleaner (Henniker Plasma, Runcorn, UK) prior to being inserted into the microscope chamber in order to remove organic residues from the grid surface. The acquisition of micrographs was performed using a Quemesa 11 MegaPixel CCD (Olympus, Shinjuku/Tokyo, Japan) camera in the bright field mode.

The dynamic light scattering (DLS) and ζ-potential measurements were carried out using a Photocor Compact-Z analyser (Photocor, Moscow, Russia) equipped with a 636.65 nm laser. The sample preparation was carried out using Milli-Q Water (18.2 MΩ/cm), and a temperature equilibrium was ensured between the sample cell and the cuvette holder. The correlation function for each of the samples was gained by averaging 10 curves, each being acquired for 20 s. Then, the data were filtered by adjusting the permissible deviation of the scattering intensity from the average value (no more than 10%), taking into account the shift of the baseline.

The optical absorption spectra were recorded using quartz cuvettes (10.0 mm optical path length) in a 200–600 nm range, at 0.1 nm steps, on an SF-2000 spectrophotometer (OKB Spectr, Saint-Petersburg, Russia) with a deuterium-halogen light source.

The IR spectra of the samples were recorded in attenuated total reflection geometry using a Spectrum 65 FT IR spectrometer (Perkin Elmer, Waltham, MA, USA) with a spectral resolution of 2 cm^{-1}, in the wavenumber range of 400–4000 cm^{-1}.

The Raman spectra were recorded using a Confotec NR500 spectrometer (SOL Instruments, Minsk, Belarus) with a 514 nm laser, using a ×40 (NA = 0.75) lens at ~30 mW laser power. The spectral resolution was 0.8 cm^{-1}.

The enzyme-like activity (peroxidase/catalase-like) of the CeO_2 sols was investigated using the chemiluminescent method with the reaction of luminol oxidation in the presence of hydrogen peroxide in a phosphate (100 mM, pH of 7.4, K_2HPO_4) or a Tris-HCl (100 mM, pH of 7.4) buffer solution, at 36 °C. The background luminescence was recorded for 60 s after mixing the solutions of hydrogen peroxide (10 mM) and luminol (50 µM). Then, an aliquot (5 µL) of the CeO_2 sol was added (250 µM), and a chemiluminescent signal was recorded for 10 min. Light sums were calculated via numerical integration of the chemiluminescence curves using the PowerGraph software v.3.3 (Moscow, Russia).

4. Conclusions

Thermohydrolysis of solutions of ceric ammonium nitrate with concentrations in the range of 0.092–0.370 M produced ceria colloidal solutions with different particle sizes in the range of 2–5 nm. All the obtained sols showed good colloid stability over three months of storage under ambient conditions. For the quantitative analysis of the enzyme-like activity of the CeO_2 sols with respect to hydrogen peroxide, a chemiluminescent method was chosen to determine reactive oxygen species based on the interaction of the probe (luminol) with hydroxyl radicals. To address the effect of chemical environment on the

CeO$_2$ nanozyme property, the analysis was conducted using two different buffers (Tris-HCl and PBS) at an identical pH (7.4). The peroxidase-like activity of cerium dioxide in the Tris-HCl buffer solution decreased with an increase in particle size. The opposite dependence was registered in the phosphate-buffered solution, while the rate of luminol oxidation in the phosphate-rich medium was significantly lower than in the Tris-HCl medium. According to the kinetic modelling, in the phosphate-rich medium, the rate of CeO$_2$ interaction with H$_2$O$_2$ was more than three orders of magnitude lower than in the Tris-HCl medium. The mechanisms of the enzyme-like activity of CeO$_2$ nanoparticles in different media are discussed.

Supplementary Materials: The following supporting information can be downloaded at https://www.mdpi.com/article/10.3390/molecules28093811/s1. Figure S1: TEM image of CeO$_2$ nanoparticles in the ceria sol obtained from 0.185 M aqueous solution of (NH$_4$)$_2$[Ce(NO$_3$)$_6$]; Figure S2: ζ-potential values of the CeO$_2$ aqueous sols prepared from (NH$_4$)$_2$[Ce(NO$_3$)$_6$] solutions after three months of storage under ambient conditions.

Author Contributions: Conceptualisation, A.E.B. and V.K.I.; methodology, A.D.F., M.M.S. and A.E.B.; validation, A.D.F. and M.M.S.; formal analysis, A.D.F.; investigation, A.D.F., S.Y.K. and K.A.C.; resources, A.E.B. and V.K.I.; data curation, A.D.F.; writing—original draft preparation, A.D.F.; writing—review and editing, A.E.B. and V.K.I.; visualisation, A.D.F. and S.Y.K.; supervision, V.K.I.; project administration, V.K.I.; funding acquisition, V.K.I. All authors have read and agreed to the published version of the manuscript.

Funding: This research was funded by the Russian Science Foundation, grant number 19-13-00416.

Institutional Review Board Statement: Not applicable.

Informed Consent Statement: Not applicable.

Data Availability Statement: Data are contained within this article.

Acknowledgments: The analysis of the composition, structure and properties of the obtained materials was carried out using the equipment of the JRC PMR IGIC RAS.

Conflicts of Interest: The authors declare no conflict of interest.

Sample Availability: Not applicable.

References

1. Montini, T.; Melchionna, M.; Monai, M.; Fornasiero, P. Fundamentals and Catalytic Applications of CeO$_2$-Based Materials. *Chem. Rev.* **2016**, *116*, 5987–6041. [CrossRef]
2. Rahman, M.M.; Zahir, M.H.; Helal, A.; Suleiman, R.K.; Haq, B.; Kumar, A.M. UV-Protected Polyurethane/f-Oil Fly Ash-CeO$_2$ Coating: Effect of Pre-Mixing f-Oil Fly Ash-CeO$_2$ with Monomers. *Polymers* **2021**, *13*, 3232. [CrossRef] [PubMed]
3. Wang, W.; Zhang, B.; Jiang, S.; Bai, H.; Zhang, S. Use of CeO$_2$ Nanoparticles to Enhance UV-Shielding of Transparent Regenerated Cellulose Films. *Polymers* **2019**, *11*, 458. [CrossRef] [PubMed]
4. Aklalouch, M.; Calleja, A.; Granados, X.; Ricart, S.; Boffa, V.; Ricci, F.; Puig, T.; Obradors, X. Hybrid sol–gel layers containing CeO$_2$ nanoparticles as UV-protection of plastic lenses for concentrated photovoltaics. *Sol. Energy Mater. Sol. Cells* **2014**, *120*, 175–182. [CrossRef]
5. Oosthuizen, D.N.; Motaung, D.E.; Swart, H.C. Gas sensors based on CeO$_2$ nanoparticles prepared by chemical precipitation method and their temperature-dependent selectivity towards H$_2$S and NO$_2$ gases. *Appl. Surf. Sci.* **2020**, *505*, 144356. [CrossRef]
6. Li, P.; Wang, B.; Qin, C.; Han, C.; Sun, L.; Wang, Y. Band-gap-tunable CeO$_2$ nanoparticles for room-temperature NH$_3$ gas sensors. *Ceram. Int.* **2020**, *46*, 19232–19240. [CrossRef]
7. Mokrushin, A.S.; Simonenko, E.P.; Simonenko, N.P.; Bukunov, K.A.; Sevastyanov, V.G.; Kuznetsov, N.T. Gas-sensing properties of nanostructured CeO$_2$-xZrO$_2$ thin films obtained by the sol-gel method. *J. Alloys Compd.* **2019**, *773*, 1023–1032. [CrossRef]
8. Wu, H.; Sun, Q.; Chen, J.; Wang, G.-Y.; Wang, D.; Zeng, X.-F.; Wang, J.-X. Citric acid-assisted ultrasmall CeO$_2$ nanoparticles for efficient photocatalytic degradation of glyphosate. *Chem. Eng. J.* **2021**, *425*, 130640. [CrossRef]
9. Torrente-Murciano, L.; Gilbank, A.; Puertolas, B.; Garcia, T.; Solsona, B.; Chadwick, D. Shape-dependency activity of nanostructured CeO$_2$ in the total oxidation of polycyclic aromatic hydrocarbons. *Appl. Catal. B Environ.* **2013**, *132*, 116–122. [CrossRef]
10. Lin, S.; Li, Z.; Li, M. Tailoring metal-support interactions via tuning CeO$_2$ particle size for enhancing CO$_2$ methanation activity over Ni/CeO$_2$ catalysts. *Fuel* **2023**, *333*, 126369. [CrossRef]
11. Adachi, G.; Imanaka, N.; Kang, Z.C. (Eds.) *Binary Rare Earth Oxides*; Springer: Dordrecht, The Netherlands, 2004; ISBN 9781402025686.

12. Zhuo, M.; Ma, J.; Quan, X. Cytotoxicity of functionalized CeO_2 nanoparticles towards Escherichia coli and adaptive response of membrane properties. *Chemosphere* **2021**, *281*, 130865. [CrossRef]
13. Leung, Y.H.; Yung, M.M.N.; Ng, A.M.C.; Ma, A.P.Y.; Wong, S.W.Y.; Chan, C.M.N.; Ng, Y.H.; Djurišić, A.B.; Guo, M.; Wong, M.T.; et al. Toxicity of CeO_2 nanoparticles—The effect of nanoparticle properties. *J. Photochem. Photobiol. B Biol.* **2015**, *145*, 48–59. [CrossRef] [PubMed]
14. Estes, L.M.; Singha, P.; Singh, S.; Sakthivel, T.S.; Garren, M.; Devine, R.; Brisbois, E.J.; Seal, S.; Handa, H. Characterization of a nitric oxide (NO) donor molecule and cerium oxide nanoparticle (CNP) interactions and their synergistic antimicrobial potential for biomedical applications. *J. Colloid Interface Sci.* **2021**, *586*, 163–177. [CrossRef]
15. Derevianko, S.; Vasylchenko, A.; Kaplunenko, V.; Kharchuk, M.; Demchenko, O.; Spivak, M. Antiviral Properties of Cerium Nanoparticles. *Acta Univ. Agric. Silvic. Mendel. Brun.* **2022**, *70*, 187–204. [CrossRef]
16. Nefedova, A.; Rausalu, K.; Zusinaite, E.; Vanetsev, A.; Rosenberg, M.; Koppel, K.; Lilla, S.; Visnapuu, M.; Smits, K.; Kisand, V.; et al. Antiviral efficacy of cerium oxide nanoparticles. *Sci. Rep.* **2022**, *12*, 18746. [CrossRef]
17. Ji, Z.; Wang, X.; Zhang, H.; Lin, S.; Meng, H.; Sun, B.; George, S.; Xia, T.; Nel, A.E.; Zink, J.I. Designed Synthesis of CeO_2 Nanorods and Nanowires for Studying Toxicological Effects of High Aspect Ratio Nanomaterials. *ACS Nano* **2012**, *6*, 5366–5380. [CrossRef]
18. Wang, L.; Ai, W.; Zhai, Y.; Li, H.; Zhou, K.; Chen, H. Effects of Nano-CeO_2 with Different Nanocrystal Morphologies on Cytotoxicity in HepG2 Cells. *Int. J. Environ. Res. Public Health* **2015**, *12*, 10806–10819. [CrossRef]
19. Lazić, V.; Živković, L.S.; Sredojević, D.; Fernandes, M.M.; Lanceros-Mendez, S.; Ahrenkiel, S.P.; Nedeljković, J.M. Tuning Properties of Cerium Dioxide Nanoparticles by Surface Modification with Catecholate-type of Ligands. *Langmuir* **2020**, *36*, 9738–9746. [CrossRef] [PubMed]
20. Ribeiro, F.M.; de Oliveira, M.M.; Singh, S.; Sakthivel, T.S.; Neal, C.J.; Seal, S.; Ueda-Nakamura, T.; de Lautenschlager, S.O.S.; Nakamura, C.V. Ceria Nanoparticles Decrease UVA-Induced Fibroblast Death through Cell Redox Regulation Leading to Cell Survival, Migration and Proliferation. *Front. Bioeng. Biotechnol.* **2020**, *8*, 577557. [CrossRef]
21. Wang, C.; Blough, E.; Dai, X.; Olajide, O.; Driscoll, H.; Leidy, J.W.; July, M.; Triest, W.E.; Wu, M. Protective Effects of Cerium Oxide Nanoparticles on MC_3T_3-E_1 Osteoblastic Cells Exposed to X-Ray Irradiation. *Cell. Physiol. Biochem.* **2016**, *38*, 1510–1519. [CrossRef]
22. Zhu, W.; Wang, L.; Li, Q.; Jiao, L.; Yu, X.; Gao, X.; Qiu, H.; Zhang, Z.; Bing, W. Will the Bacteria Survive in the CeO_2 Nanozyme-H_2O_2 System? *Molecules* **2021**, *26*, 3747. [CrossRef] [PubMed]
23. Shlapa, Y.; Solopan, S.; Sarnatskaya, V.; Siposova, K.; Garcarova, I.; Veltruska, K.; Timashkov, I.; Lykhova, O.; Kolesnik, D.; Musatov, A.; et al. Cerium dioxide nanoparticles synthesized via precipitation at constant pH: Synthesis, physical-chemical and antioxidant properties. *Colloids Surf. B Biointerfaces* **2022**, *220*, 112960. [CrossRef] [PubMed]
24. Shlapa, Y.; Timashkov, I.; Veltruska, K.; Siposova, K.; Garcarova, I.; Musatov, A.; Solopan, S.; Kubovcikova, M.; Belous, A. Structural and physical-chemical characterization of redox active CeO_2 nanoparticles synthesized by precipitation in water-alcohol solutions. *Nanotechnology* **2021**, *32*, 315706. [CrossRef]
25. Gupta, A.; Das, S.; Neal, C.J.; Seal, S. Controlling the surface chemistry of cerium oxide nanoparticles for biological applications. *J. Mater. Chem. B* **2016**, *4*, 3195–3202. [CrossRef] [PubMed]
26. Sozarukova, M.M.; Proskurnina, E.V.; Popov, A.L.; Kalinkin, A.L.; Ivanov, V.K. New facets of nanozyme activity of ceria: Lipo- and phospholipoperoxidase-like behaviour of CeO_2 nanoparticles. *RSC Adv.* **2021**, *11*, 35351–35360. [CrossRef]
27. Wu, X.; Wei, J.; Wu, C.; Lv, G.; Wu, L. ZrO_2/CeO_2/polyacrylic acid nanocomposites with alkaline phosphatase-like activity for sensing. *Spectrochim. Acta Part A Mol. Biomol. Spectrosc.* **2021**, *263*, 120165. [CrossRef]
28. Henych, J.; Šťastný, M.; Ederer, J.; Němečková, Z.; Pogorzelska, A.; Tolasz, J.; Kormunda, M.; Ryšánek, P.; Bażanów, B.; Stygar, D.; et al. How the surface chemical properties of nanoceria are related to its enzyme-like, antiviral and degradation activity. *Environ. Sci. Nano* **2022**, *9*, 3485–3501. [CrossRef]
29. Tian, Z.; Yao, T.; Qu, C.; Zhang, S.; Li, X.; Qu, Y. Photolyase-Like Catalytic Behavior of CeO_2. *Nano Lett.* **2019**, *19*, 8270–8277. [CrossRef]
30. Lang, J.; Ma, X.; Chen, P.; Serota, M.D.; Andre, N.M.; Whittaker, G.R.; Yang, R. Haloperoxidase-mimicking CeO_{2-x} nanorods for the deactivation of human coronavirus OC_{43}. *Nanoscale* **2022**, *14*, 3731–3737. [CrossRef]
31. Korschelt, K.; Schwidetzky, R.; Pfitzner, F.; Strugatchi, J.; Schilling, C.; von der Au, M.; Kirchhoff, K.; Panthöfer, M.; Lieberwirth, I.; Tahir, M.N.; et al. CeO_{2-x} nanorods with intrinsic urease-like activity. *Nanoscale* **2018**, *10*, 13074–13082. [CrossRef]
32. Ransy, C.; Vaz, C.; Lombès, A.; Bouillaud, F. Use of H_2O_2 to Cause Oxidative Stress, the Catalase Issue. *Int. J. Mol. Sci.* **2020**, *21*, 9149. [CrossRef]
33. Marquez, L.A.; Dunford, H.B. Mechanism of the Oxidation of 3,5,3′,5′-Tetramethylbenzidine by Myeloperoxidase Determined by Transient- and Steady-State Kinetics. *Biochemistry* **1997**, *36*, 9349–9355. [CrossRef] [PubMed]
34. Pütz, E.; Smales, G.J.; Jegel, O.; Emmerling, F.; Tremel, W. Tuning ceria catalysts in aqueous media at the nanoscale: How do surface charge and surface defects determine peroxidase- and haloperoxidase-like reactivity. *Nanoscale* **2022**, *14*, 13639–13650. [CrossRef] [PubMed]
35. Allen, R.C. Haloperoxidase-Catalyzed Luminol Luminescence. *Antioxidants* **2022**, *11*, 518. [CrossRef]
36. Xue, Y.; Zhai, Y.; Zhou, K.; Wang, L.; Tan, H.; Luan, Q.; Yao, X. The Vital Role of Buffer Anions in the Antioxidant Activity of CeO_2 Nanoparticles. *Chem. A Eur. J.* **2012**, *18*, 11115–11122. [CrossRef] [PubMed]

37. Ma, H.; Liu, Z.; Koshy, P.; Sorrell, C.C.; Hart, J.N. Density Functional Theory Investigation of the Biocatalytic Mechanisms of pH-Driven Biomimetic Behavior in CeO_2. *ACS Appl. Mater. Interfaces* **2022**, *14*, 11937–11949. [CrossRef]
38. Singh, S.; Dosani, T.; Karakoti, A.S.; Kumar, A.; Seal, S.; Self, W.T. A phosphate-dependent shift in redox state of cerium oxide nanoparticles and its effects on catalytic properties. *Biomaterials* **2011**, *32*, 6745–6753. [CrossRef]
39. Wang, X.; Lopez, A.; Liu, J. Adsorption of Phosphate and Polyphosphate on Nanoceria Probed by DNA Oligonucleotides. *Langmuir* **2018**, *34*, 7899–7905. [CrossRef]
40. Zhao, Y.; Li, H.; Lopez, A.; Su, H.; Liu, J. Promotion and Inhibition of the Oxidase-Mimicking Activity of Nanoceria by Phosphate, Polyphosphate, and DNA. *ChemBioChem* **2020**, *21*, 2178–2186. [CrossRef]
41. Molinari, M.; Symington, A.R.; Sayle, D.C.; Sakthivel, T.S.; Seal, S.; Parker, S.C. Computer-Aided Design of Nanoceria Structures as Enzyme Mimetic Agents: The Role of Bodily Electrolytes on Maximizing Their Activity. *ACS Appl. Bio Mater.* **2019**, *2*, 1098–1106. [CrossRef]
42. Shcherbakov, A.B.; Teplonogova, M.A.; Ivanova, O.S.; Shekunova, T.O.; Ivonin, I.V.; Baranchikov, A.Y.; Ivanov, V.K. Facile method for fabrication of surfactant-free concentrated CeO_2 sols. *Mater. Res. Express* **2017**, *4*, 055008. [CrossRef]
43. Lin, M.; Fu, Z.Y.; Tan, H.R.; Tan, J.P.Y.; Ng, S.C.; Teo, E. Hydrothermal Synthesis of CeO_2 Nanocrystals: Ostwald Ripening or Oriented Attachment? *Cryst. Growth Des.* **2012**, *12*, 3296–3303. [CrossRef]
44. Ivanov, V.K.K.; Fedorov, P.P.P.; Baranchikov, A.Y.Y.; Osiko, V.V.V. Oriented attachment of particles: 100 years of investigations of non-classical crystal growth. *Russ. Chem. Rev.* **2014**, *83*, 1204–1222. [CrossRef]
45. Tret'yakov, Y.D.; Baranchikov, A.E.; Kopitsa, G.P.; Ivanov, V.K.; Polezhaeva, O.S. Oxygen nonstoichiometry of nanocrystalline ceria. *Russ. J. Inorg. Chem.* **2010**, *55*, 325–327. [CrossRef]
46. Grabchenko, M.; Mikheeva, N.; Mamontov, G.; Salaev, M.; Liotta, L.; Vodyankina, O. Ag/CeO_2 Composites for Catalytic Abatement of CO, Soot and VOCs. *Catalysts* **2018**, *8*, 285. [CrossRef]
47. Liu, Y.; Yang, Z.; Zhang, X.; He, Y.; Feng, J.; Li, D. Shape/Crystal Facet of Ceria Induced Well-Dispersed and Stable Au Nanoparticles for the Selective Hydrogenation of Phenylacetylene. *Catal. Lett.* **2019**, *149*, 361–372. [CrossRef]
48. Sreeremya, T.S.; Krishnan, A.; Remani, K.C.; Patil, K.R.; Brougham, D.F.; Ghosh, S. Shape-Selective Oriented Cerium Oxide Nanocrystals Permit Assessment of the Effect of the Exposed Facets on Catalytic Activity and Oxygen Storage Capacity. *ACS Appl. Mater. Interfaces* **2015**, *7*, 8545–8555. [CrossRef]
49. Nakamoto, K. *Infrared and Raman Spectra of Inorganic and Coordination Compounds*; Wiley: New York, NY, USA, 1977; ISBN 0471629790.
50. Prieur, D.; Bonani, W.; Popa, K.; Walter, O.; Kriegsman, K.W.; Engelhard, M.H.; Guo, X.; Eloirdi, R.; Gouder, T.; Beck, A.; et al. Size Dependence of Lattice Parameter and Electronic Structure in CeO_2 Nanoparticles. *Inorg. Chem.* **2020**, *59*, 5760–5767. [CrossRef]
51. Deus, R.C.; Cilense, M.; Foschini, C.R.; Ramirez, M.A.; Longo, E.; Simões, A.Z. Influence of mineralizer agents on the growth of crystalline CeO_2 nanospheres by the microwave-hydrothermal method. *J. Alloys Compd.* **2013**, *550*, 245–251. [CrossRef]
52. Bellamy, L.J. *The Infra-Red Spectra of Complex Molecules*; Springer Netherlands: Dordrecht, The Netherlands, 1975; ISBN 9789401160193
53. Little, L.H. *Infrared Spectra of Adsorbed Species*; Academic Press: Cambridge, MA, USA, 1966.
54. Kang, D.; Yu, X.; Ge, M. Morphology-dependent properties and adsorption performance of CeO_2 for fluoride removal. *Chem. Eng. J.* **2017**, *330*, 36–43. [CrossRef]
55. Chen, C.; Zhan, Y.; Zhou, J.; Li, D.; Zhang, Y.; Lin, X.; Jiang, L.; Zheng, Q. Cu/CeO_2 Catalyst for Water-Gas Shift Reaction: Effect of CeO_2 Pretreatment. *ChemPhysChem* **2018**, *19*, 1448–1455. [CrossRef] [PubMed]
56. Schilling, C.; Hofmann, A.; Hess, C.; Ganduglia-Pirovano, M.V. Raman Spectra of Polycrystalline CeO_2: A Density Functional Theory Study. *J. Phys. Chem. C* **2017**, *121*, 20834–20849. [CrossRef]
57. Agarwal, R.G.; Kim, H.-J.; Mayer, J.M. Nanoparticle O–H Bond Dissociation Free Energies from Equilibrium Measurements of Cerium Oxide Colloids. *J. Am. Chem. Soc.* **2021**, *143*, 2896–2907. [CrossRef] [PubMed]
58. Vladimirov, Y.A.; Proskurnina, E.V. Free radicals and cell chemiluminescence. *Biochemistry* **2009**, *74*, 1545–1566. [CrossRef]
59. Li, C.; Shi, X.; Shen, Q.; Guo, C.; Hou, Z.; Zhang, J. Hot Topics and Challenges of Regenerative Nanoceria in Application of Antioxidant Therapy. *J. Nanomater.* **2018**, *2018*, 1–12. [CrossRef]
60. Giussani, A.; Farahani, P.; Martínez-Muñoz, D.; Lundberg, M.; Lindh, R.; Roca-Sanjuán, D. Molecular Basis of the Chemiluminescence Mechanism of Luminol. *Chem. Eur. J.* **2019**, *25*, 5202–5213. [CrossRef] [PubMed]
61. Heindl, D.; Josel, H.-P. Chemiluminescent Detection with Horseradish Peroxidase and Luminol. In *Nonradioactive Analysis of Biomolecules*; Springer: Berlin/Heidelberg, Germany, 2000; pp. 258–261.
62. Deng, M.; Xu, S.; Chen, F. Enhanced chemiluminescence of the luminol-hydrogen peroxide system by BSA-stabilized Au nanoclusters as a peroxidase mimic and its application. *Anal. Methods* **2014**, *6*, 3117–3123. [CrossRef]
63. Guan, G.; Yang, L.; Mei, Q.; Zhang, K.; Zhang, Z.; Han, M.-Y. Chemiluminescence Switching on Peroxidase-Like Fe_3O_4 Nanoparticles for Selective Detection and Simultaneous Determination of Various Pesticides. *Anal. Chem.* **2012**, *84*, 9492–9497. [CrossRef] [PubMed]

64. Liu, D.; Ju, C.; Han, C.; Shi, R.; Chen, X.; Duan, D.; Yan, J.; Yan, X. Nanozyme chemiluminescence paper test for rapid and sensitive detection of SARS-CoV-2 antigen. *Biosens. Bioelectron.* **2021**, *173*, 112817. [CrossRef]
65. Zhong, Y.; Tang, X.; Li, J.; Lan, Q.; Min, L.; Ren, C.; Hu, X.; Torrente-Rodríguez, R.M.; Gao, W.; Yang, Z. A nanozyme tag enabled chemiluminescence imaging immunoassay for multiplexed cytokine monitoring. *Chem. Commun.* **2018**, *54*, 13813–13816. [CrossRef]
66. Vlasova, I. Peroxidase Activity of Human Hemoproteins: Keeping the Fire under Control. *Molecules* **2018**, *23*, 2561. [CrossRef]
67. Sozarukova, M.M.; Proskurnina, E.V.; Ivanov, V.K. Prooxidant potential of CeO_2 nanoparticles towards hydrogen peroxide. *Nanosyst. Phys. Chem. Math.* **2021**, *12*, 283–290. [CrossRef]
68. Römer, I.; Briffa, S.M.; Arroyo Rojas Dasilva, Y.; Hapiuk, D.; Trouillet, V.; Palmer, R.E.; Valsami-Jones, E. Impact of particle size, oxidation state and capping agent of different cerium dioxide nanoparticles on the phosphate-induced transformations at different pH and concentration. *PLoS ONE* **2019**, *14*, e0217483. [CrossRef]
69. Vlasova, N.; Markitan, O. Phosphate–nucleotide–nucleic acid: Adsorption onto nanocrystalline ceria surface. *Colloids Surf. A Physicochem. Eng. Asp.* **2022**, *648*, 129214. [CrossRef]
70. Dahle, J.T.; Livi, K.; Arai, Y. Effects of pH and phosphate on CeO_2 nanoparticle dissolution. *Chemosphere* **2015**, *119*, 1365–1371. [CrossRef] [PubMed]
71. Naganuma, T. Tunable phosphate-mediated stability of Ce^{3+} ions in cerium oxide nanoparticles for enhanced switching efficiency of their anti/pro-oxidant activities. *Biomater. Sci.* **2021**, *9*, 1345. [CrossRef]
72. Singh, R.; Singh, S. Role of phosphate on stability and catalase mimetic activity of cerium oxide nanoparticles. *Colloids Surf. B Biointerfaces* **2015**, *132*, 78–84. [CrossRef] [PubMed]
73. Damatov, D.; Mayer, J.M. (Hydro)peroxide ligands on colloidal cerium oxide nanoparticles. *Chem. Commun.* **2016**, *52*, 10281–10284. [CrossRef] [PubMed]
74. Celardo, I.; Pedersen, J.Z.; Traversa, E.; Ghibelli, L. Pharmacological potential of cerium oxide nanoparticles. *Nanoscale* **2011**, *3*, 1411. [CrossRef]
75. Bashir, S.M.; Idriss, H. The reaction of propylene to propylene-oxide on CeO_2: An FTIR spectroscopy and temperature programmed desorption study. *J. Chem. Phys.* **2020**, *152*, 044712. [CrossRef] [PubMed]
76. Baldim, V.; Bedioui, F.; Mignet, N.; Margaill, I.; Berret, J.-F. The enzyme-like catalytic activity of cerium oxide nanoparticles and its dependency on Ce^{3+} surface area concentration. *Nanoscale* **2018**, *10*, 6971–6980. [CrossRef] [PubMed]
77. Kuchibhatla, S.V.N.T.; Karakoti, A.S.; Baer, D.R.; Samudrala, S.; Engelhard, M.H.; Amonette, J.E.; Thevuthasan, S.; Seal, S. Influence of Aging and Environment on Nanoparticle Chemistry: Implication to Confinement Effects in Nanoceria. *J. Phys. Chem. C* **2012**, *116*, 14108–14114. [CrossRef] [PubMed]
78. Neal, C.J.; Sakthivel, T.S.; Fu, Y.; Seal, S. Aging of Nanoscale Cerium Oxide in a Peroxide Environment: Its Influence on the Redox, Surface, and Dispersion Character. *J. Phys. Chem. C* **2021**, *125*, 27323–27334. [CrossRef]
79. Liu, X.; Wu, J.; Liu, Q.; Lin, A.; Li, S.; Zhang, Y.; Wang, Q.; Li, T.; An, X.; Zhou, Z.; et al. Synthesis-temperature-regulated multi-enzyme-mimicking activities of ceria nanozymes. *J. Mater. Chem. B* **2021**, *9*, 7238–7245. [CrossRef]
80. Seminko, V.V.; Maksimchuk, P.O.; Grygorova, G.V.; Hubenko, K.O.; Yefimova, S.L. Features of ROS generation during hydrogen peroxide decomposition by nanoceria at different pH values. *Funct. Mater.* **2021**, *28*, 420. [CrossRef]
81. Ghosalya, M.K.; Li, X.; Beck, A.; van Bokhoven, J.A.; Artiglia, L. Size of Ceria Particles Influences Surface Hydroxylation and Hydroxyl Stability. *J. Phys. Chem. C* **2021**, *125*, 9303–9309. [CrossRef]
82. Plakhova, T.V.; Romanchuk, A.Y.; Butorin, S.M.; Konyukhova, A.D.; Egorov, A.V.; Shiryaev, A.A.; Baranchikov, A.E.; Dorovatovskii, P.V.; Huthwelker, T.; Gerber, E.; et al. Towards the surface hydroxyl species in CeO_2 nanoparticles. *Nanoscale* **2019**, *11*, 18142–18149. [CrossRef]
83. Wu, B.; Lo, I.M.C. Surface Functional Group Engineering of CeO_2 Particles for Enhanced Phosphate Adsorption. *Environ. Sci. Technol.* **2020**, *54*, 4601–4608. [CrossRef]
84. McCormack, R.N.; Mendez, P.; Barkam, S.; Neal, C.J.; Das, S.; Seal, S. Inhibition of Nanoceria's Catalytic Activity due to Ce^{3+} Site-Specific Interaction with Phosphate Ions. *J. Phys. Chem. C* **2014**, *118*, 18992–19006. [CrossRef]
85. Huang, L.; Zhang, W.; Chen, K.; Zhu, W.; Liu, X.; Wang, R.; Zhang, X.; Hu, N.; Suo, Y.; Wang, J. Facet-selective response of trigger molecule to CeO_2 {1 1 0} for up-regulating oxidase-like activity. *Chem. Eng. J.* **2017**, *330*, 746–752. [CrossRef]
86. Zhang, M.; Yuan, R.; Chai, Y.; Wang, C.; Wu, X. Cerium oxide–graphene as the matrix for cholesterol sensor. *Anal. Biochem.* **2013**, *436*, 69–74. [CrossRef] [PubMed]
87. Bause, D.E.; Patterson, H.H. Enhancement of luminol chemiluminescence with halide ions. *Anal. Chem.* **1979**, *51*, 2288–2289. [CrossRef]
88. Lee, S.S.; Song, W.; Cho, M.; Puppala, H.L.; Nguyen, P.; Zhu, H.; Segatori, L.; Colvin, V.L. Antioxidant Properties of Cerium Oxide Nanocrystals as a Function of Nanocrystal Diameter and Surface Coating. *ACS Nano* **2013**, *7*, 9693–9703. [CrossRef] [PubMed]
89. Cafun, J.-D.; Kvashnina, K.O.; Casals, E.; Puntes, V.F.; Glatzel, P. Absence of Ce^{3+} Sites in Chemically Active Colloidal Ceria Nanoparticles. *ACS Nano* **2013**, *7*, 10726–10732. [CrossRef] [PubMed]

90. Ju, X.; Hubalek Kalbacova, M.; Šmíd, B.; Johánek, V.; Janata, M.; Dinhová, T.N.; Bělinová, T.; Mazur, M.; Vorokhta, M.; Strnad, L. Poly(acrylic acid)-mediated synthesis of cerium oxide nanoparticles with variable oxidation states and their effect on regulating the intracellular ROS level. *J. Mater. Chem. B* **2021**, *9*, 7386–7400. [CrossRef] [PubMed]
91. Wang, Z.; Shen, X.; Gao, X. Density Functional Theory Mechanistic Insight into the Peroxidase- and Oxidase-like Activities of Nanoceria. *J. Phys. Chem. C* **2021**, *125*, 23098–23104. [CrossRef]

Disclaimer/Publisher's Note: The statements, opinions and data contained in all publications are solely those of the individual author(s) and contributor(s) and not of MDPI and/or the editor(s). MDPI and/or the editor(s) disclaim responsibility for any injury to people or property resulting from any ideas, methods, instructions or products referred to in the content.

Article

Computational Investigation of Interactions between Carbon Nitride Dots and Doxorubicin

Mattia Bartoli [1,2], Elena Marras [3] and Alberto Tagliaferro [2,3,*]

1. Center for Sustainable Future Technologies, Italian Institute of Technology, Via Livorno 60, 10144 Torino, Italy; mattia.bartoli@polito.it
2. Consorzio Interuniversitario Nazionale per la Scienza e Tecnologia dei Materiali (INSTM), Via G. Giusti 9, 50121 Firenze, Italy
3. Politecnico di Torino, Department of Applied Science and Technology, C.so Duca Degli Abruzzi 24, 10129 Torino, Italy; elena.marras@studenti.polito.it
* Correspondence: alberto.tagliaferro@polito.it; Tel.: +39-0110907347

Abstract: The study of carbon dots is one of the frontiers of materials science due to their great structural and chemical complexity. These issues have slowed down the production of solid models that are able to describe the chemical and physical features of carbon dots. Recently, several studies have started to resolve this challenge by producing the first structural-based interpretation of several kinds of carbon dots, such as graphene and polymeric ones. Furthermore, carbon nitride dot models established their structures as being formed by heptazine and oxidized graphene layers. These advancements allowed us to study their interaction with key bioactive molecules, producing the first computational studies on this matter. In this work, we modelled the structures of carbon nitride dots and their interaction with an anticancer molecule (Doxorubicin) using semi-empirical methods, evaluating both geometrical and energetic parameters.

Keywords: carbon nitride dots; Doxorubicin; modelling

1. Introduction

Since their discovery in 2004 [1], carbon nano dots (CDs) have attracted great interest due to their unique set of properties [2]. CDs are characterized by a high fluorescence emission [3] due to their quantum confinement [4], water solubility [5] and high biocompatibility [6]. Their chemical features have spread their use across several fields with remarkable achievements as fluorescent probes in analytical procedures [7–9] and biomedical treatments [10,11] as active materials for both photocatalysis [12,13] and electrocatalysis [14,15]. The application of CDs has raised several issues related to what is and what is not a carbon dot together with serious concern about the development of a rational classification methodology [2].

The classification is of utmost relevance to properly approaching the field of CDs [16]. The most widespread classification approach is based on structural features, and it encompasses three large families of CDs [17]: (i) graphene quantum dots (GQDs), (ii) carbon nitride carbon dots (CNDs) and (iii) polymeric carbon dots (PCDs).

GQDs represent the first kind of CDs discovered and they present the simplest structures among all CDs. The chemical structure of this family is directly related to small graphene oxide clusters, and the distribution of chemical functionalities is explained simply by the Lerf–Klinowsky model [18]. Accordingly, GQDs contain few layers of oxidized graphene, with carbonyls and carboxylic residues being concentrated in the edges, producing an oxygen-rich shell that accounts for their water solubility. The oxidized graphene layers of GQDs display different morphologies and oxidation degrees [19,20] that provide different chemical platforms for further functionalization [21,22].

PCDs are a wide family of nanomaterials produced by the partial degradation of polymers [23] or through the condensation of organic precursors [24–26]. Their properties are very difficult to correlate with a unique interpretative model and each PCD requires proper characterization.

CNDs are similar to GQDs, but they present a more complex structure in which nitrogen atoms dope the graphene layers [27] or form nitrogen-rich aromatic moieties, as found by Mintz et al. [28]. This represents a very interesting possibility for synthetic chemistry, as it can tune the reactivity and properties [29], particularly in a watery medium, as mentioned by Wiśniewski [30]. As a consequence of their functional tuneability, CNDs can act as a platform for several specific tasks, such as chemotherapy for cancer treatment, by being conjugated with active molecules [31]. Among them, Doxorubicin is one of the most studied, due to its remarkable activity, biological stability and simple chemical structure [32,33]. CNDs were successfully combined with Doxorubicin for the production of conjugates (D-CNDs) for the treatment of several cancers using theragnostic approaches [34–36]. Even if the interest in the rational design of D-CNDs is great, the literature lacks computational studies devoted to quantifying the interactions between Doxorubicin and CNDs. Rashid et al. [37] investigated flutamide CNDs through DFT calculation, limiting the system to a representative portion of small CNDs. Zaboli et al. [38] went more in depth by simulating the interaction of a molecule of Doxorubicin on a single sheet of carbon nitride, taken as a representative model of CNDs. The authors were able to correctly evaluate the $\pi-\pi$ and water interaction with both a single large layer of heptazine and the Doxorubicin-conjugated material. Similarly, Shaki et al. [39] investigated the adsorption of and interaction between several anticancer species with the external and inner surfaces of carbon nanotubes. All of these studies show a common point represented by the charge transfer from the drugs to the aromatic domain-rich carrier and assume that this is the key reason for the efficacy of the systems.

These approaches are powerful, but they require a high calculation power, and they are currently limited to small chemical systems that are not sufficiently complex to properly describe CDs. Alternatively, semi-empirical quantum chemistry has been developed to resolve the computation challenges of complex systems in order to try to overcome the main limitations of the traditional Hartree–Fock approaches, the slow speed and the low accuracy [40]. The semi-empirical methods are based on the time reduction of the calculation by parameterizing or omitting certain terms based on the experimental data sources (i.e., the ionization energies of atoms, dipole moments of molecules) [41]. Accordingly, the semi-empirical quantum chemistry approaches are considerably faster and useful for modelling large molecules. The first semi-empirical quantum chemistry method was the modified neglect of the diatomic overlap (MNDO) [42] that parameterized the one-center two-electron integrals based on the spectroscopic data for the isolated atoms, evaluating other two-electron integrals using the multipole–multipole interactions of the classical electrostatics [43]. The MNDO approach was implemented using a basis set composed of only s and p orbitals even if the d orbitals set was used to describe hypervalent sulfur atoms and transition metals. Nevertheless, MNDO is characterized by several intrinsic limitations, such as the impossibility to describe both hydrogens or to predict the heat of the formation [44]. Dewar and coworkers [45] improved the MNDO approach by developing the Austin Model X (AMX) by modifying the expression of nuclear–nuclear core repulsion, emulating van der Waals interactions. This new approach required a total reparameterization of the dipole moments and ionization potentials but allowed the description of the hydrogen bond network and the heat of the formation without properly estimating the basicity.

All of these approaches were made outdated by the development of parametric method 3 (PM3) [46], which uses a Hamiltonian operator that is very similar to the AM1 Hamiltonian, but the parameterization strategy is one of molecular properties, rather than atomic ones. A consolidating method which can be used to perform the computational modelling of large organic systems with lower computational costs is the PM3. This is a semi-empirical quantum mechanical parameterization method based on the modified neglect of the differential overlap method [47].

The PM3 approach is based on the use of a set of empirical parameters to describe the electronic structure of a molecule [48]. The key feature of the PM3 is that it offers the chance to accurately predict the geometrical conformation of large molecules, including enzymes [49] and polymers [50].

Despite these advantages, the PM3 method does not perform well enough to accurately describe the properties of the transition metal complexes [51] and the effects of electron correlation [52].

In the present work, we report a computational study focused on the evaluation of the geometrical features of D-CNDs and the interaction energies (E_i) between Doxorubicin and CNDs using a PM3-based computational approach. We ran a semi-empirical simulation of four-layer CNDs with a single molecule of Doxorubicin, evaluating different geometrical interactions with the Doxorubicin of the outer layer of the CNDs and evaluating the intercalant agent in order to provide a first solid insight into the non-covalent interactions occurring within a chemotherapy agent.

2. Results
2.1. CNDs Model Structure: Definition and Modelling

The most challenging issue in the modelling of CNDs is to provide a representative species to be studied. The structural composition of CNDs is highly dependent on the synthetic method used for their production. CNDs are generally composed of a core composed of sp^2-hybridized carbons surrounded by a surface layer that contains a mixture of sp^2-/sp^3-hybridized carbon and nitrogen atoms [53]. Regarding pure carbon nitride structures, the nitrogen atoms are incorporated directly into their carbon lattice, creating structural defects and altering the electronic properties of the material. Considering CNDs, the distribution of nitrogen atoms is still a matter of discussion, and several hypotheses have been suggested to explain their several spatial arrangements.

Firstly, those conducting CND structural investigations have hypothesized a pure layered structure formed by functionalized heptazine units [27]. Mintz et al. [28] moved a step forward in the clear definition of the structure of CNDs, suggesting a more complex arrangement of the layers. Based on a detailed physical–chemical investigation, the authors proved that CNDs have a graphitic core with massive functionalization on the edges and less heteroatomic inclusion in the core. As reported by Zhou et al. [54], elucidating the differences between a pure heptazine and a realistic model was crucial in order to properly evaluate the interaction between the CNDs and the protein active site. Nevertheless, the great variability of their functionalization prevented the realization of a general model compound that is able to describe all CNDs species. Accordingly, we tentatively propose a model that considers both the heptazine and functionalized graphene layers, as shown in Figure 1.

The heptazine structure (L_h^n) was assembled following the method used by Zhou et al. [54] to produce CNDs composed of pure heptazine. The graphene layer L_g^n could be modeled using at least four geometrical models, as reported by Mandal et al. [19]. We selected type-4 as it showed the lowest energy band gap. According to the Lerf–Klinowski model of graphene oxide [18], oxygen-containing residues symmetrically tailored the L_g^n layer on the edges with oxygen-based functionalities. The oxygen-based functions of the CND edges are still a matter of debate, but Kirbas et al. [29] have reported how they can be tuned by varying the amounts of urea and organic acids used for their production. We selected, as oxygen-based functionalities, hydroxyl and carboxylic groups to evaluate the effects of the presence of a strong network of hydrogen bonds inside the structure. This condition could promote the dismutation of carbonyl groups under harsh synthetic environments while avoiding the presence of carbonyl residues [23]. The other key point to be defined was related to the molecular weight of L_g^n. We limited the L_g^n size to 81 carbon atoms in order to achieve a four-layer structure with two each of L_h^n and L_g^n, respectively, for an average molecular weight higher than 4000 g/mol but with an expected size of below 3 nm, in accordance with the data reported in the literature regarding the common size of CNDs [55]. The layered structure was optimized in vacuo at a temperature of 303.15 K, and the graphical results are shown in Figure 2.

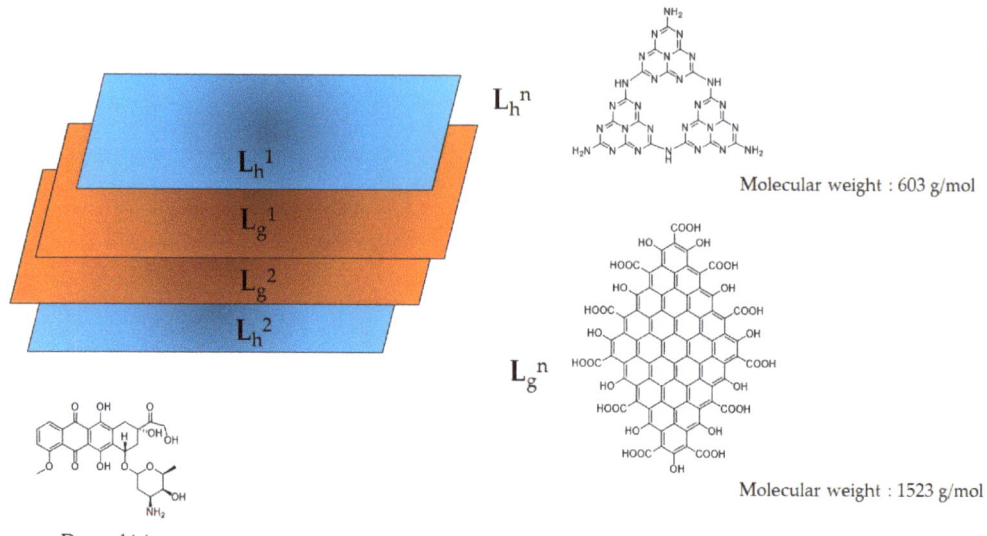

Figure 1. Scheme of representative species used to model Doxorubicin, CNDs and D-CNDs.

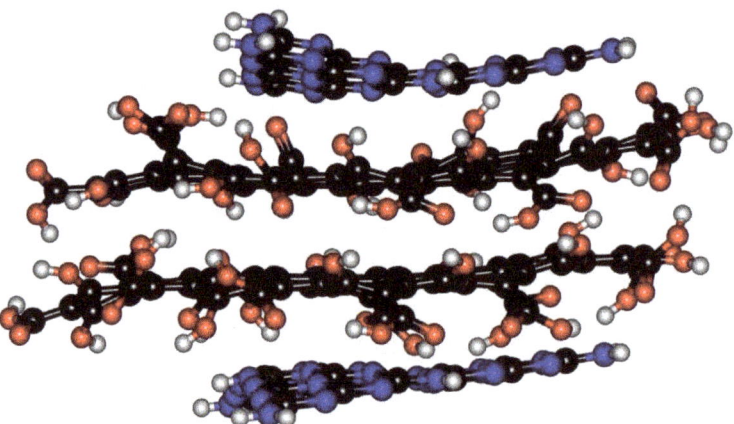

Figure 2. In vacuo optimized CNDs structural model using the PM3-based semi-empirical approach.

The optimized structure was characterized by a free energy of 180.4 kcal/mol and a maximum size of 2.0 × 1.3 nm, with different layer distances related to the intrinsic asymmetry of the model used. The L_g^1–L_g^2 distance was found to be up to 0.41 nm, while the maximum distance between the L_g^n and L_h^n was lower: 0.39–0.34. The L_g^1–L_g^2 interlayer distances were considerably higher than that of the neat graphite structure (0.335 nm) [56], which is in agreement with the massive functionalization that induced this defectiveness. The presence of functional groups on the edge of the L_g^n induced a deformation due to the simultaneous effects of steric hindrance and the attraction/repulsion between the oxygen-based functions [57,58]. The L_h^1 and L_h^2 layer was distorted, with dihedral angles of 1.4° and 7.8° due to the interaction between the nitrogen residues and oxygen-based functionalities. Interestingly, L_h^n was non-centered on the L_g^n, leaving a more oxygen-rich region on one of the CNDs' extremities. This was due to the asymmetry of the L_g^n, which can promote the favoring of the formation of a hydrogen bonds network on the edges, rather than π−π stacking.

As a matter of fact, this simple model was a rough approximation of CNDs, neglecting a real distribution of the oxygen and nitrogen functions between L_g^n and L_h^n, but it was worth considering it due to the simple approach proposed to introduce the graphene-oxidized core into the CNDs. We discuss the limitations and applications of our approach in the dedicated section below.

2.2. Doxo@CNDs Model Structures Modelling

The four-layer model adopted to enable us to describe the CNDs allowed several kinds of interaction with Doxorubicin, forming several D-CNDs. We investigated the structures reported in Figure 3, in which Doxorubicin interacted only with L_h^n (Figure 3a, D-CNDs1), intercalated between L_h^1 and L_g^1 (Figure 3b, D-CNDs2), intercalated between L_g^1 and L_g^2 (Figure 3c, D-CNDs3) and interacted with all L_h^n and L_g^n (Figure 3d, D-CNDs4) values. In Table 1, the main structural and energetic features in are summarized.

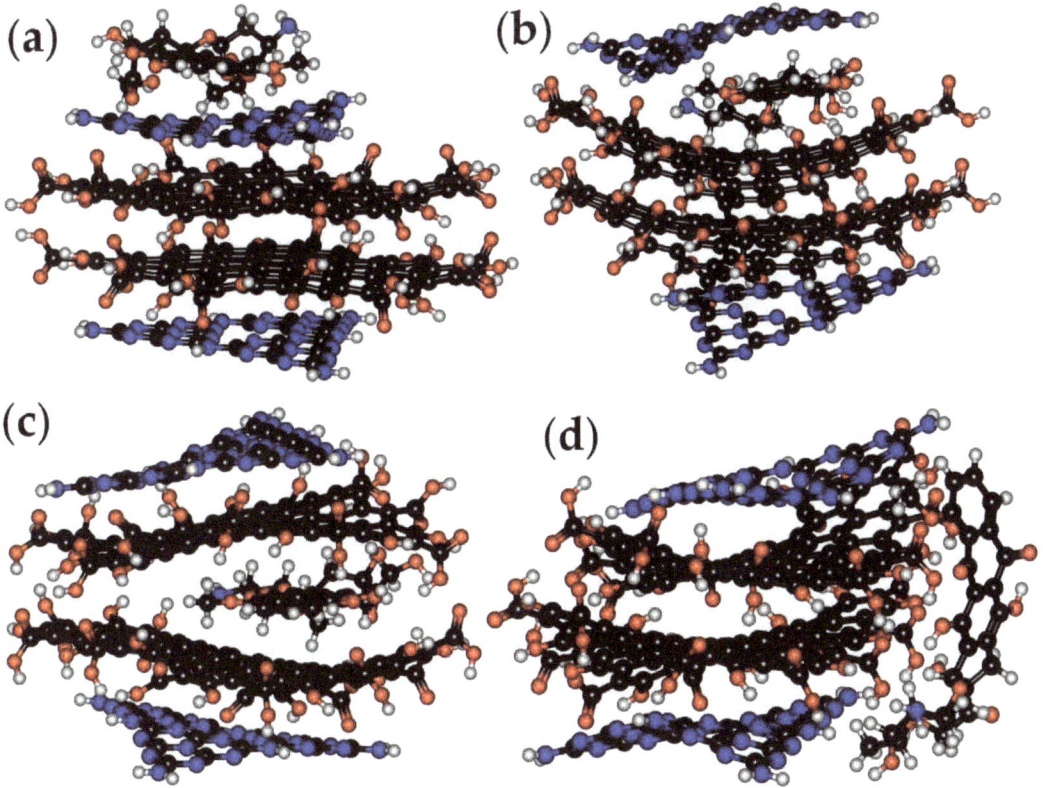

Figure 3. Optimized molecular structure of (**a**) D-CNDs1, (**b**) D-CNDs2, (**c**) D-CNDs3 and (**d**) D-CNDs 3.

Table 1. Summary of main structural features and energetic outcomes of Doxorubicin, CNDs and D-CNDs species.

Specie	Distances (nm)				Angular Strain (°)			Free Energy (kcal/mol)	E_i (kcal/mol)	Δ_{HL} [e] (eV)
	L_h^1–L_g^1	L_g^1–L_g^2	L_g^2–L_h^2	Doxorubicin-Layer	L_h^1	L_h^2	L_g^n			
Doxorubicin	--	--	--	--	--	--	--	128.7	--	0.33
CNDs	0.39	0.41	0.34	--	1.4	7.8	9.1	180.4	--	1.56
D-CNDs1	0.55	0.39	0.67	0.25 [a]	0.3	0.1	15.9	280.6	−28.5	1.38
D-CNDs2	0.70	0.50	0.42	0.40 [a] /0.36 [b]	12.3	66.1	26.1	316.2	7.1	1.71
D-CNDs3	0.35	0.73	0.38	0.36 [b] /0.37 [c]	0.2	0.1	63.2	269.6	−39.5	1.58
D-CNDs4	0.42	0.36	0.34	0.50 [d]	21.4	58.2	42.6	325.6	16.5	1.46

[a] Minimum distance between Doxorubicin and L_h^1. [b] Minimum distance between Doxorubicin and L_g^1. [c] Minimum distance between Doxorubicin and L_g^2. [d] Minimum distance between Doxorubicin and CNDs. [e] Δ_{HL} = energy value of LUMO − energy value of HOMO.

D-CNDs1 showed a consistent increase in terms of L_h^1–L_g^1 distance compared with CNDs up to 0.55 nm with a decrement of the L_h^n strains and significant increase in L_g^n up to 15.9°. This modification occurred due to the strong interaction of Doxorubicin with L_h^1 at a distance of 0.25 nm. As it clearly emerged, the decrement of the L_h^n strain compacted the L_g^n, promoting the interactions through weak forces. When the Doxorubicin intercalated the CNDs layers, greater changes occurred in the CNDs' structure. D-CNDs2 showed an increase in all layers' distance and the strains. Interestingly, Doxorubicin was closer to L_g^1 (0.36 nm) than L_h^1 (0.40 nm), probably due to a better interaction with the graphene layer as a consequence of its greater spatial dimension. Similar distances were observed in D-CNDs3, where the distance between the Doxorubicin and L_g^n layer was the same (0.36–0.37 nm). In this case, we observed the maximum strain of L_g^n, reaching 63.2° due to the formation of a task-like structure in which Doxorubicin found its place. D-CNDs4 proved that the interaction with all the CNDs layers also induced a significant structural modification, with the increase in L_h^1–L_g^1 distance being up to 0.42 nm and the layers being distorted. These computational results open an interesting pathway to evaluating the microscopic analysis of D-CNDs in which structural modification could be correlated with the kind of conjugation that has occurred in the specie. Considering the values of E_i, the more stable structure was D-CNDs3, in which Doxorubicin is fully contained in the CNDs structures and stabilized by the π–π interactions of L_g^n (E_i −39.5 kcal/mol), while the D-CNDs2 and D-CNDs4 were less stable (E_i 7.1 and 16.6 kcal/mol, respectively). D-CNDs1 was also stabilized by the ordering of the piling-up of Doxorubicin on L_h^1 with an E_i −28.5 kcal/mol. These results suggest that the π–π interactions were the main driving force for the interaction between the Doxorubicin and CNDs, rather than the hydrogen bond ones.

Doxorubicin loading to the CND also affected the HOMO–LUMO gap (Δ_{HL}). The calculated Doxorubicin Δ_{HL} was in agreement with the one reported by Lopez-Chavez et al. [59] using DFT approaches. The CNDs showed an Δ_{HL} of up to 1.56 eV, while the presence of Doxorubicin altered the Δ_{HL} by both increasing or decreased it based on its position. D-CNDs1 and D-DCNDs4 (1.38 and 1.46 eV, respectively) showed an Δ_{HL} lower than CNDs, while both D-CNDs2 and D-CNDs3 showed sensible increments of Δ_{HL} (1.71 and 1.58 eV, respectively). As reported by Bharathy et al. [60], a decrease in Δ_{HL} could be correlated to an increase in the reactivity of the systems, and this suggested that Doxorubicin-containing CNDs were less reactive when Doxorubicin was inserted between the layers. Particularly, the insertion between L_h^1 and L_g^1 seemed to be the best configuration among the ones tested due to the interaction with two layers without a massive alteration, as in the case of D-CNDs4. The insertion between L_g^1 and L_g^2 was not so effective, probably due to the exposure of a consistent part of Doxorubicin to the external environment. Even if Δ_{HL} is a powerful tool for discussing the stability of a system, it is not entirely sufficient to describe the kind of interaction occurring in the system. According to Johnson et al. [61], the evaluation of interactions could be properly conducted using only NBO analysis to evaluate the atom–atom contribution to the stabilization. Nonetheless, this approach has a very high computational cost for large systems such as the one that is very representative of those used in the interpretation of CDs. Alternatively, it is possible to evaluate the electron density and use electron maps such those reported in Figure 4.

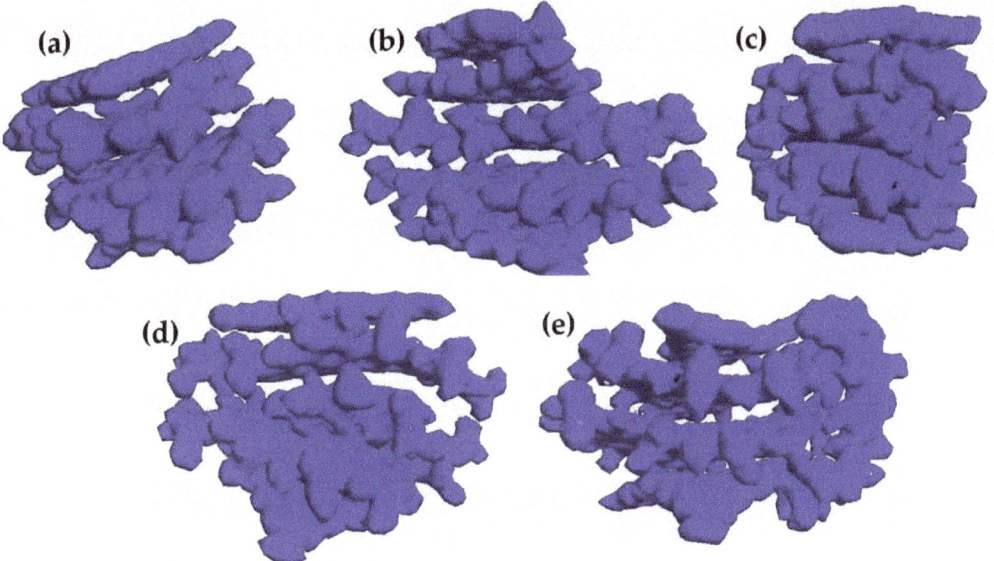

Figure 4. Electron density maps of (**a**) CNDs, (**b**) D-CNDs1, (**c**) D-CNDs2, (**d**) D-CNDs3 and (**e**) D-CNDs 3.

The electron density map showed a rich region between L_g^1 and L_g^2 in CNDs that was reduced in the D-CNDs, with the exception of D-CNDs2. In this case, the insertion of Doxorubicin between L_h^1 and L_g^1 induced the concentration of electron density between the reducing L_g^n layers. The localization of high electron density on Doxorubicin in D-CNDs3 and D-CNDs4 suggested that it can react more easily in these two configurations than in the others.

3. Computational Methods

The Doxorubicin, CND and D-CND structures were drawn using Chem Sketch (ACD Lab, Toronto, CA, USA), and they were modelled using ArgusLab 4.0.1 [62,63] software without considering the discrete solvent medium. The calculations were performed using a personal computer with the Windows 10 operating system (built 19045) equipped with an Intel(R) Core(TM) i5-10210U CPU @ 1.60 GHz, 2112 Mhz, 4 cores, 8 logic gates and 16 GB of RAM installed. Conformational analysis was carried out through geometrical optimization using the PM3 semi-empirical quantum mechanical parameterization of the Hartree–Fock calculation method.

The geometry of Doxorubicin, CND and D-CND structures were considered optimized when the converged set point of 0.001 kcal/(Åmol) was reached using the Polak–Ribiere conjugate gradient algorithm for the optimization process [64] using up to 50,000 cycles. The interaction energies between the Doxorubicin and CNDs were calculated as follows:

$$E_i = E_{D-CNDs} - (E_{CNDs} + E_D) \qquad (1)$$

where E_i is the interaction energy, $E_{D\text{-}CNDs}$ is the total energy of the optimized Doxorubicin–CNDs structure, $E_{D\text{-}CNDs}$ is the total energy of the free optimized CNDs structure and E_D is the total energy of the free optimized Doxorubicin.

The optimized structures were also used for the numerical evaluation of the highest occupied molecular orbital (HOMO) and lower unoccupied molecular orbital (LUMO) energetic values.

We also report the electron density map produced by the calculation ran using Argus-Lab 4.0.1 with a 40-slice grid.

The visualization of the molecules and geometrical parameters of the optimized structures were evaluated using Avogadro 1.2 software [65]. The layer distances are reported as the maximum distance, and the angular strain is defined as the dihedral angle formed by the layers.

4. The Interpretation of CDs Structures and Their Simulations: A Critical Discussion

The use of a computation model for the interpretation of CDs can be a powerful tool, but it should be used with great attention. The first limitation that the computational routes showed is related to the nature of what is simulated. As reported by Zhou et al. [54], the simple synthesis of CNDs led to the production of mixtures of great complexity, even using high-performance purification systems such as dialysis. The authors identified at least five well-characterized species inside the CNDs, with different distributions of functionalities. The authors proposed a data-driven interpretation model in which a layered structure was highly functionalized and variable. Nevertheless, the authors clearly stated that each fraction produced was not composed of a single specie but of a narrow distribution of very similar species. According to this capital finding, the computational approaches to CDs interpretation are intrinsically limited due to the mismatch between the simulated input and real materials. Nonetheless, there is a limited number of cases in which the gap between reality and the simulation is narrow enough to be neglected. GQDs can be investigated using the computation tool together with a complex set of electron microscopy and spectroscopic techniques, allowing for good agreement between the empirical data and computational outputs. On the contrary, PCDs are out of the range of possibility due to their random shapes. CNDs lay in between PCDs and GCDs, showing simple features to be integrated in a simulation such as a heptazine structure, and in more complex ones, distribution functions, as reported in Figure 5.

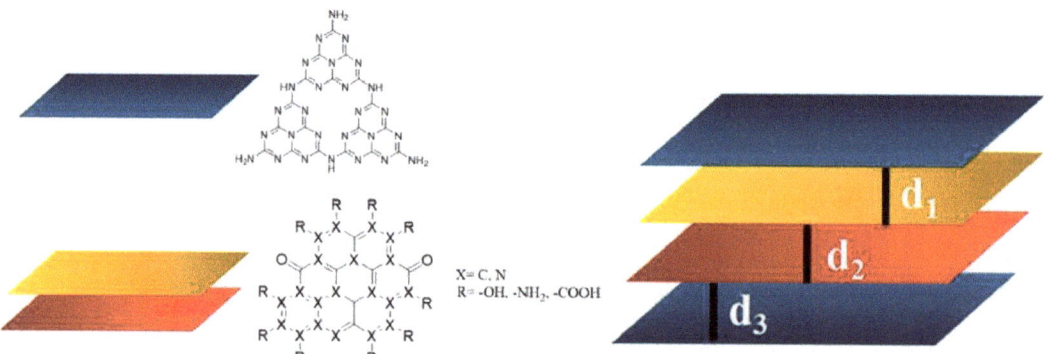

Figure 5. Real data-driven interpretative scheme of CNDs. Reprinted with all permission from Zhou et al. [54].

As clearly emerged, the L_h^n layers have been defined, and their identification can be made using a combination of NMR and mass spectroscopy investigation. The definition of the L_g^n layers is practically impossible and there is no solution to their unique identification. Furthermore, the interaction between L_h^n and L_g^n is only considered to be a weak interaction, but it is impossible to exclude the formation of a chemical bond between them, which could probably form as a consequence of the opening of epoxy functions. Therefore, the trade-off between representability and the reality of CND models should be evaluated considering how the model is able to predict the properties, but also, this approach could be misleading because the distribution of the particles could show the same properties as one of the particles. Nowadays, the computational study of CNDs could provide interesting insight into how these species interact with other molecules, but they are still far from being used as a predictive instrument with which to rationally design the synthesis and applications.

The customizable design of CNDs is of utmost relevance for all of those applications that require the fine tuning of the bot size and chemical functionalities [66].

Gao et al. [67] studied the use of the ab initio approach to CND hybrid materials coupled with CDs while assuming a layered structured for species. The authors showed the relevance of both the size and the shape of the layers in driving both of the relative positions and the alignment of bands. The authors proved that a considerably small enlargement size of the CNDs was able to enhance the visible light response of the species, forming a proper type-II van der Waals heterojunction between the CNDs and CDs [68]. The tuning of the band gap is also crucial for all of those catalytic applications, ranging from electrochemical to photochemical ones [69,70].

Li et al. [71] evaluated the effect of CNDs on the expression of protein kinases for the regulation of the cell signaling pathway. The authors showed the crucial role played by the phosphorylation of CNDs and how this is related to the edge functionalities.

The results discussed showed that the cutting edges applications of CNDs required a proper integration of empirical data with a solid interpretation that only a computational approach could provide. Nevertheless, some key issue remain unsolved together with opportunity as summarized in Table 2.

Table 2. Summary of advantages and unresolved issue related to CNDs modelling for custom-designed applications.

Advantages	Disadvantages
■ Comprehensive interpretations of the spatial arrangement of the CNDs structures.	■ Required a great analytic efforts to define the chemical functionalities.
■ Evaluation of band gap.	■ Limitation to GQDs and CNDs.
■ Description of electrochemical properties.	■ Required long times.
■ Model the interaction between CNDs and biological species such as protein and nucleic acids.	■ For several application is necessary only the knowledge of physiochemical properties.

5. Conclusions

The computational studies of CDs have made the first steps and already provided key information that have driven the research in new directions. Here, we describe a simple, representative and solid model of CNDs useful for evaluating their interaction with Doxorubicin, a widely used anticancer molecule. The modelled D-CNDs species showed various unique structures, suggesting that the intercalated species are more stable if the intercalation occurred between L_g^n or through adsorption directly onto L_h^n. Interestingly, $\pi-\pi$ interactions seemed to play a greater role compared to hydrogen bond, but this result is quite sensitive to the model used. Nevertheless, this first computational study on three-dimensional D-CND models enlightens the critical role of the modelling approach in order to fully understand the CDs' physiochemical behavior.

In the future, we aim to refine this model using a synthetic approach oriented towards the production of well-defined CNDs with reproducible and uniquely identifiable features. This will allow us to definitively prove the robustness of the approach in the prediction of the CDs' chemical, optical and geometrical properties.

Author Contributions: Conceptualization, M.B. and A.T.; methodology, M.B.; formal analysis, E.M., M.B.; investigation, E.M., M.B. and A.T.; resources, A.T.; writing—original draft preparation, E.M., M.B. and A.T.; writing—review and editing, E.M., M.B. and A.T.; supervision, A.T. All authors have read and agreed to the published version of the manuscript.

Funding: This research received no external funding.

Conflicts of Interest: The authors declare no conflict of interest.

References

1. Xu, X.; Ray, R.; Gu, Y.; Ploehn, H.J.; Gearheart, L.; Raker, K.; Scrivens, W.A. Electrophoretic analysis and purification of fluorescent single-walled carbon nanotube fragments. *J. Am. Chem. Soc.* **2004**, *126*, 12736–12737. [CrossRef] [PubMed]
2. Giordano, M.G.; Seganti, G.; Bartoli, M.; Tagliaferro, A. An Overview on Carbon Quantum Dots Optical and Chemical Features. *Molecules* **2023**, *28*, 2772. [CrossRef]
3. Yan, F.; Sun, Z.; Zhang, H.; Sun, X.; Jiang, Y.; Bai, Z. The fluorescence mechanism of carbon dots, and methods for tuning their emission color: A review. *Microchim. Acta* **2019**, *186*, 583. [CrossRef] [PubMed]
4. Connerade, J.P. A review of quantum confinement. In *AIP Conference Proceedings*; American Institute of Physics: College Park, MD, USA, 2009; pp. 1–33.
5. Yan, X.; Cui, X.; Li, B.; Li, L.-S. Large, solution-processable graphene quantum dots as light absorbers for photovoltaics. *Nano Lett.* **2010**, *10*, 1869–1873. [CrossRef]
6. Zhu, L.; Shen, D.; Wu, C.; Gu, S. State-of-the-Art on the Preparation, Modification, and Application of Biomass-Derived Carbon Quantum Dots. *Ind. Eng. Chem. Res.* **2020**, *59*, 22017–22039. [CrossRef]
7. Chen, B.-B.; Liu, M.-L.; Gao, Y.-T.; Chang, S.; Qian, R.-C.; Li, D.-W. Design and applications of carbon dots-based ratiometric fluorescent probes: A review. *Nano Res.* **2023**, *16*, 1064–1083. [CrossRef]
8. Gallareta-Olivares, G.; Rivas-Sanchez, A.; Cruz-Cruz, A.; Hussain, S.M.; González-González, R.B.; Cárdenas-Alcaide, M.F.; Iqbal, H.M.; Parra-Saldívar, R. Metal-doped carbon dots as robust nanomaterials for the monitoring and degradation of water pollutants. *Chemosphere* **2023**, *312*, 137190. [CrossRef] [PubMed]
9. Lo Bello, G.; Bartoli, M.; Giorcelli, M.; Rovere, M.; Tagliaferro, A. A Review on the Use of Biochar Derived Carbon Quantum Dots Production for Sensing Applications. *Chemosensors* **2022**, *10*, 117. [CrossRef]
10. Jing, H.H.; Bardakci, F.; Akgöl, S.; Kusat, K.; Adnan, M.; Alam, M.J.; Gupta, R.; Sahreen, S.; Chen, Y.; Gopinath, S.C. Green Carbon Dots: Synthesis, Characterization, Properties and Biomedical Applications. *J. Funct. Biomater.* **2023**, *14*, 27. [CrossRef]
11. Hui, S. Carbon dots (CDs): Basics, recent potential biomedical applications, challenges, and future perspectives. *J. Nanopart. Res.* **2023**, *25*, 68. [CrossRef]
12. Sendão, R.M.; Esteves da Silva, J.C.; Pinto da Silva, L. Applications of Fluorescent Carbon Dots as Photocatalysts: A Review. *Catalysts* **2023**, *13*, 179. [CrossRef]
13. Sun, P.; Xing, Z.; Li, Z.; Zhou, W. Recent Advances in Quantum Dots Photocatalysts. *Chem. Eng. J.* **2023**, *9*, 141399. [CrossRef]
14. Sikiru, S.; Oladosu, T.L.; Kolawole, S.Y.; Mubarak, L.A.; Soleimani, H.; Afolabi, L.O.; Toyin, A.-O.O. Advance and prospect of carbon quantum dots synthesis for energy conversion and storage application: A comprehensive review. *J. Energy Storage* **2023**, *60*, 106556. [CrossRef]
15. Zhai, Y.; Zhang, B.; Shi, R.; Zhang, S.; Liu, Y.; Wang, B.; Zhang, K.; Waterhouse, G.I.; Zhang, T.; Lu, S. Carbon dots as new building blocks for electrochemical energy storage and electrocatalysis. *Adv. Energy Mater.* **2022**, *12*, 2103426. [CrossRef]
16. Georgakilas, V.; Perman, J.A.; Tucek, J.; Zboril, R. Broad family of carbon nanoallotropes: Classification, chemistry, and applications of fullerenes, carbon dots, nanotubes, graphene, nanodiamonds, and combined superstructures. *Chem. Rev.* **2015**, *115*, 4744–4822. [CrossRef]
17. Mansuriya, B.D.; Altintas, Z. Carbon Dots: Classification, properties, synthesis, characterization, and applications in health care—An updated review (2018–2021). *Nanomaterials* **2021**, *11*, 2525. [CrossRef]
18. Lerf, A.; He, H.; Forster, M.; Klinowski, J. Structure of graphite oxide revisited. *J. Phys. Chem. B* **1998**, *102*, 4477–4482. [CrossRef]
19. Mandal, B.; Sarkar, S.; Sarkar, P. Exploring the electronic structure of graphene quantum dots. *J. Nanopart. Res.* **2012**, *14*, 1317. [CrossRef]
20. Yan, X.; Li, B.; Cui, X.; Wei, Q.; Tajima, K.; Li, L.-s. Independent tuning of the band gap and redox potential of graphene quantum dots. *J. Phys. Chem. Lett.* **2011**, *2*, 1119–1124. [CrossRef]
21. Yan, X.; Cui, X.; Li, L.-s. Synthesis of large, stable colloidal graphene quantum dots with tunable size. *J. Am. Chem. Soc.* **2010**, *132*, 5944–5945. [CrossRef]
22. Hai, X.; Mao, Q.-X.; Wang, W.-J.; Wang, X.-F.; Chen, X.-W.; Wang, J.-H. An acid-free microwave approach to prepare highly luminescent boron-doped graphene quantum dots for cell imaging. *J. Mater. Chem. B* **2015**, *3*, 9109–9114. [CrossRef]
23. Seven, E.S.; Kirbas Cilingir, E.; Bartoli, M.; Zhou, Y.; Sampson, R.; Shi, W.; Peng, Z.; Ram Pandey, R.; Chusuei, C.C.; Tagliaferro, A.; et al. Hydrothermal vs microwave nanoarchitechtonics of carbon dots significantly affects the structure, physicochemical properties, and anti-cancer activity against a specific neuroblastoma cell line. *J. Colloid Interface Sci.* **2023**, *630*, 306–321. [CrossRef]
24. Xia, C.; Zhu, S.; Feng, T.; Yang, M.; Yang, B. Evolution and synthesis of carbon dots: From carbon dots to carbonized polymer dots. *Adv. Sci.* **2019**, *6*, 1901316. [CrossRef]

25. Chen, J.; Li, F.; Gu, J.; Zhang, X.; Bartoli, M.; Domena, J.B.; Zhou, Y.; Zhang, W.; Paulino, V.; Ferreira, B.C.; et al. Cancer cells inhibition by cationic carbon dots targeting the cellular nucleus. *J. Colloid Interface Sci.* **2023**, *637*, 193–206. [CrossRef]
26. Zhang, W.; Chen, J.; Gu, J.; Bartoli, M.; Domena, J.B.; Zhou, Y.; Ferreira, B.C.; Kirbas Cilingir, E.; McGee, C.M.; Sampson, R.; et al. Nano-carrier for gene delivery and bioimaging based on pentaetheylenehexamine modified carbon dots. *J. Colloid Interface Sci.* **2023**, *639*, 180–192. [CrossRef]
27. Liyanage, P.Y.; Graham, R.M.; Pandey, R.R.; Chusuei, C.C.; Mintz, K.J.; Zhou, Y.; Harper, J.K.; Wu, W.; Wikramanayake, A.H.; Vanni, S.; et al. Carbon Nitride Dots: A Selective Bioimaging Nanomaterial. *Bioconjugate Chem.* **2019**, *30*, 111–123. [CrossRef] [PubMed]
28. Mintz, K.J.; Bartoli, M.; Rovere, M.; Zhou, Y.; Hettiarachchi, S.D.; Paudyal, S.; Chen, J.; Domena, J.B.; Liyanage, P.Y.; Sampson, R.; et al. A deep investigation into the structure of carbon dots. *Carbon* **2021**, *173*, 433–447. [CrossRef]
29. Kirbas, E.C.; Sankaran, M.; Garber, J.M.; Vallejo, F.A.; Bartoli, M.; Tagliaferro, A.; Vanni, S.; Graham, R.; Leblanc, R.M. Surface Modification Nanoarchitectonics of Carbon Nitride Dots for Better Drug Loading and Higher Cancer Selectivity. *Nanoscale* **2022**, *14*, 9686–9701. [CrossRef] [PubMed]
30. Wiśniewski, M. The Consequences of Water Interactions with Nitrogen-Containing Carbonaceous Quantum Dots—The Mechanistic Studies. *Int. J. Mol. Sci.* **2022**, *23*, 14292. [CrossRef]
31. Liyanage, P.Y.; Zhou, Y.; Al-Youbi, A.O.; Bashammakh, A.S.; El-Shahawi, M.S.; Vanni, S.; Graham, R.M.; Leblanc, R.M. Pediatric glioblastoma target-specific efficient delivery of gemcitabine across the blood–brain barrier via carbon nitride dots. *Nanoscale* **2020**, *12*, 7927–7938. [CrossRef]
32. Carvalho, C.; Santos, R.X.; Cardoso, S.; Correia, S.; Oliveira, P.J.; Santos, M.S.; Moreira, P.I. Doxorubicin: The good, the bad and the ugly effect. *Curr. Med. Chem.* **2009**, *16*, 3267–3285. [CrossRef]
33. Rivankar, S. An overview of doxorubicin formulations in cancer therapy. *J. Cancer Res. Ther.* **2014**, *10*, 853–858. [CrossRef] [PubMed]
34. Zhang, W.; Dang, G.; Dong, J.; Li, Y.; Jiao, P.; Yang, M.; Zou, X.; Cao, Y.; Ji, H.; Dong, L. A multifunctional nanoplatform based on graphitic carbon nitride quantum dots for imaging-guided and tumor-targeted chemo-photodynamic combination therapy. *Colloids Surf. B Biointerfaces* **2021**, *199*, 111549. [CrossRef]
35. Dong, J.; Zhao, Y.; Chen, H.; Liu, L.; Zhang, W.; Sun, B.; Yang, M.; Wang, Y.; Dong, L. Fabrication of PEGylated graphitic carbon nitride quantum dots as traceable, pH-sensitive drug delivery systems. *New J. Chem.* **2018**, *42*, 14263–14270. [CrossRef]
36. Dong, J.; Zhao, Y.; Wang, K.; Chen, H.; Liu, L.; Sun, B.; Yang, M.; Sun, L.; Wang, Y.; Yu, X. Fabrication of graphitic carbon nitride quantum dots and their application for simultaneous fluorescence imaging and pH-responsive drug release. *ChemistrySelect* **2018**, *3*, 12696–12703. [CrossRef]
37. Rashid, A.; Perveen, M.; Khera, R.A.; Asif, K.; Munir, I.; Noreen, L.; Nazir, S.; Iqbal, J. A DFT study of graphitic carbon nitride as drug delivery carrier for flutamide (anticancer drug). *J. Comput. Biophys. Chem.* **2021**, *20*, 347–358. [CrossRef]
38. Zaboli, A.; Raissi, H.; Farzad, F. Molecular interpretation of the carbon nitride performance as a template for the transport of anti-cancer drug into the biological membrane. *Sci. Rep.* **2021**, *11*, 18981. [CrossRef]
39. Shaki, H.; Raissi, H.; Mollania, F.; Hashemzadeh, H. Modeling the interaction between anti-cancer drug penicillamine and pristine and functionalized carbon nanotubes for medical applications: Density functional theory investigation and a molecular dynamics simulation. *J. Biomol. Struct. Dyn.* **2020**, *38*, 1322–1334. [CrossRef]
40. Valatin, J. Generalized hartree-fock method. *Phys. Rev.* **1961**, *122*, 1012. [CrossRef]
41. Bartlett, R.J.; Stanton, J.F. Applications of Post-Hartree—Fock Methods: A Tutorial. *Rev. Comput. Chem.* **1994**, *2*, 65–169.
42. Dewar, M.J.; Thiel, W. Ground states of molecules. 38. The MNDO method. Approximations and parameters. *J. Am. Chem. Soc.* **1977**, *99*, 4899–4907. [CrossRef]
43. Dewar, M.J.; Thiel, W. A semiempirical model for the two-center repulsion integrals in the NDDO approximation. *Theor. Chim. Acta* **1977**, *46*, 89–104. [CrossRef]
44. Engelke, R. Limitations on mndo and mndo/ci computations of activation barriers. *Chem. Phys. Lett.* **1981**, *83*, 151–155. [CrossRef]
45. Dewar, M.J.; Zoebisch, E.G.; Healy, E.F.; Stewart, J.J. Development and use of quantum mechanical molecular models. 76. AM1: A new general purpose quantum mechanical molecular model. *J. Am. Chem. Soc.* **1985**, *107*, 3902–3909. [CrossRef]
46. Stewart, J.J. Optimization of parameters for semiempirical methods II. Applications. *J. Comput. Chem.* **1989**, *10*, 221–264. [CrossRef]
47. Cavasotto, C.N.; Aucar, M.G.; Adler, N.S. Computational chemistry in drug lead discovery and design. *Int. J. Quantum Chem.* **2019**, *119*, e25678. [CrossRef]
48. Repasky, M.P.; Chandrasekhar, J.; Jorgensen, W.L. PDDG/PM3 and PDDG/MNDO: Improved semiempirical methods. *J. Comput. Chem.* **2002**, *23*, 1601–1622. [CrossRef] [PubMed]
49. Toledo, M.V.; Briand, L.E.; Ferreira, M.L. A Simple Molecular Model to Study the Substrate Diffusion into the Active Site of a Lipase-Catalyzed Esterification of Ibuprofen and Ketoprofen with Glycerol. *Top. Catal.* **2022**, *65*, 944–956. [CrossRef]
50. Bystrov, V.; Paramonova, E.; Meng, X.; Shen, H.; Wang, J.; Fridkin, V. Polarization switching in nanoscale ferroelectric composites containing PVDF polymer film and graphene layers. *Ferroelectrics* **2022**, *590*, 27–40. [CrossRef]
51. Stewart, J.J. Optimization of parameters for semiempirical methods. III extension of pm3 to be, mg, zn, ga, ge, as, se, cd, in, sn, sb, te, hg, tl, pb, and bi. *J. Comput. Chem.* **1991**, *12*, 320–341. [CrossRef]
52. Ignatov, S.; Razuvaev, A.; Kokorev, V.; Alexandrov, Y.A. Extension of the PM3 Method on s, p, d Basis. Test Calculations on Organochromium Compounds. *J. Phys. Chem.* **1996**, *100*, 6354–6358. [CrossRef]

53. Rono, N.; Kibet, J.K.; Martincigh, B.S.; Nyamori, V.O. A review of the current status of graphitic carbon nitride. *Crit. Rev. Solid State Mater. Sci.* **2021**, *46*, 189–217. [CrossRef]
54. Zhou, Y.; Kandel, N.; Bartoli, M.; Serafim, L.F.; ElMetwally, A.E.; Falkenberg, S.M.; Paredes, X.E.; Nelson, C.J.; Smith, N.; Padovano, E.; et al. Structure-activity relationship of carbon nitride dots in inhibiting Tau aggregation. *Carbon* **2022**, *193*, 1–16. [CrossRef] [PubMed]
55. Liu, H.; Wang, X.; Wang, H.; Nie, R. Synthesis and biomedical applications of graphitic carbon nitride quantum dots. *J. Mater. Chem. B* **2019**, *7*, 5432–5448. [CrossRef] [PubMed]
56. Lipson, H.S.; Stokes, A. The structure of graphite. *Proc. R. Soc. London. Ser. A. Math. Phys. Sci.* **1942**, *181*, 101–105.
57. Sakorikar, T.; Vayalamkuzhi, P.; Jaiswal, M. Geometry dependent performance limits of stretchable reduced graphene oxide interconnects: The role of wrinkles. *Carbon* **2020**, *158*, 864–872. [CrossRef]
58. Gómez-Navarro, C.; Burghard, M.; Kern, K. Elastic properties of chemically derived single graphene sheets. *Nano Lett.* **2008**, *8*, 2045–2049. [CrossRef]
59. Lopez-Chavez, E.; Garcia-Quiroz, A.; Santiago-Jiménez, J.C.; Díaz-Góngora, J.A.; Díaz-López, R.; de Landa Castillo-Alvarado, F. Quantum–mechanical characterization of the doxorubicin molecule to improve its anticancer functions. *MRS Adv.* **2021**, *6*, 897–902. [CrossRef]
60. Bharathy, G.; Prasana, J.C.; Muthu, S. Molecular conformational analysis, vibrational spectra, NBO, HOMO–LUMO and molecular docking of modafinil based on density functional theory. *Int. J. Cur. Res. Rev* **2018**, *10*, 36–45. [CrossRef]
61. Johnson, E.R.; Keinan, S.; Mori-Sánchez, P.; Contreras-García, J.; Cohen, A.J.; Yang, W. Revealing Noncovalent Interactions. *J. Am. Chem. Soc.* **2010**, *132*, 6498–6506. [CrossRef]
62. Chaudhary, N.K.; Mishra, P. Metal complexes of a novel Schiff base based on penicillin: Characterization, molecular modeling, and antibacterial activity study. *Bioinorg. Chem. Appl.* **2017**, *2017*, 6927675. [CrossRef] [PubMed]
63. Agrahari, A.K. A computational approach to identify a potential alternative drug with its positive impact toward PMP22. *J. Cell. Biochem.* **2017**, *118*, 3730–3743. [CrossRef] [PubMed]
64. Zhang, L.; Zhou, W.; Li, D.-H. A descent modified Polak–Ribière–Polyak conjugate gradient method and its global convergence. *IMA J. Numer. Anal.* **2006**, *26*, 629–640. [CrossRef]
65. Hanwell, M.D.; Curtis, D.E.; Lonie, D.C.; Vandermeersch, T.; Zurek, E.; Hutchison, G.R. Avogadro: An advanced semantic chemical editor, visualization, and analysis platform. *J. Cheminform.* **2012**, *4*, 17. [CrossRef]
66. Liu, S.; Li, D.; Sun, H.; Ang, H.M.; Tadé, M.O.; Wang, S. Oxygen functional groups in graphitic carbon nitride for enhanced photocatalysis. *J. Colloid Interface Sci.* **2016**, *468*, 176–182. [CrossRef] [PubMed]
67. Gao, G.; Jiao, Y.; Ma, F.; Jiao, Y.; Waclawik, E.; Du, A. Carbon nanodot decorated graphitic carbon nitride: New insights into the enhanced photocatalytic water splitting from ab initio studies. *Phys. Chem. Chem. Phys.* **2015**, *17*, 31140–31144. [CrossRef]
68. Jin, C.; Ma, E.Y.; Karni, O.; Regan, E.C.; Wang, F.; Heinz, T.F. Ultrafast dynamics in van der Waals heterostructures. *Nat. Nanotechnol.* **2018**, *13*, 994–1003. [CrossRef]
69. Feng, J.; Liu, G.; Yuan, S.; Ma, Y. Influence of functional groups on water splitting in carbon nanodot and graphitic carbon nitride composites: A theoretical mechanism study. *Phys. Chem. Chem. Phys.* **2017**, *19*, 4997–5003. [CrossRef]
70. Gao, Y.; Hou, F.; Hu, S.; Wu, B.; Wang, Y.; Zhang, H.; Jiang, B.; Fu, H. Graphene quantum-dot-modified hexagonal tubular carbon nitride for visible-light photocatalytic hydrogen evolution. *ChemCatChem* **2018**, *10*, 1330–1335. [CrossRef]
71. Li, X.; Zhou, Y.; Xu, Y.; Xu, H.; Wang, M.; Yin, H.; Ai, S. A novel photoelectrochemical biosensor for protein kinase activity assay based on phosphorylated graphite-like carbon nitride. *Anal. Chim. Acta* **2016**, *934*, 36–43. [CrossRef]

Disclaimer/Publisher's Note: The statements, opinions and data contained in all publications are solely those of the individual author(s) and contributor(s) and not of MDPI and/or the editor(s). MDPI and/or the editor(s) disclaim responsibility for any injury to people or property resulting from any ideas, methods, instructions or products referred to in the content.

MDPI
St. Alban-Anlage 66
4052 Basel
Switzerland
www.mdpi.com

Molecules Editorial Office
E-mail: molecules@mdpi.com
www.mdpi.com/journal/molecules

Disclaimer/Publisher's Note: The statements, opinions and data contained in all publications are solely those of the individual author(s) and contributor(s) and not of MDPI and/or the editor(s). MDPI and/or the editor(s) disclaim responsibility for any injury to people or property resulting from any ideas, methods, instructions or products referred to in the content.

www.ingramcontent.com/pod-product-compliance
Lightning Source LLC
LaVergne TN
LVHW070052120526
838202LV00102B/2052